9th International Conference on X-Ray Microscopy 2008

Journal of Physics: Conference Series Volume 186

Zurich, Switzerland
21-25 July 2008

ISBN: 978-1-61738-290-1

Printed from e-media with permission by:

Curran Associates, Inc.
57 Morehouse Lane
Red Hook, NY 12571
www.proceedings.com

Some format issues inherent in the e-media version may also appear in this print version.

Printed by Curran Associates, Inc. (2010)

Institute of Physics
Dirac House, Temple Back
Bristol BS1 6BE UK

Phone: 44 1 17 929 7481
Fax: 44 1 17 920 0979

techtracking@iop.org

TABLE OF CONTENTS

INTRODUCTION

METHODS: SR-BASED MICROSCOPY

METHODS: LAB-BASED MICROSCOPY

METHODS: TOMOGRAPHY

METHODS: COHERENCE

X-RAY OPTICS

APPLICATIONS: BIOLOGY AND ENVIRONMENTAL SCIENCE

APPLICATIONS: CONDENSED MATTER AND MATERIALS SCIENCE

XRM 2008

xrm2008.web.psi.ch

This volume compiles the contributions to the International Conference on X-Ray Microscopy (XRM2008) held on 20–25 July 2008 in Zurich, Switzerland. The conference was the ninth in a series which started in Göttingen in 1984. Over the years the XRM conference series has served as a forum bringing together all relevant players working on the development of methods, building instrumentation, and applying x-ray microscopy to challenging issues in materials science, condensed matter research, environmental science and biology.

XRM2008 was attended by about 300 participants who followed 44 oral presentations and presented 220 posters.

Figure 1: Participants of the XRM2008 conference gathered in front of the main building of the ETH-Zurich.

The conference showed that x-ray microscopy has become a mature field resting on three pillars. The first are workhorse instruments available even to nonspecialist users. These exist at synchrotron sources world-wide as well as in laboratories. They allow the application of established microscopy methods to solve scientific projects in areas as diverse as soil science, the investigation of cometary dust particles, magnetic materials, and the analysis of ancient parchments. Examples of all of these projects can be found in this volume. These instruments have become so well understood that now they are also commercially available.

The second pillar is the continued development of methods. Methods like stroboscopic imaging, wet cells or high and low temperature environments add versatility to the experiments. Methods like phase retrieval and ptychographic imaging allow the retrieval of information which hitero was thought to be inaccessible.

The third pillar is the extension of such instruments and methods to new photon sources. With x-ray free electron lasers on the horizon the XRM community is working to transfer their know-how to these novel sources which will offer unprecedented brightness and time structure, but which at the same time require unprecedented effort to perform the experiment and to extract meaningful information from the data.

Resting on these three pillars, the XRM community seems well prepared to solve the scientific questions of today and to help solve even more challenging scientific questions in the future.

Many people contributed to the success of XRM2008, first and foremost were the participants with their excellent contributions and through their lively discussions. Organizing the event was made possible due to many helping hands and brains at the organizing institution, the Paul Scherrer Institut. It is our pleasure to thank all of these people. Financial support was given by the Gold Sponsor, XRADIA, by the European Round Table for Synchrotrons and Free Electron Lasers, the Center for Imaging Science and Technology at Zurich, and by the Swiss National Science Foundation. The conference was kindly hosted by ETH Zurich which provided a perfect setting for this venue.

We thank all the participants of XRM2008, everybody who helped in the organization and all financial supporters and are looking forward to hearing about further progress during XRM2010 which will be organized by Argonne National Laboratory in Chicago.

C Quitmann, C David, F Nolting, F Pfeiffer and M Stampanoni
Proceedings Editors

Figure 2 View over Zurich and into the alps from the terrace of the ETHZ main building during the XRM2008 reception.

Figure 3 Flag tosser and Alphorn blowers in front of the ETH Zurich main building during the XRM2008 opening reception.

Figure 4 Industry exhibition in the entrance hall of the ETHZ main building.

Figure 5 Lively discussions amongst the conference participants took place during the two poster sessions.

Figure 6 Lecture hall with the participants busily discussing between talks.

Figure 7 Past and present winners of the Werner Meyer-Ilse Award are joined by Andrea Meyer-Ilse, the wife of the deceased x-ray microscopy pioneer Werner Meyer-Ilse. From left to right: Weilun Chao (winner 2005), Ann Sakdinawat (winner 2008), Piere Thibault (winner 2008), Andrea Meyer-Ilse and M Feser (winner 2002).

Figure 8 Reception in the court of Schloss Rapperswil during the XRM2008 conference excursion.

Figure 9 Conference participants walking towards the castle of Rapperswil during the conference excursion.

Figure 10 Accompanying persons group gathered in the port of Lucerne.

Dates

Issue 1 (2009)

All papers published in this volume of *Journal of Physics: Conference Series* have been peer reviewed through processes administered by the proceedings Editors. Reviews were conducted by expert referees to the professional and scientific standards expected of a proceedings journal published by IOP Publishing.

Dates

Issue 1 (2009)

International Programm Committee (IPC)

Harald Ade	NCSU
Sadao Aoki	University Tsukuba
David Attwood	Lawrence Berkeley National Laboratory/CXRO
Christian David	Paul Scherrer Institut
Peter Fischer	Lawrence Berkeley National Laboratory
Adam Hitchcock	McMaster University
Chris Jacobsen	SUNY, Stony Brook
Denis Joyeux	Lab Charles Fabry de l'Institut d'Optique
Yasushi Kagoshima	University of Hyogo
Hiroshi Kihara	Kansai Medical University
Janos Kirz	SUNY Stony Brook
Maya Kiskinova	ELETTRA
Ian McNulty	Argonne National Lab/APS
Alan Michette	Kings College London
Graeme Morrison	Kings College London
Keith Nugent	University of Melbourne
Zhu Peiping	BSRF Institute of High Energy Physics
Francois Polack	Soleil
Christoph Quitmann	Paul Scherrer Institut
Günther Schmahl	University Göttingen
Gerd Schneider	Bessy
Hyun-Joon Shin	Pohang Accelerator Lab
Jean Susini	ESRF
Mau-Tsu Tang	NSRRC
Tony Warwick	Lawrence Berkeley Lab/ALS

Local Organizing Committee

Christoph Quitmann	Chair, Scientific Program
Charlotte Heer	Secretary
Christian David	Scientific Program
Frithjof Nolting	Scientific Program
Franz Pfeiffer	Scientific Program

Marco Stampanoni	Scientific Program
Robert Rudolph	Sponsoring, Financials
Alfred Waser	Industry Exhibition
Robert Keller	Public Relation
Markus Knecht	Computing and WWW
Annick Cavedon	Proceedings and Excursions and Accompanying Persons Program
Margrit Eichler	Excursions and Accompanying Persons Program
Kathy Eikenberry	Excursions and Accompanying Persons Program
Marlies Locher	Excursions and Accompanying Persons Program

Dates

Issue 1 (2009)

The 9th International Conference on X-Ray Microscopy was graciously supported by the following institutions:

Gold sponsor

Paul Scherrer Institut

European Round Table for Synchrotron Radiation and Free Electron Laser

Eidgenössische Technische Hochschule Zürich
Swiss Federal Institute of Technology Zurich

Swiss Federal Institute of Technology Zurich

Center for Imaging Science and Technology

Swiss National Science Foundation

Dates

Issue 1 (2009)

9th International Conference on X-Ray Microscopy IOP Publishing
Journal of Physics: Conference Series **186** (2009) 012001 doi:10.1088/1742-6596/186/1/012001

The History and Future of X-ray Microscopy

J Kirz[1,2] and C Jacobsen[2]

[1] Advanced Light Source, Lawrence Berkeley National Laboratory, 1 Cyclotron Road, Berkeley, CA 94720, USA

[2] Department of Physics and Astronomy, Stony Brook University, Stony Brook, NY 11794, USA

E-mail: jkirz@lbl.gov, Chris.Jacobsen@stonybrook.edu

Abstract. We take a somewhat whimsical look at the history of X-ray microscopy, and extrapolate some trends into the future.

1. History

This is a story of spies, heroes, villains, false starts, and a brush with real fame. We divide the history into seven epochs: 1) Ancient history 1895–1945; 2) The Classical Period 1946–1960; 3) The Dark Ages 1961–1971; 4) The Renaissance 1972–1982; 5) Romanticism 1983–1993; 6) The Age of Reason 1994–2002; and 7) The Industrial Revolution 2003–2008.

1.1. Ancient History 1895–1945

This was the time of the pioneers who, following Röntgen's 1895 discovery of X-rays [1], established point-projection microscopy with a resolution of a few microns [2], understood the nature of X rays, and developed absorption and emission spectroscopy. It was, coincidentally, during this period that R.W. Wood demonstrated the use of phase zone plates as focusing elements with visible light [3]. The theoretical starting point of X-ray optics can be traced to the brief paper by Einstein [4] in which he suggested that the index of refraction for X rays in most materials should be just slightly less than 1.

1.2. The Classical Period 1946–1960

Right after World War II several groups became interested in X-ray microscopy. Arne Engstrom in Sweden developed the technique of quantitative elemental imaging [5]. Paul Kirkpatrick at Stanford worked on grazing incidence optics. With Albert Baez he developed the crossed-mirror system that bears their names [6] (see Fig. 1). He convinced himself that this optical element will lead to a very high resolution imaging microscope [7], having neglected to consider off-axis aberrations; additional examples were constructed with Howard Pattee. In Cambridge, Cosslett and Nixon first carried out work in point projection x-ray microscopy (Fig. 2), and then went on to work primarily on x-ray analysis in the electron microscope [8]. These three groups took turns organizing international conferences on X-ray microscopy: Cosslett hosted the first one in 1956 [9], Engstrom the second in 1959 [10], with Kirkpatrick taking responsibility for the third in 1962 [11]. This series of conferences shifted focus from microscopy to microanalysis around the 1962 meeting.

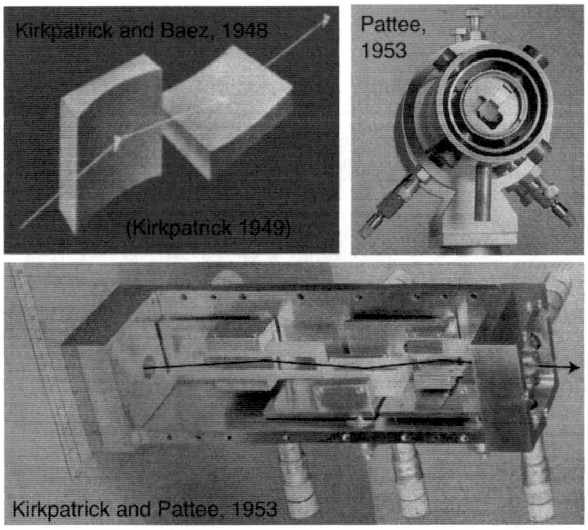

Figure 1. The crossed-1D-lenses focusing scheme developed by Kirkpatrick and Baez using elliptical profile mirrors [6]. At top left is an illustration from Kirkpatrick's 1949 contribution to *Scientific American* [7], where the editor's blurb in the table of contents stated "The X-ray Microscope: It would be a big improvement on microscopes using light or electrons, for X-rays combine short wavelengths, giving fine resolution, and penetration. The main problems standing in the way have now been solved." At bottom is a three-mirror system constructed by Kirkpatrick and Pattee [12], and at top right is an improved, concentric-cylinder mounting scheme developed by Pattee (1953; picture from a 1983 letter by Pattee to M. Howells). More recent focusing schemes involving multilayer Laue lenses [13] return to the crossed-1D-lenses focusing geometry.

It was during this period that Baez published his suggestion that zone plates should be used as X-ray lenses [14, 15], and the book by Cosslett and Nixon [16] summarized everything that was known about X-ray microscopy.

1.3. The Dark Ages 1961–1971

The sixties were a relatively quiet period in X-ray microscopy. The Cambridge group continued to be active, especially Theodore Hall and his collaborators [18]. It was revealed only decades later that Hall was the second Soviet spy in Los Alamos [19], and that Cosslett was also involved in clandestine activities.

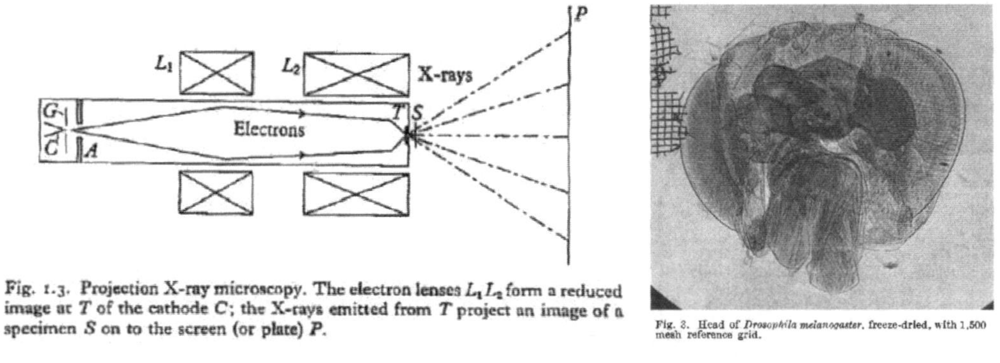

Figure 2. Point projection microscopy scheme developed by Cosslett and Nixon. The scheme is shown at left (from [16]), while at right is shown a point projection image of an insect head (from [17]).

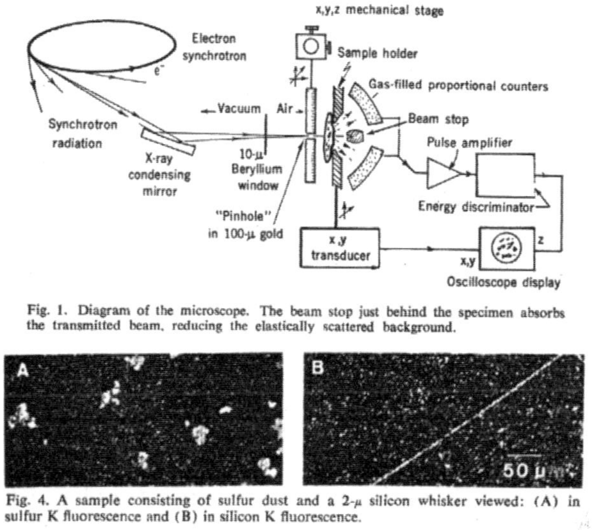

Figure 3. The first synchrotron-based x-ray microprobe was developed by Horowitz and Howell at the Cambridge Electron Accelerator in 1972 [21]. The microprobe used a deep pinhole fabricated by plating around a silicon whisker. Unfortunately this promising instrument was short-lived, as the accelerator was shut down in 1973 due to the high energy physics community moving on to bigger machines.

Of course the seeds for the renaissance were planted during the dark ages: it was toward the end of the decade that Schmahl and Rudolph turned their attention to holographic fabrication of zone plates [20].

1.4. The Renaissance 1972–1982

The decade of the seventies saw the first lightsource-based X-ray microscopes: Horowitz and Howell demonstrated both transmission and fluorescence microscopy at the Cambridge Electron Accelerator [21] (Fig. 3), Aoki and Kikuta did the first holographic experiments in Japan [22], and Schmahl *et al.* built the first TXMs that used zone plates, first using a laboratory source [23] and soon after using synchrotron radiation [24, 25] (Fig. 4). Stony Brook activities in developing STXM started at SSRL [26], and moved to the NSLS [27] (Fig. 5).

1.5. Romanticism 1983–1993

Following the first uses of synchrotron light sources, a lot of new avenues were tried, some more successful than others. The King's College group built a STXM at Daresbury [28] (using

Figure 4. The first zone plate TXMs (transmission x-ray microscopes) developed by Schmahl, Rudolph, and Niemann. At left is shown an instrument operated at DESY in Hamburg in 1976 [24], while an instrument operated at ACO in Orsay is shown in the middle image. At right is shown an image of a diatom obtained at Orsay [25]. Images courtesy G. Schmahl.

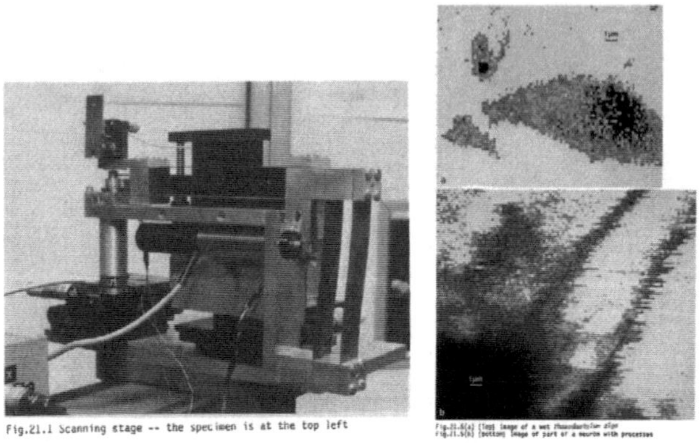

Figure 5. The first STXM (scanning transmission x-ray microscope) using zone plate optics was constructed by Rarback, Kirz, and Kenney at U-15 at the National Synchrotron Light Source at Brookhaven [27]. It used 300 nm zone plates and operated with $E/\Delta E \simeq 300$; the images shown took nearly an hour each to acquire.

a source with poor performance for coherent illumination), and so did the Göttingen group at BESSY [29]. Kunz's group at DESY built a microscope based on a commercial ellipsoid with terrible aberrations [30]. The Stony Brook group (Ade *et al.*) at NSLS built a scanning photoemission microscope two decades before the technology was ready for such an instrument [31]. MAXIMUM, a somewhat similar device at Wisconsin, was not much more successful [32]. Efforts at further development of holography at the NSLS achieved some successful demonstrations [33, 34, 35], but they also illustrated just how difficult it was to perform coherence-based experiments on second generation light sources. The first X-ray lasers were demonstrated at Livermore and Princeton [36, 37, 38], but at only a few pulses per day they were not very user-friendly for microscopy experiments.

The novelty of X-ray microscopy also led some people astray. Prof. Baldini from Harvard teamed up with scientists at IBM for flash-contact-microscopy of blood platelets [39]. One of the images even appeared on the front page of the Science section of *The New York Times* (Jan. 15, 1985). Unfortunately it was not an image, but rather the platelet stuck onto the photoresist "detector," as the authors later discovered. The euphoria over the development of x-ray lasers led to another cover article (April 2, 1985) in the Science section of *The New York Times* that gushed "But aside from its weapons applications, the X-ray laser has excited biologists, chemists and physicists because of its possible use in a super microscope, an instrument that will perhaps be capable of taking holographic three-dimensional movies of the genetic code of a living cell." The 1992 NASA Scientist of the Year Award was given to Richard Hoover for inventing (US Patent 5,107,526 on April 21, 1992), according to NASA, a "revolutionary new microscope [that] should enable researchers to see in great detail high contrast x-ray images of proteins, chromosomes and other tiny carbon structures inside living cells. Resolution of the microscope could be so high that it may produce detailed images of the building blocks of life—tiny DNA molecules." The instrument was to use a carbon K X-ray source and a multilayer-coated Schwartzschild objective, similar to a scanning microscope proposed by Eberhard Spiller some years before [40]. Several of us got together and sent a letter to the head of NASA, pointing out that radiation damage would make it impossible to do what was being claimed. We received an answer from the top lawyer at the agency, who assured us that no laws were violated...

It was also during this period that the present series of X-ray Microscopy Conferences started (see Fig. 6A), with the first one in Göttingen in 1983 hosted by Günter Schmahl and his group [41]. Major steps during this period included:

- The invention of Zernike phase contrast microscopy by Schmahl & Rudolph [42, 43].
- Following a proposal by Sayre [44], the development of e-beam fabricated zone plates at MIT [45] and more substantially at IBM [46]. IBM also introduced the community to silicon

nitride windows [47].
- The first undulator beamline for microscopy at the NSLS [48].
- The establishment of the Center for X-ray Optics in Berkeley by David Attwood.
- Spectromicroscopy using XANES [31, 49, 50, 51, 52].

Most of these conference proceedings are relatively easy to come by [53, 54, 55, 56, 57, 58]. The proceedings of the Sep. 20–24, 1993 conference held in Chernogolovka, Russia are a bit harder to find [59], but the conference was memorable: the Congress of People's Deputies was dissolved by President Boris Yeltsin on Sep. 21, and rumors were rampant during the meeting. Most foreign participants had returned home before street riots and battles took place over Sep. 28-Oct. 5.

1.6. The Age of Reason 1994–2002

X-ray microscopy started a major expansion during this period, with new instruments at new light sources, such as ALS, APS, ESRF, ELETTRA, NSRRC, Spring-8, Aarhus, Ritsumeikan, etc. Tomography [60, 61], cryo [62, 63], and cryo-tomography [64, 65, 66, 67] were demonstrated. The range of applications grew rapidly, including soil science, geochemistry, polymer science, magnetism, etc. Groups in Göttingen, Stockholm, London, Tsukuba and elsewhere were designing and building laboratory-based instruments. David Sayre's old dream of diffraction microscopy (recording the diffraction pattern of a non-crystal, and reconstructing it) had its first successful realization at the NSLS [68].

1.7. The Industrial Revolution 2003–2008

In the last five years X-ray microscopy has entered the mainstream. We are no longer working with an esoteric, new, unproven technique. What brought about this change is the rapidly growing list of successful and highly visible applications in environmental and soil science, geo- and cosmo-chemistry, polymer science, biology, magnetism, energy research, materials and surface science, among others. Without applications we are just a curiosity.

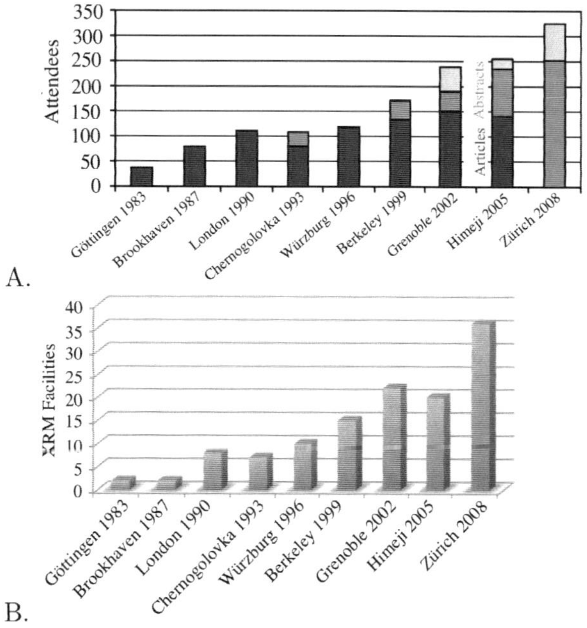

Figure 6. Growth in the x-ray microscopy community. A) Number of attendees, articles with abstracts, and abstracts only at the modern series of x-ray microscopy conferences. The number of articles at the 2008 meeting is not indicated in this figure. B) X-ray microscopy facilities at synchrotron light sources worldwide. The count is based on papers presented at each x-ray microscopy conference.

New microscopes are springing up everywhere, at new and old lightsources (CLS, SSRL, APS, SLS, ALS, BESSY, etc.; see Fig. 6B), and in laboratories without lightsources. The demand has resulted in the success of new commercial enterprises that sell microscopes (Xradia, ACCEL), zone plates (Xradia, zoneplates.com), detectors (DECTRIS), substrates (Silson), etc, and these enterprises have spurred the spread of microscopes. The growth has been particularly stunning in hard X-ray microscopy, where its early promise [21, 69] has been realized with new zone plates, Kirkpatrick-Baez optics, compound refractive lenses and multilayer Laue lenses [13, 70] getting to the point where they rival the resolution and efficiency of soft X-ray optics.

2. The future

It is foolish to try to prognosticate, yet here we go! What shapes the future are the new ideas. If we knew what they were, well ... So we are reduced to extrapolating from what we know, what is already in the pipeline, and especially what has been presented at this Conference. Among the areas that we feel are set for rapid growth is multi-dimensional microscopy, where the added dimension may be space, time, energy, or a combination of these. Since the added dimension generally requires multiple exposures, dealing with the issue of radiation damage becomes particularly important.

2.1. Time resolved microscopy

Until recently we have been largely restricted to the investigation of slowly varying phenomena. In many cases these are of tremendous technical importance, as with the destruction of interconnects in computer chips by electromigration [71, 72], or the curing of concrete [73]. More recently the very fast switching of magnetic vortices has been demonstrated by groups using both STXM and TXM based techniques. With the soft X-ray FEL source at Hamburg (FLASH) coming into operation, time resolved imaging has entered a new area. The time for an exposure is now on the order of 10 fs, and with clever arrangement the group of Chapman *et al.* have succeeded in observing phenomena with a time resolution that is comparable with this exposure time [74]. With the upcoming upgrade of FLASH and the completion of LCLS, XFEL and SCSS, we can confidently predict that very high time resolution and flash imaging will be a major growth area.

2.2. Tomography

Because one of the important advantages of X-rays is their penetrating power, much of the demand in applications is to examine samples too thick for electron microscopes [75, 76, 77]. Here one often encounters the problem of overlapping structures. 2D images are therefore often too complex and not informative, making 3D imaging particularly desirable. 3D microscopy with submicron resolution has been demonstrated over the past decade at several sources, with earlier results noted above and more recent results presented in talks at this conference by Brennan, Cloetens, Feser, Heim, and Larabell, among others. One major development of the past few years has been the success of the tomographic TXM instruments built by Xradia, which make hard X-ray tomography at < 60 nm resolution routine at lightsources [78, 79, 80] and in laboratory [81] instruments.

New, specialized tomographic instruments are also coming on line. At this Conference we saw the first stunning results from the National Center for X-ray Tomography, dedicated to cryo-tomography of biological specimens. The hard X-ray microtomography program at the ESRF has been evolving in the direction toward nanotomography based on projection geometry using a very intense nanoprobe [82].

Even higher resolution tomography has been demonstrated using diffraction microscopy [83, 84, 85, 86]. At this point, collecting and reconstructing the tomographic data set by this technique requires a great deal of effort, but as the technique matures, this should also become more routine.

We expect rapid growth in tomographic applications!

2.3. Elemental and chemical mapping

Spectromicroscopy, where a stack of images [52] is collected at closely spaced steps in energy near an absorption edge (especially near the carbon K edge) is now the main application of scanning microscopes operating at synchrotron light sources. The results provide spatially resolved information on the chemical state of the element in question. Here the incident X-ray energy is the third dimension in the data-set.

In a growing number of instruments it is the elemental constituents that are being mapped by X-ray fluorescence. This involves a hard-X-ray scanning microscope, where the third dimension of the data collected is the X-ray fluorescence spectrum (either in its entirety or sampled at key emission lines) for each pixel. Interest is growing rapidly in mapping trace metals in biological and environmental samples because of the importance of these in health, disease, nutrition and pharmacology. We expect a rapid growth in this area, and in a combination of X-ray fluorescence with spectromicroscopy to map the oxidation state of the elements found, to the extent allowed by radiation damage.

2.4. More than three dimensions

In samples that are less radiation sensitive, we can foresee four dimensional imaging. This may take the form of tomographic spectromicroscopy [87], tomographic movies, etc.

Other added dimensions may be in the form of correlative microscopy, be it visible, infrared, electron, or other. In fact this has always been a part of X-ray microscopy, but it is moving into a level of sophistication that has not been seen in the past. For example Larabell *et al.* developed a capability of looking at the same frozen hydrated specimen in their X-ray microscope, and in a visible light microscope equipped with a custom cryo-stage. This clearly points in a promising direction for the future!

2.5. The spatial resolution frontier

As the spatial resolution of x-ray optics improves (Fig. 7), an increasing number of scientifically important problems can be addressed. Taking this concept to the limit, the stated specification of the NSLS II facility is that it has to demonstrate 1 nm spatial resolution. This is an ambitious goal, though it raises interesting questions regarding radiation damage, depth of field, contrast, and the like. There is broader consensus among practicing x-ray microscopists that sub-10 nm resolution in practical applications is a worthy and ultimately reachable goal.

Zone plates are likely to demonstrate better than 10 nm resolution in the near future. They are most useful as microscope lenses, or to focus the radiation to form a nanoprobe as in a STXM. Multilayer Laue lenses are likely to get us to 5 nm in nanoprobe applications.

Diffraction microscopy is a different animal. It is an image forming technique that has already demonstrated very high resolution, though how exactly its resolution is determined is still a topic of some discussion. However, its lack of probe-forming optics rules out fluorescence mapping, and it does not give instant images. Plans for single molecule imaging [99] using X-ray lasers such as the LCLS call for high enough resolution to determine the structure of large protein molecules by diffraction microscopy, but this requires many identical copies of the molecule, each being exposed to the ultrashort and ultrastrong beam. Clearly if this, or something similar can be made to work, it will have tremendous impact!

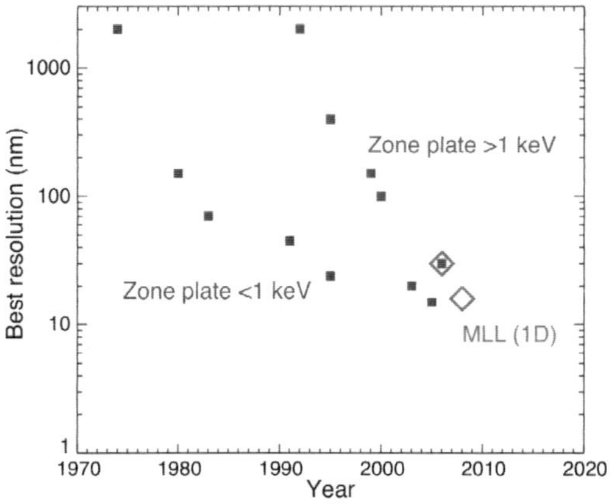

Figure 7. Best focusing values for diffractive focusing optics by year. Only those results that represent a new improvement from previous results are shown. Results are separated between those at ¡1 keV photon energies [23, 88, 25, 89, 90, 91, 92], and at ¿1 keV photon energies [93, 94, 95, 96, 97]. Also shown are 1D focusing results of multilayer Laue lenses (MLLs) [70, 98]. Results for refractive and reflective optics are not shown, though they have also experienced dramatic resolution improvements over time.

2.6. New ideas

A number of new ideas aim to extend the resolution that can be achieved in scanning microscopy, or in Fourier transform holography. The basic scheme is to record the diffraction pattern to high angles. In the approach by the Swiss Light Source investigators [100, 101], the detector in the STXM is the fast pixel-array detector PILATUS, and the full diffraction pattern is recorded at overlapping pixels of the scan. A four-fold resolution enhancement has been demonstrated. Williams *et al.* [102] use a beam diverging from a small focus to record the diffraction pattern. In this case the central part of the pattern is basically just the Fourier transform hologram of the object, and the resolution is set by the size of the small focus. But by detecting and reconstructing the large angle diffraction pattern, the resolution is extended substantially. The use of spherical, as opposed to planar, illumination facilitates the reconstruction.

3. Closing remarks

We have emphasized new developments. These often take heroic efforts, and take a while to mature, if they ever do. They are of great importance for the future of the field. It is equally, or may be even more important to have user-friendly workhorse microscopes that churn out results, and are available to the novice. We are entering a period where we have both. This guarantees the continued growth and health of X-ray microscopy.

Acknowledgments

We are very grateful to the organizers of XRM 2008 for a very successful and enjoyable meeting, and acknowledge the generous support of DOE, NSF and NIH for the efforts at Stony Brook. The Advanced Light Source is supported by the Director, Office of Science, Office of Basic Energy Sciences, of the U.S. Department of Energy under Contract No. DE-AC02-05CH11231.

References

[1] Röntgen W C 1895 *Sitzungsberichte der Würzburger Physik-medic. Gesellschaft* **137** 41
[2] Malsch F 1939 *Naturwissenschaften* **27** 854–855
[3] Wood R W 1898 *Philosiphical Magazine* **45** 511–522
[4] Einstein A 1918 *Verhandlungen der Deutschen Physikalischen Gesellschaft* **9** 86–87
[5] Engström A 1946 *Acta Radiologica (Supplement)* **63** 1–106
[6] Kirkpatrick P and Baez A V 1948 *Journal of the Optical Society of America* **38** 766–774
[7] Kirkpatrick P 1949 *Scientific American* **180** 44–47
[8] Cosslett V E 1959 *Nature* **184** 860–862 ISSN 0028-0836
[9] Cosslett V E, Engström A and Pattee Jr H H (eds) 1957 *International Symposium on X-ray Optics and X-ray Microanalysis* (New York: Academic Press) cavendish Laboratory, Cambridge, 1956

[10] Engström A, Cosslett V E and Pattee Jr H H (eds) 1960 *X-ray Microscopy and X-ray Microanalysis* (Amsterdam: Elsevier) stockholm, 1959
[11] Pattee Jr H H, Cosslett V E and Engström A (eds) 1963 *X-ray Optics and X-ray Microanalysis* (New York: Academic Press) stanford, 1962
[12] Kirkpatrick P and Pattee H 1953 *Advances in Biological and Medical Physics* **3** 247–283
[13] Maser J, Stephenson G, Vogt S, Yun W, Macrander A, Kang H, Liu C and Conley R 2004 *Design and Microfabrication of Novel X-ray Optics II* vol 5539 ed Snigirev A and Mancini D (SPIE) pp 185–194
[14] Baez A V 1952 *Nature* **169** 963–964
[15] Baez A V 1952 *Journal of the Optical Society of America* **42** 756–762
[16] Cosslett V E and Nixon W C 1960 *X-ray Microscopy* (London: Cambridge University Press)
[17] Cosslett V and Nixon W 1952 *Nature* **170** 436–438 ISSN 0028-0836
[18] Hall T, Rockert H and Saunders R d H 1972 *X-Ray Microscopy in Clinical and Experimental Medicine* (Springfield, Illinois: C. C. Thomas)
[19] Albright J and Kunstel M 1997 *Bombshell: The Secret Story of America's Unknown Atomic Spy Conspiracy* (Crown)
[20] Schmahl G and Rudolph D 1969 *Optik* **29** 577–585
[21] Horowitz P and Howell J A 1972 *Science* **178** 608–611
[22] Aoki S and Kikuta S 1974 *Japanese Journal of Applied Physics* **13** 1385–1392
[23] Niemann B, Rudolph D and Schmahl G 1974 *Optics Communications* **12** 160–163
[24] Niemann B, Rudolph D and Schmahl G 1976 *Applied Optics* **15** 1883–1884
[25] Niemann B, Rudolph D and Schmahl G 1983 *Nuclear Instruments and Methods in Physics Research* **208** 367–372
[26] Rarback H, Kenney J, Kirz J and Xie X S 1980 in Ash [103] pp 449–456
[27] Rarback H, Shu D, Feng S C, Ade H, Jacobsen C, Kirz J, McNulty I, Vladimirsky Y, Kern D P and Chang T H P 1988 in Sayre *et al.* [53] pp 194–200
[28] Morrison G R, Bridgwater S, Browne M T, Burge R E, Cave R C, Charalambous P S, Foster G F, Hare A R, Michette A G, Morris D, Taguchi T and Duke P 1989 *Review of Scientific Instruments* **60** 2464–2467
[29] Rudolph D, Niemann B, Schmahl G and Christ O 1984 in Schmahl and Rudolph [41] pp 192–202
[30] Voss J and Kunz C 1992 in Michette *et al.* [54] pp 235–237
[31] Ade H, Kirz J, Hulbert S L, Johnson E, Anderson E and Kern D 1990 *Applied Physics Letters* **56** 1841–1843
[32] Ng W, Ray-Chaudhuri A K, Crossley S, Crossley D, Gong C, Guo J, Hansen R W C, Margaritondo G, Cerrina F, Underwood J H, Perera R C C and Kortright J 1990 *Journal of Vacuum Science and Technology A* **8** 2563–2565
[33] Howells M, Jacobsen C, Kirz J, Feder R, McQuaid K and Rothman S 1987 *Science* **238** 514–517
[34] McNulty I, Kirz J, Jacobsen C, Anderson E, Kern D and Howells M 1992 *Science* **256** 1009–1012
[35] Lindaas S, Howells M, Jacobsen C and Kalinovsky A 1996 *Journal of the Optical Society of America A* **13** 1788–1800
[36] Trebes J E, Brown S B, Campbell E M, Matthews D L, Nilson D G, Stone G F and Whelan D A 1987 *Science* **238** 517–519
[37] Skinner C H, DiCicco D S, Kim D, Rosser R J, SSuckewer, Gupta A P and Hirschberg J G 1990 *Journal of Microscopy* **159** 51–60
[38] DaSilva L B, Trebes J E, Balhorn R, Mrowka S, Anderson E, Attwood D T, Barbee Jr T W, Brase J, Corzett M, Gray J, Koch J A, Lee C, Kern D, London R A, MacGowan B J, Matthews D L and Stone G 1992 *Science* **258** 269–271
[39] Feder R, Banton V, Sayre D, Costa J, Baldini M and Kim B K 1985 *Science* **227** 63–64
[40] Spiller E 1980 in Ash [103] pp 365–391
[41] Schmahl G and Rudolph D (eds) 1984 *X-ray Microscopy* (*Springer Series in Optical Sciences* vol 43) (Berlin: Springer-Verlag)
[42] Schmahl G and Rudolph D 1987 in Cheng and Jan [104] pp 231–238
[43] Schmahl G, Rudolph D, Schneider G, Guttmann P and Niemann B 1994 *Optik* **97** 181–182
[44] Sayre D 1972 Proposal for the utilization of electron beam technology in the fabrication of an image forming device for the soft x-ray region Tech. Rep. RC 3974 (#17965) IBM T. J. Watson Research Laboratory Yorktown Heights, New York
[45] Shaver D C, Flanders D C, Ceglio N M and Smith H I 1980 *Journal of Vacuum Science and Technology* **16** 1626–1630
[46] Kern D, Coane P, Acosta R, Chang T H P, Feder R, Houzego P, Molzen W, Powers J, Speth A, Viswanathan R, Kirz J, Rarback H and Kenney J 1984 *Science with Soft X-Rays* vol 447 ed Himpsel F J and Klaffky R W (Bellingham, Washington: Society of Photo-Optical Instrumentation Engineers (SPIE)) pp 204–213
[47] Pawlak J, Cheng P C and Shinozaki D M 1987 in Cheng and Jan [104] pp 336–345

[48] Rarback H, Buckley C, Ade H, Camilo F, DiGennaro R, Hellman S, Howells M, Iskander N, Jacobsen C, Kirz J, Krinsky S, Lindaas S, McNulty I, Oversluizen M, Rothman S, Sayre D, Sharnoff M and Shu D 1990 *Journal of X-ray Science and Technology* **2** 274–296
[49] Ade H, Zhang X, Cameron S, Costello C, Kirz J and Williams S 1992 *Science* **258** 972–975
[50] Zhang X, Ade H, Jacobsen C, Kirz J, Lindaas S, Williams S and Wirick S 1994 *Nuclear Instruments and Methods in Physics Research A* **347** 431–435
[51] De Stasio G, Hardcastle S, Koranda S F, Tonner B P, Mercanti D, Teresa Ciotti M, Perfetti P and Margaritondo G 1993 *Physical Review* **E 47** 2117–2121
[52] Jacobsen C, Flynn G, Wirick S and Zimba C 2000 *Journal of Microscopy* **197** 173–184
[53] Sayre D, Howells M R, Kirz J and Rarback H (eds) 1988 *X-ray Microscopy II* (*Springer Series in Optical Sciences* vol 56) (Berlin: Springer-Verlag)
[54] Michette A G, Morrison G R and Buckley C J (eds) 1992 *X-ray Microscopy III* (*Springer Series in Optical Sciences* vol 67) (Berlin: Springer-Verlag)
[55] Thieme J, Schmahl G, Umbach E and Rudolph D (eds) 1998 *X-ray Microscopy and Spectromicroscopy* (Berlin: Springer-Verlag)
[56] Meyer-Ilse W, Warwick T and Attwood D (eds) 2000 *X-ray Microscopy: Proceedings of the Sixth International Conference* (Melville, NY: American Institute of Physics) ISBN 1-56396-926-2
[57] Susini J, Joyeux D and Polack F 2003 *Journal de Physique IV* **104** V
[58] Aoki S, Kagoshima Y and Suzuki Y (eds) 2006 *X-ray Microscopy: Proceedings of the 8th International Conference* IPAP Conference Series 7 (Tokyo, Japan: The Institute of Pure and Applied Physics (IPAP))
[59] Aristov V V and Erko A I (eds) 1994 *X-ray Microscopy IV* Proceedings of the 4th International Conference, Chernogolovka, Russia, September 20–24, 1993 (Chernogolovka, Russia: Bogorodskii Pechatnik) ISBN 5-900846-01-6
[60] Haddad W S, McNulty I, Trebes J E, Anderson E H, Levesque R A and Yang L 1994 *Science* **266** 1213–1215
[61] Lehr J 1997 *Optik* **104** 166–170
[62] Schneider G and Niemann B 1998 in Thieme *et al.* [55] pp I–25–34
[63] Maser J, Jacobsen C, Kirz J, Osanna A, Spector S, Wang S and Warnking J 1998 in Thieme *et al.* [55] pp I–35–44
[64] Weiß D, Schneider G, Niemann B, Guttmann P, Rudolph D and Schmahl G 2000 *Ultramicroscopy* **84** 185–197
[65] Wang Y, Jacobsen C, Maser J and Osanna A 2000 *Journal of Microscopy* **197** 80–93
[66] Schneider G, Anderson E, Vogt S, Knöchel C, Weiss D, Legros M and Larabell C 2002 *Surface Review and Letters* **9** 177–183
[67] Larabell C and Le Gros M 2004 *Molecular Biology of the Cell* **15** 957–962
[68] Miao J, Charalambous P, Kirz J and Sayre D 1999 *Nature* **400** 342–344
[69] Jacobsen C 1992 in Michette *et al.* [54] pp 274–277
[70] Kang H, Maser J, Stephenson G, Liu C, Conley R, Macrander A and Vogt S 2006 *Physical Review Letters* **96** 127401
[71] Schneider G, Hambach D, Niemann B, Kaulich B, Susini J, Hoffmann N and Hasse W 2001 *Applied Physics Letters* **78** 1936–1938
[72] Zschech E, Geisler H, Rinderknecht J, Schneider G, Spolenak R and Schmeisser D 2008 *Current Nanoscience* **4** 256–266 ISSN 1573-4137
[73] Kurtis K E, Monteiro P J M, Brown J T and Meyer-Ilse W 1998 *Cement and Concrete Research* **28** 411–421
[74] Chapman H, Hau-Riege S, Bogan M, Bajt S, Barty A, Boutet S, Marchesini S, Frank M, Woods B, Benner W, London R, Rohner U, Szoeke A, Spiller E, Moeller T, Bostedt C, Shapiro D, Kuhlmann M, Treusch R, Ploenjes E, Burmeister F, Bergh M, Caleman C, Huldt G, Seibert M and Hajdu J 2007 *Nature* **448** 676–679 ISSN 0028-0836
[75] Sayre D, Kirz J, Feder R, Kim D M and Spiller E 1977 *Science* **196** 1339–1340
[76] Jacobsen C, Medenwaldt R and Williams S 1998 in Thieme *et al.* [55] pp II–93–102
[77] Grimm R, Singh H, Rachel R, Typke D, Zillig W and Baumeister W 1998 *Biophysical Journal* **74** 1031–1042
[78] Yin G C, Tang M T, Song Y F, Chen F R, Liang K, Duewer F, Yun W, Ko C H and Shieh H P 2006 *Applied Physics Letters* **88** 241115 ISSN 0003-6951
[79] Chu Y, Yi J, De Carlo F, Shen Q, Lee W K, Wu H, Wang C, Wang J, Liu C, Wang C, Wu S, Chien C, Hwu Y, Tkachuk A, Yun W, Feser M, Liang K, Yang C, Je J and Margaritondo G 2008 *Applied Physics Letters* **92** 103119 ISSN 0003-6951
[80] Chen Y T, Lo T N, Chu Y, Yi J, Liu C J, Wang J Y, Wang C L, Chiu C W, Hua T E, Hwu Y, Shen Q, Yin G C, Liang K, Lin H M, Je J and Margaritondo G 2008 *Nanotechnology* **19** 395302 ISSN 0957-4484
[81] Tkachuk A, Duewer F, Cui H, Feser M, Wang S and Yun W 2007 *Zeitschrift Kristallographie* **222** 650–655
[82] Mokso R, Cloetens P, Maire E, Ludwig W and Buffiere J Y 2007 *Applied Physics Letters* **90** 144104 ISSN

0003-6951

[83] Miao J, Ishikawa T, Johnson B, Anderson E, Lai B and Hodgson K 2002 *Physical Review Letters* **89** 088303

[84] Chapman H, Barty A, Marchesini S, Noy A, Hau-Riege S P, Cui C, Howells M, Rosen R, He H, Spence J, Weierstall U, Beetz T, Jacobsen C and Shapiro D 2006 *Journal of the Optical Society of America A* **23** 1179–1200

[85] Pfeifer M, Williams G, Vartanyants I, Harder R and Robinson I 2006 *Nature* **442** 63–66

[86] Barty A, Marchesini S, Chapman H, Cui C, Howells M, Shapiro D, Minor A, Spence J, Weierstall U, Ilavsky J, Noy A, Hau Riege S, Artyukhin A, Baumann T, Willey T, Stolken J, van Buuren T and Kinney J 2008 *Physical Review Letters* **101** 055501

[87] Johansson G, Tyliszczak T, Mitchell G, Keefe M and Hitchcock A 2007 *Journal of Synchrotron Radiation* **14** 395–402

[88] Schmahl G, Rudolph D and Niemann B 1980 in Ash [103] pp 393–412

[89] Jacobsen C, Williams S, Anderson E, Browne M T, Buckley C J, Kern D, Kirz J, Rivers M and Zhang X 1991 *Optics Communications* **86** 351–364

[90] Schneider G, Schliebe T and Aschoff H 1995 *Journal of Vacuum Science and Technology B* **13** 2809–2812

[91] Chao W *et al.* 2003 *Optics Letters* **28** 2019–2021

[92] Chao W, Harteneck B, Liddle J, Anderson E and Attwood D 2005 *Nature* **435** 1210–1213

[93] Yun W B, Viccaro P J, Chrzas J and Lai B 1992 *Review of Scientific Instruments* **63** 582–585

[94] Lai B, Yun W, Xiao Y, Yang L, Legnini D, Cai Z, Krasnoperova A, Cerrina F, DiFabrizio E, Grella L and Gentili M 1995 *Review of Scientific Instruments* **66** 2287–2289

[95] Kaulich B, Oestreich S, Salome M, Barrett R, Susini J, Wilhein T, Di Fabrizio E, Gentili M and Charalambous P 1999 *Applied Physics Letters* **75** 4061–4063

[96] David C, Kaulich B, Barrett R, Salome M and Susini J 2000 *Applied Physics Letters* **77** 3851–3853

[97] Feng Y, Feser M, Lyon A, Rishton S, Zeng X, Chen S, Sassolini S and Yun W 2007 *Journal of Vacuum Science and Technology B* **25** 2004–2007

[98] Kang H, Yan H, Winarski R, Holt M, Maser J, Liu C, Conley R, Vogt S, Macrander A and Stephenson G 2008 *Applied Physics Letters* **92** 221114 ISSN 0003-6951

[99] Neutze R, Wouts R, van der Spoel D, Weckert E and Hajdu J 2000 *Nature* **406** 752–757

[100] Rodenburg J, Hurst A, Cullis A, Dobson B, Pfeiffer F, Bunk O, David C, Jefimovs K and Johnson I 2007 *Physical Review Letters* **98** 034801–1–4

[101] Thibault P, Dierolf M, Menzel A, Bunk O, David C and Pfeiffer F 2008 *Science* **321** 379–382

[102] Williams G J, Quiney H M, Dhal B B, Tran C Q, Nugent K A, Peele A G, Paterson D and de Jonge M D 2006 *Physical Review Letters* **97** 025506 ISSN 0031-9007

[103] Ash E A (ed) 1980 *Scanned Image Microscopy* (London: Academic Press)

[104] Cheng P C and Jan G J (eds) 1987 *X-ray Microscopy: Instrumentation and Biological Applications* (Berlin: Springer-Verlag)

Full-field transmission x-ray microscopy at SSRL

J C Andrews[1,4], S Brennan[1], P Pianetta[1], H Ishii[2], J Gelb[3], M Feser[3], J Rudati[3], A Tkachuk[3] and W Yun[3]

[1] Stanford Synchrotron Radiation Laboratory, Menlo Park CA 94025, USA
[2] Institute of Geophysics and Planetary Physics, Lawrence Livermore National Laboratory, Livermore CA 94550, USA
[3] Xradia Inc., Concord CA 94520, USA

Abstract. A full-field hard-x-ray microscope at SSRL with 15 micron field of view and 40 nm resolution, as well as 3D tomographic capabilities, has successfully imaged samples of biological, environmental and astronomical origin. Spectroscopic imaging of a particle of comet dust from the NASA Stardust mission showed significantly more absorption contrast above the Fe K-edge, and tomography revealed the three-dimensional structure of the terminal particle and track through the aerogel.

1. Introduction

The imaging of biological, environmental and materials samples to capture structural detail and chemical composition has been attempted by many methods. X-ray microscopy in the soft x-ray region has been used extensively for biological imaging [1], and full-field hard x-ray microscopy has also been used successfully at several synchrotron sources [2]. The Xradia full-field hard x-ray microscope [3] was the first commercial instrument capable of sub-40 nm resolution. Using high resolution zone plates [4], it provides 3D tomographic capabilities [5], enabling spatial resolution in the 30-60 nm range. The transmission x-ray microscope (TXM) implemented at SSRL is capable of 2-dimensional imaging in phase and absorption contrast with 40 nm resolution, and 3D tomography and spectroscopic imaging from 4-14 keV [6]. Therefore, this instrument is an excellent tool for nondestructive high-resolution imaging as demonstrated across a wide range of samples, including mouse bone, plant roots, yeast cells [7]; and comet dust, as described below.

2. Microscope setup and capabilities

The x-ray microscope is located on beam line 6-2 at the Stanford Synchrotron Radiation Laboratory, with a 54-pole wiggler. The beam passes through vertical and horizontal slits, a vertically collimating mirror (M_0) followed by vertical slits, monochromator, and toroidal mirror (M_1) to focus the beam at the virtual source (S3) (figure 1a). A diffuser 3 meters before S3 and vertical stabilizer immediately after S3 (not shown) help to provide uniform illumination. A capillary condenser (C) provides hollow-cone illumination that is focused through a pinhole (P) onto the sample (figure 1b). A central stop (not shown) eliminates the direct beam and the remaining undeflected light is removed by the pinhole. The image is projected by a micro zone plate (MZP) onto the CCD (1024 x 1024). Zernike phase contrast can be provided by a gold phase ring (PR) that is centered prior to imaging by inserting a Bertran lens (BL).

[4] Corresponding author. E-mail: jandrews@slac.stanford.edu

The microscope is capable of 40 nm resolution in phase contrast at 8 keV (and soon at 5 keV), and absorption contrast from 4-14 keV over a 14 μm field of view. Because of the relatively large depth of focus (~20-40 μm), the samples can be imaged without sectioning as is routinely required for electron microscopy (EM). This depth of focus also provides full transmission imaging as opposed to surface only for EM. The sample stage allows for xyz and theta motion for 3D tomography. Spectroscopic imaging above and below the absorption edge facilitates studies of nanoscale trace element distribution. A modulation transfer function of an image of a Siemens star with 30 nm minimum feature size and analysis of the finer region confirmed 40 nm resolution (figure 2).

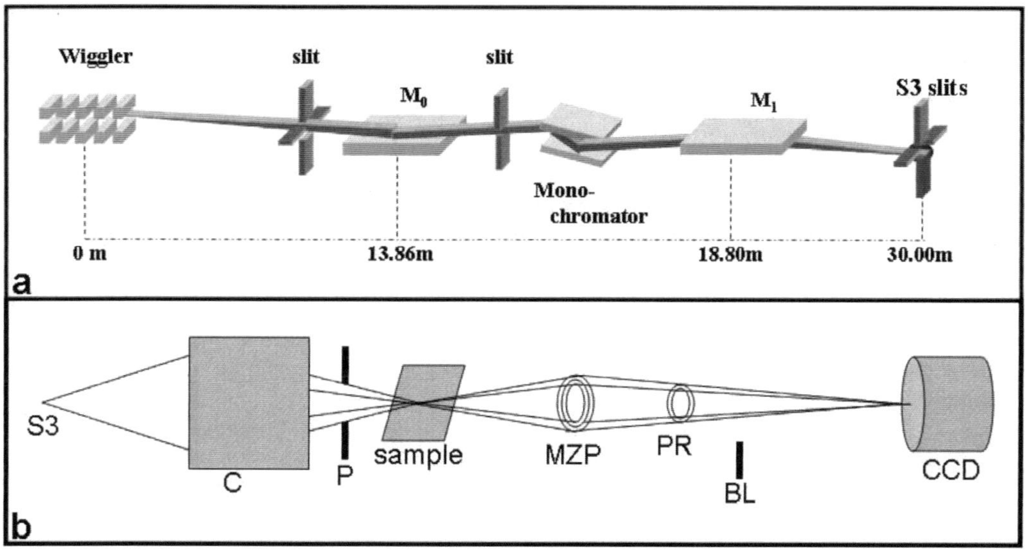

Figure 1: Setup of TXM on beamline 6-2 at SSRL (a) before and (b) after virtual source. (a) The beam passes through vertical and horizontal slits, a vertically collimating mirror (M_0) followed by vertical slits, monochromator, and toroidal mirror (M_1) to focus the beam at the virtual source (S3). (b) Capillary condenser (C) provides hollow cone illumination and pinhole (P) removes undeflected light. Sample image is focused by micro zone plate (MZP) onto CCD detector, with optional phase ring (PR) for use at 8 keV and 5 keV. A Bertran lens (BL) is used to position the phase ring, and is removed from the beam for imaging. Drawings are not to scale.

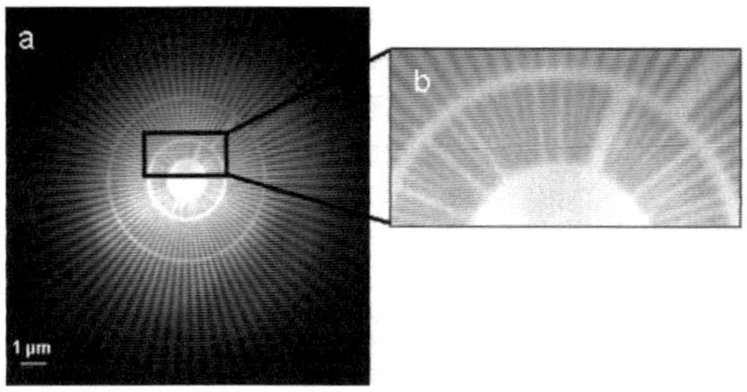

Figure 2: Phase contrast image at 8 keV of Siemens star with 30 nm minimum feature size (a) and detail of region with smallest features (b). Modulation transfer function (not shown) confirms 40 nm resolution.

3. Spectroscopic and tomographic imaging of comet dust

In collaboration with NASA and the Institute for Geophysics and Planetary Physics at Lawrence Livermore National Laboratory, we obtained TXM images of dust gathered from comet 81-P/Wild2 by the Stardust spacecraft. Microprobe measurements (not shown) revealed the presence of Fe, Ni and Cu in the particle [8]. 3D tomography (not shown) portrayed the shape of the terminal particle with a portion of the trail through the aerogel. Spectroscopic imaging through the iron K-edge (figure 6 a-c) in absorption contrast shows the location of iron within the particle, which grows darker above the iron K-edge (at 7112 eV).

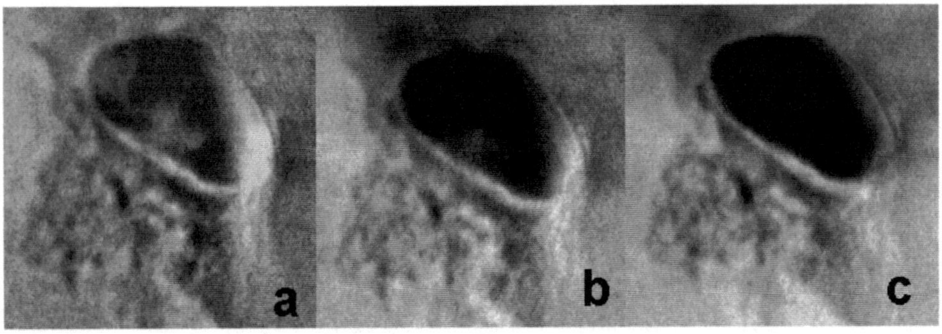

Figure 6: Imaging of comet track T12 from Comet 81-P/Wild2 gathered in the Stardust mission. Absorption contrast images at (a) 7100 eV, (b) 7115 eV and (c) 7150 eV show darkening of the particle due to absorption above the Fe K-edge.

4. Conclusion

In summary, the TXM at beam line 6-2 at SSRL has successfully imaged samples with biological, environmental and materials applications. With this method we have been able to discern features not previously apparent with other methods. In the immediate future we will be upgrading the CCD to 2K x 2K pixels to increase the field of view, and adding a vortex detector for fluorescence detection of trace elements within a portion of the field of view. Plans are also being drawn up for a liquid nitrogen cryostat, and for a more sensitive sample stage which will improve tomographic resolution.

Acknowledgments

This work has been supported by NIH/NIBIB grant number R01-EB004321. SSRL is supported by the Department of Energy, Office of Basic Energy Sciences.

References

[1] Parkinson D Y, McDermott G, Etkin L D, LeGros M A and Larabell C 2008 *J. Struct. Biol.* **162** 380
[2] Kim G B, Yoon Y J, Shin T J, Youn H S, Gho Y S and Lee S J 2009 *Microsc. Res. Technique* **71** 639
Neuhausler U and Schneider G 2006 *Microelectr. Eng.* **83** 1043
Rau C et al. 2007 *Nuclear Instr. Meth. Phys. Res. B* **261** 850
[3] Yin G C, Tang M T, Song Y F, Chen F R, Liang K S, Duewer F W, Yun W, Ko C H and Shieh H P D 2006 *Appl. Phys. Lett.* **88** 241115
[4] Feng Y, Feser M, Lyon A, Rishton S, Zeng X, Chen S, Sassolini S and Yun W 2007 *J. Vac. Sci. Technol.* B. **25** 2004
Chu Y S *et al.* 2008 *Appl. Phys. Lett.* **92** 103119
[5] Tkachuk A, Duewer F, Cui H, Feser M, Wang S and Yun W 2007 *Z. Kristallogr.* **222** 650
[6] Andrews J C, Brennan S, Patty C, Luening K, Pianetta P, Almeida E, van der Meulen M C H, Feser M, Gelb J, Rudati J, Tkachuk A and Yun W 2008 *Synchr. Rad. News* **21** 17
[7] Andrews J C *et al.*, this proceedings
[8] J.G. Flynn *et al.* 2006 *Science* **314** 1731

Journal of Physics: Conference Series **186** (2009) 012003 doi:10.1088/1742-6596/186/1/012003

The microXAS beamline at the Swiss Light source: towards nano-scale imaging

C.N. Borca[1], D. Grolimund[1], M. Willimann[1], B. Meyer[1], K. Jefimovs[1,2], J. Vila-Comamala[1], C. David[1]

[1]Paul Scherrer Institut, Villigen PSI, CH-5232, Switzerland,
[2]EMPA, Dübendorf, CH-8600, Switzerland

camelia.borca@psi.ch

Abstract. The microXAS beamline is a dedicated hard X-ray microprobe facility allowing a combination of fluorescence, spectroscopy and diffraction techniques in an energy scale from 4 to 23 keV. This paper presents a short review of the beamline, including the recent developments towards nano-scale imaging.

1. General description of beamline

The small source size and low emittance of the new 3rd generation synchrotron sources in combination with recent advances in the design and production of hard X-ray focusing optics opens up new possibilities for high-flux hard X-ray microprobe beamlines.

The microXAS beamline is located at the long straight section X05L of the SLS ring. A minigap in-vacuum undulator (U19) serves as radiation source providing high brightness X-rays in the energy range from <4 to 23 keV. The photon flux delivered at 12 keV is 3×10^{12} photons/sec, while the optical elements used ensure an energy resolution of $\Delta E/E < 10^{-4}$.

The optical layout of the beamline is composed of several pairs of slits, a bendable toroidal, horizontally deflecting mirror and a monochromator. The vertical mirror unit serves three main purposes: (i) to collimate the beam in the vertical dimension, (ii) to allow for dynamic demagnification in horizontal dimension, and to act as a low-pass filter with an energy cut-off of ~23 keV given by the Rh coating. The horizontal focusing corresponds to the first part of a two-step focusing strategy that offers two main advantages: a secondary source with flexible size adjustment by precision slits (the capability of dynamical focusing and the possibility of optimizing the overall acceptance of the subsequent microfocusing optical system. The fixed–exit double-crystal monochromator is equipped with three different pairs of crystals: Si(111), Si(311) for higher energy resolution and Ge(111) for higher flux throughput.

2. Micro-probe setup

The entire micro-probe setup is installed on a stable optical table. For achromatic focusing in the entire energy range of 4-23 keV we use an elliptical shape mirror pair in the Kirkpatrick-Baez (KB) geometry, commercially available from Xradia. The mirrors are 10 cm in length, coated with Rh and enclosed in a He chamber that ensures their thermal stability. The acceptance of the horizontal and vertical mirrors at a nominal incident angle of 3.1 mrad is ~ 0.3 mm for both directions. The respective nominal focal distances are 150 mm and 260 mm for the centers of horizontally and vertically focusing mirrors, respectively. Since the source is strongly asymmetric, the horizontal and vertical dimensions of the focal spot are not the same. Highest total photon flux density is obtained for a 6×1 μm^2 spot. A more symmetric spot size near 1 μm^2 can be obtained by dynamically reducing the secondary source size. Figure 1 shows typical beam profiles as measured by means of knife edge scans. 'Reasonable effort' spot size values measured using monochromatic beam are 1.3 (H) × 0.9 (V) μm^2, while using a white beam the spot size is 1.3 (H) × 0.75 (V) μm^2. The monochromatic photon flux measured in the focal spot is in the order of 10^{11} photons/sec.

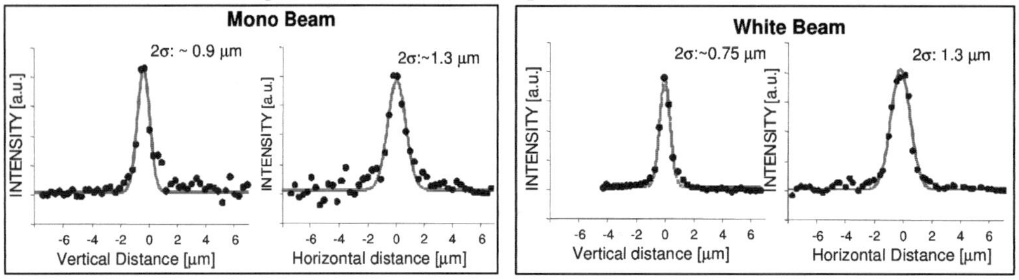

Figure 1. Beam size measured by knife-edge scans for micro-focused white light (left panel) and a monochromatic micro-beam (right panel).

Several types of detectors and beam monitors are available at the beamline corresponding to the needs of the different experimental setups available. For the micro-spectroscopy setup we use a single element Si(Li) detector (Ketek) with an active area of 50 mm^2. The XRF map in Figure 2 shows the spatial distribution of Ni fluorescence intensity obtained from an 80 nm thick Ni patterned film fabricated by electron beam lithography using lift-off technique. The size of the grid lines was 1 μm in both horizontal and vertical directions, with a period of 10 μm. The sample is mounted on a fully encoded XYZ manipulator, with encoder resolution of 50 nm for the horizontal axis and 100 nm for the other two directions. Additional X-Y piezo nanopositioners, as well as 2 sets of rotation stages, can be mounted on the sample manipulator in case precise sample positioning is needed (for example in the micro-XRD setup). For monitoring the incoming micro-beam we use an ion chamber of reduced dimensions, developed in house, adapting the type of flowing gas with the energy needed.

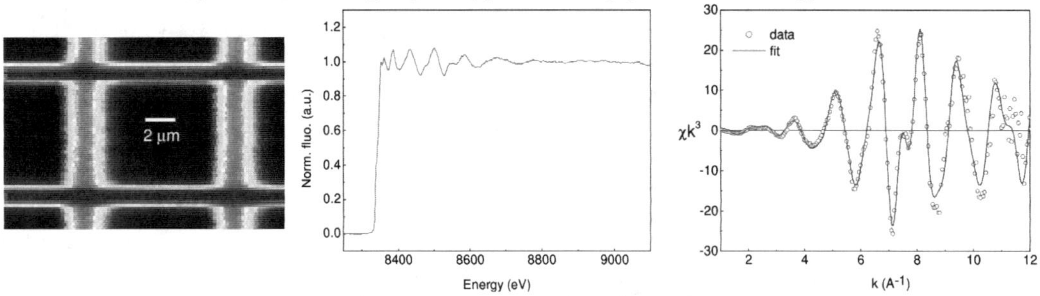

Figure 2. XRF map and μ-EXAFS for Ni K-edge. The size of the Ni grid lines is 1 μm and a period of 10 μm. The middle panel shows an average of 5 unsmoothed μ-EXAFS scans acquired at the intersection of two Ni lines. The third panel shows the corresponding EXAFS oscillations and the corresponding fit.

The μ-EXAFS at the Ni K-edge was performed at the intersection of two grid lines, using a beam size of 1.3 μm (H) x 0.5 μm (V). The average of 5 unsmoothed scans is shown in Figure 2 (middle panel), while the corresponding EXAFS oscillations in k-space are shown in the right panel. The fit shown with continuous line was performed using the IFEFFIT package, indicating the presence of 6 Ni nearest neighbors at a distance of 2.45(7) Å, as expected for bulk Ni^0.

3. Nano-probe setup

Focusing of soft X-ray beams with Frenel zone plates (FZP) has proven to give focal spots with excellent spatial resolution combined with low background [1]. The diffraction limit for the resolution of a FZP is on the order of the smallest, outermost zone width, meaning that nanolithography processes with very high resolution is required. Efficient focusing of hard X-rays is more difficult, as additional problems must be overcome: to achieve acceptable diffraction efficiencies, the zone plate structures should be made from heavy materials and the required height of the structures needs to be a micron or more. This means that the structures have to be 10 or even 20 times higher than their width, when both high resolution and efficiency is needed. Lensless X-ray microscopy has been successfully demonstrated at our beamline at 6-8 keV [2]. Here we describe recent results on the focusing of 15 keV X-rays using gold FZPs with an outermost zone width of 100 nm and a profile height of 1 μm. The fabrication process of gold FZPs with an aspect ratio of more than 10 is described in reference [3].

A schematic drawing of key components of the setup used at the micro-XAS beamline is shown in Figure 3. Experiments were performed in air at atmospheric pressure using an X-ray energy of 15 keV. The X-ray beam passes first through an aperture and then it is focused by the FZP. In order to reduce the background caused by the divergent radiation of the higher diffraction orders, a 5 μm order selecting aperture (OSA) is put 3 mm downstream of the FZP. The coherence length of the X-ray beam at this energy is ideally about 100 μm in vertical direction and 20 μm in horizontal direction. In order to cut the incoherent part of the radiation we used a 20 μm coherence defining aperture in front of the FZP. The FZP has a diameter of 30 μm and an outermost zone-width of $\Delta r \cdot N$ = 100 nm and therefore, only zones down to $\Delta r \cdot N$ = 150 nm were illuminated through the aperture. In case of fully coherent illumination, a focal spot of $\delta = 1.22 \times \Delta r_N$ = 182 nm in the first diffraction order could be achieved in our case.

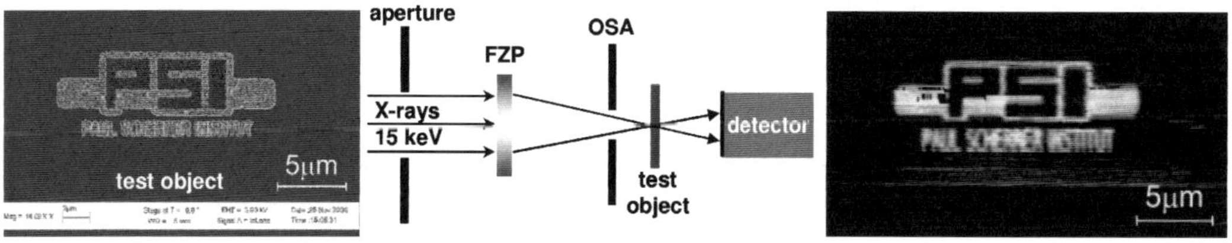

Figure 3. Schematic drawing of the focusing setup and SEM-image of the test pattern used for imaging (right) and of the reconstructed fluorescence map (left). Only zones down to 150 nm were illuminated through a 20 μm aperture in front of the zone plate.

A number of improvements can be done to increase the aspect ratio of the structures of the FZPs, and therefore further reduce spatial resolution at higher energies, as further described in reference [4]. The future development plans at the microXAS beamline include full implementation of hard X-ray nano-imaging capabilities, starting with fluorescence imaging and going to nano-spectroscopy in a later stage.

References

[1] W. Chao, B.D. Harteneck, J.A. Liddle, E.H. Anderson, D.T. Attwood, Nature 435 (2005) 1210,
[2] J. M. Rodenburg, A.C. Hurst, A.G. Cullis, B.R. Dobson, F. Pfeiffer, O. Bunk, C. David, Phys. Rev. Lett. 98 (2007) 034801-3; I. Johnson, K. Jefimovs, O. Bunk, C. David, M. Dierolf, J. Gray, D. Renker, F. Pfeiffer, Phys. Rev. Lett 100 (2008) 155503-4,
[3] K. Jefimovs , O. Bunk, F. Pfeiffer, D. Grolimund, J.F. van der Veen, C. David, Microelectronic Engineering 84 (2007) 1467–1470,
[4] Vila-Comamala et al., this volume.

Modeling of XANES-Spectra with the FEFF-Program

E. Bosman and J. Thieme

Institute for X-Ray Physics, Georg-August-University Göttingen, Friedrich-Hund-Platz 1, 37077 Göttingen, Germany

E-mail: e.bosman@gmx.de, jthieme@gwdg.de

Abstract. The aim of this project is the calculation of the absorption coefficient μ of x-ray absorption spectra as a function of energy and a comparison with experimental data. A characteristic fine structure can be found in x-ray absorption spectra (XAS) consisting of the XANES (X-Ray Absorption Near Edge Structure) and the EXAFS (Extended X-Ray Absorption Fine Structure) region. XANES is characterized by multiple scattering and provides information about coordination chemistry and bonding angles of the irradiated sample. The program FEFF 8.4 was used for the calculations of the absorption K-edge spectra. FEFF was generated for ab initio multiple scattering calculations of X-ray Absorption Fine Structure (XAFS) of atom-clusters. The code yields scattering amplitudes, phases and other quantities. We computed μ at the K-edge of several elements like Ti, S and Fe. For this purpose, clusters of Na_2SO_4, Ba_2TiO_4, FeS_2, $CaSO_4$ $2(H_2O)$ were used, working with several space groups. Some of the calculations are consistent with the results of the experiments, but others show energy shifts in the range of some eV. In summary, the FEFF calculations and the experimental data exhibit similarities as well as deviations. By using trimming parameters, deviations could be eliminated to a certain extent, which will be presented.

1. Introduction

The aim of this project is the calculation of the absorption coefficient μ of x-ray absorption spectra as a function of energy and a comparision with experimental data. If the experimental data agree well with the simulated data, additional knowledge about the sample could be obtained, or one could predict the characteristics of a sample to a certain extent [7], [8].
X-ray absorption spectroscopy is the measurement of the absorption of x-rays as a function of x-ray energy. In a plot of the absorption coefficient μ as a function of the x-ray energy the characteristical oscillations above the absorption edge (fig. 2, solid line) known as fine structure can be distinguished.

2. Theory

The fine structure of the XANES region can be explained as follows. An incident x-ray photon is absorbed by an atom (fig. 1, grey) and causes either fluorescence- or the Auger effect. After interaction the photoelectron wave of the emitted photoelectron propagates. It is scattered at the neighbouring atoms (fig. 1, white), which cause multiple scattering. The multiple scattering paths are illustrated with arrows. The outgoing and the backscattered photoelectron wave result in interference, which is observable as fine structure near the absorption edge [1]. Such XANES

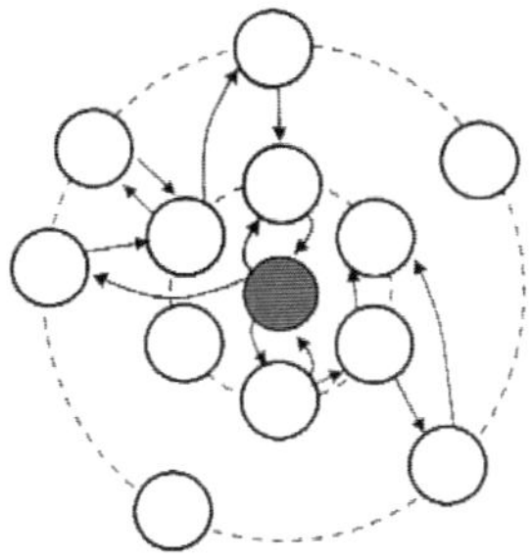

Figure 1. The absorber atom (grey) is surrounded by several neighbouring atoms (white) on the first and second coordination shell (dashed lines). The arrows describe the direction of the propagation of the photoelectron wave along the multiple scattering paths [1].

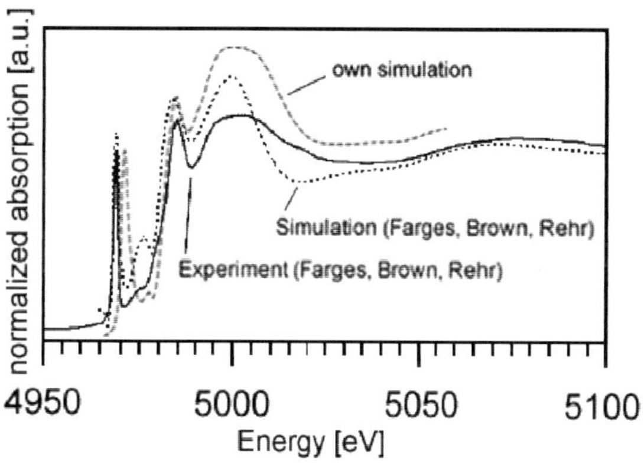

Figure 2. Measured (solid line) [6] and calculated (dashed and dotted [6]) XANES-spectra of the Ti K-edge of Ba_2TiO_4.

spectra provide information about coordination chemistry, bonding angles, bonding types (e.g. C-C, C=C, C≡C) and oxidation number of the irradiated sample.

3. FEFF 8.4

FEFF 8.4 [2] is a program for the calculation of multiple scattering in order to simulate XAFS-spectra. The potential geometry of the crystal lattice is approximated by a muffin-tin like shape. FEFF calculates with the Hartree-Fock method self-consistent orbitals to obtain the scattering potentials. Some quantities, which are calculated by FEFF, are the scattering amplitude and the wave number of the photoelectron wave, the phase shift between outgoing and backscattered wave, the absorption coefficient and the multiple scattering paths.

4. Results and comparison with measured data

With the calculated data the absorption coefficient μ as a function of the photon energy E can be plotted for comparison with measured spectra. Fig. 3 shows spectra taken at the sulfur K-edge of FeS_2 where the number of scatteres has been increased. In Fig. 4 spectra of $CaSO_4\ 2(H_2O)$ are shown where by the use of a so-called correction card the peak positions could be manipulated. In the diagrams the calculated absorption is normalized to the measured absorption.

For FeS_2 the calculated K-edge (peak A) of S agrees with the measured data. The fine structure (C' and D') above the edge first appears at a cluster size of 21 atoms and the B' peak first at 103 atoms (not shown here). Calculations with more than 167 atoms did not lead to better conformity with the experiment. The shoulder B of the measured XANES-spectrum is recognizable as B' in the spectra of the two largest atom clusters. The peaks C' and D' are approximately at the same position as in the measured spectrum. However, the heights of the peaks C' and D' do not agree with the experiment very well. The proportion of A to C is not verified in the calculations, but the proportion of C to D (reduction of about 1/5) is. With FEFF the relative positions (on the energy scale) of the peaks and the relative absorption maxima can be calculated quite reliable, but the absolute energy values are often inconsistent as can be seen in the figures.

Due to an energy shift of about 11 eV regarding the K-edge position, the computed XANES-spectrum of $CaSO_4\ 2(H_2O)$ (dashed line) does not agree with the measured data. Energy shifts of the calculated K-edge were frequently observed, independently of the analyzed substance or the element of the absorber atom. In addition, a double peak feature (A') is noticeable, which

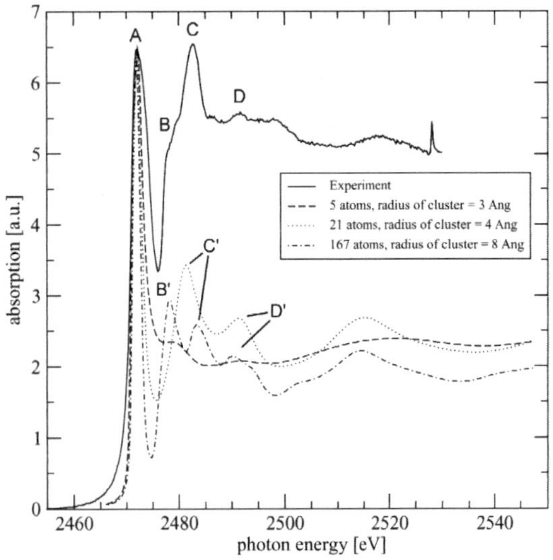

Figure 3. Measured (solid line) [5] and calculated (dashed and dotted) XANES-spectra of the S K-edge of FeS_2

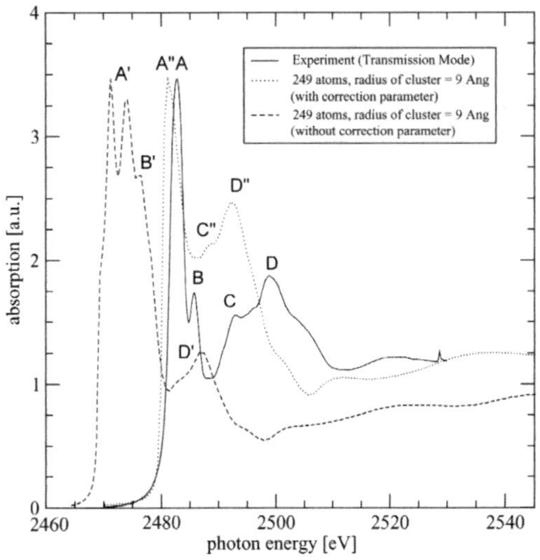

Figure 4. Measured (solid line) [5] and calculated (dashed and dotted) XANES-spectra of the S K-edge of $CaSO_4\ 2(H_2O)$

can be generally assessed as a numerical artefact. With correction parameters these deviations could be eliminated to a certain extent (dotted line). In this case the absorption edge was shifted by 10 eV to higher energies and the atomic coordinates were multiplied by the value 0.9 to reduce the interatomic distances, which means that the cluster shrinks to 90% of its original size. Fig. 2 shows spectra taken at the titanium K-edge of Ba_2TiO_4 where the calculated spectra agree well with the measured data.

In general, there is always a difference between the well known model substance and the measured sample, because a real sample never has the perfect structure of the model and this results in diverse spectra. To sum up, in some cases the computed spectra verify the experiment without further manipulations and in other cases correction parameters are necessary to reduce differences to the measured spectra to a certain extent. Thus, FEFF might not be able to predict characteristics of a sample from the environment.

Acknowledgments

This work has been supported by BMBF under contract number 05KS4MGB7 and by the DFG within the Collaborative Research Center SFB 755 Nanoscale Photonic Imaging.

References

[1] B. K. Agarwal, "X-Ray Spectroscopy", Springer-Verlag, 1991

[2] A. Ankudinov, B. Ravel, J.J. Rehr, "FEFF 8.4" program, 2002, `www./leonardo.phys.washington.edu/feff/`

[3] J.J. Rehr, R.C. Albers, "Theoretical approaches to x-ray absorption fine structure", Reviews of Modern Physics, Vol. 72, No. 3, July 2000

[4] M. Newville, "Fundamentals of X-Ray Absorption Fine Structure", APS XAFS-Summer School, July 2005

[5] J. Thieme et al., 2007

[6] F. Farges, G. E. Brown and J. J. Rehr, "Ti K-edge XANES studies of Ti coordination and disorder in oxide compounds: Comparison between theory and experiment", Phys. Rev. B, 56, 1809, 1997

[7] S. W. Goh, A. N. Buckley et al., "Sulfur electronic environments in α-NiS and β-NiS: examination of the relationship between coordination number and core electron binding energies", Phys. Chem. Minerals, 98, 2006

[8] S. W. Goh, A. N. Buckley et al., "The oxidation states of copper and iron in mineral sulfides, and the oxides formed on initial exposure of chalcopyrite and bornite to air", Geochimica et Cosmochimica Acta 70, 2006

Fresnel zone-plate based X-ray microscopy in Zernike phase contrast with sub-50 nm resolution at NSRL

Jie Chen[1], Wenjie Li[1], Yijin Liu[2], Zhengbo Yue[3], Jinping Tian[1], Longhua Liu[1], Ying Xiong[1], Gang Liu[1], Chunru Wang[4], Ziyu Wu[1], Hanqing Yu[3] and Yangchao Tian[1,a]

[1]National Synchrotron Radiation Laboratory, [2]School of Physics, [3]Laboratory of Environmental Engineering, School of Chemistry, University of Science and Technology of China, Hefei Anhui 230029, People's Republic of China
[4]Institute of Chemistry, The Chinese Academy of Sciences, Beijing 10060, People's Republic of China

[a]E-mail: ychtian@ustc.edu.cn

Abstract. A transmission X-ray microscope using Fresnel zone-plates (FZPs) has been installed at U7A beamline of National Synchrotron Radiation Laboratory (NSRL). The objective FZP with 45 nm outermost zone width delivers a sub-50 nm resolution. A gold phase ring with 2.5 μm thickness and 4 μm width was placed at the focal plane of the objective FZP at 8 keV to produce a negative Zernike phase contrast. A series of samples were used to test the performance of the Zernike phase contrast X-ray microscopy.

1. Introduction

The key power of the FZP based multi-keV X-ray microscopy is the large penetration of the X-rays and the several tens of nanometers resolution. It is potential to image the interior nanoscale features in thick objects, such as microprocessor chip, micro/nano-sized materials, and biological samples. However, the absorption contrast of the image diminished rapidly with the photon energy, especially for objects consisted of light elements. Therefore, methods are desired to improve the image contrast. One method is the Zernike phase contrast microscopy proposed by Zernike [1]. A similar method was pioneered in the soft X-ray region [2] and was then applied to multi-keV X-ray microscopy.

Here we report a FZP based transmission X-ray microscopy in Zernike phase contrast at 8keV with sub-50 nm resolution at NSRL. Samples such as Au spoke pattern, Au lines-and-spaces pattern, carbon microtube, and wheat straw were used to test the performance of the microscope.

2. Optical layout

The transmission X-ray microscope was constructed at beamline U7A of NSRL. The detailed schematic experimental setup of this x-ray microscope is described elsewhere [3, 4].

3. Experimental results

The performance of the microscope was first tested with an Au spoke pattern with 600 nm thickness and 50 nm finest line widths at the center. Figure 1(a) presents the Zernike phase contrast image of the pattern. The measured Zernike phase contrast is about 30%. It is better than the calculated absorption contrast 12% of 600 nm thick gold at 8 keV. Figure 1(b) and (c) show the absorption contrast image and Zernike phase contrast image of an Au lines-and-spaces pattern respectively. The Au lines appear dark. The measure absorption contrast is about 22% while the measured Zernike phase contrast is about 33%. We found bright boundaries of the lines in Fig.1 (c), which is the so-called halo effect corresponding to negative phase contrast. More details about the halo effect can be found in [5]. The halo effect is especially apparent around the lower spatial frequency lines, since the halos would overlap around the high spatial frequency features such as the spoke pattern in Fig. 1, which is little blurred by the halos.

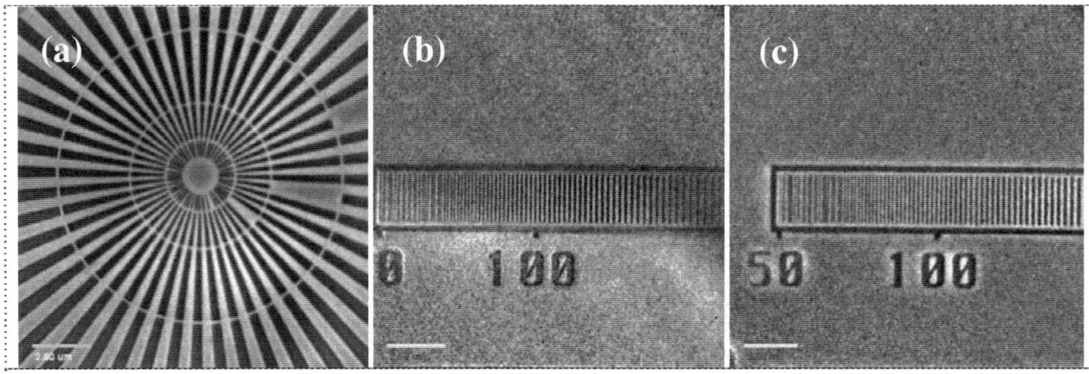

Figure 1. (a) Zernike phase contrast image of spoke pattern. (b) Absorption contrast and (c) Zernike phase contrast images of lines-and-spaces pattern.

The halo effect is pronounced when the indices of refraction between the neighbouring regions are highly different. Therefore, Zernike phase contrast would be more suitable for low-Z materials and thin specimens. Carbon microtubes with 1 μm diameter were used to test at 8 keV. The carbon microtubes were prepared by optimal floating chemical vapor deposition (CVD) method [6]. Compared with the absorption contrast image of a carbon microtube in Fig. 2(a), the phase contrast image in Fig. 2(b) shows better contrast. The dark spheres are reference gold particles.

Concerns about global warming, the soaring cost of gasoline and national security issues have rekindled interest in producing liquid transportation fuels from renewable resources, particularly those derived from biomass (cellulose, lignin, and hemicellulose). Recalcitrance to saccharification is a major limitation for conversion of lignocellulosic biomass to ethanol. Anaerobic fermentation of lignocellulosic wastes is attractive due to environmental benefits, such as reducing the greenhouse effect, generating useful products, and increasing resource availability. Wheat straw was collected from a farm in the suburb of Hefei city, China. After it was pretreatment, mixed rumen microorganisms, composed of bacteria, fungi, and protozoa (all needed enzyme components with high enzyme activity), were used for the degradation of lignocellulosic wastes. More details can be found elsewhere [7]. Figure 2(c) is a phase contrast 3x3 mosaic image of wheat straw before anaerobic degradation. It shows detailed microstructures of lignin and cellulose fibers in the straw. X-ray tomography had been performed on the red square area in Fig. 2(c). The dark spheres in the images are gold particles used as alignment mark. The total 71 sequential tomographic images were automatically collected from -70° to +70° in 2° intervals at 8.0 keV. Subsequently, these projections were aligned and reconstructed by a standard filtered-back-projection algorithm. Figure 2(d) presents the three-dimensional (3D) rendering of the straw. Figure 2(e-f) are two single phase contrast images of wheat straw after anaerobic degradation of lignin in waste straw by ruminal microbes after 12 days. Tunnelings can be clearly seen perpendicular to the lignins and fibers, which might be one of the ways for rumen microorganisms to attack the straw. It is very interesting that the conversion processes can be observed on nanoscale by the x-ray microscopy. Further investigation will be focused on the correlation between the microstructures and the conversion mechanism.

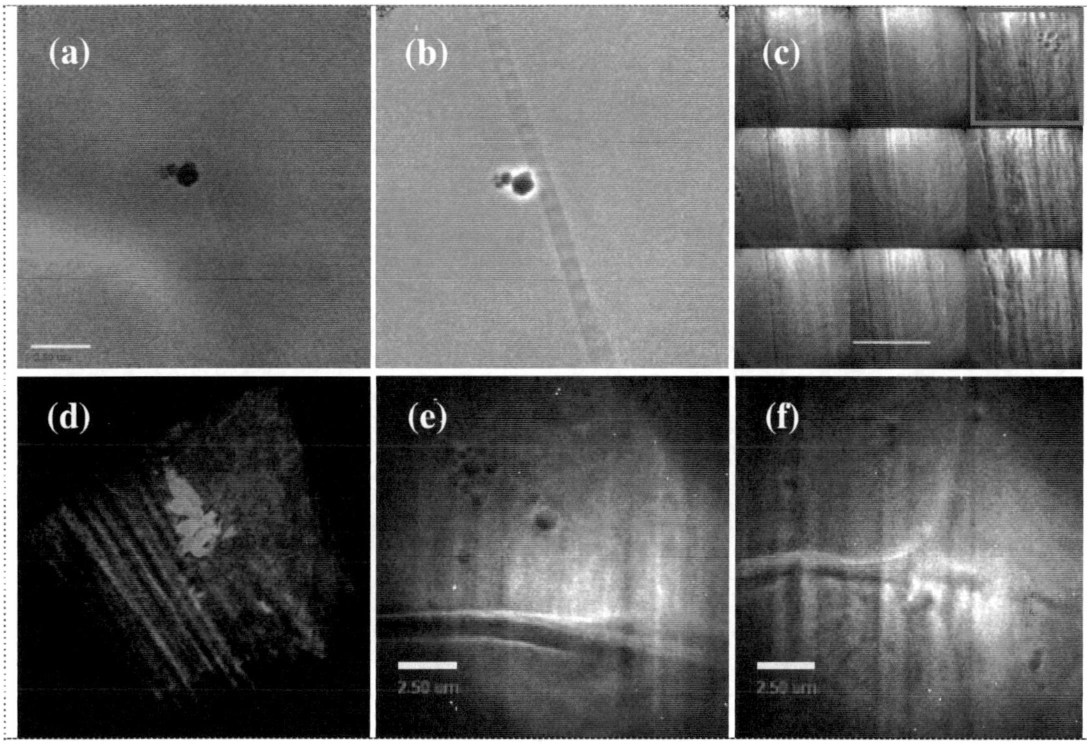

Figure 2. (a) Absorption contrast and (b) Zernike phase contrast images of a carbon microtube; (c) Phase contrast 3x3 mosaic image of wheat straw before anaerobic degradation; (d) 3D rendering of the straw in the red square in (c); (e) and (f) are phase contrast images of wheat straw after anaerobic degradation of lignin in waste straw by ruminal microbes after 12 days.

4. Conclusions

We have demonstrated a FZP based transmission X-ray microscope in Zernike phase contrast at 8 keV. A 50 nm resolution test spoke pattern was imaged. Halo effect was found in the lines-and-spaces pattern. Low-Z materials, such as carbon microtube, could be observed. In addition to the results of the wheat straw and nanoscale materials in [3], all these performance highlights the potential of its applications in environmental science, and materials science.

Acknowledgment

The authors are grateful for the financial support from the 985 project of the State Ministry of Education, and the National Science Foundation of China (10675113, 10734070).

References

[1] F. Zernike, *Z. Tech. Phys.* **11**, 454-457 (1935).
[2] G. Schmahl, D. Rudolph, P. Guttmann, G. Schneider, J. Thieme, and B. Niemann, *Rev. Sci. Instrum.* **66**, 1282-1286 (1995).
[3] J. Chen, C. Wu, J. Tian, W. Li, S. Yu, and Y. Tian, *Appl. Phys. Lett.* **92**, 233104 (2008).
[4] Y. Tian, W. Li, J. Chen, L. Liu, G. Liu, A. Tkachuk, J. Tian, Y. Xiong, J. Gelb, G. Hsu, W. Yun, *Rev. Sci. Instrum.* **79**, 103708 (2008).
[5] http://www.microscopyu.com/tutorials/java/phasecontrast/shadeoff/index.html
[6] L. Song, L. Ci, L. Lv, Z. Zhou, X. Yan, D. Liu, H. Yuan, Y. Gao, J. Wang, L. Liu, X. Zhao, Z. Zhang, X. Dou, W. Zhou, G. Wang, C. Wang, S. Xie, *Adv. Mater.* **16**, 1529–1534 (2004).
[7] Z. Hu, S. Liu, Z. Yue, L. Yan, M. Yang, H. Yu, *Environ. Sci. Technol.* **42**, 276-281 (2008).

Journal of Physics: Conference Series **186** (2009) 012006 doi:10.1088/1742-6596/186/1/012006

Quantitative scanning differential phase contrast microscopy

M D de Jonge,[1,2] B Hornberger,[3,4] C Holzner,[4] B Twining,[5] D Paterson,[1,2] I McNulty,[2] C Jacobsen,[4] S Vogt[2]

1-Australian Synchrotron, 2-Advanced Photon Source, 3-Xradia, Inc, 4-Stony Brook University , 5- University of South Carolina

martin.dejonge@synchrotron.org.au

Abstract. Differential phase contrast can be realised in the scanning transmission x-ray microscope by use of a detector with an appropriate configuration of detector elements. Use of such a configured detector significantly improves the ability to image specimens with low absorption contrast. A natural application is found in scanning fluorescence x-ray microprobe studies of biological specimens. We discuss the technique and several applications.

1. Introduction

The scanning fluorescence x-ray microprobe can be used to map elemental distributions, to obtain spatially-resolved chemical information by means of XANES measurements, and to obtain longer-range bonding and co-ordination information from XAFS measurements. The spatial resolution of such measurements ranges from several micrometres down to around 50 nm using present synchrotron facilities. Virtually any format of specimen can be measured, with limitations coming mainly from mechanical constraints and the desire to avoid imaging artifacts. Studies have been performed on geological, environmental, and soil specimens and a wide range of biological specimens, including cells, plants, and whole organisms.

While the microprobe can be used to investigate distributions on one side of a bulk specimen, the high penetration of hard x-rays recommends the application to a particular class of specimens with low absorption. In this regime the incident probe beam maintains its intensity as it traverses the specimen. Quantitative studies are possible when fluorescence x-rays escape with only minimal absorption [1]. While such studies can be performed on thin sections of specimens with high linear attenuation coefficient, a popular alternate regime is the measurement of thick specimens with low linear attenuation. This regime is particularly applicable to the imaging of trace metals in a biological context, and is ideally suited to measure almost any specimen coming from the biological community.

The high penetration of x-rays is required to produce high-fidelity and quantitative fluorescence maps; however, the same condition leads to transmission images having extremely low contrast. The framework of a cell or organism is largely composed of lipid, protein, and water. These components absorb high-energy x-rays only weakly, and so studies involving biological specimens are usually blind to the biological ultrastructure in which the trace metals are situated. Detailed ultrastructural knowledge can help an investigator identify various organelles within the organism and thereby contextualise the elemental maps. While it has long been known that phase offers stronger contrast at high x-ray energies, phase contrast methods have only recently been developed for the STXM.

Journal of Physics: Conference Series **186** (2009) 012006 doi:10.1088/1742-6596/186/1/012006

2. Differential phase contrast imaging in the STXM

Sensitivity to phase structure is possible in the scanning microprobe simply by employing a detector with spatial sensitivity [2–4]. Two different methods for reconstructing quantitative values of the phase for scanning x-ray microprobe have been reported recently [5, 6]. Although the detector configurations used for these demonstrations were identical, the analytical approaches are completely different. The method of Hornberger [5] determines the CTF appropriate to the detector elements, and inverts this CTF to determine the phase advance and absorption of the specimen. This method is most general, and can be applied to any configuration of detector. The method of de Jonge [6] expands the specimen function in a Taylor series and demonstrates that the first-order effect of a specimen on the intensity distribution in the detector plane is (1) to attenuate according to the mean optical thickness under the beam footprint (according to the usual assumptions of STXM imaging) and (2) to shift the x-ray beam in proportion to the mean refractive index gradient under the beam footprint. The next most significant term in the Taylor expansion describes a novel differential interference contrast effect, but this is small when the focal spot is small, when the absorption gradient is modest, and when the specimen is fully within the focal plane. The centre-of-mass of the intensity distribution can be used to gauge the shift of the intensity distribution when that distribution is well defined. We determine the centre of mass by examining the normalized differences in the detector halves (see Fig. 1); for example, the vertical component of the phase derivative DPC_V is gauged using $DPC_V = (I_{top} - I_{bottom}) / (I_{top} + I_{bottom})$.In this article we use this method to visualize the DPC.

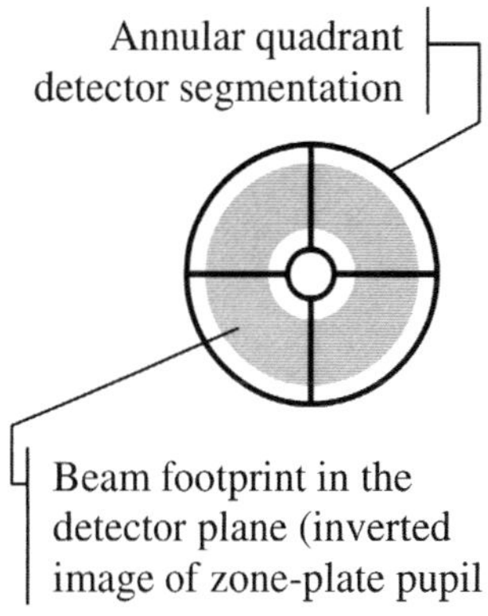

Figure 1. Detector segmentation and intensity distribution in the detector plane.

3. Applications

3.1. Specimen visualisation

The low absorption contrast of biological specimens makes them difficult to find in the x-ray beam. One of the immediate uses of DPC imaging is to facilitate this process by providing high contrast images in real time. Figure 2 shows a scan of pennate diatoms. The absorption image (left) shows very low contrast; the diatoms are barely visible. The DPC image (right) shows very clearly the diatoms, including periodic structures along their length and potentially some internal organs. Fine hairs, possibly flagella, connect the two diatoms; these are likely to be protein fibres with estimated thickness of around 5 nm. Both of these images are generated form the one scan; the absorption image is generated from the sum of all detector segments, the DPC image from the normalized differences defined in Section 2.

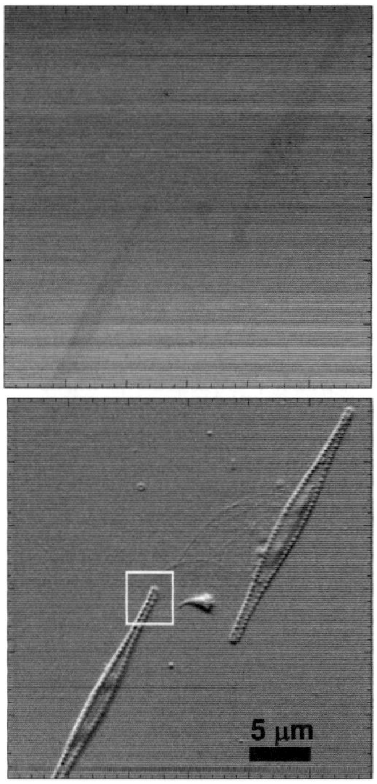

Figure 2. The absorption image (upper) barely resolves these pennate diatoms, but the vertical component of the DPC (lower) shows excellent structural detail. Thin flagella between the two diatoms have an estimated phase shift of π/300 radians, corresponding to around 5 nm of protein. The several 'doughnut' features are the result of beam damage caused by dwelling the focussed x-ray beam on the EM grid membrane. Theoretical sensitivity is estimated at better than 1 nm of protein for the STXM at 2-ID-B of the APS. Scan parameters: x-ray energy = 2.5 keV, step size = 30 nm, dwell = 5 ms, zone plate: radius = 80 µm, finest zone width = 50 nm.

3.2. Quantitative phase reconstruction

Quantitative phase reconstructions require a well characterised optical system because current reconstruction algorithms make assumptions about the optical parameters. For example, our use of a quadrant diode to measure the location of the centre of mass of the intensity requires assumptions about the source, the uniformity of the zone-plate illumination, and the uniformity of the zone-plate efficiency. Zone-plate and optical parameters are used to scale the phase gradient, and so these must be accurately known.

Figure 3 shows a reconstruction of the phase of the tip of a diatom (Fig. 2). Tooth-like structures are clearly visible. A line-out of the reconstructed phase along these structures shows that the phase sensitivity is easily better than $\pi/100$ radians; we estimate that the phase of the flagella to be of order $\pi/300$ – corresponding to about 1 nm of protein – demonstrating the extreme sensitivity of this imaging modality.

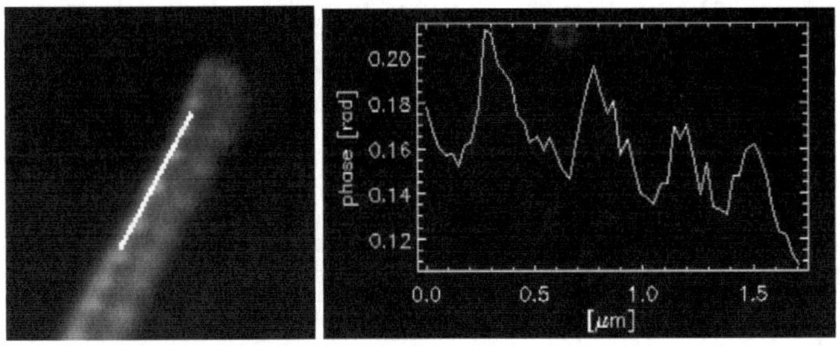

Figure 3. Phase reconstruction (left) of the tip of the pennate diatom indicated in Fig. 2. The frustule (or shell) of the diatom is comprised largely of SiO_2. A lineout along the indicated path is shown (right), clearly showing phase structures of order $\pi/100$ radians, corresponding to around 30 nm of SiO_2.

References

[1] BS Twining, SB Baines, NS Fisher, J Maser, S Vogt, C Jacobsen, A Tovar-Sanchez, and SA Sañudo-Wilhelmy, Anal. Chem. 2003, 75, 3806-3816

[2] NH Dekkers and H de Lang, Optik (Jena) **41**, 452 (1974).

[3] BC McCallum, MN Landauer, and JM Rodenburg, Optik (Jena) **101**, 53 (1995).

[4] GR Morrison and B Niemann in: J Thieme, G Schmahl, E Umbach, D Rudolph (Eds.), X-ray Microscopy and Spectromicroscopy, Springer, Berlin, 1998, pp. I-85–I-94.

[5] B Hornberger, M Feser, and C Jacobsen, Ultramicroscopy **107**, 644 (2007).

[6] MD de Jonge, B Hornberger, C Holzner, D Legnini, D Paterson, I McNulty, C Jacobsen, and S Vogt, Phys. Rev. Letts., **100,** 163902 (2008).

Development of a low-energy X-ray fluorescence system combined with X-ray microscopy

A. Gianoncelli[1], T. Klatka[2], R. Alberti[2], D. Bacescu[1], A. De Marco[3], A. Marcello[3], A. Longoni[2], B. Kaulich[1], and M. Kiskinova[1]

[1] Sincrotrone Trieste S.C.p.A., ELETTRA, s.s. 14 km 163.5 in Area Science Park, 34012 Trieste, Italy
[2] Politecnico di Milano, Dipartimento di Elettronica e Informazione, Piazza Leonardo da Vinci, 23100 Milano, Italy and INFN sezione di Milano, via Celoria 16, 20133 Milano, Italy
[3] Laboratory of Molecular Virology of the International Centre for Genetic Engineering and Biotechnology (ICGEB), Padriciano 99, 34012 Trieste, Italy

alessandra.gianoncelli@elettra.trieste.it

Abstract. A system for Low Energy X-ray Fluorescence (LEXRF) has been implemented in the TwinMic [1] microscope station operating at Elettra synchrotron facility. The integrated LEXRF is coupled with the STXM operation mode. This allows simultaneous collection of XRF maps, brightfield images and phase contrast images [2], providing complementary information about the sample's chemical specificity and morphology with high spatial resolution. The system is based on a modular setup of multiple Be windowless Silicon Drift Detectors (SDDs), read-out by a custom designed fast multichannel acquisition system [3]. The number of the SDDs (up-to 8 detectors) will be increased in the future in order to minimize the XRF acquisition time (higher count rate and solid angle) and to increase the XRF detection limits, therefore allowing us to acquire XRF maps with sub-micron spatial resolution.

1. Introduction

X-ray Microscopy (XRM) responds the increasing demands in material research for characterization with lateral resolution at sub-micrometric length scales. Combining XRM transmission imaging with spectroscopy, based on X-ray absorption (XANES) or X-ray Fluorescence (XRF), allows complementing morphological information with chemical specificity. While XANES requires stack of images (acquired scanning the photon energy) to be correlated and computationally analyzed, the XRF provides with a single scan at selected photon energy simultaneous identification of many chemical elements and their lateral distribution.

Almost all synchrotron light sources have dedicated experimental stations for XRF microscopy using photon energies in the hard x-ray range (usually > 5 keV). Low energy XRF is still poorly exploited, since it requires in-vacuum operation and short working distances of the zone plate focusing optics. The advantage of LEXRF is that it is very adequate for identifying the presence of light elements, e.g. C, N, O, Mg, Na etc, that are the main constituents of the life matter. To our knowledge the only report for LEXRF setup, which allows fluorescence detection down to the C edge, showed only elemental maps of Ca, Na, Si, K, Mn, Fe and Cu lines [4].

Further-on, by implementing LEXRF in a Scanning Transmission X-ray Microscope (STXM) the acquisition of the XRF spectrum for each pixel in the raster-scan can be combined with simultaneous X-ray absorption and phase contrast maps, which provide complementary information of the specimen morphology, allowing the location of the elements to be unambiguously established.

Most of the synchrotron facilities performing XRF analysis typically use Ge or Si(Li) detectors, the developed system reported here is based on the use of SDDs. The advantage of the SDDs is a very low output capacitance (of the order of 100 fF), independent of the active area of the device, resulting in a much lower electronic noise with respect to conventional silicon diodes of equivalent area and thickness [5]. Moreover, the integration of the first stage of the front-end electronics [6] on the detector chip allows high energy resolution even at high counting rates with the SDD operated near room temperature. The combination of low noise performance and an optimized entrance window [7] for low-energy photon detection makes the SDD ideal for detection of low energy X-rays down to B-Kα line (183eV). SDDs have also the advantage of non requiring liquid nitrogen cooling, which makes them more compact and thus more suitable for a set-up with limited space and mechanical constraints.

In the following we present the LEXRF set-up for TwinMic and the first preliminary results, obtained in the commissioning period, illustrating the performances of the developed system.

2. Experimental setup

The developed LEXRF system has been successfully installed in the TwinMic station, operated in the 250-2200 eV energy range at the Elettra synchrotron facility, in Trieste, Italy [1]. TwinMic is an X-ray Transmission Microscope which hosts Scanning Transmission (STXM) and full-field (TXM) microscopes, with easy switch between the two modes. In STXM mode the sample is raster-scanned allowing simultaneous detection of transmission (absorption, phase) images and emitted signals (photons and electrons). The optical resolution of the STXM is typically chosen as a compromise between photon flux and lateral resolution required for the specific experiment, ranging from 0.05 – 1 μm.

The LEXRF setup has been coupled with the STXM operation mode: the SDDs are acquiring the fluorescence signal emitted by the sample while absorption and phase contrast images are collected by a special configured electron-multiplied fast read-out CCD camera from Andor Technology [2]. The optical scheme of the microscope setup is shown in Figure 1.

The main limitations implementing LEXRF in a STXM are the geometrical constraints due to the short working distance (a few mm) between the sample and the zone plate-based focusing system. In TwinMic the SDDs are arranged in a symmetric concentric configuration as illustrated in Figure 1 with an angle of 70° between incident beam and detector axis. The SDDs-sample distance is around 28mm. The detectors are mounted on a copper water-cooling base plate to remove the heat dissipation from the Peltier cooling of each SDD. Customized pre-amplifying and biasing electronics (provided by Politecnico di Milano/INFN) are located in close proximity to the detectors. The charge preamplifier is based on pulsed reset technique [3,8-9] and is characterized by extraordinary energy resolution performance, high energy resolution stability (resolution worsening is about 3.5 eV at Mn Kα with input photon rate of 150 kcps) and a very low peak shift (below 0.1% over the range of incoming events from 1kcps up to 150kcps).

3. Results

TwinMic was operated in STXM mode using photon energy of 1440eV, chosen to excite the light chemical elements of interest. The pilot experiments were carried out with only two Be windowless 30mm^2 active area SDDs. The SDDs have a nominal FWHM of 135 eV at Mn Kα cooled down to -20°C. For the pilot experiment the X-ray beam was focused to 1 μm spot size with a photon flux of 10^{10}-10^{11} ph/s in the probe. The microprobe size was a compromise to achieve sufficient flux for these pilot experiments in the commissioning phase of the LEXRF set-up. Some XRF maps were also collected recently with sub-micron resolution.

Fig. 2 shows the first results demonstrating the expanded potential of the TwinMic station. The transmission and XRF images are of an hepatocyte from human liver, which represent one of the important applications in human biology and virology. The aim of such measurements are detection of changes of sub-cellular processes and cell bio-functionality due to virus replication by externally administered probes, and these commissioning results confirm the feasibility. The XRF maps show the spatial distribution of C, Na and Mg, constructed by considering the total counts corresponding to the C, Na and Mg Kα lines. They can be directly correlated to the sub-cell constituents, distinguished in the transmission absorption and phase contrast maps. For good statistics we used a dwell time of 20 s per XRF spectrum, requiring a total acquisition time of about 10 hours, which will become four times faster implementing all 8 detectors.

4. Conclusive remarks

With the development and the implementation of the LEXRF system TwinMic has become the first microscope in the soft X-ray energy range (250-2200 eV) capable to combine XRF elemental mapping with simultaneous acquisition of absorption and phase contrast images, proving complementary morphological information. For the pilot experiment a spatial resolution of 1 μm was chosen in order to have a reasonable count rate and acquisition time. The upgrade of the LEXRF set-up to 8 SDDs will allow pushing the spatial resolution sub-100 nm range by increasing the acceptance solid angle as well as reducing the measurement time when using modest sub-μm resolution sufficient for probing a great variety of specimen.

In conclusion, the LEXRF setup have expanded the potentials of the TwinMic station, opening new application opportunities in material science, human and marine biology, virology, environmental science and most recently in food science and cultural heritage.

References

[1] B. Kaulich et al. 2006 Proc. 8th Int. Conf. X-ray Microscopy IPAP Conf. Series 7 22.
[2] A. Gianoncelli et al. 2006 Applied Physics Letters **89** 251117.
[3] R. Alberti et al. 2007 IEEE Trans on Nucl. Sci. **54**, 751.
[4] A.-M. Flank et al. 2006 Nuclear Instruments and Methods in Physics Research B **246**.
[5] E. Gatti, P. Rehak, 1984 Nucl. Instr. and Meth. A **225** 608.
[6] P.Lechner et al.1996 Nucl. Instr. and Meth. A **377** 346.
[7] R. Hartmann, K.-H. Stephan, L. Strüder, 2000 Nucl. Instrum. Meth. A **439** 216.
[8] Niculae A, Lechner P, Fiorini C, Longoni A, 2006 Nucl. Instr. and Meth., A **568** 336.
[9] R Alberti, C. Guazzoni, T. Klatka 2008 JINST 3 P03003.

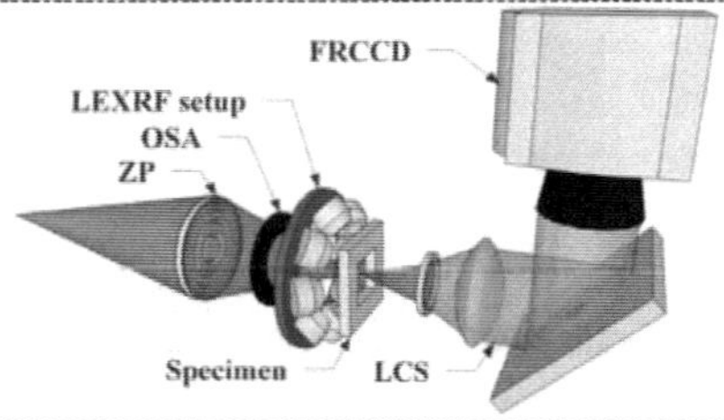

Fig. 1: Optical scheme of the TwinMic STXM mode operation with the LEXRF setup implemented.

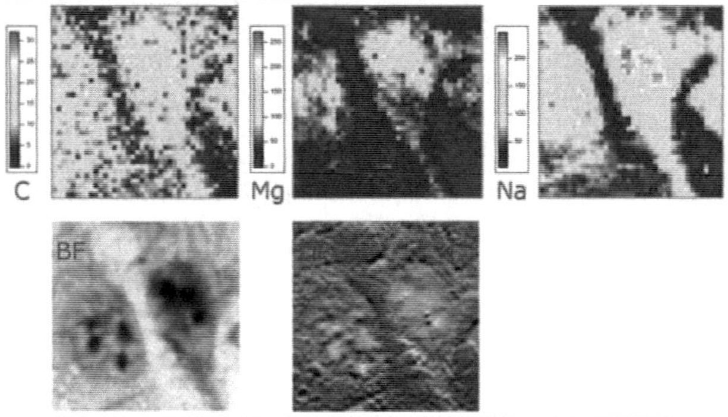

Figure 2. Scanning transmission and LEXRF maps of an hepatocyte from human liver simultaneously acquired: BF is the brightfield or absorption map, DPC is the corresponding differential phase contrast map. Image size of 40 x 40 μm^2 with 40 x 40 pixels raster scan and XRF acquisition time of 20s per pixel.

Implementation of soft x-ray microscopy with several tens nanometers spatial resolution at NSRL

Shiping Jiang[1], Liang Chen

National Synchrotron Radiation Laboratory, the University of Science and Technology of China, Hefei230029, China

E-mail: spjiang@ustc.edu.cn

Abstract.A transmission soft x-ray microscope (TXM), which is similar to the full-field x-ray microscopes installed on other synchrotron radiation sources in the world, was developed at National Synchrotron Radiation Laboratory (NSRL) in Hefei. An x-ray image taken with the microscope was acquired and its spatial resolution was estimated to be better than 70nm.

1. Introduction

Soft x-ray microscopy has many advantages for biological imaging, especially in "water window" region (2.34-4.38 nm), where the absorption coefficients of organic materials are about one order of magnitude than that of water and organic materials show strong absorption while water shows almost transparent. Compared to charged particles (such as electrons), soft x-ray has higher penetration power and thick samples (up to 10 μm) can be imaged with no need to be sectioned. Some kinds of soft x-ray microscopes have been developed in the world. Among these, transmission soft x-ray microscope s (TXM) based on zone plates and synchrotron radiation have been attracted with much attention because of its short exposure time, real time imaging and high spatial resolution. At the same time, the low numerical aperture x-ray lenses of TXM, which lead to the large depth of field, makes it true to achieve microtomography on whole cells. This depth of field is the basic reason for the development of TXM in water window region, which matches the demand for whole cell 3D imaging [1].

The project of updating the beamline and endstation of soft x-ray microscopy was completed and a transmission full-field soft x-ray microscope working at water window region was built at National Synchrotron Radiation Lab (NSRL) in 2007 [2].

2. Instrumentation

[1] To whom any correspondence should be addressed.

The beamline of TXM at NSRL was basically based on a fully dedicated 800 MeV electron storage ring. It consists of a typical front end, a plane mirror chamber, a linear monochromator and a micro zone plate imaging system. The main components of the beamline are shown in figure 1. Synchrotron radiation from a bending magnet was irradiated on a plane mirror with the grazing incidence angle of 2.5 degree. The use of plane mirror was to suppress high energies of synchrotron radiation. The plane mirror used SiO_2 as the basis and the surface coated with 30 nm nickel.

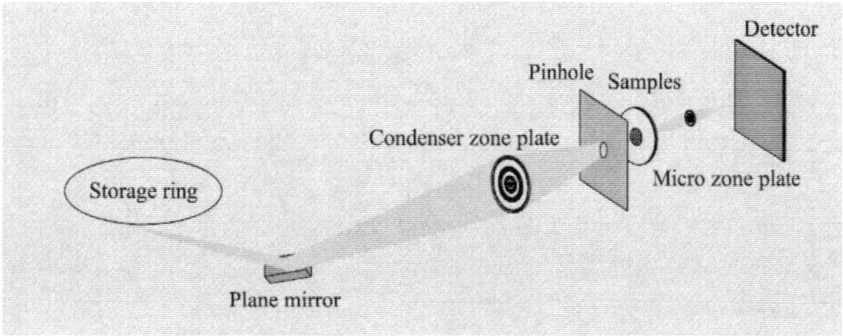

Figure 1. Schematic layout of the TXM

The reflected synchrotron radiation was then irradiated on a condenser zone plate (CZP). There was an annular aperture with a diameter of 10 mm in front of the CZP. The aperture was to restrict the radiation on the CZP and just made all of the CZP irradiated. The CZP together with a 15 μm pinhole consists of a linear monochromator which provide monochromatic soft x-rays. The diameter of CZP is 9 mm, with the outmost ring width of 50 nm. There was a central stop of CZP with diameter of 4 mm which used to prevent the zero-order radiation illuminating on the sample and reduce the background. The spectrum resolution ($\lambda/\Delta\lambda$) of x-ray beam behind the pinhole reach to about 600

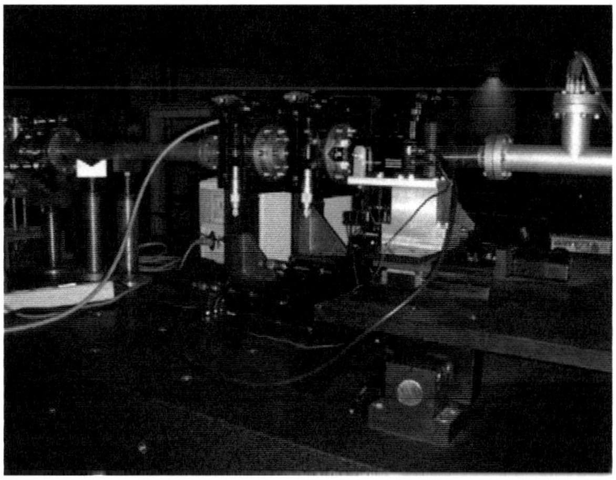

Figure 2. Photo of the TXM at NSRL (in part)

The far left is the Condenser zone plate and the center is the pinhole, sample and objective zone plate in the photo. CCD detector of the TXM is beyond the photo

The selected monochromatic soft x-rays were then irradiated on the samples. The sample stage was a three-dimensional-driven stage and repeated precision was 200 nm. There was a micro zone plate for magnifying with 800× the imaging on the detector. The detector was a back-thinned illuminated open CCD (Andor, DO434) and the quantum efficiency was about 70% at the wavelength of 2.4 nm. There used a photo diode (AXUV-100G) to measure the intensity of the spot at the sample position and was about 3.4 × 109 photons/s.100mA at 2.4 nm, which could be calculated from the photo-current read from the AXUV photo diode.

3. Result

The first image taken with the TXM was used to measure its spatial resolution. The x-ray image of a resolution pattern is shown in Fig 3. It is obvious that the spatial resolution of the image is better than 70 nm. Time of the exposure was 60s when the electronic current in storage ring was about 90mA. The sample for imaging was part of a siemens star test pattern.

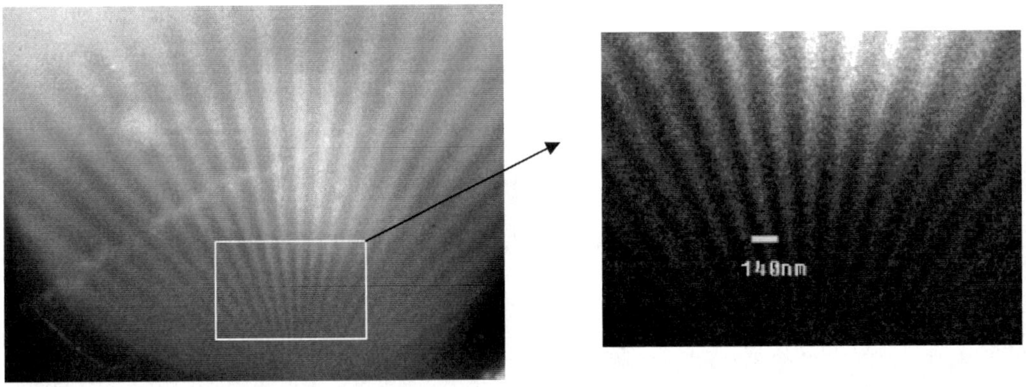

Figure 3. the first image taken with the TXM at 2.4nm

4. Conclusion

The X-ray microscope at NSRL will mainly apply in life science. The spatial resolution of the imaging can reach less than 70 nm. With the microscope, it comes true for single cell microtomography with soft x-rays at "water window" region. There is no doubt that it will become a new platform for cell biology researchers.

[1] Gerd Schneider. 2003, *Anal Bioanal Chem* , **376** 558–561

[2] S.P.Jiang, L.Chen, L.B.Wan et al. 2006, *Proc.8th.Int.Conf.X-ray microscopy, Conf. Series 7*:,edited by S.Aoki, Y.Kagoshima and Y.Suzuki, IPAP,Hyogo,Japan, 30-31.

Image enhancement of x-ray microscope using frequency spectrum analysis

Wenjie Li[1], Jie Chen[1], Yijin Liu[2], Jinping Tian[1], Xiaobo Zhang[1], Gang Liu[1], Ziyu Wu[2,*] and Yangchao Tian[1,*]

[1] National Synchrotron Radiation Laboratory, University of Science and Technology of China, Hefei, Anhui 230029, P. R. China
[2] Institute of High Energy Physics, Chinese Academy of Science, Beijing 100049, P. R. China

[*] E-mail: wuzy@ihep.ac.cn (Ziyu Wu), ychtian@ustc.edu.cn (Yangchao Tian)

Abstract. We demonstrate a new method for x-ray microscope image enhancement using frequency spectrum analysis. Fine sample characteristics are well enhanced with homogeneous visibility and better contrast from single image. This method is easy to implement and really helps to improve the quality of image taken by our imaging system.

1. Introduction

A full-field, high-resolution transmission hard x-ray microscope (TXM) based on zone plate and operated at the photon energy from 7 to 11 keV has been installed at the U7A beamline at National Synchrotron Radiation Laboratory (NSRL) [1]. This facility can provide images with 50 nm spatial resolution and be able to image some weak absorption material at 8 keV by using Zernike phase ring to enhance the contrast. Although phase contrast imaging is powerful, the image contrast sometimes is still not so satisfactory. Another defect of the images are caused by the nonuniform illumination. The common solution is to divide the sample image by a flat field image. But since the illumination cannot be stationary in this sensitive system all the time, the divided images sometimes might not be so homogeneous. In addition, it's hard to take each flat field image immediately after taking the projection image during tomography. So other complementary method would be needed.

Image enhancement techniques have been applied in many kinds of fields to improve image quality. Special frequency enhancement is one of the effective methods [2]. In this paper, we propose an approach for image enhancement of x-ray microscope, which can achieve homogeneous visibility and enhanced contrast from single image on our imaging system.

2. Method

By the Fourier transform theory, the frequency scale of original signal is based on the spatial scale of the signal [3]. Based on this theory, we propose an image enhancement approach for our x ray microscope: according to the priori knowledge of the characteristic scale range of the sample, the frequency special can be separated into three regions, high-frequency (HF), intermediate-frequency (IF) and low-frequency (LF). Each frequency band corresponds to the different spatial scale. In order to improve image quality and contrast, the power of IF and HF will be increased, which represents the sample information and detail information; and the power of LF will be decreased, which represents the nonuniform background.

The steps of proposed approach can be followed as:

a) Assume that the normalized original image is *I*. At first, let *I'=ln(I+1)*. This step can change multiplicative noise to additive noise, which is advantageous for further process [4].
b) Secondly, two dimensional Fast Fourier Transform (FFT2) is applied to *I'* to obtain frequency special *F* of the image: *F=FFT2{I'}*. Based on the Gaussian low-pass filter and high-pass filter [3], *F* can be separated into three frequency region: *LF*, *IF* and *HF*. The cut-off frequency of Gaussian filters can be designed by the characteristic scale range of the sample and parameters of our imaging system.
c) Thirdly, set *F'=a*LF+b*IF+c*HF*, in which *a, b, c* are called enhancement coefficients. Via changing these three coefficients, different enhanced images can be achieved.
d) At last, two dimensional inverse Fast Fourier Transform (IFFT2) and exponent arithmetic are applied to obtain the final enhancement image *EI*: *EI=exp(IFFT2{F'})*

The total procedure is summarized in figure 1.

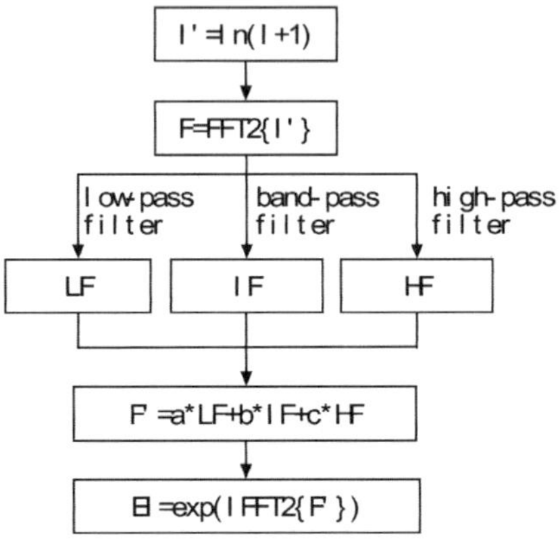

Figure 1. The flow chart of the frequency enhancement algorithm

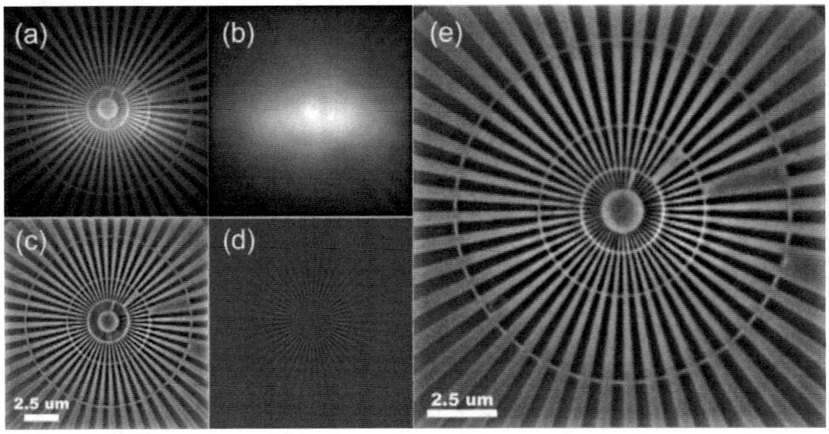

Figure 2. (a) is the original image; (b)-(d) are regarded as the LF, IF and HF images separately; (e) is an enhancement image obtained by the proposed approach. (a)-(d) are with same scale bar.

3. Experiments and Results

Here a spoke pattern image (figure 2a) taken by our system with the size of 512×512 was used to deal with the frequency enhancement algorithm. The pixel size of the image is 29nm and the characteristic scale range of the spoke pattern is about 50nm-500nm, which can be considered as the priori knowledge to design the three filters. Figure 2(b)-(d) show the intermediate results, which represents the nonuniform background, main information and the detail information of the spoke. Finally an enhanced image shown in figure 2(e) was obtained by setting the enhancement coefficients as *a*=0.05, *b*=0.4, *c*=0.55. The coefficients *a* reduces the nonuniform background while the coefficients *b* and *c* increase the main information and detail information of the spoke sample.

Another example of carbon microtube [5] shown in figure 3 is used to evaluate the enhancement effect between enhanced and original images (the dark spheres are golden dots used for alignment of tomography). The enhancement coefficients can be chosen as a=0.05, b=0.8, c=0.15 and the filter was designed by the characteristic scale range of the carbon microtube (~100nm-1um). Figure 3(a) is the original image and figure 3(b) is the enhanced image obtained by our approach. Figure 3(c) and 3(d) are two uniform images obtained by the common approach mentioned above. The flat filed image used in figure 3(d) was acquired several hours later than the one used in figure 3(c). Compared with the image in figure 3(c) and 3(d), the enhanced image in figure 3(b) has better homogeneous visibility. The contrasts of carbon microtube in the enhanced image and original image were also calculated. Figure 3(e) and 3(f) show the line profile indicated by the yellow line of two normalized images in figure 3(b) and 3(c) separately. The contrast of carbon microtube is about 18% in figure 3(e) and 13% in figure 3(f). These results reveal that the enhanced image of carbon microtube has better visibility and contrast than original one.

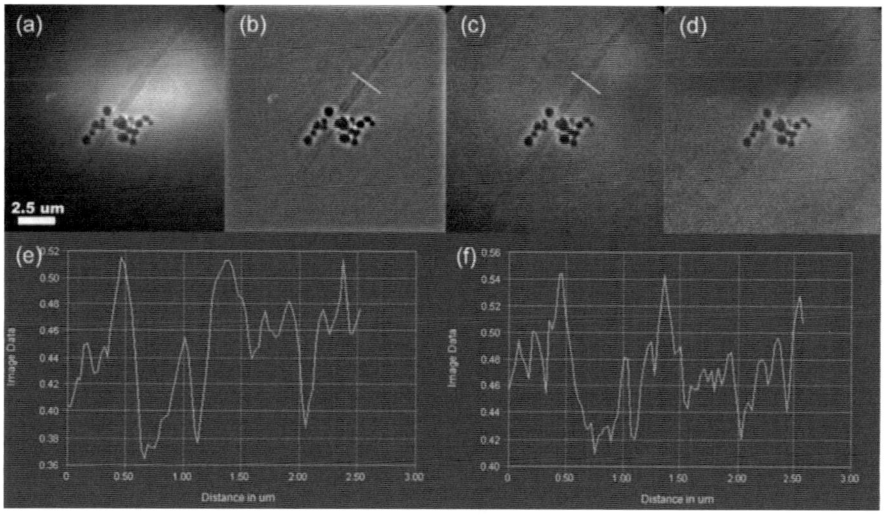

Figure 3. (a) is the original image of carbon microtube; (b) is the enhanced image; (c) and (d) are the images divided (a) by two different flat field images; (e) and (f) are the line profiles indicated in (b) and (c) separately.

4. Conclusion

To improve image quality of our TXM, we propose an approach for image enhancement using frequency spectrum analysis. The algorithm suggested in this study is mainly to reconstruct an image with better visibilities and contrast. It really helps to enhance the sample characteristics of TXM images. In future more features would be applied to develop the algorithm with better effect.

Acknowledgments

The authors are grateful for the financial support from the 985 project of the State Ministry of Education, and the National Natural Science Foundation of China (10675113, 10734070).

References

[1] Tian Y, Li W, Chen J, Liu L, Liu G, Tkachuk A, Tian J, Xiong Y, Gelb J, Hsu G, and Yun W 2008 *Rev. Sci. Instrum.* **79** 103708

[2] Yang C C 2008 *OPTIK* **119** 143-146

[3] Gonzalez R C and Woods R E 2002 *Digital Image Processing, Second Edition* (Prentice Hall)

[4] Dougherty E R and Astola J 1994 *An Introduction to Nonlinear Image Processing* (SPIE Optical Engineering Press)

[5] Song L, Ci L, Lv L, Zhou Z, Yan X, Liu D, Yuan H, Gao Y, Wang J, Liu L, Zhao X, Zhang Z, Dou X, Zhou W, Wang G, Wang C, Xie S 2004 *Adv. Mater.* **16**, 1529–1534

Journal of Physics: Conference Series **186** (2009) 012010 doi:10.1088/1742-6596/186/1/012010

High resolution projection X-ray microscope equipped with fluorescent X-ray analyzer and its applications

K Minami, Y Saito, H Kai, K Shirota and K Yada

TOHKEN CO., LTD., X-ray R&D Dept.,
2-27-7 Tamagawa Chofu, Tokyo 182-0025, Japan

E-mail: kminami@tohken.co.jp

Abstract. We have newly developed an open type fine-focus X-ray tube "TX-510" to realize a spatial resolution of 50nm and to radiate low energy characteristic X-rays for giving high absorption contrast to images of microscopic organisms. The "TX-510" employs a ZrO/W(100) Schottky emitter and an "In-Lens Field Emission Gun". The key points of the improvements are (1) reduced spherical aberration coefficient of magnetic objective lens, (2) easy and accurate focusing, (3) newly designed astigmatism compensator, (4) segmented thin film target for interchanging the target materials by electron beam shift and (5) fluorescent X-ray analysis system.

1. Introduction

Today, projection X-ray microscopes are widely utilized as inspecting systems in the fields of electrical and electronic industries. But they have been unable to resolve sub-micron meter inner structures, because of their modest focusing ability of the "micro-focus X-ray tube". Our electron optical technologies have made it possible to produce a superior open type nano-focus X-ray tube, named "TX-500", which employs a ZrO/W(100) Schottky emitter and an "In-Lens Field Emission Gun". The ZrO/W(100) Schottky emitter is located at the center of a strong magnetic lens field to get a high brightness electron beam [1]. We reported a projection X-ray inspection system "TUX-5000F" equipped with the "TX-500" tube and showed a resolution of 100nm, several years ago. However, it was difficult to resolve extremely minute structures as in LSI (large scale integrated circuits). Higher resolution and higher contrast are strongly demanded from the markets such as electronic devices, semiconductor devices, micro electro mechanical systems (MEMS), new material industries and so forth. And now, other application fields, such as biology, medical science etc..., are emerging. It is also required to have element analysis function in combination with projection X-ray microscope. To meet those demands, we have developed a new nano-focus X-ray tube "TX-510" and applied it to a highly resolving projection X-ray microscope "TUX-5000FS". Details are described below.

2. Apparatus

The nano-focus X-ray tube "TX-510" needs to be evacuated to 10^{-8}Pa range and so a combination of turbo-molecular and sputter ion pumps are used. Figure 1 shows a schematic diagram of "TX-510". This X-ray tube "TX-510" has a reflection electron detector near the target in order to get a reflection electron image of the target surface. This function makes the following four features possible: (a) easy and accurate focusing, (b) precise astigmatism compensation, (c) easy interchange of target materials

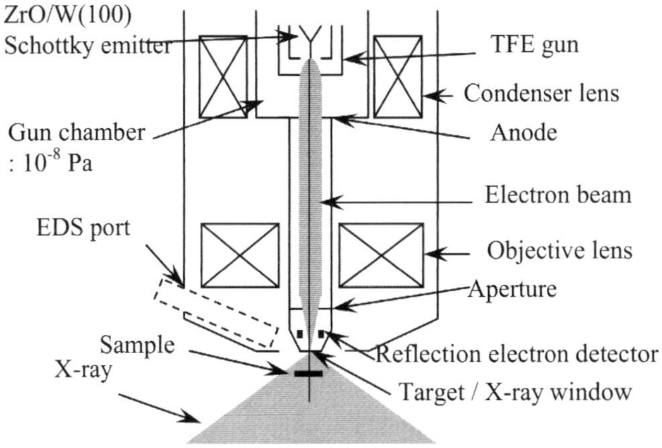

Figure 1. Schematics of nano-focus X-ray tube "TX-510".

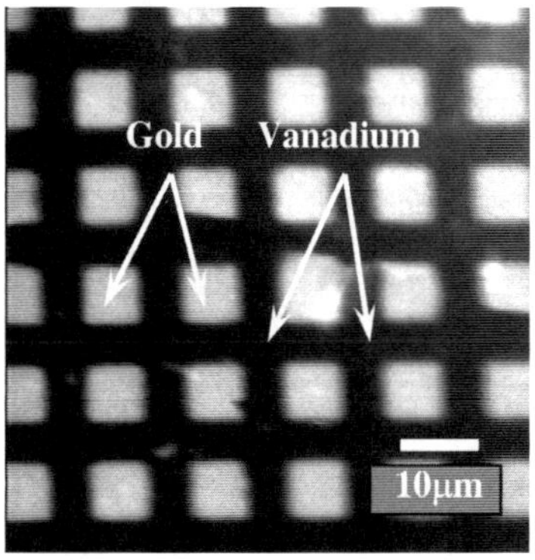

Figure 2. A reflected electron image of two kinds of target materials: vanadium and gold.

by electron beam shift and also (d) elongation of X-ray target life time. Figure 2 shows a reflected electron image of the target surface. In this case, V-K_α (4.95keV) and Au-L_α (9.71keV), -M_α (2.12keV) are useful to get a high contrast X-ray image for minute and/or light element specimens. It needs about two or three times acceleration voltage of a characteristic X-rays excitation energy for the high contrast X-ray imaging. Next, we have realized an element analysis function by installing an energy dispersive X-ray spectrometer (EDS) without any deterioration of resolution. As shown in figure 1, it is possible to set an EDS probe inside the objective lens to achieve high detection efficiency. The shortest distance to the specimen is 19.6mm, the maximum take off angle is 35 degrees and detection solid angle is 0.13 steradians when a Si-Li detector of 50mm^2 is used. Apertures are set between the target and the specimen to limit the analyzing area on the specimen. The minimum size of the analyzing area is about 10μm in diameter. Figure 3 shows an analyzed spectrum of a gold mesh. Peaks for Au-M_α(2.12keV), -L_α(9.71keV) and -L_β(11.44keV) are clearly detected, but some unnecessary ghost peaks are also included, which are corresponding to lead, iron and nickel. These ghost spectra are due to the system noise caused by fluorescent radiation from the limiting apertures, so we need to solve this problem. Third, we have succeeded to decrease spherical aberration coefficient of the objective lens down to two-fifth compared with our older model "TX-500", thereby

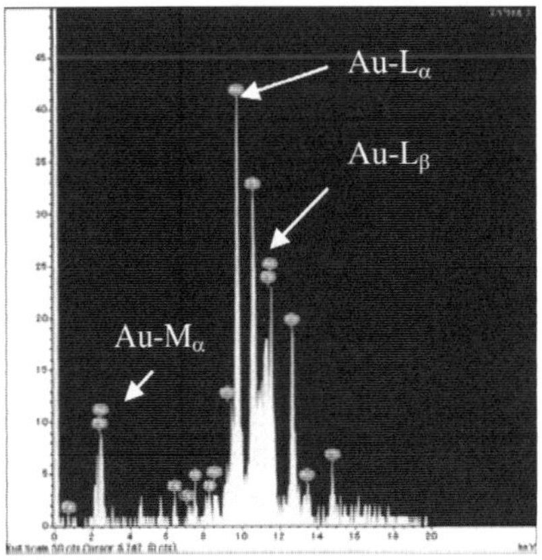

Figure 3. An example of fluorescence X-ray spectra taken for a gold mesh. Process time of EDS system is 120sec. Accelerating voltage is 20kV. Emission current form TFE gun is 80μA.

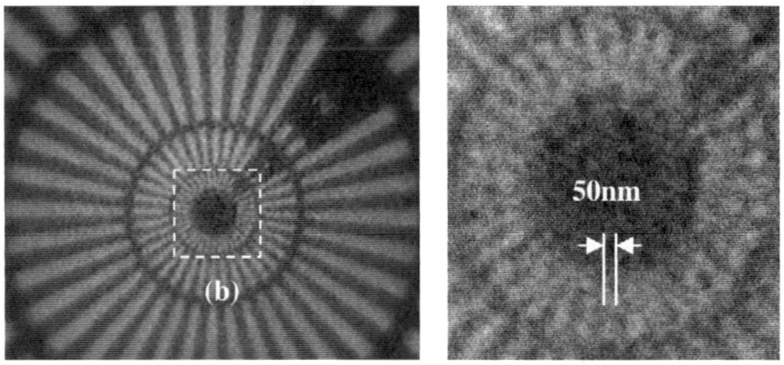

(a) (b)

Figure 4. X-ray images of X-ray resolution test chart taken by using TUX-5000FS. A low magnification image (a). The minimum width of radial line and space of the inner circle is 50nm in digital magnified image (b). Accelerating voltage is 30kV and emission current from TFE gun is 160μA. Target material is platinum of 0.6 μm thick.

we have achieved a resolution of 50 nm. Figure 4 are images of X-ray resolution test chart taken by using new type of "TUX-5000FS". This test chart "ATN/XRESO-50HC" is made by NTT-AT nanofabrication Co., Japan. To get an improved absorption contrast for low energy characteristic X-rays, a vacuum or helium gas atmosphere is needed. The "TUX-5000FS" has a vacuum tight specimen chamber equipped with a helium introducing system. Both an X-ray CCD camera "XC+605HT" (X-ray Precision, Inc., Japan) and an X-ray imaging system with an intensifier "E5889BE-P1K" (TOSHIBA ELECTRON TUBE & DEVICES Co., Ltd., Japan,) are available, but only the X-ray CCD camera is able to work in the vacuum.

3. Application

Figure 5 shows an X-ray image (a) and a scanning electron microscope (SEM) image (b) of *Paramecium caudatum*, which is fixed by osmium-tetroxide acid and freeze-dried. Size of *Paramecium* is about 200 μm long and the thickness is about 30 μm. The X-ray image was taken with vanadium target in helium gas atmosphere. Some internal organelles like the macronucleus and vacuoles are clearly seen inside the *Paramecium* in figure 5 (a).

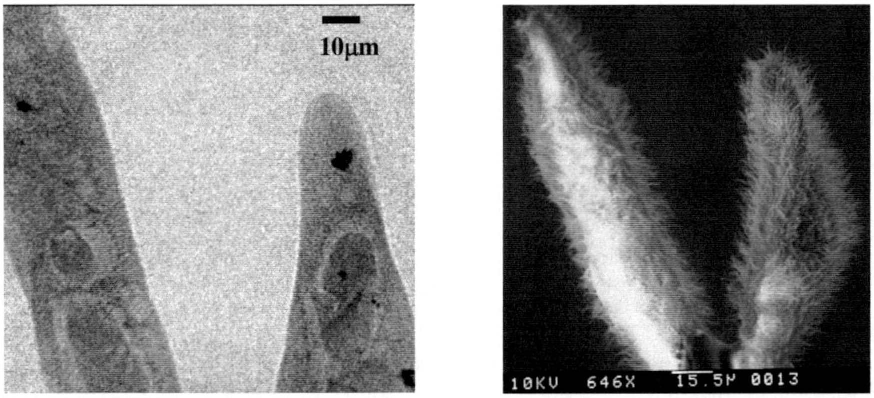

(a) (b)

Figure 5. A X-ray image of *Paramecium* taken by using "TUX-5000FS" in He gas atmosphere (a) and an identical SEM image (b). Accelerating voltage is 15kV and emission current from TFE gun is 364μA. Target material is vanadium of 0.7μm thick.

4. Summary

We have succeeded to develop a new type nano-focus X-ray tube "TX-510", which is able to form extremely small and bright X-ray source. Segmented thin film targets with various materials for low energy characteristic X-rays can be equipped in combination with EDS system. Thus, the new projection X-ray microscope "TUX-5000FS" installed with "TX-510" tube has realized high resolution of 50nm and high absorption contrast to a minute and light element specimen like *Paramecium.*

5. Acknowledgment

This development was supported by a grant from the Ministry of Economy, the Japanese Government. Grant number is Heisei18.02.08Kantou No.150 and Heisei19.02.26Kantou No.77.

Reference

[1] Y Saito, H Kai, K Shirota, and K Yada 2005 *Proc. 8th Int. Conf.X-Ray Microscopy (Himeji)* IPAP Conf. Series **7**, ed S Aoki, Y Kagoshima, Y Suzuki (Tokyo: IPAP) pp35-37

Journal of Physics: Conference Series **186** (2009) 012011 doi:10.1088/1742-6596/186/1/012011

Image reconstruction using a configurable detector in STXM

G R Morrison[1], A Gianoncelli[2], B Kaulich[2]
[1]King's College London, Dept. of Physics, Strand, London WC2R 2LS, UK
[2]Sincrotrone Trieste, S.S. 14, km 163.5, Area Science Park, I-34012 Trieste-Basovizza, Italy

E-mail: graeme.morrison@kcl.ac.uk

Abstract. The use of a fast-readout CCD x-ray detector in a scanning transmission x-ray microscope (STXM) allows a range of simultaneous imaging modes to be realised. The CCD records the full 2-D distribution of x-rays transmitted by the sample for every pixel in the raster scan that generates the STXM image, potentially generating large volumes of data. The data in each CCD frame can then be processed in a variety of ways to produce useful image signals, including both absorption and phase contrast. Since the detector effectively records a convergent-beam micro-diffraction pattern for every position in the STXM raster other processing algorithms can be used to provide additional information to the user.

1. Introduction

The use of a segmented detector for the transmitted x-ray detector in the scanning transmission x-ray microscope (STXM) is increasing, as it allows a range of different forms of image contrast to be made available simultaneously, and more effectively exploits the information available about the sample structure. There are two main approaches: the first makes use of a large 2-D array of detector elements to record a full 2-D intensity distribution for every pixel in the raster scan of the STXM [1–3], while the other uses a few custom-designed detector segments, usually in the form of circular arcs and quadrants [4, 5]. Both types of detector allow the conventional incoherent brightfield image to be formed by summing the signals from all the detector elements, and both also provide information about the redistribution of the x-ray transmitted x-ray signal across the detector plane as a result of the x-ray interaction with the sample. It is this redistribution of the intensity that allows phase contrast image signals to be produced, usually in the form of differential phase contrast, when an antisymmetric detector response function is used.

The use of customised segmentation means that only a few image signals need to be recorded for each pixel in the image scan, and the detector can be optimised for fast and efficient data collection. The disadvantages are that the choice of detector configuration is fixed at the hardware design stage, and that the detector has a well-defined central axis that must be aligned carefully with optical axis of the STXM to ensure that signal differences across the detector plane arise only from the sample interaction, and not from beam misalignment. On the other hand, the use of a pixellated 2-D array detector, as shown schematically in Figure 1, means that a great variety of detector configurations can be chosen in software after the image data are acquired, so that different imaging modes can be considered retrospectively. The disadvantage in this case is that a much larger volume of raw data is acquired in the first instance, and 2-D

Journal of Physics: Conference Series **186** (2009) 012011 doi:10.1088/1742-6596/186/1/012011

array detectors tend to have somewhat longer readout times. In practice an array of 128^2 pixels allows quite fine control over the detector configuration, and a charge-coupled device (CCD) array of this size can be read out in only a few milliseconds [2].

In this paper we describe experiments made using the fast-readout CCD that is now in standard use as the transmitted x-ray detector on the Twinmic STXM on beamline 1.1L [6] at the Elettra synchrotron.

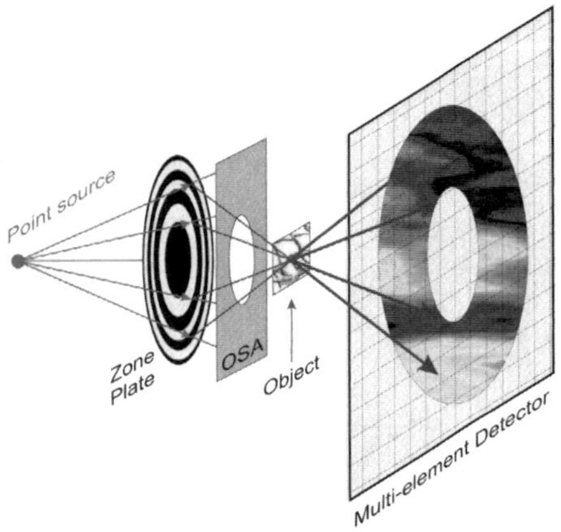

Figure 1. Schenatic representation of an STXM fitted with a pixellated detector, that can record a 2-D image of the transmitted x-ray signal.

2. Imaging with the Twinmic STXM

In the STXM the transmitted x-ray detector can be assumed to lie in the far-field (Fourier transform plane) relative to the specimen plane, and the response of a configurable detector is described by the function $R(\mathbf{k})$, where $\mathbf{k}$ is a spatial frequency vector. The image signal when the x-ray probe $\psi(\mathbf{r}, \mathbf{r}_o)$ is centred at position $\mathbf{r}_o$ on the object is then

$$s(\mathbf{r}_o) = \int |\Psi(\mathbf{k}, \mathbf{r}_o)|^2 R(\mathbf{k})\, \mathrm{d}\mathbf{k} \tag{1}$$

and the choice of imaging mode is determined by $R(\mathbf{k})$. Setting $R(\mathbf{k}) = 1$ gives the incoherent brightfield image, while $R(\mathbf{k}) = \mathbf{k}$ yields a first-moment detector response that produces differential phase contrast, with the image signal linearly related to the phase gradient of the object transmittance [7].

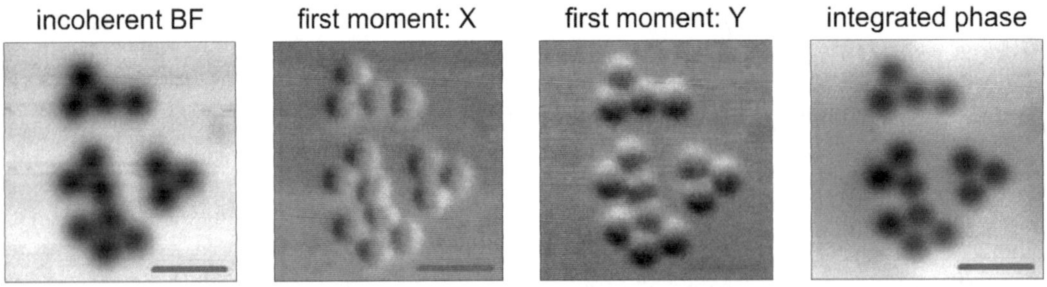

Figure 2. Images of 1 μm diameter polystyrene spheres taken at a photon energy $\sim$ 653eV. The scale bar is 2 μm. The contrast in the brightfield and integrated phase images has been stretched to fill the available grey levels.

It is possible to reconstruct the object phase $\phi(x, y)$ by integrating the phase gradient signals; one approach that seems particularly quick and effective, even in the presence of image noise, is a Fourier method [8, 9]. With this method, the first-moment signals shown in Figure 2 are used to create a complex array

$$g(x, y) = a\left(\frac{\partial \phi}{\partial x} + i\frac{\partial \phi}{\partial y}\right) \quad \Rightarrow \quad \Phi(k_x, k_y) = \frac{G(k_x, k_y)}{2\pi i a(k_x + ik_y)} \tag{2}$$

where $\Phi(k_x, k_y)$ and $G(k_x, k_y)$ are the Fourier transforms of $\phi(x, y)$ and $g(x, y)$ respectively, and a is a constant. The object phase can be found by an inverse transform of $\Phi(k_x, k_y)$. An example of this phase integration for images of 1 μm diameter polystyrene spheres is shown in Figure 2.

3. Discussion

The phase integration shown here is a very rapid process (typically less than 1 s) that can be applied directly to the image data from the STXM scan. However, there are some implicit assumptions: first, that the moment signals represent only a phase gradient, not a mixture of amplitude and phase information, and second, that the size of the focused x-ray probe is much smaller than any features of interest in the object field. In practice, for high resolution imaging it may often be necessary to correct for these effects.

For this reason, more sophisticated reconstruction algorithms are of considerable interest, and hybrids of conventional STXM and coherent diffraction imaging methods look particularly attractive. Ptychographic methods [10, 11] offer the ability to image extended objects and retrieve the complex amplitude transmitted by the object at a resolution that is not limited by the size of the x-ray probe, but they still require a good estimate of the form of the probe to ensure rapid convergence. The recent demonstration of scanning x-ray diffraction microscopy [3] looks even more promising, since the reconstruction algorithm can provide not only a complex reconstruction at high resolution, but also a measure of the illuminating probe itself.

Acknowledgment
The Twinmic X-ray microscope development was funded by the EU "Access to Research Infrastructure action of Improving Potential" Programme, under contract HPRI-CT-2001-50024.

References

[1] Morrison G R, Eaton W J, Barrett R and Charalambous P S 2003 *J. Phys. IV* **104** 547–550
[2] Gianoncelli A, Morrison G R, Kaulich B, Bacescu D and Kovac J 2006 *Appl. Phys. Lett.* **89** 251117
[3] Thibault P, Dierolf M, Menzel A, Bunk O, David C and Pfeiffer F 2008 *Science* **321** 379–382
[4] Feser M, Hornberger B, Jacobsen C, De Geronimo G, Rehak P, Holl P and Struder L 2006 *Nucl. Instrum. Methods Phys. Res., Sect. A* **565** 841–854
[5] Hornberger B, Feser M and Jacobsen C 2007 *Ultramicroscopy* **107** 644–655
[6] Kaulich B *et al.* 2006 *Proc. 8th Int. Conf. X-ray Microscopy* (*IPAP Conference Series* vol 7) ed Aoki S, Kagoshima Y and Suzuki Y (Tokyo: Institute of Pure and Applied Physics) pp 22–25
[7] Waddell E M and Chapman J N 1979 *Optik* **54** 83–96
[8] Arnison M R, Larkin K G, Sheppard C J R, Smith N I and Cogswell C J 2004 *J. Microsc.* **214** 7–12
[9] Kottler C, David C, Pfeiffer F and Bunk O 2007 *Opt. Express* **15** 1175–1181
[10] Rodenburg J M, Hurst A C, Cullis A G, Dobson B R, Pfeiffer F, Bunk O, David C, Jefimovs K and Johnson I 2007 *Phys. Rev. Lett.* **98** 034801
[11] Rodenburg J M, Hurst A C and Cullis A G 2007 *Ultramicroscopy* **107** 227–231

Journal of Physics: Conference Series **186** (2009) 012012 doi:10.1088/1742-6596/186/1/012012

First differential phase contrast results from PolLux

J. Raabe [1], B. Watts [1], G. Tzvetkov [1, 2], R. H. Fink [2], C. Quitmann [1]

[1] Swiss Light Source (SLS), Paul Scherrer Institut, 5232 Villigen-PSI, Switzerland

[2] Department Chemistry and Pharmacy & Interdisciplinary Center for Molecular Materials (ICMM), Egerlandstraße 3, 91058 Erlangen, Germany

Corresponding author: joerg.raabe@psi.ch

Abstract. We report on the design and performance of the PolLux soft x-ray scanning transmission x-ray microscopy (STXM) instrument, which has been installed at the X07DA bending magnet beamline of the Swiss Light Source. Although PolLux started routine user operation in 2006, new features like differential phase contrast imaging have recently been implemented to broaden the experimental possibilities.

1. Motivation

X-ray microscopy has developed into a routine technique at many synchrotron facilities. In the soft x-ray regime, scanning-transmission x-ray microspectroscopy (STXM) allows combined spectroscopy and microscopy of soft matter, environmental samples and mesoscopic magnetic structures. In addition to elemental and chemical imaging, linear or circular polarization is routinely used to visualize molecular orientation, and magnetic order.

Differential phase contrast (DPC) is an additional feature now available at the PolLux STXM. This provides significant enhancement of image quality, reduced radiation damage and allows observation of minute differences in the refractive index that are not detectable in pure absorption experiments. Although DPC is quite common in hard x-ray imaging with coherent beams, it is rarely used in soft x-ray microscopy, mainly due to complications in the detection of soft x-rays. DPC is implemented in the STXM X1A at Brookhaven [1] using a segmented x-ray detector, and in the TWINMIC end station at ELETTRA via a fast read-out CCD camera [2]. We have realized a DPC setup similar to that at TWINMIC for the PolLux instrument and present here the current performance of PolLux with first examples of DPC detection.

2. PolLux Beamline and Experimental Station

In the PolLux STXM a Fresnel zone plate (FZP) focuses the x-rays to a diffraction limited focus ($\varnothing_{\text{focus}}$ ~ tens of *nm*). Images are acquired by raster-scanning the sample through the focus while measuring the transmitted intensity, I_{trans}, with a relatively large x-ray detector ($\varnothing_{\text{detector}}$ ~ hundreds of *μm*). Variation of the photon energy, *E*, the polarization, $\vec{P}$, or sample parameters such as the temperature, *T*, allows recording multidimensional datasets, e.g. $I_{trans}(x, y, E, \vec{P}, T, \ldots)$. These are analyzed by comparison to the well-known absorption edges of the elements and to reference samples, thus providing quantitative information about the elemental composition, chemical bonding, sample morphology and magnetic order.

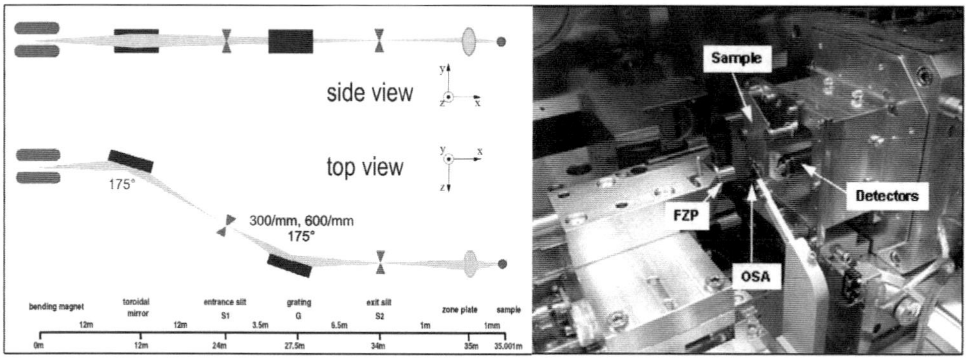

Figure 1. Left: Sketch of the beamline layout, based on a spherical grating monochromator. Right: Image of the microscope interior with the major components labelled.

The PolLux instrument is installed at bending magnet beamline X07DA of the Swiss Light Source (Villigen, Switzerland). Its design is based on the STXM installed at BL 5.3.2 of the Advanced Light Source (Berkeley, USA) but includes several substantial improvements. The beamline allows a choice of two monochromator gratings in order to provide an extended photon energy range (200 - 1400 *eV*), while still providing high flux and resolution for the study of transition metal samples and their magnetism. The SLS provides excellent stability of the electron beam in terms of intensity, through its top-up mode, and in terms of position, via a fast orbit-feedback system. The user can introduce local vertical tilt of the storage ring orbit to provide circularly polarized light, without disturbingother beamlines. The helicity can presently be switched (both left or right) within less than 1 *sec* which greatly reduces long term drift effects between images.

The optical layout of the beamline is shown in *Fig. 1 (left).* A horizontally deflecting toroidal mirror creates a horizontal focus of the source at the entrance slit, *S1*. In the vertical direction the source is imaged by the toroidal mirror onto the exit slit, *S2*. The monochromator, with either of ist two spherical gratings (300 *lines/mm* and 600 *lines/mm*), creates dispersion in the horizontal plane. The exit slit (*S2*) is the source for the Fresnel zone plate (FZP). Details of the design and performance are provided in refs. [3],[4].

A photograph of the PolLux instruments sample environment is shown in *Fig. 1 (right).* The FZP focuses the light into a diffraction limited spot. Higher diffraction orders are filtered out using an order-sorting aperture (OSA), located between the FZP and the sample. Samples are usually prepared on either semitransparent silicon nitride membranes, or transmission electron microscope grids. The intensity of the transmitted x-ray beam is measured by a detector behind the sample. For standard absorption measurements two detectors are available, a photodiode and a photomultiplier (coupled to a scintillator). The instrument has already demonstrated spatial resolution below 20 nm [5].

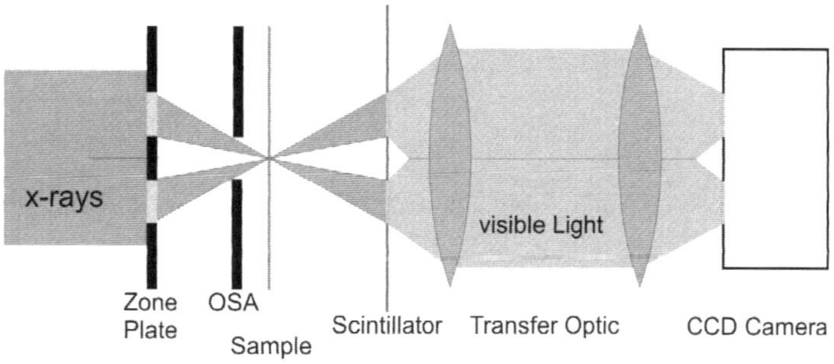

Figure 2. Optical layout of the DPC mode showing the scintillator and the transfer optics used to image the x-ray image on a CCD camera with high read out speed.

3. Differential Phase Contrast Imaging at PolLux

In the standard mode of operation, a STXM records the integrated intensity of the transmitted x-rays, thus providing the local absorption coefficient of the sample at each pixel. Extending the detection system to a spatially resolved detector provides access to phase contrast. At PolLux this is realized by

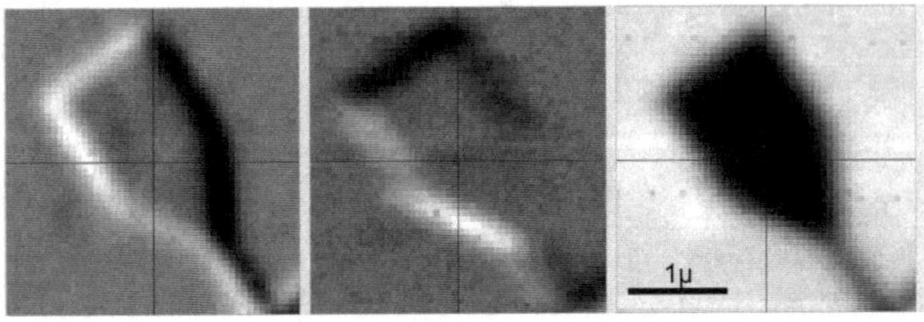

Figure 3. Image of a test pattern (100 nm thick Gold on a silicon nitride membrane), recorded at E = 1 keV. Images on the left and in the centre show differential phase contrast in the horizontal and vertical direction, respectively. The right image shows the conventional absorption contrast.

first converting x-ray image transmitted by the sample into visible light via a scintillator. This image is then transferred onto a CCD camera using intermediate optics (Fig. 2). For a homogenous sample the image on the CCD has the shape of a bright ring, representing the zone plate with its central stop. In the presence of a phase gradient in the sample, this ring is deflected. We calculate this deflection by defining four areas on the detector and by performing a four-quadrant detection in software. The two asymmetry signals measure the horizontal and the vertical phase gradient of the sample, respectively. Details on the realization of a segmented detector and the calculation of the asymmetry signals can be found in [1] for a segmented diode, and in [2] for segmentation in software using a CCD detector. The standard absorption signal is given by the integrated intensity of the CCD detector (Fig. 3).

It should be noted that the CCD image contains even more information about the sample structures in the x-ray focus. This diffraction information can be obtained by application of phase retrieval algorithms [7] and will be the focus of further work on the PolLux instrument.

Acknowledgment:
Financial support by the BMBF (contract 05KS7WE1) is gratefully acknowledged.

References

[1] B. Hornberger, M. D. de Jonge, M. Feser, P. Holl, C. Holzner, C. Jacobsen, D. Legnini, D. Paterson, P. Rehak, L. Strüder, and S. Vogt. J. Synchrotron Rad. **15**, 355 (2008)
[2] A. Ginoincelli, G.R. Morrison, B. Kaulich, D. Bacesu, and J. Kovac, Appl. Phys. Lett **89**, 251117 (2006).
[3] U. Flechsig, C. Quitmann, J. Raabe, M. Böge, R. Fink, and H. Ade, Ninth International Conference on Synchrotron Radiation Instrumentation, AIP Conf. Proc. **879**, 505 (2007)
[4] J. Raabe, G. Tzvetkov, U. Flechsig, M. Böge, A. Jaggi, B. Sarafimov, M.G.C. Vernooij, T. Huthwelker, H. Ade, D. Kilcoyne, T. Tyliszczak, R. H. Fink, and C. Quitmann, Rev. Sci. Instrum. **79**, 113704 (2008)
[5] K. Jefimovs, J. Vila-Comamala, T. Pilvi, J. Raabe, M. Ritala, C. David, Phys. Rev. Lett. **99**, 264801 (2007)
[6] Andor DV860 with electron multiplying CCD, resolution: 128 x 128 pixels, frame rate up to 3500Hz (http://www.andor.com/)
[7] P. Thibault, M. Dierolf, A. Menzel, O. Bunk, C. David, and F. Pfeiffer, Science **321**, 379-382 (2008)

High-throughput X-ray fluorescence imaging using a massively parallel detector array, integrated scanning and real-time spectral deconvolution

C.G. Ryan[1,6,7], D.P. Siddons[2], G. Moorhead[3,6], R. Kirkham[3], G. De Geronimo[4], B.E. Etschmann[1,7], A. Dragone[2], P.A. Dunn[3], A. Kuczewski[2], P. Davey[3], M. Jensen[3], J.M. Ablett[2,8], J. Kuczewski[2], R. Hough[1] and D. Paterson[5]

[1] CSIRO, Exploration and Mining, Monash University, Clayton VIC, Australia
[2] National Synchrotron Light Source, Brookhaven National Laboratory, NY, USA
[3] CSIRO Materials Science and Engineering, Clayton VIC, Australia
[4] Instrumentation Division, Brookhaven National Laboratory, Brookhaven NY, USA.
[5] Australian Synchrotron, Clayton VIC, Australia
[6] School of Physics, University of Melbourne, Parkville VIC, Australia
[7] CODES Centre of Excellence, University of Tasmania, Hobart TAS, Australia

Email: Chris.Ryan@csiro.au

Abstract. A step improvement in X-ray fluorescence imaging performance is demonstrated through close integration of a large detector array, dedicated data acquisition, stage control and real-time parallel data processing, to achieve efficient elemental imaging with <1 ms per pixel, image sizes in excess of 4 megapixels, full-spectral data collection and spectral deconvolution, at detected photon rates up to 6 M/s, in prototype tests at the NSLS using a 96 detector array.

1. Introduction

X-ray fluorescence microprobe imaging is commonly hampered by read-out delays. An alternative data acquisition strategy uses a multiparameter approach in which each photon event is tagged by sample XY position and detector identity, and these 'events' are streamed for further processing or storage. This strategy exhibits zero read-out overhead and enables faster raster scanning. Complex full-spectral SXRF data can be decomposed into elemental components using a matrix transform method called Dynamic Analysis (DA) [1,2] originally developed for PIXE imaging [3]. The simplicity of the method permits its execution on this event stream in real-time [4].

Large planar monolithic silicon detector arrays and dedicated application specific integrated circuits (ASICs) for signal processing offer the potential for large solid-angle collection, with the resulting high count rates spread over a large number of detectors. The speed of field-programmable gate array (FPGA) computation permits real-time processing of each event from such an array.

A new spectroscopy detector system is under development that combines a large detector array and custom ASICs development by BNL and a pipelined, parallel processor with embedded DA image projection developed at CSIRO. It subtends a solid-angle of ~1.5 sr, is closely coupled with sample

[8] Present address: Synchrotron Soleil, L'orme des Merisiers, Saint-Aubin, BP 48, F-91192, Gif-sur-Yvette, France

scanning and performs real-time processing and projection of elemental images [5,6]. The project aims for a 384 detector array. This work reports on tests using a 96 detector prototype, called Maia-96.

2. Detector Concept

The detector comprises a low-leakage Si pad array developed at BNL [5], Peltier cooled to -35 °C. It is directly wire-bonded to CMOS ASICs each providing 32 low-noise preamplifiers with high order shapers and baseline stabilizers (HERMES [7]). Shaped pulses go to 32-channel peak-detector derandomizer ASICs, which capture pulse height and time-over-threshold (SCEPTER [8]).

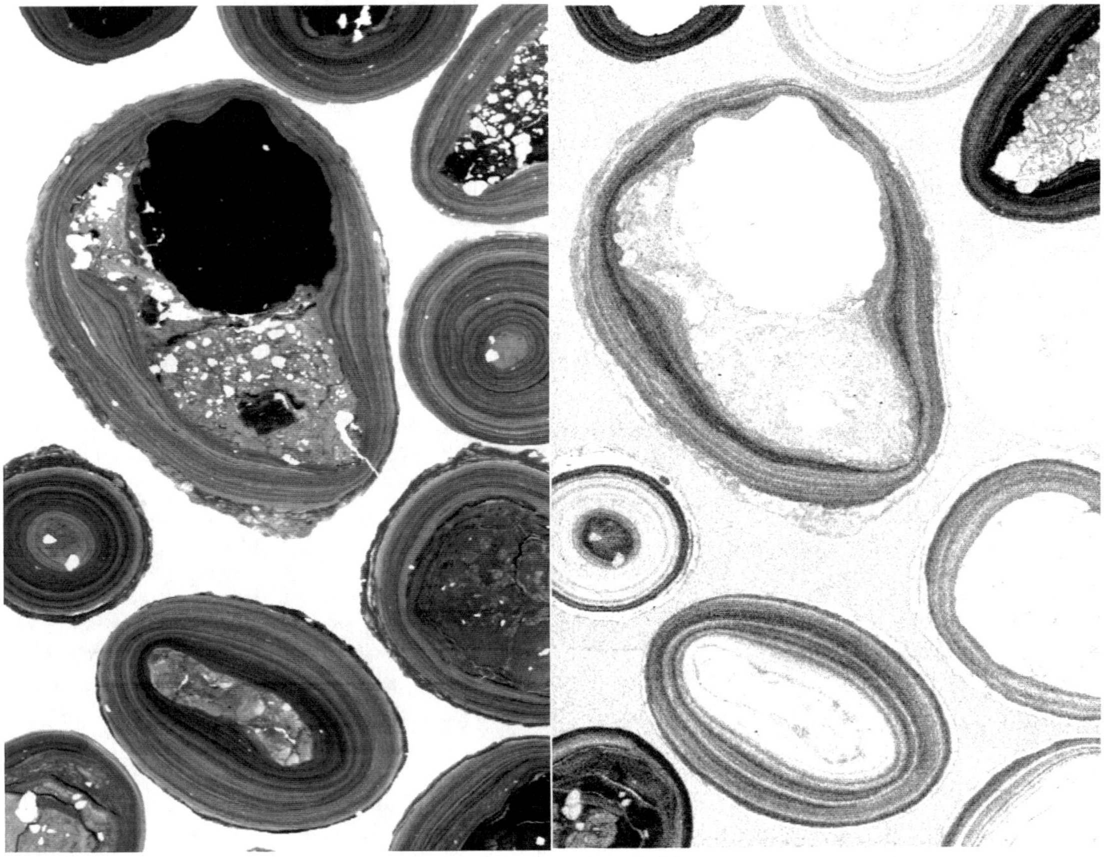

Figure 1. Major element Fe (left) and trace Y (right) grey-scale images of iron-oxide nodules, Rose Dam; 1625 x 2625 pixels (13 x 21 mm^2, 5 ms dwell per pixel), 7.5 μm pixels, using 17.2 keV X-rays.

Outputs from SCEPTER representing both energy (E) and time-over-threshold (T) are digitized by dual 14-bit fast synchronous ADCs interfaced to the parallel processing engine HYMOD, developed at CSIRO for high speed instrumentation. HYMOD consists of a FPGA connected to 6 large static RAMs, a PowerPC processor and fast serial (12 x 3.125 Gb/s) and Ethernet ports (2 x 1Gb/s). FPGA code is written using a CSIRO-developed pipelined, parallel processing language called 3PL [9].

The processing pipeline comprises: (i) pile-up rejection, (ii) energy calibration, (iii) 4K spectrum accumulation, (iv) DA image accumulation, and optionally (v) event pre-scaling and logging. Each ET event is tagged by detector number and XY position. Pile-up rejection compares E with that expected for its T, with outliers discarded. The detector number selects energy calibration tables to map channel number onto DA matrix column, which provides the increments for elemental images for the current XY pixel [2]. DA accounts for changing elemental sensitivity across the array due to varying take-off angle, absorption and efficiency [10]. DA is implemented as 32 parallel pipelines, one per chemical element, read out and displayed at XY pixel rate. In parallel, events can be pre-scaled to suppress intense lines when logging to disk. HYMOD also controls the stage, driving stepper motors directly.

3. Prototype Tests

Earlier tests demonstrated the concept using a 32 detector prototype with real-time DA projection into elemental images [11,6]. Tests at NSLS beam-line X27A [12] reported here used a 96 element prototype (called Maia-96) with the array bonded to three HERMES-SCEPTER pairs, and direct drive of a stage capable of 10 mm/s slew speed. These tests collectively demonstrated (i) count rates up to 6 M/s, (ii) image sizes up to 4.3 megapixels, (iii) dwell times as short as 0.8 ms (limited by stage speed for 7.5 μm pixels), (iv) image frames (560 x 280 pixels) as short as 6.5 minutes, (v) DA processing using HYMOD, using a second HYMOD to simulate a detector data-source, at 10^8 events per second and (vi) compression schemes designed to filter major element components in the raw data stream.

Figure 1 illustrates performance showing images of Fe and trace Y in iron-oxide nodules from Rose Dam, Western Australia, revealing an amazing diversity of zonation reflecting contrasting growth history and sedimentary sorting processes; Y concentrations vary up to ~100 ppm and average ~10 ppm. The pixel size was chosen to slightly over-sample given the beam size. The desired running time of ~6 hours determined the required stage speed of 1.5 mm/s and a dwell per pixel of 5 ms.

4. Conclusions

Development of a 384 detector system, integrating full-spectral data collection, high speed stage control and real-time data processing and image display are well advanced with successful tests of a 96 detector prototype at the NSLS, which features megapixel imaging at dwell times as short as 0.8 ms and count rates up to 6 M/s. Given these capabilities, the choice of pixel size and dwell time are now dictated by the needs of the application, and not by data acquisition limitations. Two 384 detector systems are under construction for installation at the Australian Synchrotron and the NSLS.

Acknowledgements

We acknowledge financial support from the *Access to Major Research Facilities Programme.* Research carried out at the National Synchrotron Light Source, Brookhaven National Laboratory, which is supported by the U.S. Department of Energy, under Contract No. DE-AC02-98CH10886.

References

[1] Ryan C G, Etschmann B E, Vogt S, Maser J, Harland C L, Van Achterbergh E and Legnini D 2005, *Nucl. Instr. Meth.* B **231** 183
[2] Ryan C G 2000, *Int. J. Imaging Systems and Technology* **11** 219
[3] Ryan C G and Jamieson D N 1993, *Nucl. Instr. and Meth.* B **77** 203
[4] Ryan C G, Jamieson D N, Churms C L and Pilcher J V 1995, *Nucl. Instr. and Meth.* B **104** 157
[5] Siddons D P, Beuttenmuller R H, O'Connor P, Kuczewski A J, Li Z 2004, *AIP Conference Proceedings* **705** 953
[6] Ryan C G, Siddons D P, Moorhead G, Kirkham R, Dunn P A, Dragone A, De Geronimo G 2007, *Nucl. Instr. Meth.* B **260** 1
[7] De Geronimo G, O'Connor P, Beuttenmuller R H, Li Z, Kuczewski A J, and Siddons D P 2003, *IEEE Trans. Nucl. Sci.* **50** 885
[8] Dragone A, De Geronimo G, Fried J, Kandasamy A, O'Connor P and Vernon E 2005, *IEEE Nuclear Science Symposium Conference Record* **2** 914
[9] Dunn P, Ryan C G, Kirkham R, Moorhead G, Davey P, submitted to *Nucl. Instr. Meth.* A
[10] GeoPIXE software for PIXE/SXRF imaging, http://nmp.csiro.au/GeoPIXE.html
[11] Siddons D P, Dragone A, De Geronimo G, Kuczewski A, Kuczewski J, O'Connor P, Li Z, Ryan C G, Moorhead G, Kirkham R, Dunn P 2006, *proc. of IEEE Nuclear Science Symposium, Medical Imaging and Room Temperature Detector Workshop*, San Diego, October 2006
[12] Ablett J M, Kao C C, Reeder R J, Tang Y, Lanzirotti A 2006, *Nucl. Instr. Meth.* A **562** 487

9th International Conference on X-Ray Microscopy IOP Publishing
Journal of Physics: Conference Series **186** (2009) 012014 doi:10.1088/1742-6596/186/1/012014

Fluorescence X-ray micro-spectroscopy activities at ESRF

M Salomé[1], P Bleuet[1], S Bohic[2,1], J Cauzid[3,1], E Chalmin[1], P Cloetens[1], M Cotte[4,1], V De Andrade[1], G Martinez-Criado[1], S Petitgirard[1,5], M Rak[1], J A Sans Tresserras[1], J Szlachetko[6,1], R Tucoulou[1] and J Susini[1]

[1]European Synchrotron Radiation Facility, X-ray Imaging Group, BP 220, F-38043 Grenoble Cedex, France
[2]INSERM U-836, Institut des Neurosciences Grenoble, Université Joseph Fourier UMR-S 836, F-38042 Grenoble, France
[3]CREGU & UMR G2R 7566, Université Poincaré, BP 23, F-54506 Vandoeuvre-Les-Nancy Cedex, France
[4]Centre de Recherche et de Restauration des Musées de France, CNRS-UMR 171, Palais du Louvre, 14, quai François Mitterrand, F-7501 Paris, France
[5]Laboratoire des Sciences de la Terre, Ecole Normale Supérieure de Lyon, Lyon F-69007, France
[6]Institute of Physics, Jan Kochanowski University, 25-406 Kielce, Poland

E-mail: salome@esrf.fr

Abstract. The X-ray Microscopy and Micro-analysis beamlines at ESRF operate complementary state-of-the-art instruments at ID21, ID22, ID18F and more recently ID22NI. Within a multi-modal strategy, these beamlines develop micro-imaging techniques with various contrast mechanisms (µXRF, µXANES, µXRD and phase contrast) and host experiments with scientific topics ranging from Geochemistry to Archeology, Environmental sciences, Biology and Material sciences. Future challenges include pushing spatial resolution down to the nano-scale and the development of innovative 3D micro-analysis techniques.

1. Introduction

The X-ray Microscopy and Micro-analysis beamlines of the European Synchrotron Radiation Facility form a very complete platform to investigate samples at the micro- and nano-scale with different imaging modalities. ID21 is dedicated to scanning micro-spectroscopy (µXANES) in transmission and fluorescence modes in the "low" energy range between 2 and 7.2 keV [1]. ID22 covers the complementary "high" energy range from 6 keV to 30 keV [2] and proposes a large panel of imaging techniques, combining X-ray micro-fluorescence (µXRF), micro-spectroscopy, micro-diffraction (µXRD), 3D micro-tomography and X-ray Excited Optical Luminescence (XEOL) [3]. ID22 is also extremely modular in terms of controlled sample environments. The ID18f end-station which has been developed in collaboration with MITAC (University of Antwerp, Belgium) is dedicated to µXRF and µXRD at fixed energy [4]. A pilot project towards nano-scale imaging is presently under development at ID22NI. Last, the platform includes a Fourier Transform Infra-Red (SR-FTIR) microscope located at ID21 and providing complementary molecular information on the samples. The combination of all these techniques accessible in a coordinated platform is a clear asset for the correlative investigation of complex heterogeneous samples. A typical example is the analysis of ancient paintings using μXRF/μXANES/μXRD and SR-FTIR for the characterization of the mineral phases and organic binders respectively [5]. Examples of recently published experiments can be found in the following references related to scientific applications like Archeometry [6], Earth and Planetary sciences [7][8][9][10][11] or Environment sciences [12][13]. This paper will focus more on the major technical developments undertaken since the last XRM conference.

2. A new SXM at ID21

The ID21 Scanning X-ray Microscope (SXM) has recently been refurbished to include enhanced functionalities. Zone plates (ZonePlates Ltd, UK) are used as focusing optics and a typical spot size of $0.3x0.7\mu m^2$ with a photon flux of 10^8 -10^9 photons/s is routinely achieved in the 2-7 keV range. The optics stage has been fully redesigned to improve the guiding accuracy of the ZP along the optical axis, which is essential to ensure micro-beam stability during μXANES scans. A Kirkpatrick-Baez system (KB) based on elliptically shaped fixed-focus Ni coated mirrors is presently under construction for the SXM. The design of this KB is very compact due to mechanical/thermal stability requirements as well as to ensure a short focal length for efficient source demagnification. Advantages of the KB configuration will be a higher photon flux (a gain x30 is expected) and achromaticity which is preferable for μXANES. The modular optics stage can accommodate either the ZP or KB configuration. The SXM is now equipped with a HpGe 7-element fluorescence detector (Princeton Gamma-Tech, US), which offers an increased solid angle for an optimized fluorescence photons collection. A load-lock system allows faster exchange of sample under vacuum, and will greatly facilitate operation under cryogenic conditions. Control of the SXM is now performed through a fully integrated graphical user interface (GUI) developed by the ESRF Beamline Instrumentation Software Support (BLISS) Group. This GUI includes on-line visualization of the sample through a video-microscope and direct grabbing of regions to be scanned from the video images. XRF data analysis is also facilitated by the development at ESRF of the freely distributable PyMca package [14], which batch processing capabilities allow for treatment of large data sets to produce elemental maps. All these developments contribute to a convenient operation by external users.

3. Towards nano-probes: the ID22 Nano-imaging end-station

Development of hard X-ray nano-probes not only requires outstanding quality optics, but also high precision mechanics and ultimate control of many environmental parameters (e.g. temperature, hygrometry, vibrations…). In this context, the ID22NI nano-imaging project aims at deepening our expertise in all instrumental aspects related to hard X-ray nano-probes in view of the future nano-focussing activities within the ESRF Upgrade Programme. ID22NI is dedicated to 3D imaging with a target routine spatial resolution of the order of 50nm. The microscope combines different imaging schemes to provide both high resolution morphological information in Projection X-ray Microscopy mode and fluorescence micro-analysis in Scanning X-ray Microscopy mode. The instrument is located on the ID22 high-β section. A secondary source is created using high heat-load slits to increase the transverse coherence of the source. The microscope is operated in pink beam. Multilayer coated mirrors in Kirkpatrick-Baez geometry are used as focusing optics [15]. The multilayer coating increases the numerical aperture and beam flux, and is acting as a monochromator with an energy bandwidth compatible with fluorescence and projection imaging. The smallest focus achieved so far at 17keV exhibits a full width half maximum of $76x84nm^2$ (VxH) with a photon flux of $6x10^9$ photons/s. A higher beam flux of 10^{12} photons/s in a slightly larger spot size of $100x150nm^2$ can be achieved by increasing the slits opening. This spatial resolution in the hard energy range opens up new possibilities in biology and more particularly for the mapping of trace metals at sub-cellular level, as illustrated by the 2D fluorescence imaging of iron distribution in neuron cells performed at ID22NI [16].

4. From 2D to 3D

A natural evolution of XRF elemental mapping is the extension towards in-depth third dimension. Compared to standard X-ray absorption tomography, fluorescence tomography is more challenging since it is limited by self-absorption and matrix effects. Two strategies have been developed to overcome these physical limitations. The first strategy, pioneered at ID18f by the MITAC group (University of Antwerp, Belgium), consists of a geometrical solution, called confocal, and is based on the use of two lens optics, focusing the X-ray beam and restricting the detector angular acceptance to a small volume [17]. The intersection of the coinciding focii determines the analyzed volume in the sample. The second approach, developed at ID22, is algorithmic and relies on the combination of several signals (transmission, fluorescence and Compton) to derive the 3D distribution of elements from a pencil-beam tomographic scan [18]. The latter technique has recently been extended with the possibility to acquire simultaneously X-ray diffraction data and reconstruct a 3D map of the crystalline phases in a sample [19].

Acknowledgments
The authors would like to acknowledge the outstanding support provided by the beamline technical staff, S. Labouré, F. DiChiaro, O. Beraldin and the involvement of our collaborators from the ESRF support groups R. Baker, R. Barrett, G. Berruyer, Y. Dabin, E. Gagliardini, C. Guilloud, O. Hignette, R. Hino, J.C. Labiche, A. Solé, M. Soulier and H.P. Van Der Kleij.

References

[1] Susini J, Salomé M, Fayard B, Ortega R and Kaulich B 2002 *Surf. Rev. Lett.* **9** 203
[2] Somogyi A, Tucoulou R, Martínez-Criado G, Homs A, Cauzid J, Bleuet P, Bohic S, Simionovici A 2005 *J. Synchrotron Radiat.* **12** 208
[3] Martínez-Criado G, Alén B, Homs A, Somogyi A, Miskys C, Susini J, Pereira-Lachataignerais J and Martínez-Pastor J 2006 *Appl. Phys. Lett.* **89** 221913-1-221913-3
[4] Somogyi A, Drakopoulos M, Vincze L, Vekemans B, Camerani C, Janssens K, Snigirev A, Adams F 2001 *X-ray spectrometry* **30**(4) 242
[5] Cotte M, Susini J, Solé V A, Taniguchi Y, Chillida J, Checroun E, Walter P 2008 *Journal of Analytical Atomic Spectrometry* **23** 820
[6] Cotte M, Susini J, Metrich N, Moscato A, Gratziu C, Bertagnini A and Pagano M 2006 *Anal. Chem.* **78** 7484
[7] Cauzid J, Philippot P, Bleuet P, Simionovici A, Somogyi A and Golosio B 2007 *Spectrochim. Acta B* **62** 799
[8] Ménez B, Rommevaux-Jestin C, Salomé M, Wang Y, Philippot P, Bonneville A and Gérard E 2007 *Chem. Geol.* **240**, 182
[9] Lemelle L, Labrot P, Salomé M, Simionovici A, Viso M and Westall F 2008 *Organic Geochemistry* **39** 188
[10] Flynn G J et al 2006 *Science* **314** 1731
[11] Bleuet P, Simionovici A, Lemelle L, Ferroir T, Cloetens P, Tucoulou R and Susini J 2008 *Appl. Phys. Lett.* **92** 213111-1
[12] Isaure MP, Fayard B, Sarret G, Pairis S and Bourguignon J 2006 *Spectrochim. Acta B* **61** 1242
[13] Prietzel J, Thieme J, Salomé M and Knicker H 2007 *Soil Biology and Biochemistry* **39** 877
[14] Solé V A, Papillon E, Cotte M, Walter Ph and Susini J 2007 *Spectrochim. Acta B* **62**(1):63
[15] Hignette O, Cloetens P, Rostaing G, Bernard P and Morawe C 2005 *Rev. Sci. Instrum.* **76** 063709
[16] Ortega R , Cloetens P, Devès G, Carmona A and Bohic S 2007 *Plos ONE* 2(9):e925 doi:10.1371/journal.pone.0000925
[17] Vincze L, Vekemans B, Brenker F E, Falkenberg G, Rickers K, Somogyi A, Kersten M, Adams F 2004 *Anal. Chem.* **76**(22) 6786
[18] Golosio B, Somogyi A, Simionovici A, Bleuet P, Susini J, Lemelle L 2004 *Appl. Phys. Lett.* **84**(12) 2199
[19] Bleuet P, Welcomme E, Dooryhée E, Susini J, Hodeau JL and Walter P 2008 *Nature Mater.* **7** 468

NanoXAS, a novel concept for high resolution microscopy

I Schmid[1], J Raabe[1], C Quitmann[1], S Vranjkovic[2], H J Hug[2] and R H Fink[3]

[1] Swiss Light Source, Paul Scherrer Institut, 5232 Villigen - PSI, Switzerland
[2] Empa, Federal Lab for Materials Testing and Research, 8600 Dübendorf, Switzerland
[3] Universität Erlangen-Nürnberg, 91058 Erlangen, Germany

E-mail: iris.schmid@psi.ch

Abstract. We currently develop a novel type of scanning x-ray microscope. This instrument will combine the chemical specificity of x-ray absorption spectroscopy with the very high spatial resolution of scanning probe microscopy . In a fundamentally new instrumental approach, the instrument can be used as a conventional scanning transmission x-ray microscope (STXM), as a conventional scanning probe microscope (SPM) or in a mode combining these two techniques. In the latter case, the sample is placed in the focus generated by the fresnel zone plate of the STXM. The SPM-tip placed downstream acts as as a local detector of the emitted photoelectrons. Simulations and experiments have shown that the use of shielded SPM-tips is crucial to obtain a strongly increased chemical resolution. In contrast to similar projects underway at other synchrotrons we use a coaxial geometry. This should greatly enhance the flux density and reduce background signals caused by straylight illuminating the tip.

1. Introduction

In materials science there are two different techniques dominating the field: X-ray microscopy and scanning probe microscopy (SPM). While x-ray microscopy provides chemical sensitivity, but is currently limited to x $\sim$ 15nm resolution [1], SPM provides sub-nm resolution but with no or only little chemical information. Some combination of the two techniques would be advantageous if it could combine the spatial resolution of SPM with chemical sensitivity of X-ray microscopy. The basic principle of combining x-ray absorption spectroscopy (XAS) and scanning tunneling microscopy (STM) has been previously demonstrated by Eguchi et al. [2] and Saito et al. [3], who reported element specific images acquired by an STM tip rastered over a x-ray illuminated surface. Thus, acquiring two-dimensional chemical images as well as localized spectral analysis through a STM tip is possible, although the measured photocurrents are weak (pA) compared to the tunneling currents (nA). For these experiments the sample was illuminated by the X-ray beam at grazing incidence, which results in a very large area of the sample as well as the tip being illuminated to create a very large background signal. This approach has the further difficulty of possible intermixing of the distance feedback and the photoelectron signal. S. Suzuki et al. [4] avoided signal intermixing by combining XAS with non-contact atomic force microscopy (NC-AFM) [5]. They detected a peak in the measured frequency shift signal of the oscillating cantilever when the photon energy was scanned over the absorption edge of the underlying sample. However, images have not yet been produced by this technique so the spatial resolution has not been assessed.

We propose a new technical approach for combining XAS with SPM: A conventional scanning transmission x-ray microscope (STXM) setup using a fresnel zone plate (FZP) to focus the x-ray beam to a spot size < 50 nm in diameter and thus increasing the emitted photoelectron density, is one part of the instrument. On the other side of the sample, a coaxially insulated cantilever tip is placed in the center of the focused beam in order to locally collect the emitted photoelectrons. Note that this arrangement, illustrated in Fig. 1 (a), requires that the tip be downstream of the sample and thus the sample is illuminated through the back-side. The use of the scanning force microscopy (SFM) instead of an STM-technique has the advantage that the tip-sample distance can be controlled by the resonance frequency of the cantilever, which is clearly separated from the photocurrent signal of interest. Furthermore, the instrument can also be used in normal STXM- or SFM-mode as complementary techniques. The instrument, named NanoXAS, will be operational at a dedicated beamline of the Swiss Light Source (SLS) by July 2009.

2. Description of Instrument and Beamline

The NanoXAS instrument is designed to combine three experimental techniques: STXM, SFM and their combination. NanoXAS will be built in a UHV-chamber, attached to bend magnet beamline X07DB. The optical layout of the beamline very similar to the PolLux [6] beamline (X07DA) of the SLS, but will have an extended energy range of 200-1600 eV. A load-lock chamber attached to the main UHV-chamber and two manipulators will allow for fast sample and cantilever exchange under UHV-conditions. The principle of the NanoXAS instrument is shown in Fig. 1 (a). The FZP focuses the soft x-ray light to a diffraction limited spot. The focal length of the FZP depends on the photon energy. Higher diffraction orders are filtered out using an order sorting aperture (OSA), located between the FZP and the sample. The OSA is positioned so that it provides maximum rejection of higher diffraction orders and maximum transmission of first order light through its aperture. The distance between OSA and the sample is typically less than 1 mm, depending on the photon energy. The sample is deposited on a semitransparent SiN-membrane or mouted on a transmission electron microscope grid. To allow conventional STXM experiments a large ($\o \sim 1$ mm) photodiode detector can measure the transmitted x-ray intensity. If the photodiode is retracted, the SPM-tip can approach the sample from the downstream side where it collects the emitted photoelectrons. This tip current, I_{tip}, is preamplified using a low noise I/V converter. In addition, a lock-in amplifier can be

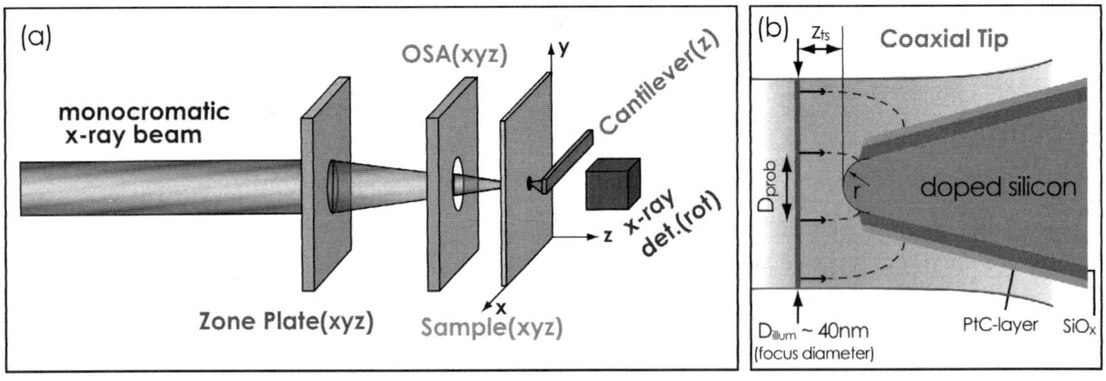

Figure 1. Sketch of the NanoXAS instrument (a) and schematic view of coaxial SPM-tip located in the focus of the zone plate with arrows indicating the photoelectrons

used to filter out unwanted background as caused, for example, by direct illumination of the tip. For lock-in amplification, the incoming x-ray beam can be chopped at about 1 kHz and the monochromator allows modulation of the photon energy by $\sim$ 50 eV/s.

Imaging is accomplished by scanning the sample, mounted on a fine- (closed-loop scanner 40 x 40 x 5 μm) and coarse-stage (positioner), across the x-ray focus. The positioner allows positioning the relevant sample region in the x-ray beam. In order to control the tip-sample distance, z_{ts}, dynamic scanning force microscopy operation modes will be used. In the constant frequency shift mode, z_{ts} can be controlled to fractions of nanometers whereas in the constant height mode the average z_{ts} is kept constant and can be adjusted to about 5-50 nm, depending on the sample surface.

3. Coaxial Cantilever Design and approximation of measured Tip current

The scientific and technological challenge of the NanoXAS project lies in controlling the diameter of the probed area, D_{probe}, from which the SPM-tip collects photoelectrons and measures them as a current, I_{tip}. Simulations performed by Chiu et al. [7] discuss the resolution limits of insulated STM-tips irradiated by x-ray light. They define D_{prob} as the diameter of a circular region on the sample surface within which emitted electrons contribute to 90 % of I_{tip}. Obviously, D_{prob} will limit the spatial resolution of the instrument $\Delta x \sim D_{prob}$.

Since photo electrons are emitted in the entire illuminated area A_{illum} having a diameter $D_{illum} \sim 40$ nm, the challenge is in preventing electrons produced far from the tip from reaching the tip and contributing to the signal. This requires a small tip radius, r, a very effective shielding of the tip [8, 9, 10], a small tip-sample distance (z_{ts}) as well as an adequate bias voltage on the tip, U_{tip}, and the shield, U_{shield}. Fig. 1 (b) shows the planed layout of a coaxial SPM-tip. It consists of an inner conductor (red) made of doped Silicon, an insulating SiO_x layer and a thin conducting PtC shield layer of several nm each. For the fabrication of such tips we plan to use high aspect ratio silicon cantilevers, produce a thin oxide layer by thermal oxidation and then to deposit a thin, smooth PtC layer by e-beam evaporation. Finally, the shield and the insulation layers are removed from the tip apex by focused ion beam (FIB).

In order to get an idea about the size of I_{tip} that can be expected, we first measured the photoelectron density, i.e. the number of emitted photoelectrons per area I_{sample}/A_{illum}, where I_{sample} is the drain current of the sample and A_{illum} is the total illuminated sample area. This is experimentally done by measuring the drain current of an illuminated cobalt test sample using the STXM [6] at the PolLux beamline of the SLS. For a focus diameter of $\sim$ 35 nm the measured drain current at the Co L_3 absorption edge (778.4 eV) was about 3 pA. This results in a photocurrent density of about 4000 A/m^2, so compared to a typical photocurrent density of a non-focused undulator source (U5 at the NSRRC [7]) of about 0.7 A/m^2, we gain a factor of more than 1000.

References

[1] Jefimovs K, Vila-Comamala J, Pilvi T, Raabe J, Ritala M and David C 2007 *Phys. Rev. Lett.* **99** 264801
[2] Eguchi T and et al 2006 *Appl. Phys. Lett.* **89** 243119 (pages 3)
[3] Saito A and et al 2006 *J. Synchrotron Rad.* **13** 216–220
[4] Suzuki S, Koike Y, Fujikawa K, Matsudaira N, Nakamura M, Chun W J, Nomura M and Asakura K 2006 *Catalysis Today* **117** 80–83
[5] Morita Sand Wiesendanger R and Meyer E 2002 *Noncontact Atomic Force Microscopy* (Berlin: Springer)
[6] Raabe J and et al Pollux: A new instrument for soft x-ray spectromicroscopy at the sls submitted to Rev. Sci. Instr.
[7] Chiu C Y, Chan Y L, Hsu Y J and Wei D H 2008 *Appl. Phys. Lett.* **92** 103101 (pages 3)
[8] Akiyama K, Eguchi T, An T, Hasegawa Y, Okuda T, Harasawa A and Kinoshita T 2005 *Rev. Sci. Instrum.* **76** 083711 (pages 3)
[9] Saito A and et al 2007 *Surf. Sci.* **601** 5294–5299
[10] Rose V, Freeland J W, Gray K E and Streiffer S K 2008 *Appl. Phys. Lett.* **92** 193510 (pages 3)

Hard X-Ray Scanning Microscopy with Fluorescence and Diffraction Contrast

Christian G. Schroer[1], Pit Boye[1], Jan M. Feldkamp[1], Jens Patommel[1], Andreas Schropp[1], Andreas Schwab[1], Sandra Stephan[1], Manfred Burghammer[2], Sebastian Schöder[2], Christian Riekel[2], and Walter H. Schröder[3]

[1] Institute of Structural Physics, Technische Universität Dresden, D-01062 Dresden, Germany
[2] European Synchrotron Radiation Facility ESRF, B. P. 220, F-38043 Grenoble Cedex, France
[3] Phytosphere Institute, ICGIII, Research Center Jülich, D-52425 Jülich, Germany

E-mail: schroer@xray-lens.de

Abstract. Based on nanofocusing parabolic refractive x-ray lenses we have developed and built a hard x-ray scanning microscope that was tested and put to use at beamline ID13 of the ESRF. It can provide a monochromatic hard x-ray nanobeam with lateral extension below 100 nm (down to 50 nm) and a flux up to 10^9 ph/s in the energy range from 15 to 25 keV. The microscope exploits transmission, fluorescence, and diffraction contrast to obtain local elemental and nanostructural information from the sample. Tomographic scanning yields high resolution elemental maps from the inside of an object. Coherent x-ray diffraction imaging with nanofocused illumination yields images of objects with highest spatial resolution, e. g., 5 nm in a given example.

1. Introduction

One of the key strengths of hard x-ray microscopy is the large penetration depth in matter that allows one to investigate the inner structure of an object, also inside special sample environments. Scanning microscopy can exploit different x-ray analytical techniques as contrast mechanisms, such as diffraction, fluorescence, and absorption spectroscopy. In this way, one can obtain local structural, elemental, and chemical information from a specimen, respectively. This has applications in a variety of scientific fields, for instance in biomedicine, chemistry, materials, earth, and environmental science, and nanotechnology.

Based on refractive x-ray optics [1, 2], a hard x-ray scanning microscope was developed, tested, and put to use at the European Synchrotron Radiation Facility (ESRF) in Grenoble, France.

2. Hard X-Ray Scanning Microscope

At the heart of a scanning x-ray microscope lies a high-resolution focusing optic that images the source onto the sample position in a strongly demagnifying geometry. The smaller the source and the more photons are emitted into the aperture of the optic, the more intensive is the microbeam. Therefore, highly brilliant sources are needed for this technique, such as undulator sources at third-generation storage rings.

In the hard x-ray range, refractive x-ray lenses are well suited to generate intensive microbeams [2]. Reducing the focal length of these optics is advantageous in two ways: the geometric image of the source decreases with decreasing focal length and so does the size of the diffraction limited Airy disc. The short focal length (10 to 30 mm in the hard x-ray range) is realized with so-called nanofocusing lenses (NFLs) [3]. With these optics, lateral beam sizes down to 50 nm have been reached [4]. Currently, they are made out of silicon by nanofabrication techniques, generating a line focus. In the future, their performance in terms of transmission and diffraction limit can be improved by using more transparent lens materials, such as boron or diamond. Theoretically, beam sizes down to below 20 nm are conceivable with these optics [3]. Further improvements of the focal size can be achieved with adiabatically focusing lenses [5].

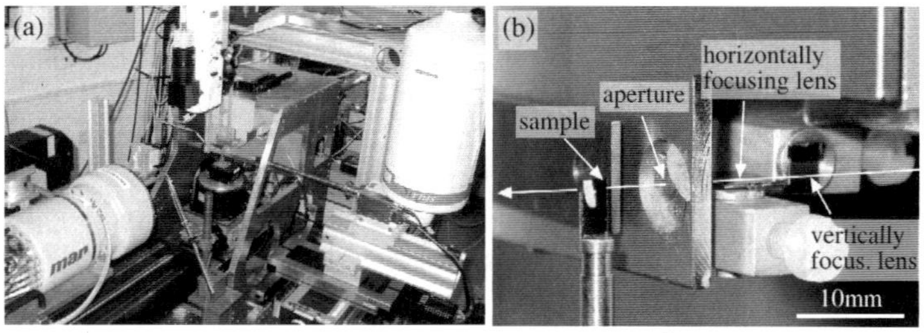

Figure 1. (a) Hard x-ray nanoprobe installed at beamline ID13 of the ESRF. (b) Detailed view of optics and sample.

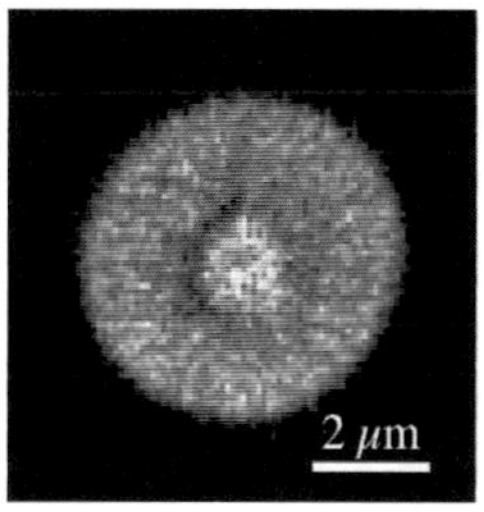

Figure 2. Calcium distribution on a virtual section through a leaf hair of *Arabidopsis thaliana.*

Fig. 1(a) shows a hard x-ray scanning microscope developed at Aachen University and TU Dresden in collaboration with beamline ID13 of the ESRF. In this microscope, two nanofocusing lenses are aligned behind each other in crossed geometry to generate a point focus at the sample position [Fig. 1(b)]. By appropriately prefocusing the beam using parabolic refractive x-ray lenses made of Be [2], efficient diffraction limited focusing is achieved. In the microscope the sample can be scanned along three translations and one rotation, allowing also for scanning tomography. Fluorescence, transmission, and diffraction contrast can be exploited.

3. Applications

The scanning microscope described in the previous section was tested and put to use at beamline ID13 of the ESRF. A variety of experiments were carried out, taking advantage of the different x-ray analytical contrasts.

Fig. 2 shows a tomographic reconstruction of the calcium distribution on a virtual section through a leaf hair (tichome) of *Arabidopsis thaliana*. To obtain this tomogram, the sample was scanned in translation (step size 100 nm) and rotation over 180 degrees in 60 steps, recording at each position in the scan a full x-ray fluorescence spectrum of the sample. After extraction of the calcium Kα line, the tomogram was reconstructed using a self-consistent tomographic reconstruction scheme that takes absorption of the incident and fluorescence radiation into account [6]. As the sample is very small, all corrections are small, too, and multiple scattering and secondary fluorescence effects are negligible.

Besides x-ray fluorescence, diffraction can be used as contrast, measuring, for example, the locally resolved diffraction pattern of individual SiGe Stranski-Krastanow islands [7].

In view of future applications of coherent x-ray diffraction imaging [9] at storage ring based or x-ray free-electron laser sources, we have investigated the coherence preservation in the

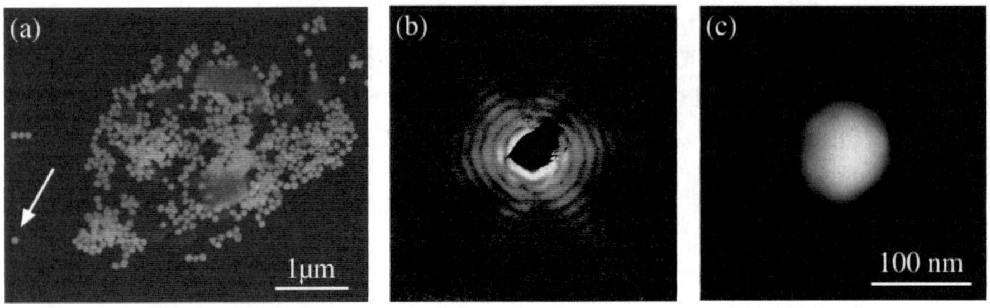

Figure 3. (a) Scanning electron micrograph of a cluster of gold nanoparticles on a silicon nitride membrane. (b) Coherent x-ray diffraction pattern recorded of the single gold particle pointed to by the arrow in (a). (c) Reconstruction of the gold particle from the diffraction pattern in (b), using phase retrieval techniques [8].

nanofocused beam in a recent experiment [8]. A gold nanoparticle, pointed to by the arrow in Fig. 3(a) was illuminated by a nanobeam in the diffraction limited focus of the x-ray microscope at $E = 15.25$ keV. The lateral extension of the beam in terms of field amplitude was about 150×150 nm^2 (FWHM), fully illuminating the sample of about 90 nm diameter. The diffraction pattern shown in Fig. 3(b) was recorded in the forward direction in ten one minute exposures on a diffraction camera (FReLoN 4M, 50 μm pixel size, sample-detector distance 1.25 m). From this diffraction pattern, the gold particle in Fig. 3(c) was reconstructed with the unprecedented spatial resolution of 5 nm by phase retrieval techniques. This experiment shows that high resolution coherent diffraction imaging is possible with nanofocused illumination by refractive x-ray lenses. This is important for single particle experiments at future x-ray free-electron laser sources [10] and for high resolution scanning diffraction microscopy [11].

Acknowledgments
This work is supported by the German Ministry of Education and Research (BMBF) under grant number 05KS7OD1 and by VI-203 of the Impuls- und Vernetzungsfonds (IVF) of the Helmholtz-Society.

References
[1] Snigirev A, Kohn V, Snigireva I and Lengeler B 1996 *Nature (London)* **384** 49
[2] Lengeler B, Schroer C G, Kuhlmann M, Benner B, Günzler T F, Kurapova O, Zontone F, Snigirev A and Snigireva I 2005 *J. Phys. D: Appl. Phys.* **38** A218–A222
[3] Schroer C G, Kuhlmann M, Hunger U T, Günzler T F, Kurapova O, Feste S, Frehse F, Lengeler B, Drakopoulos M, Somogyi A, Simionovici A S, Snigirev A, Snigireva I, Schug C and Schröder W H 2003 *Appl. Phys. Lett.* **82** 1485–1487
[4] Schroer C G, Kurapova O, Patommel J, Boye P, Feldkamp J, Lengeler B, Burghammer M, Riekel C, Vincze L, van der Hart A and Küchler M 2005 *Appl. Phys. Lett.* **87** 124103
[5] Schroer C G and Lengeler B 2005 *Phys. Rev. Lett.* **94** 054802
[6] Schroer C G 2001 *Appl. Phys. Lett.* **79** 1912–1914
[7] Hanke M, Dubslaff M, Schmidbauer M, Boeck T, Schöder S, Burghammer M, Riekel C, Patommel J and Schroer C G 2008 *Appl. Phys. Lett.* **92** 193109
[8] Schroer C G, Boye P, Feldkamp J, Patommel J, Schropp A, Schwab A, Stephan S, Burghammer M, Schöder S and Riekel C 2008 *Phys. Rev. Lett.* **101** 090801
[9] Miao J, Charalambous P, Kirz J and Sayre D 1999 *Nature* **400** 342–344
[10] Neutze R, Wouts R, van der Spoel D, Weckert E and Hajdu J 2000 *Nature* **406** 752–757
[11] Thibault P, Dierolf M, Menzel A, Bunk O, David C and Pfeiffer F 2008 *Science* **321** 379–382

Mechanical design of ultraprecision weak-link stages for nanometer-scale x-ray imaging

D. Shu[1] **and J. Maser**[2,3]

[1]APS Engineering Support Division, Argonne National Laboratory, Argonne, IL 60439, USA
[2]Center for Nanoscale Materials, Argonne National Laboratory, Argonne, IL 60439, USA
[3]X-ray Sciences Division, Argonne National Laboratory, Argonne, IL 60439, USA

shu@aps.anl.gov

Abstract. A nanopositioning diagnostic setup has been built to support the Argonne Center for Nanoscale Materials (CNM) nanoprobe instrument commissioning process at the APS. Its laser Doppler interferometer system provides subnanometer positioning diagnostic resolution with large dynamic range. A set of original APS designed ultraprecision PZT-driven weak-link stages with high-stiffness motor-driven stages has been tested with this diagnostic setup.

In this paper we present a preliminary test result of the ultraprecision weak-link stage system developed for the CNM hard x-ray nanoprobe instrument at APS sector 26. A test result for a novel laminar weak-link mechanism with sub-centimeter travel range and sub-nanometer positioning resolution is also introduced in this paper as a future work.

1. Introduction

The CNM hard x-ray nanoprobe instrument at APS sector 26 combines a scanning probe mode with a full-field transmission mode. It uses x-ray fluorescence for trace element mapping and spectroscopy; x-ray diffraction to obtain local structural information such as crystallographic phase, strain texture, and x-ray transmission in phase; and absorption to image internal structures of complex devices [1]. The nanoprobe instrument is designed for operating with photon energies between 3 keV and 30 keV with 30-nm spatial resolution. Imaging and spectroscopy at this resolution level require staging of x-ray optics and specimens with a mechanical repeatability of better than 10 nm [1].

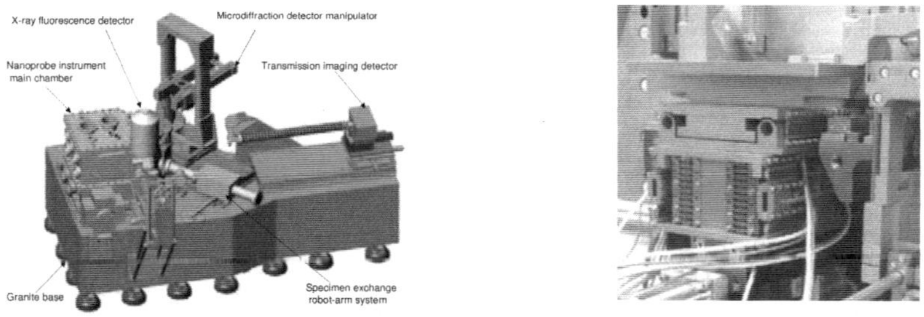

Fig. 1. Left, a 3-D model of the the optomechanical structure for the hard x-ray nanoprobe instrument. Right, photograph of the ultraprecision weak-link stage system for FOM.

As shown in Figure 1 (left side), the optomechanical structure for the CNM hard x-ray nanoprobe instrument consists of the following major subassemblies: a granite base, a nanoprobe instrument main chamber with scanning stages for x-ray focusing optics and sample stages, an industrial robot-arm-based microdiffraction detector manipulator, a stage system for an x-ray transmission imaging detector, a stage system for an x-ray fluorescence detector, and a precision specimen exchange robot-arm system to provide automated high-throughput operation capability [2]. Inside the instrument main chamber, the positioning stages for the optics are grouped into several subcomponents that are engineered to move in or out of the beam to allow configuration of the instrument for scanning probe mode and full-field transmission mode. Fast feedback for differential vibration control between the zone-plate x-ray optics and the sample holder has been implemented in this design. A specially designed, custom-built laser Doppler displacement meter (LDDM) system provides two-dimensional differential displacement measurement with subnanometer resolution between the focusing optics module (FOM) and the specimen module (SM) [3].

2. Linear stages with laminar overconstrained weak-link mechanisms

Unlike traditional kinematic flexure mechanisms, laminar overconstrained weak-link mechanisms provide much higher structure stiffness and stability. Using photochemical machining processes with lithography techniques we are able to construct a strain-limited overconstrained mechanism on the thin metal sheets. By stacking these thin-metal weak-link sheets with alignment-pins, we can construct a solid complex laminar weak-link structure with ultrahigh positioning sensitivity and stability [4-6]. More than forty sets of rotary laminar weak-link mechanisms have been made for APS users in synchrotron radiation instrumentation applications, such as high-energy-resolution monochromators for inelastic x-ray scattering, and x-ray analyzers for ultra-small-angle scattering and powder-diffraction experiments. Linear laminar weak-link mechanisms also demonstrated high resolution with high-stiffness for high-precision linear stage applications, such as x-ray microscopy applications.

Figure 1 (right side) shows the stages for the nanoprobe FOM manufactured by Xradia, Inc. [7], which provide a three-dimensional positioning capability with 0.125-nm LDDM measuring resolution in 12 mm x 12 mm x 12 mm range. It includes three stepping-motor-driven translation stages and two PZT-driven high-stiffness stages using overconstrained weak-link parallelogram mechanisms for ultraprecision motion control.

As shown in Figure 2 (left side), two PZT drivers are applied for the horizontal PZT-driven stage (APS T8-22) to guarantee the linear motion trajectory accuracy in a specific direction. The resolution of the stage is 0.2-nm with a travel range of 15 microns. To gain a higher positioning resolution, two sets of precision motion reduction mechanisms using laminar weak-link structures are implemented in the APS PZT-driven vertical stage T8-23 as shown in Figure 2 (middle-left side). Each set of precision motion-reduction mechanism was driven by a pair of PZT actuators to provide a 0.05-nm positioning resolution in a 4-micron travel range with a 5-kg load capacity. Physik Instrumente PI-841.10 PZT actuators with E-509 strain gauge sensor servo-control modules [8] were used to drive both APS T8-22 and T8-23 weak-link stages.

Fig. 2. Left, a 3-D model for the APS T8-22 PZT-driven horizontal stage. Middle-left, a 3-D model for the APS T8-23 PZT-driven vertical stage. Middle-right, a 3-D model for the APS T2-24 stepping-motor-driven vertical stage. Right, a series of 1-nm steps generated by the APS-designed T8-23 ultraprecision weak-link stage on top of the original T2-24 high-stiffness motor-driven stage with a 4-kg load under LDDM closed-loop control. The test is performed with a nanopositioning diagnostic setup on an optical table at the APS nanopositioning lab.

The structural stiffness of the stepping-motor-driven stage (APS T2-24), the base of two motorized horizontal stages and two PZT-driven stages, is critical to the final performance of the FOM stages group. Argonne has developed a special crossed roller linear guiding system with sub micron positioning sensitivity and 2-D heavy preloading capability. The coupling structure between the T2-24 guiding system and its driver's leading screw is also unique. As shown in Figure 2 (middle-right side), it can decouple any misalignment between the two motion components and ensure the positioning sensitivity and structural stiffness performances of both stages [9].

3. Test of the weak-link stages with the APS nanopositioning diagnostic setup

Although the single horizontal weak-link stage has demonstrated Angstrom level positioning sensitivity and stability at the APS since 2002 [10], many construction and commissioning details needed to be optimized to ensure reliable nanometer-scale operation for a multi-dimensional scanning stage system with a few-kilogram load capacity for the CNM nanoprobe FOM.

A nanopositioning diagnostic setup has been built to support the CNM nanoprobe instrument commissioning process at the APS. Its laser Doppler interferometer system provides subnanometer positioning diagnostic resolution with large dynamic range. A set of original APS-designed ultraprecision PZT-driven weak-link stages T8-22 and T8-23 with a high-stiffness vertical motor-driven stage T2-24 has been tested with this diagnostic setup. A series of 1-nm steps generated by the APS T8-23 ultraprecision weak-link stage on top of the T2-24 vertical motorized stage with a 4-kg load under LDDM closed-loop control is demonstrated as shown in Figure 2 (right side). To support the CNM nanoprobe instrument commissioning process, more tests will be performed for motorized vertical stages with different construction and assembly conditions.

4. Future work for ultraprecision linear weak-link stages

For scanning microscope instrumentation developers it is always a dream to have a compact single stage to cover large travel range with very high positioning resolution. Based on an advanced structure design with a laminar overconstrained weak-link technique, we are exploring the feasibility of building a weak-link-based precision linear guiding system with sub-nanometer resolution and sub-centimeter travel range with a reasonably compact size [11]. A test stage for a single weak-link module has been constructed with sizes of 120 mm x 130 mm x 25 mm. A 3-mm travel range is demonstrated in this test with 3- to 4-μrad tilt errors in the middle of 2.4 mm travel range. Using four such modules, a vertical weak-link stage has been designed for a scanning sample stage system for a hard x-ray test setup with multilayer Laue lenses (MLLs) [12].

Acknowledgments

The authors would like to thank D. Nocher and R. Ranay from Argonne National Laboratory for their help in the development of laminar weak-link mechanisms . This work was supported by the U.S. Department of Energy, Office of Science, Office of Basic Energy Sciences, under Contract No. DE-AC02-06CH11357.

References

[1] J. Maser et al., Proc. 8th Int. Conf. X-ray Microscopy, IPAP Conf. Series 7 (July 2006) 26-29.
[2] D. Shu et al., Proc. 8th Int. Conf. X-ray Microscopy, IPAP Conf. Series 7 (July 2006) 56-58.
[3] U.S. Patent granted No. 7,331,714, D. Shu, J. Maser, B. Lai, F. S. Vogt, M. V. Holt, C. A. Preissner, R. Winarski, and G. B. Stephenson, 2008.
[4] U.S. Patent granted No. 6,607,840, D. Shu, T. S. Toellner, and E. E. Alp, 2003.
[5] U.S. Patent granted No. 6,984,335, D. Shu, T. S. Toellner, and E. E. Alp, 2006.
[6] D. Shu, T. S. Toellner, and E. E. Alp, Nucl. Instrum. and Methods A 467-468 (2001) 771-774.
[7] Xradia is a trademark of the Xradia Inc. California, U.S.A.
[8] Physik Instrumente (PI) is a trademark of the Physik Instrumente GmbH & Co. Germany.
[9] U.S. Patent granted No. 5,526,903, D. Shu, and J. Barraza, 1996.
[10] D. Shu, Y. Han, T. S. Toellner, and E. E. Alp, Proc. SPIE Vol. 4771 (2002) 78-90.
[11] D. Shu and J. Maser, U.S. Patent application in progress for ANL-IN-08-013.
[12] J. Maser et al., Proc. SPIE-Int. Soc. Opt. Eng. 5539 (2004) 185.

Broadband X-ray full field microscopy at a superbend

M Stampanoni[1,2], F Marone[1], G Mikuljan[1], K Jefimovs[3], P Trtik[3], J Vila-Comamala[1], C David[1], R Abela[1]

[1]Paul Scherrer Institut, Villigen, Switzerland
[2]Institute for Biomedical Engineering, University and ETH Zürich, Zürich, Switzerland
[3]EMPA, Materials Science & Technology, Dübendorf, Switzerland

marco.stampanoni@psi.ch

Abstract. Over the last decade, synchrotron-radiation based X-ray Tomographic Microscopy (SRXTM) has established itself as a fundamental tool for non-invasive, quantitative investigations of a broad variety of samples, with application ranging from space research and materials science to biology and medicine. The beamline for **TO**mographic **M**icroscopy and **C**oherent r**A**diology experimen**T**s (TOMCAT) has been recently equipped with a full field, hard X-ray microscope with a theoretical pixel size down to 30 nm and a field of view of 50 microns. The nanoscope performs well at X-ray energies between 8 and 12 keV: here we illustrate the experimental setup and the performance of the instrument in both microscopy and tomography mode.

1. Introduction

The combination of the unique properties of X-ray microscopy (high-resolution, large penetration depth and chemical speciation) with advanced tomographic methods as well as the exceptional properties of third-generation synchrotron radiation sources allow to obtain volumetric information of a specimen with voxel size in the micrometer range.

Even rookie synchrotron-CT users require today flexible endstations, capable to cover field of views (FOV) varying up to three orders of magnitude with corresponding spatial resolution and/or acquisition mode. Frequently, such users also demand an easy and fast swap between antipodal configurations, for instance between a 10 mm FOV with 10 microns resolution in phase contrast and a 50 microns FOV with 100 nm resolution in pure absorption.

Reaching and trespassing the isotropic barrier of 100 nm remains a challenge and only few instruments, mostly in the soft X-ray range, i.e. for energies well below 10 keV have this capability [1]. The most common approach to achieve sub 100 nm (2D) resolution is full-field X-ray microscopy. The key optical elements are a condenser to provide illumination of the sample and an objective lens to produce a magnified image of the sample on the detector. The illumination should be as homogeneous and intense as possible and the numerical aperture should be matched to that of the objective lens in order to obtain optimum resolution [2]. Instead of the more conventional Fresnel zone plates, tapered capillaries or mirrors, we made use of a beam-shaper with complex optical functionality as condenser in the X-ray regime. The idea of the design is to divide a conventional Fresnel zone plate (FZP) into sectors, keeping the local spatial frequency within each sector constant (in our case square gratings). As a result, each sector will produce a flat-top illumination in the focal plane [3].

Journal of Physics: Conference Series **186** (2009) 012018 doi:10.1088/1742-6596/186/1/012018

2. Experimental Setup

The nanoscope installed at the TOMCAT beamline is approximately 12 meters long. The beamshaper, is placed just after the exit window of the beamline, at 17 meters from the source. The X-ray beam at that position is tailored to a size of 2x2 mm^2 by the exit slits of the beamline. A motorized slits pair (JJ-Xray, Type 3 Slit Air Version) is used as an adjustable order sorting aperture (OSA). The sample is mounted on an air-bearing rotation stage with and xy-centering device. As an objective lens, a gold FZP with a 100 nm outermost zone and a diameter of 100 microns is used [4]. Immediately downstream the zone plate, the high-resolution microscope (normally used at TOMCAT for routine experiments in the micron range) can be inserted into the beampath. This allows a very quick alignment of beamshaper, beamstopper and zone plate. The focal length of the FZP used for this experiment is 64.5 mm and 80.6 mm at 8 keV and 10 keV respectively. By positioning the 2D detector at distances larger than 10 m (Photonic Science Hystar Camera, 2048x2048 pixels, true 14 bit dynamic range, 1:3 magnifying fiber optic taper and 1.5 mg/cm^2 Gadox scintillator resulting in a optical resolution of 4.7 microns) we can obtain X-ray magnifications up to 180x.

3. System performance

The radiographic imaging properties of the system have been characterized at 8 keV. We were targeting a resolution of 100 nm. For this purpose, as a test sample, we used a 20 microns diameter gold zoneplate with an outermost zone width of 100 nm. Figure 1 shows a resolution measurement for an X-ray projection. Figure 1b depicts the corresponding X-ray radiographic image (absorption image) of the test sample shown in figure 1a. A line profile over the central region of the zone plate (Fig. 1c) shows that intensity oscillations at the frequency of 5 lp/micron could be detected with sufficient contrast (Fig. 1d). The detector has been used in 2x binning mode, giving a theoretical pixel size of approximately 60 nm.

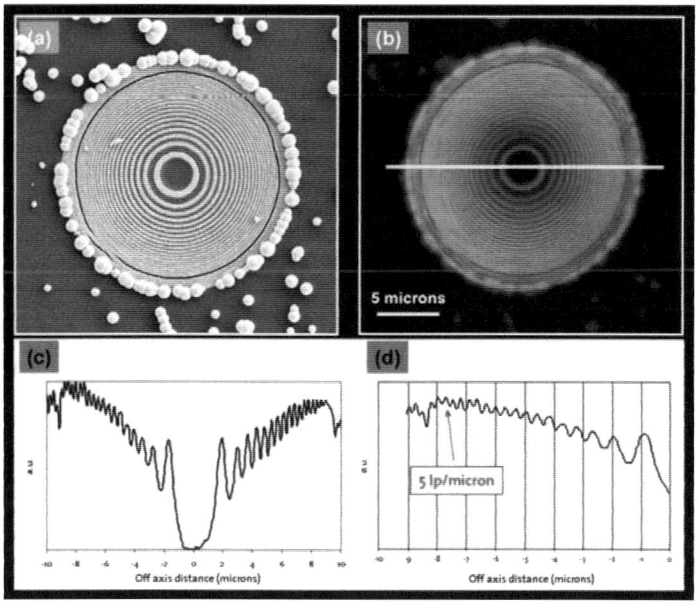

Figure 1. (a) SEM of the reference sample: 20 microns diameter gold FZP with 100 nm outermost zone width. (b) Corresponding X-ray projection acquired at 8 keV. (c) Line profile and (d) zoom in into the profile, showing contrast at the target resolution of 100 nm.

We developed a test sample which is easy to align and ideally suited for tomographic investigations. It consists of a $BaTiO_3$ sphere of 22 microns diameter with calibration holes of 5, 2, 1, 0.5 and 0.25 microns diameter, milled with focus ion beam, see Figure 2a. The sphere has been coated by a thin gold layer to avoid charging during FIB milling. In the corresponding radiographic projection, see Fig. 2b, the gold coating is visible on the right side, while in the center the typical conical shape of the FIB holes are recognizable. After tomographic data acquisition (501 projections over 180°), we generated an isosurface rendering which clearly shows the entrance point of the reference apertures, see Fig. 2c. Holes with diameters of 5, 2, 1 and 0.5 microns on the surface of the sphere are clearly visible. The position of the 0.25 microns hole can be grasped as well.

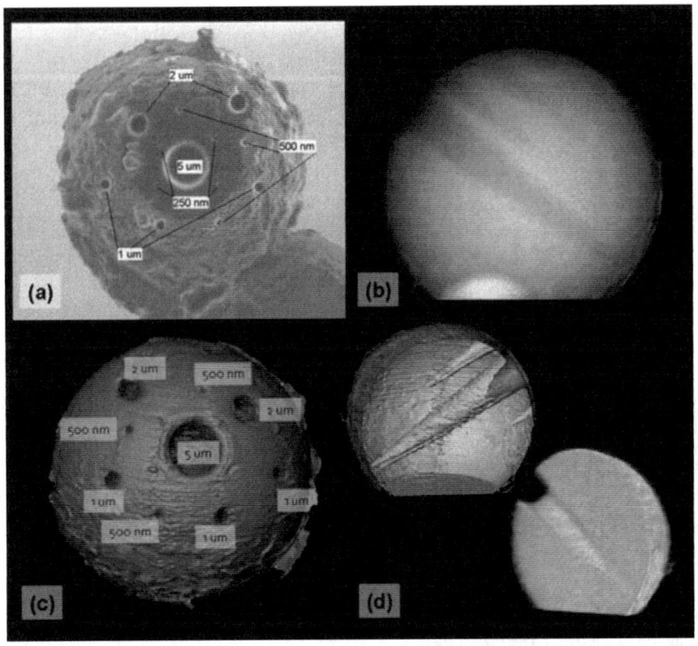

Figure 2. (a) SEM of the reference sample used for nanotomography: a $BaTiO_3$ sphere of 22 microns diameter with calibration holes of 5, 2, 1, 0.5 and 0.25 microns. (b) Corresponding X-ray projection acquired at 10 keV. (c) Surface rendering showing the entrance points of the reference holes. (d) Virtual cuts through the test sample, showing the typical cone shape of the FIB-milled holes. The FIB-milling in deep recesses resulted in holes of aspect ratio of approximately 5:1. Therefore the 5 microns hole was milled all the way through the sphere, while the 2 microns only reached the center of the sphere.

4. Outlook

In the next future we will further improve the stability and fine tune our registration algorithms. A new, larger beamshaper will collect a beam area of up to 3x3 mm^2 and tailor it down to 50x50 $microns^2$, therefore augmenting the flux at the sample position by a factor of 4. We also plan to update the present detector with a more efficient system which should result in a reduction of the exposure time by at least a factor of 2.

Finally, our efforts will be focused in the tentative of producing phase contrast images with the nanoscope.

References

[1] Meyer-Ilse W, Hamamoto D, Nair A, Lelievre SA, Denbeaux G, Johnson L, Pearson AL, Yager D, LeGros MA, Larabell, "High resolution protein localization using soft X-ray microscopy", Journal of Microscopy **201**, 395-403, (2001)

[2] Born M & Wolf E, "Principles of Optics", 7th ed., Cambridge University Press, 595 (1999)

[3] Jefimovs K, Vila-Comamala J, Stampanoni M, Kaulich B, David C, "Beam-shaping condenser lenses for full-field transmission X-ray microscopy", Journal of Synchrotron Radiation **15**, 106-108 (2008)

[4] Jefimovs K, Bunk O, Pfeiffer F, Grolimund D, van der Veen J F, David C, Fabrication of Fresnel zone plates for hard x-rays, Microelectronic Engineering **84**, 1467–1470 (2007)

Journal of Physics: Conference Series **186** (2009) 012019 doi:10.1088/1742-6596/186/1/012019

Development of an Auto-focusing Imaging System in the Soft X-Ray Microscope Beamline of the SR Center in Ritsumeikan University

K Takemoto[1], H Namba[2], M Mukai[1], K Okuno[3], T Ohta[3] and H Kihara[1]
[1] Department of Physics, Kansai Medical University, 18-89 Uyamahigashi, Hirakata, Osaka, 573-1136, Japan
[2] Department of Physical Science, Ritsumeikan University, 1-1-1, Noji-Higashi Kusatsu, Shiga, 525-8577, Japan
[3] SR Center, Ritsumeikan University, 1-1-1, Noji-Higashi Kusatsu, Shiga, 525-8577, Japan

E-mail: takemoto@makino.kmu.ac.jp

Abstract. An X-ray microscope beamline (BL-12) was installed in the SR center, Ritsumeikan University, in 1996 and has been operated for 12 years. After XRM2005, an automated CZP positioning system has been used for daily observation. Using this system, several unique biological samples were observed. In 2007, we started a new project to develop an auto-focusing imaging system in the beamline. The system can be controlled by a single finger and makes possible to perform the multi-wavelength imaging in a wide wavelength region from 1.73 to 4.73 nm. With this system, an edge absorption imaging can be easily applied to many elements.

1. Introduction

At Ritsumeikan University (Kusatsu, Japan), the world smallest electron storage ring with one-body superconducting magnet, 'AURORA' has been operated since 1996 (the electron energy of 575 MeV and the ring current of 300 mA). An X-ray microscope beamline (BL-12) was installed also in 1996 and has been operated for 12 years. The present optical configuration of the X-ray microscope is the same with the previous one [1-2]. The optical system consists of two zone plates. In conjunction with a pinhole a condenser zone plate (CZP) acts as a dispersive and focusing element. A CZP is a Göttingen KZP 7 type (diameter: 9 mm, outermost zone width: 53.7 nm, number of the zones: 41,890) [3]. Pinhole diameter is 20 μm. An objective zone plate (OZP) was fabricated at IBM/LBNL (diameter: 50 μm, outermost zone width: 45 nm, number of the zones: 277) [4]. The images are detected by a back-illuminated CCD (C4880-21-24WD, Hamamatsu K.K.). In ray tracing calculation, the wavelength resolution was estimated to be 200 at full-width at half maximum when a beam size was 0.28×2.6 μm^2 (2σ) and distance between the source and the CZP of 7.1m. Its resolution was estimated to be about 70 nm (20-80%) from the intensity gradient of the knife edge of the mesh [1, 2]. We reported several improvements in the previous paper [5]. In 2004, an automated CZP positioning system was designed and manufactured. This system covers 1.6 - 3.3 nm wavelength range and a number of specimens have been observed since X-ray microscopy conference in Himeji 2005 (XRM2005). Energy calibration was performed by using nitrogen and oxygen absorption edges. Although multi-wavelength observation was possible, it was necessary to adjust the system to an appropriate energy at each time. Fully automated wavelength scanning system was highly demanded. In 2007, we started a new project to develop a full auto-focusing imaging system in our beamline. This system can be controlled by a single finger. It makes possible to perform the multi-wavelength imaging in a wide wavelength region from 1.73 to 4.73 nm. With this system, an edge absorption imaging is easily applied to many elements. In this paper, details of this project and some recent results after XRM2005 are presented.

2. Vanadium Binding in *Dictyostelium discoideum*

Vanadium is particularly attractive as a contrast enhancement reagent in soft X-ray microscopy because it has a strong absorption L-edge (2.43 nm) in the water window. In order to evaluate the capability as a labelling agent of vanadium, a labelling operation using vanadium was performed. *Dictyostelium discoideum* Ax-2 amoebae, cultured in axenic medium (HL5), were collected into centrifuge tubes and washed with 0.01M phosphate buffer. After 1 % glutaraldehyde fixation, the cells were stained with 0.2M $VOSO_4$ dissolved in distilled water for 16 h.

Figure 1 is an image of *Dictyostelium* cell stained with vanadium, 0.2 M. Optical microscopic observation indicates that the cell size is approximately 7.5 μm in diameter, and there is a core of about 4 μm. Figure 1b and c show X-ray micrographs taken at 2.3 and 2.5 nm, which correspond to the energies above and below vanadium L-edge respectively. Figure 1d shows the vanadium signal which was obtained by subtracting figure 1c from figure 1b. Figure 1d obviously shows that there is a high vanadium dense core, whose size is 4 μm in diameter. Surrounding the core, filamentous structure is observed in the cytoplasm, which may correspond to the cytoskeleton. This result indicates that the nucleus and cytoskeleton are a kind of vanadyl binding sites.

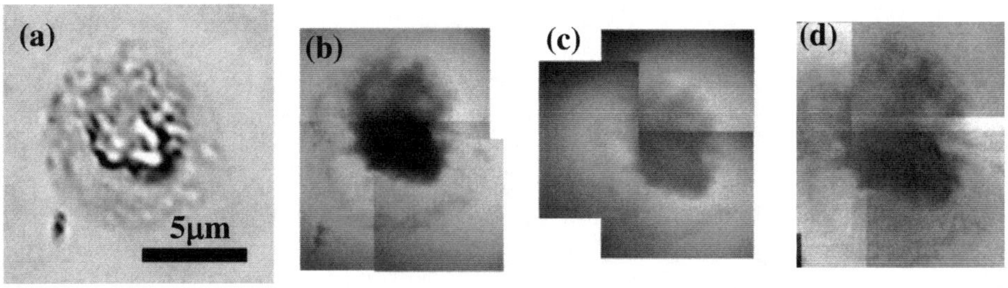

Figure 1. Optical microscopic and X-ray microscopic images of *Dictyostelium* cell stained with 0.2M $VOSO_4$ solution. (a) Optical microscopic image, (b) X-ray image at 2.3 nm and (c) at 2.5 nm. (d) Vanadium image which was obtained by subtracting the image (c). Exposure time was 60 s.

3. Improvement of Soft X-ray Microscope

The previous system was a half-automatic observation system [5]. Since the CZP moves together with the CZP chamber by a stepping motor, the usable wavelength range is restricted by a working stroke of a vacuum bellows which connected to the CZP chamber [1, 2, 5].

In the new system, a CZP chamber is newly manufactured. A particular wavelength is selected by moving the CZP with a new stepping motor along the optical axis of the synchrotron beam in it (see Figure 2). Figure 3 shows schematics and photographs of the mounting assembly for the CZP. The CZP is mounted on piezo positioner stages PP-30 for vacuum (MICOS, Germany). A central stop is also mounted on a 2-axis positioning stage with a tiny SQL Series SQUIGGLE motor SQ-100V for vacuum (New Scale Technologies, Inc, USA). The central stop stage is attached to a CZP holder plate and dependent moving to the CZP is possible. The new CZP chamber covers 1.73 - 4.73 nm wavelength range. Linear actuators are newly attached to the OZP and sample stages. As the results, every stage moving is computer-controlled by b using a LabVIEW (National Instruments (NI)) system. With this system, user can control the system by a single finger and perform the multi-wavelength imaging in a wide wavelength region from 1.73 to 4.73 nm.

4. Summary

After XRM2005, an automated CZP positioning system has been used for daily observation and shows good performance. Using the soft X-ray microscope at BL12, we have evaluated the effect of the contrast enhancement based on vanadium labelling for *Dictyostelium discoideum*. Using 0.2M $VOSO_4$ staining solution, a cell nucleus and cytoplasm were visible and especially details of cytoskeleton were clearly seen.

In order to offer more user-friendly equipment, a full automatic observation system has been designed and manufactured. Pulse motor is attached to all moving stages and every movement is controlled by pulse. In addition, a new motor control system is developed using LabVIEW. Using this system, user can control the system by a single finger and perform the multi-wavelength imaging in a wide wavelength region from 1.73 to 4.73 nm. During the 2008 fiscal year, we look forward to observing bio-specimens with this new user-friendly X-ray microscope.

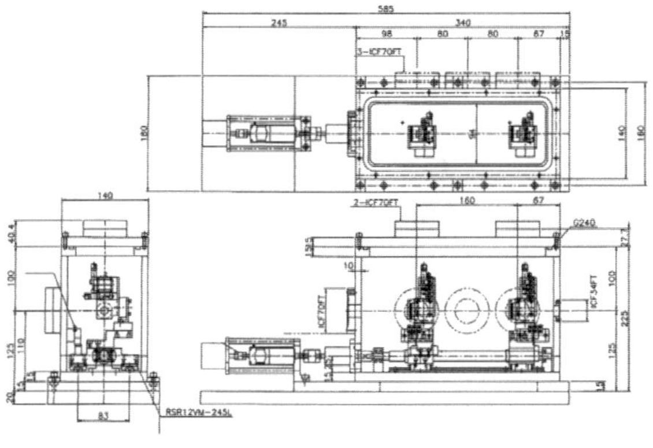

Figure 2. Schematic of new CZP chamber with the CZP stage.

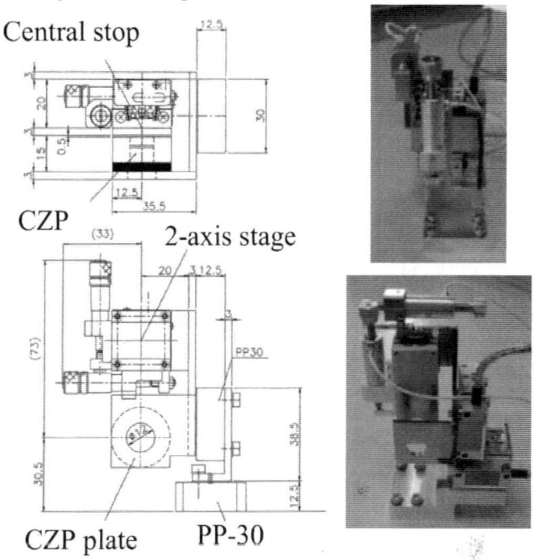

Figure 3. Schematic and photograph of mounting assembly for the CZP.

Acknowledgments

The authors are grateful to Dr. Ohigasi for his collaboration and discussion on the new microscopic system. We are grateful to Ritsumeikan University SR Center staffs for their support. Dr. D. Rudolph *et al.* at Göttingen group and Dr. D. Attwood *et al.* at Lawrence Berkeley National Laboratory are appreciated for the use of a condenser zone plate and an objective zone plate, respectively.

References

[1] Hirai A, Takemoto K, Nishino K, Niemann B, Hettwer M, Rudolph D, Anderson E, Attwood D, Kern D P, Nakayama Yand Kihara H 1999 *Jpn. J. Appl. Phys.* **38** 274-278

[2] Hirai A, Takemoto K, Nishino K, Watanabe N, Anderson E, Attwood D, Kern D, Hettwer M, Rudolph D, Aoki S, Nakayama Y and Kihara H 1998 *J. Synchrotron Rad* **5** 1102-1104

[3] Schmahl G, Rudolph D, Niemann B, Guttmann P, Thieme J, Schneider G, David D, Diehl M, and Wilhein T 1995 *Optik* **93** (1995) 95-102

[4] Anderson E and Kern D, "X-Ray Microscopy III", edited by A.G. Michette, G.R. Morrison, C.J. Buckley (Springer-Verlag, Berlin, 1998) 75-78.

[5] Kimura K, Takemoto T and Kihara H, "X-Ray Microscopy", edited by S. Aoki, Y. Kagoshima, Y. Suzuki (Institute of Pure and Applied Physics, Tokyo, 2006) 219-221

Zernike phase-contrast x-ray microscope with pseudo-Kohler illumination generated by sectored (polygon) condenser plate

Akihisa Takeuchi, Kentaro Uesugi, and Yoshio Suzuki

JASRI / SPring-8, Kouto 1-1-1, Sayo, Hyogo 679-5198, Japan

take@spring8.or.jp

Abstract. Zernike phase contrast x-ray microscope has been developed at the undulator beamline 20XU and 47XU of SPring-8. The system consists of a pseudo-Köhler-illuminating system, a Fresnel zone plate objective with outermost zone width of 100 nm, a Zernike phase plate (0.96-μm-thick tantalum, $\lambda/4$ or $3\lambda/4$ phase-shifter at 8 keV) installed at the back-focal plane of the objective, and a visible-light conversion type cooled CCD camera as an image detector. A sectored (polygon) condenser plate is employed as the condenser in order to secure a large and flat field of view. Details and experimental results of the system will be shown.

1. Introduction

Full-field transmission x-ray microscope (TXM) with Fresnel zone plate (FZP) objective has been developed at the undulator beamline 20XU and 47XU of SPring-8 aimed for 100-nm-resolution x-ray CT [1-4]. Since the field of view of CT system determines the measurable size of object, large and flat image field is required for the TXM optics as well as the spatial resolution. From this point of view, employing of a Köhler-illumination is one of the best choices. Therefore, in order to produce a (pseudo-) Köhler-illumination, we have previously developed a circle condenser plate (CCP) which has concentric diffractive grating patterns with even pitch (Fig. 1(a)). The spatial resolution of the system has been reached to 160 nm, and the CT using this system is now in operation constantly showing the spatial resolution of better than 300 nm in three-dimension [4].

However, obtained field of view had an extremely inhomogeneous intensity distribution proportional to the inverse of transverse distance from optical axis (gray line in Fig. 2). In order to solve this problem, we newly developed a sectored (polygon) condenser plate (SCP) which consists of

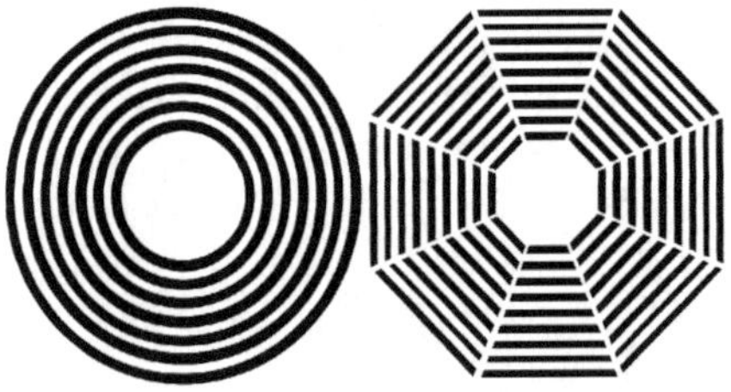

(a) (b)

Figure 1. Schematic drawing of condensers, (a) circle condenser plate (CCP) and (b) sectored condenser plate (SCP).

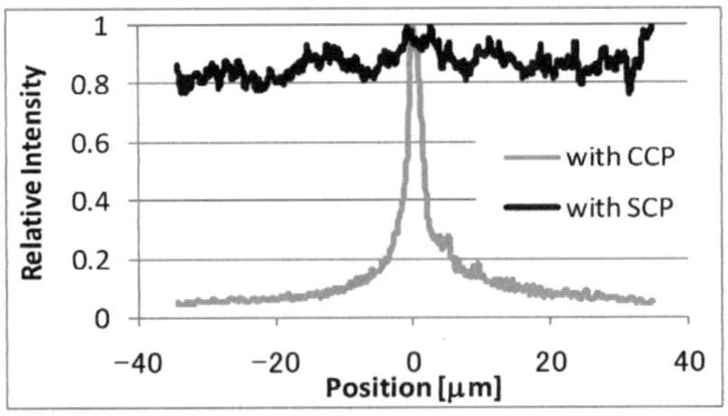

Figure 2. Intensity profiles of image field, gray line: with CCP, black line: with SCP.

plural diffractive gratings with even pitch (Fig. 1(b)). As shown in the figure, these gratings are arranged in a radial manner so that all the +1st diffractions are gathered at object plane. The use of SCP successfully produces even intensity distribution over the field of view of 50-100 μm (black line in Fig. 2). Calculated modulation transfer functions (MTF) of the TXM system using a CCP and using an octagon SCP with same optical parameters are shown in Fig. 3. Because the illumination system with the SCP gives eight oblique illuminating beams, the MTF shows a step-function. Therefore, the MTF curve has different shape from that of CCP system. However, they have the same diffraction-limit being superior to the case of parallel illumination.

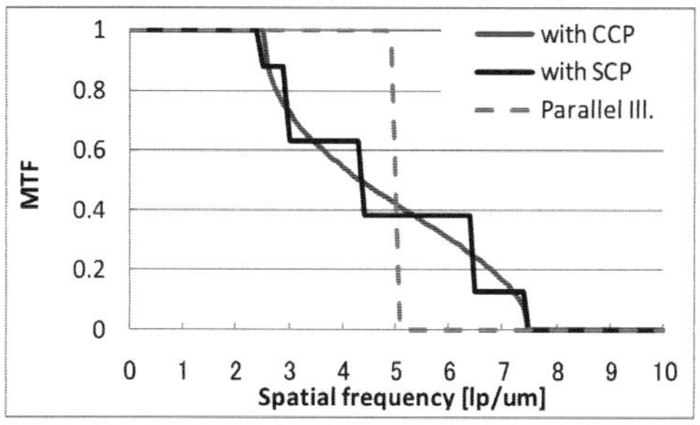

Figure 3. Calculated MTFs of TXM system, gray line: with CCP, black line: with SCP, broken gray line: parallel illumination.

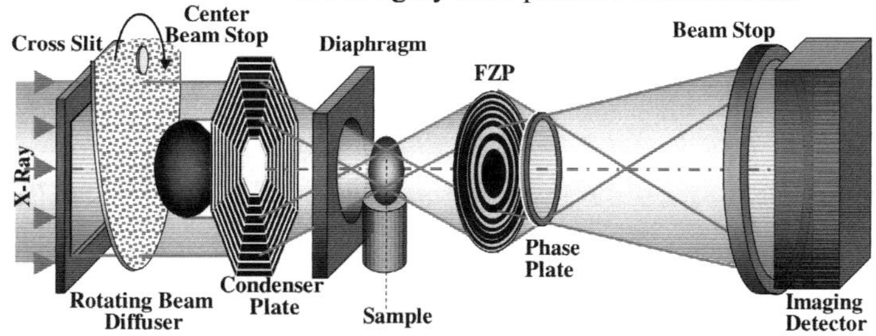

Figure 4. Schematic diagram of Zernike phase-contrast x-ray microscope (ZXM) optical system.

On the other hand, in the user-experiment, requirement of use of phase-contrast imaging has become higher and higher because of the miniaturization of volume of object related to the improvement of spatial resolution. As an optional use of the TXM system, a Zernike phase contrast x-ray microscope (ZXM) coupled with the Köhler-illuminating system has been developed (Fig. 4). Principle of the Zernike phase contrast method is that the filtering of Fourier spectrum of the object formed at the back-focal plane. It is well-known that the Köhler-illumination is very suitable for the Zernike phase contrast method because the diffracted rays of the object and non-diffracted rays (0th-order) are spatially sorted at the back-focal plane.

2. Experimental setup

A SCP consists of 8 sectors of diffractive gratings with even pitch of 400 nm (200 nm line width) is used as the condenser. Total width of the SCP is 1000 μm, and the pattern is drawn with tantalum (1.65 μm thick). A FZP with outermost zone width of 100 nm is used as the objective. The zone material is tantalum with 1 μm thick. A Zernike phase plate (ZPP) made from tantalum with 0.96 μm thick is placed at the back-focal plane of the objective FZP. In this case, the ZPP works as a $\lambda/4$ phase shifter for 8 keV x-rays. A pair of positive and negative ring patterns is drawn in the ZPP, and one can choose between the bright phase contrast mode and dark phase contrast mode by selecting these ring patterns. Line width of each ring pattern is 4 μm. Used SCP, FZP and ZPP are fabricated by NTT-AT, Japan. A cooled CCD camera (C4880-41S, Hamamatsu Photonics, Japan) coupled with relay lens and scintillator (P43) screen is placed at the image plane. The CCD camera is used under the 2×2 binning mode and in this case the effective pixel size and the camera format are 3.4 μm and 2000×1312 pixels, respectively. A rotating beam diffuser (letter paper) is installed in front of the SCP in order to eliminate speckle noise. Angular spread of the beam behind the diffuser is about 13 μrad in FWHM.

Black line in Fig. 5 shows a calculated contrast transfer function (CTF) of the ZXM system under the dark phase contrast mode. Periodic rectangular tantalum pattern with 0.5 μm thick is assumed as the object. Although the CTF in high frequency region is the same as that of TXM mode, as shown in Fig. 3 and Fig. 5, it drastically decreases in the low frequency region because of the finite width of the phase ring (in this case, the line width is 4 μm). It is because there is no phase shift if the distance at the Fourier space between the diffracted beam and the non-diffracted beam is smaller than the line width of phase ring. Therefore, image contrast of the lower frequency region is almost same as the TXM mode. In this case, the image contrast at the frequency region lower than 0.129 μm^{-1} corresponding to the 4 μm in real space is approximately 0.067, that is just same as transmission image contrast of tantalum 0.5 μm thick. Since the CTF of ZXM mode has characteristics of a band-pass filter as shown in Fig. 5, therefore, edge-enhance-like contrast modulation is observed in the image.

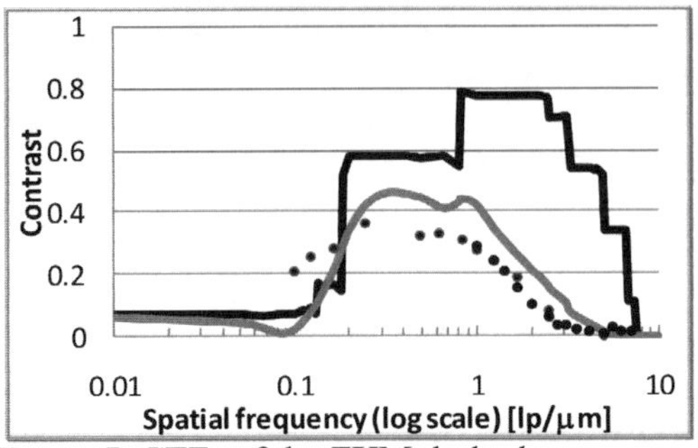

Figure 5. CTFs of the ZXM dark phase contrast mode. Black line: calculated value, gray line: calculated value considering with beam spread with diffuser, detector MTF and system magnification, dots: experimental data.

3. Results

Figures 6(a)-6(c) show x-ray images (above) and intensity profiles (below) of tantalum test pattern with 0.5 μm thick observed with TXM mode, ZXM bright phase contrast mode, and ZXM dark phase contrast mode, respectively. Although clear image is obtained even with the TXM mode, the ZXM mode images show several times higher contrast than the TXM mode. There is no difference in spatial resolution between three modes, however, edge-enhance-like contrast modulation is seen in the ZXM mode images due to the band-pass filtering effect. Image contrast under the ZXM dark phase contrast mode is plotted in Fig. 5. This shows a good agreement with the calculated CTF (gray line in Fig. 5).

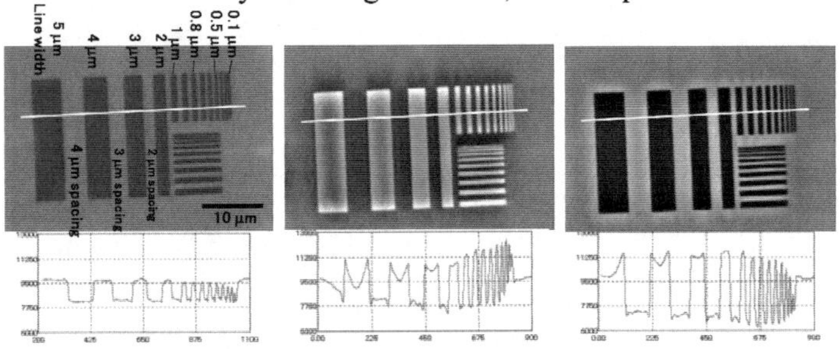

(a) (b) (c)

Figure 6. X-ray images and intensity profiles of tantalum test pattern (thickness: 0.5 μm), (a) TXM image, (b) ZXM bright phase contrast, and (c) ZXM dark phase contrast image. Beamline: BL47XU, x-ray energy: 8 keV, exposure time: 0.15 sec, pixel size: 47 nm.

Figure 7 shows x-ray images of human cervical cancer HeLa cells. The samples are cultured on a Si_3N_4 window, fixed with gultaraldehyde and then dried by critical point drying. Although the sample is hardly seen with the TXM mode, detailed structures such as the nucleolus in the nucleus are clearly observed with both of ZXM modes.

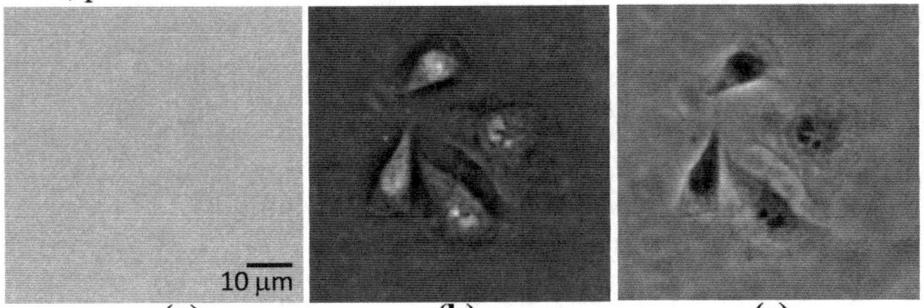

(a) (b) (c)

Figure 7. X-ray images of human cervical cancer HeLa cells (dry), (a) TXM image, (b) ZXM bright phase contrast, and (c) ZXM dark phase contrast image. Beamline: BL20XU, x-ray energy: 8 keV, exposure time: 30 sec, pixel size: 130 nm.

The experiments were performed under the approval of SPring-8 committee 2007A1738 and 2007B1584. We thank Prof. A. Itou for providing HeLa cells and their helpful advice.

References

[1] A. Takeuchi, H. Takano, K. Uesugi and Y. Suzuki 2001 *Proc. SPIE* **4499** 29.
[2] A. Takeuchi, K. Uesugi, H. Takano and Y. Suzuki 2002 *Rev. Sci. Instrum.* **73** 4246-4249.
[3] A. Takeuchi, K. Uesugi, Y. Suzuki, S. Tamura and N. Kamijo 2006 *IPAP Conf.* Series **7** 360.
[4] K. Uesugi, A. Takeuchi and Y. Suzuki 2006 *Proc. SPIE* **6318** 63181F.

Journal of Physics: Conference Series **186** (2009) 012021 doi:10.1088/1742-6596/186/1/012021

X-ray phase micro-tomography using an interference microscope with zone plates

N Watanabe[1], M Hoshino[1], K Yamamoto[1], S Aoki[1], A Takeuchi[2], and Y Suzuki[2]

[1]Institute of Applied Physics, University of Tsukuba, Tsukuba, Ibaraki 305-8573, Japan.
[2]SPring-8, JASRI, Mikazuki, Hyogo 679-5198, Japan

watanabe@bk.tsukuba.ac.jp

Abstract. Using an x-ray common-path interference microscope with two zone plates at SPring-8 BL20XU, quantitative phase imaging was investigated by the fringe-scanning method. The phase sensitivity was estimated to be λ/62. Tomographic three dimensional phase image of diamond particles and polystyrene beads of 2.8 μm in diameter could be clearly reconstructed.

1. Introduction

In x-ray region, phase-contrast is much higher than absorption contrast especially for a specimen which consists of light elements, such as biological specimens. Various phase-contrast microscopes, such as an interference microscope, a Zernike-type phase-contrast microscope, a deferential interference microscope, and holography have been developed in the x-ray region. Using an x-ray interference microscope, phase shift of a specimen can be imaged directly and it is suitable for phase tomography.

One possible solution for an x-ray interference microscope is a combination of an x-ray microscope and an off-axis reference beam. A phase shift image was obtained by using fringe-scan technique. Disadvantage of such an off-axis interference microscope is that the spacing of interference fringe is about 10 μm or less and it is necessary to use fringe-scanning technique to observe a specimen. This problem can be solved by using a co-axial reference beam. This type of microscope was discussed by Wilhein et al. [1]. The authors was developed a common-path interference microscope with two zone plates. A fringe-less phase image of a 300 nm line and 300 nm space patterns could be resolved at 10 keV [2]. A different type of such a microscope was also developed by Koyama et al. [3].

In this paper, we show the performance of phase imaging and phase tomography using the common-path interference microscope.

2. Common-path interference microscope with two zone plates

The optical system is shown in Fig. 1. The two zone plates (ZP1, ZP2) had the same specifics. The diameter and the outermost zone width were 155 μm and 100 nm, respectively. The zone material was tantalum of 1.0 μm in thickness. The zone plates were fabricated by NTT Advanced Technology Inc. Two zone plates were placed twice of the focal length apart from each other. Illumination and reference beams were separated by a double pinhole of 50 μm in diameter. The (+1, 0) order x-rays were transmitted through a specimen and were focused on a CCD camera (Hamamatsu Photonics K.K.,

3.14 μm/pixel at the phosphor screen) with the magnification ratio of 50. The (0, −1) order x-rays of the reference beam were overlapped with the image of a specimen. The two beams interfered with the same phase difference over the image plane. Then, the interference image was observed without any interference fringes. The beam stop 1 prevents the -1st order reference beam of ZP1 from overlapping the image plane. The beam stop 2 prevents the 1st order of ZP2 from overlapping the image plane. The phase of the reference beam could be changed by rotating the phase shifter of 70-μm quartz sheet.

10 keV x-rays of SPring-8 BL20XU were used. This beamline has an in-vacuum undulator and a Si (111) double-crystal monochromator. The monochromator is placed at 46 m from the source. The distance between the source and the experimental hutch was 240 m. In this optical system, illuminating x-rays interfere with x-rays transmitted through the first zone plate at the symmetrical position about the optical axis. Then, spatial coherence area must be larger than the diameter of the zone plate. This relatively long beamline length is suitable to obtain a large coherence area.

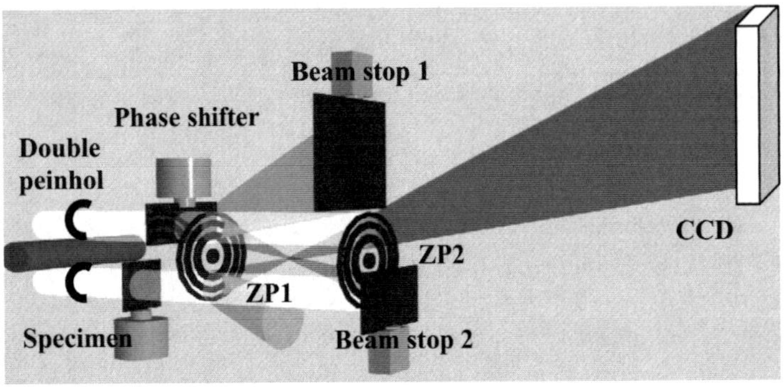

Figure 1. Optical system of the interference microscope.

3. 2D phase image

Figure 2 shows a phase image of an aluminum foil obtained by 4-step fringe-scanning method. The standard deviation of the phase shift was 0.13 rad at the area A in the foil (100 × 50 pixels, 62 nm/pixel), and 0.034 rad at the area B. Suppose that the phase sensitivity was three times of this background, it was calculated to be λ/62. Figure 3 shows the phase shift variation with the thickness of the aluminum foil. The result of the 18-μm foil agreed very closely with the calculated value from the Henke table [4]. But the results of the 5 μm and 12.5 μm were slightly larger than the calculations.

Figure 5 shows the phase shift profile of overlapped 2-μm copper foils (total 4 μm). The solid line is the experimental data and the dashed line is calculated one. The obtained phase shift near the copper absorption edge was smaller than the calculated one. But the complicated structure was observed in the obtained profile.

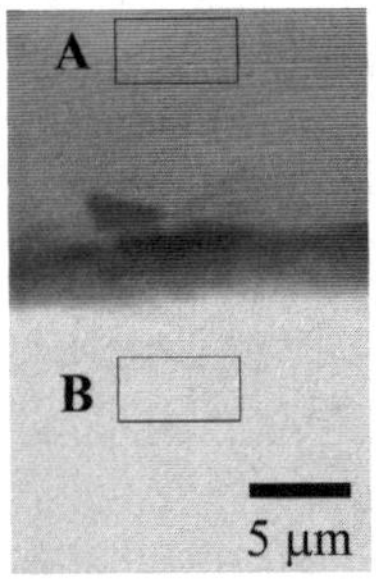

Figure 2. Phase image of the edge of an aluminum foil. The standard deviations of the area A and B are 0.13 rad and 0.034 rad, respectively.

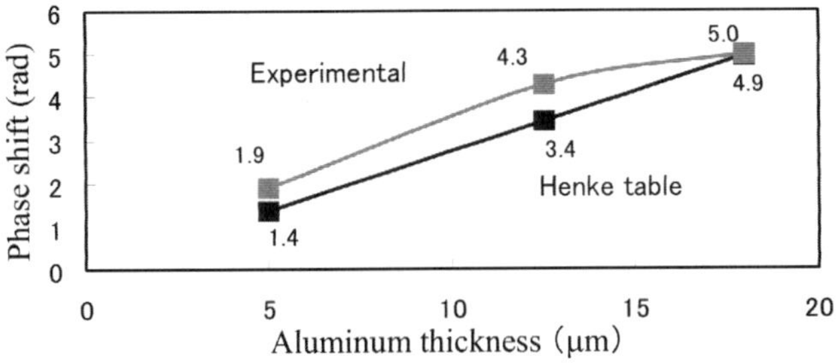

Figure 3. Phase shift variation with the thickness of the aluminum foil. The upper line shows the experimental values and the lower line shows the calculation from the Henke table.

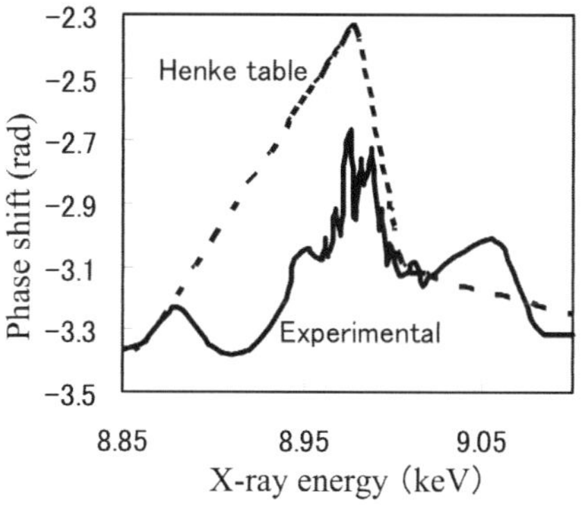

Figure 4. phase shift profile of overlapped 2-μm copper foils (total 4 μm).

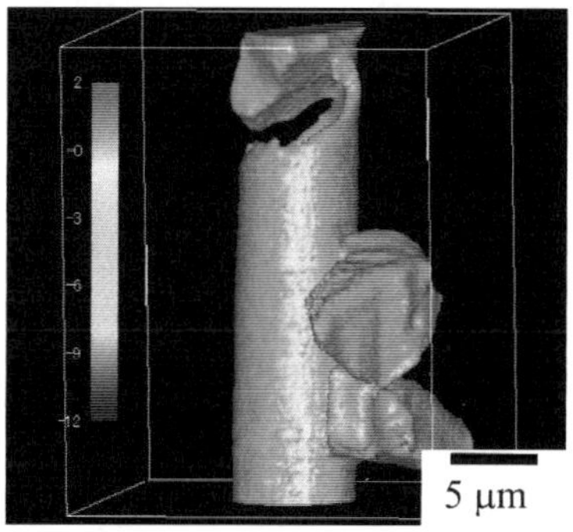

Figure 5. Phase tomography of diamond particles attached to a glass capillary of 7 μm in outer diameter (rendering view).

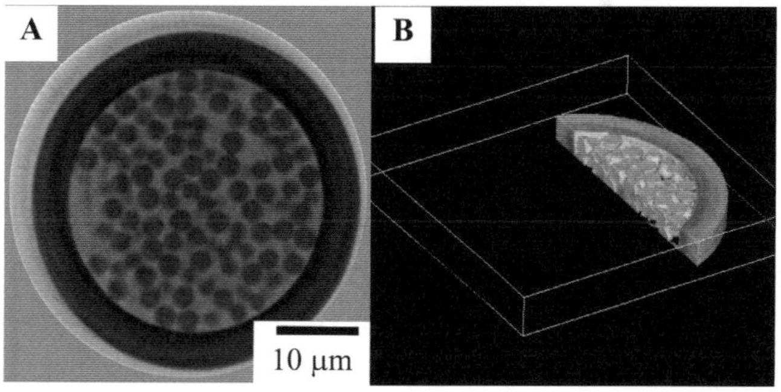

Figure 6. Phase tomography of polystyrene beads of 2.8-μm diameter in a glass capillary. A: the section image, B: the three dimensional rendering view.

4. Phase tomography

Phase tomography was investigated using test samples, such as diamond particles attached to a glass capillary and polystyrene beads of 2.8 μm in diameter packed in a glass capillary. The phase projections were obtained by 4-step fringe-scanning method.The three-dimensional phase images were reconstructed from the 90 phase projections of different angles of view over the range of 180 degrees. The exposure time of each image was 20 s.

Figure 5 shows the diamond image and Fig. 6 shows the polystyrene image. The almost transparent samples, such as the diamond particles and the polystyrene beads could be clearly observed three dimensionally.

This work was supported by Grant-in-Aid for Scientific Research (B) (18360044) from the Ministry of Education, Culture, Sports, Science and Technology of Japan.

References

[1] Wilhein T, Kaulich B and Susini J 2001 Opt. Commun. 193, 19.
[2] Watanabe N, Hoshino M, Sato M, Takeda Y, Namiki T, Aoki S, Takeuchi A and Suzuki Y 2006 Proc. 8th Int. Conf. X-ray Microscopy, IPAP Conf. Series 7 pp.372-374.
[3] Koyama T, Tsuji T, Yoshida K, Takano H, Tsusaka Y and Kagoshima Y 2006 Jpn. J. Appl. Phys. 45, L1159
[4] Henke, B L, Gullikson, E M, Davis, J C, Atomic Data Nucl. Data Tables 54, 181-342 (1993). (http://henke.lbl.gov/optical_constants/)

Journal of Physics: Conference Series **186** (2009) 012022 doi:10.1088/1742-6596/186/1/012022

X-ray fluorescence micro-tomography and laminography using an x-ray scanning microscope

N Watanabe[1], M Hoshino[1], K Yamamoto[1], S Aoki[1], A Takeuchi[2] and Y Suzuki[2]

[1]Institute of Applied Physics, University of Tsukuba, Tsukuba, Ibaraki 305-8573, Japan
[2]SPring-8, JASRI, Mikazuki, Hyogo 679-5198, Japan

E-mail: watanabe@bk.tsukuba.ac.jp

Abstract. Using a scanning microscope with a zone plate, x-ray fluorescence micro-tomography was investigated at SPring-8 BL20XU. A 120 nm-thick zinc layer could be resolved in the reconstructed section image. A frozen phytoplankton and a iron impurity of a diamond could be also reconstructed. X-ray fluorescence laminography was tested at SPring-8 BL47XU. A tantalum line pattern of 3 μm line width could be reconstructed.

1. Introduction

X-ray fluorescence microscopy is a powerful method to map elemental distributions in biological and material samples. However, this method is usually two dimensional. X-ray fluorescence tomography enables observation of three dimensional elemental distributions of an internal structure [1]. A combination of a third generation x-ray undulator and a zone plate enabled us to make an x-ray scanning microscope with a resolution around 100 nm. Using the x-ray scanning microscope, x-ray fluorescence micro-tomography and laminography were investigated at SPring-8.

2. X-ray scanning microscope with a zone plate

Figure 1 shows the scanning microscope system. The x-ray beam-line BL20XU of SPring-8 was used. This beam-line has an x-ray undulator and a double-crystal monochromator cooled by liquid nitrogen. X-rays at 9.8 keV were focused by a zone plate. The diameter and the outermost zone width were 155 μm and 100 nm, respectively. The zone plate was placed about 250 m downstream from the undulator source. This long distance enabled us to obtain the focused spot size of about 200 nm. The first order focus was used and the other order x-rays were eliminated by an order sorting aperture (OSA). The zone plate and the OSA were raster scanned synchronously to obtain an image with high resolution. For micro-tomography, a specimen was set on a precision rotary stage. Transmission and fluorescence x-rays were recorded by an ionization chamber and a silicon drift detector (SDD), respectively.

3. X-ray fluorescence tomography

A test sample of iron (140 nm), zinc (120 nm), and copper (200 nm) layers evaporated on a tungsten wire (figure 2 (a)) was used for evaluation of the micro-tomography system. By rotating the specimen by 5.0° intervals, x-ray fluorescence line projections from 72 different angles of view were recorded. The scan pitch was 106 nm and the exposure time was 0.1 s/pixel. The convolution back projection method with Shepp-Logan filter was used without correcting for attenuation [2].

Journal of Physics: Conference Series **186** (2009) 012022 doi:10.1088/1742-6596/186/1/012022

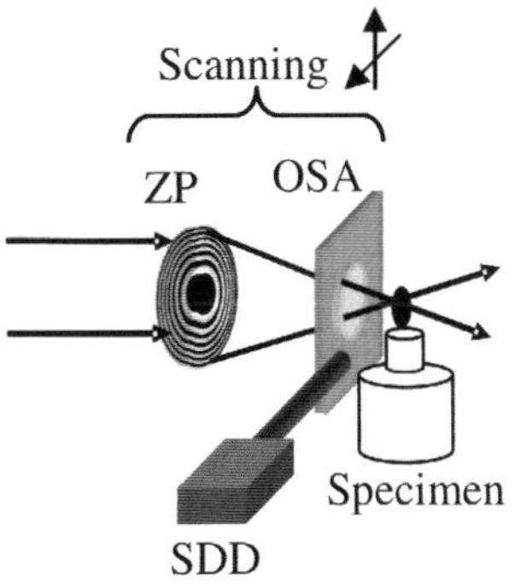

Figure 1. Optical system.

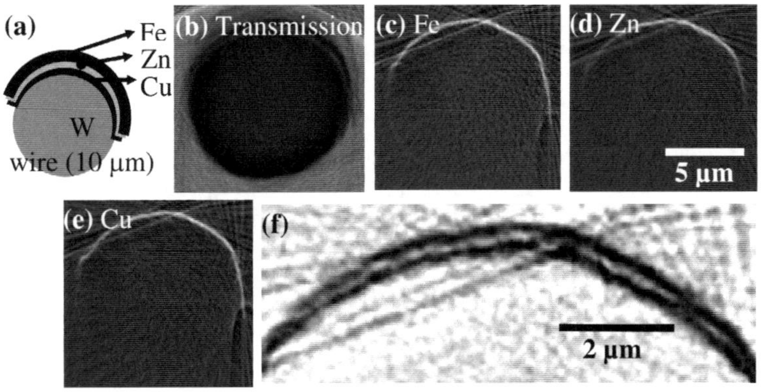

Figure 2. (a) Schematic of the test sample. (b) Transmission CT image. (c) - (e) X-ray fluorescence CT image. (e) Composite image of Iron and Copper, after gradation reversal.

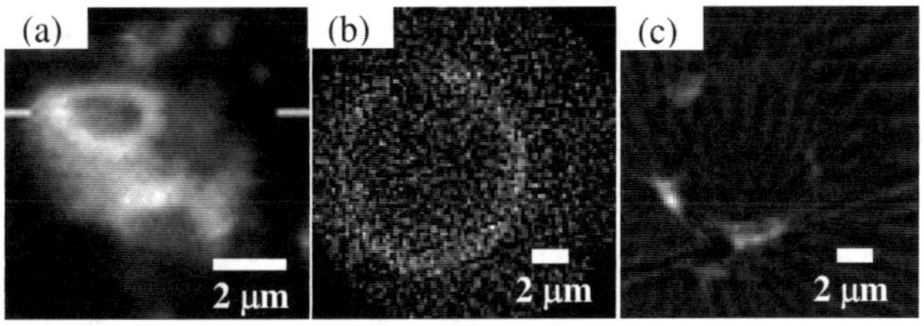

Figure 3. X-ray fluorescence images of E. huxleyi. (a) Calcium fluorescence image. (b) Transmission CT image at 13.4 keV. (c) Calcium fluorescence CT image.

Figure 2 (b) is the transmission section image and figure 2 (c) – (e) are the x-ray fluorescence section images of iron, zinc and copper, respectively, Figure 2 (f) is the composite image of (c) and (e). The zinc layer (120 nm) could be distinguished, which can be seen as a gap of the two lines.

Figure 3 shows x-ray fluorescence tomography of an oceanic phytoplankton, Coccolithophorid (Emiliania huxleyi) [3]. Coccolithophorids are unicellular, calcifying algae. For a hydrated biological specimen, radiation damage is serious especially in tomography. Then, Cryojet system (Oxford Instruments) was introduced to keep a specimen at about 150 K. Figure 2 (a) is the Ca fluorescence image. Figure 2 (b) and (c) are the tomographic section images of transmission and Ca fluorescence. The reconstructions were calculated from 45 projections over 360° (0.18 μm/pixel, 1.0 s/pixel). Fig. 2 (c) shows the distribution of coccolith.

Using an x-ray scanning microscope, it was frequently difficult to determine where the reconstruction area was located in a sample. Then, a full-field x-ray microscope with a zone plate was incorporated into an x-ray fluorescence microscope to obtain the overall 3D image of a region of interest area as shown in figure 4 (a). An impurity in a synthetic diamond of about 0.5 mm in diameter was observed. The impurity mainly consisted of iron and nickel. Figure 4 (b) shows the reconstructed iso-surface image from 180 projections / 360° recorded by the full-field microscope. Figure 4 (c) shows the corresponding section images of x-ray fluorescence tomography along the white dash lines in figure 4 (b). The reconstructions were calculated from 120 projections / 360° (0.45 μm/pixel, 0.1 s/pixel).

4. X-ray fluorescence laminography

X-ray transmission computed laminography is successfully developed and has come to be used for imaging of a laterally extended flat specimen three-dimensionally [4]. If x-ray fluorescence is used for imaging, it is expectable that it is applicable to elemental analysis of extended flat specimen three-dimensionally. Then, x-ray fluorescence laminography was tested. The optical system is the same as the tomography system except the specimen stage. The rotation axis of a specimen was tilted 30° from the perpendicular to the optical axis as shown in figure 5 (a). 10 keV x-rays of SPring-8 BL47XU were used. Figure 5 (b) is the tantalum fluorescence image of a tantalum line pattern of 0.5 μm thickness. Figure 5 (c) and (d) are the section and the three dimensional rendering images, respectively. The reconstruction was calculated by the convolution back projection method from 18 projections at the angle intervals of 20°. The reconstruction quality was much worse than the previous experiments. This is due to the inclination of the rotation axis and the small number of projection. The other reconstruction method, such as ART, is considered to give a better image.

Acknowledgement
The authors are grateful to Prof. Y. Shiraiwa, Univ. Tsukuba for the sample of Emiliania huxleyi.

References
[1] Simionovici A, Chukalina M, Günzler F, Schroer Ch, Snigirev A, Snigireva I, Tümmler J and Weitkamp T 2001 Nucl. Instr. Meth. **A 467-468** 889.
[2] Kak A C and Slaney M 1988 Principles of Computerized Tomographic Imaging (New York: IEEE).
[3] Obata T, Araie H and Shiraiwa Y 2004 Plant Cell Physiol. **45** 1434.
[4] Helfen L, Baumbach T, Mikulik P, Kiel D, Pernot P, Cloetens P and Baruchel J 2005 Appl. Phys. Lett. 86 71915.

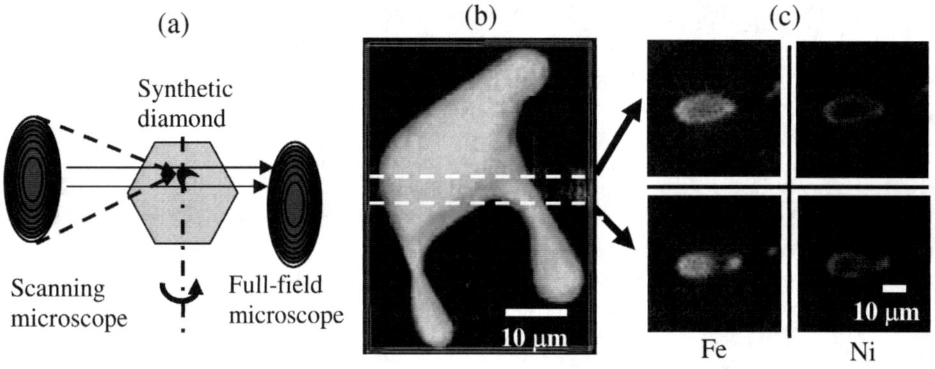

Figure 4. (a) A transmission x-ray microscope was incorporated. (b) The iso-surface image of x-ray transmission CT at 9.8 keV. (c) The section images of iron and nickel fluorescence CT along the white dash lines of (b).

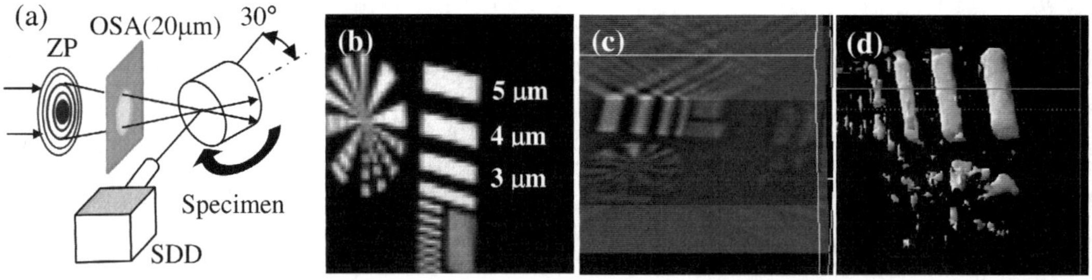

Figure 5. (a) X-ray laminography system. (b) The x-ray fluorescence image of the tantalum line pattern. (c) The vertical and horizontal section images of x-ray fluorescence laminography image, (d) The three-dimensional rendering image. The tantalum line of 3 μm in width could be imaged.

Journal of Physics: Conference Series **186** (2009) 012023 doi:10.1088/1742-6596/186/1/012023

Soft X-Ray Beam Induced Current Technique

B Watts[1,4], D Queen[2], A L D Kilcoyne[3], T Tyliszczak[3], F Hellman[2] and H Ade[1]

1 Department of Physics, North Carolina State University, Raleigh, NC 27695, USA
2 Department of Physics, University of California, Berkeley, CA 94720, USA
3 Advanced Light Source, Lawrence Berkeley Nat. Lab., Berkeley, CA 94720, USA
4 Paul Scherrer Institute, Villigen 5232, Switzerland
benjamin.watts@gmail.com

Abstract. Direct mapping of the charge transport efficiency of polymer solar cell devices using a soft X-ray beam induced current (SoXBIC) method is described. By fabricating a polymer solar cell on an x-ray transparent substrate, we demonstrate the ability to map polymer composition and nanoscale structure within an operating solar cell device and to simultaneously measure the local charge transport efficiency via the short-circuit current. A simple model is calculated and compared to experimental SoXBIC data of a PFB:F8BT bulk-heterojunction device in order to gain greater insight into the device operation and physics.

1. Introduction

It is well known that the nanoscale structure determines the exciton splitting and charge transport properties of bulk-heterojunction polymer photovoltaic devices. However the precise details of these effects hide in the entangling of the different steps of energy capture in the device. Comparison of nanoscale composition images from scanning transmission X-ray microscopy (STXM) with conventional macroscopic efficiency testing of polymer solar cells has proven that such testing is insufficient for determining the influence of nanoscale structures [1]. A more direct way to measure nanoscale device efficiency is required to further understanding in this area. To this end, we have developed a method to produce charge transport efficiency images via X-ray beam induced current (XBIC) [2] in a STXM instrument, simultaneously measuring of conventional STXM transmission images for the calculation of composition maps.

2. SoXBIC Principle

Traditional STXM spectromicroscopy uses a zone plate to focus soft X-rays onto a thin sample, measuring the transmitted intensity. In the soft X-ray beam induced current (SoXBIC) technique, the transmitted light is still utilized to map composition as in normal STXM operation, however the absorbed soft X-ray photons also create charge carriers of the active polymer layer, and produce a measureable current between the device electrodes. Therefore, measuring the short-circuit current, provides a current signal for each pixel in the recorded image to complement the transmission data.

The soft X-ray photons of the SoXBIC beam have considerably more energy than the visible photons absorbed in normal solar cell operation. Hence, rather than the formation of an exciton that must be split before the charges can be transported to the device electrodes, the soft X-ray photon produces one or more high energy photo- and Auger electrons, which then creates a cascade of lower energy, secondary electrons and a corresponding number of positively charged molecules. The large number of free charges created by each absorbed soft X-ray photon renders the nature of the original absorption event insignificant. Since SoXBIC does not involve the creation nor splitting of excitons, the technique is a measure of only the soft X-ray absorption and charge transport properties of the device. Further, since the soft X-ray absorption properties of materials are easily characterized and modeled, SoXBIC allows the charge transport properties to be isolated. The spatial spread of this secondary electron cascade is limited to about 10 nm [3] and so SoXBIC charge transport maps have the same spatial resolution as the simultaneous STXM composition maps.

3. Experiment

Poly(9,9'-dioctylfluorene-co-bis-N,N'-(4,butylphenyl)-bis-N,N'-phenyl-1,4-phenylenediamine) (PFB) and poly(9,9'-dioctylfluorene-co-benzothiadiazole) (F8BT) were purchased from American Dye Source with molecular weights (polydispersity) of 23k (2.7) and 38k (2.3) respectively. PEDOT:PSS was purchased as Baytron P from H.C. Starck.

The lower indium-tin oxide (ITO) electrode was magnetron sputtered onto a silicon nitride membrane in a 5 mTorr Ar atmosphere at ~0.5 As^{-1} with a base pressure of 10^{-7} Torr. Polymer layers were spincast; PEDOT:PSS at 4 krpm and then 1:1 PFB:F8BT from p-xylene at 3 krpm. 30 nm of Al were then thermally evaporated 1-2 As^{-1} (in a base pressure of 10^{-6} Torr) to form the top electrode and complete the device. The connection pads of the ITO layer were exposed by wiping away the polymer with acetone and then isopropanol and electrical connections were made to the ITO and AL electrodes by wedge-bonding fine gold wires.

The device was mounted (oriented such that the probing soft X-rays enter the device through the aluminium electrode) in the 5.3.2. STXM at the Advanced Light Source in Berkeley, USA, and the electrodes connected to a Keithley 6517A Electrometer in short-circuit. The output of the electrometer was connected to a voltage-to-frequency converter and the resulting pulses counted and recorded by a PC. STXM images were recorded with a dwell time of 500 ms and a pixel size of 50 nm at 280, 284.7, 285.2, 285.7, 286.6, 289 and 320 eV (in order of ascending photon energy) with both the transmitted intensity and current signals recorded simultaneously.

4. Results and Discussion

The composition map of the segregated PFB:F8BT layer of the device, shown in figure 1, exhibits a domain structure ~0.5 μm in size and domain purities of about 59% and 65% for the PFB- and F8BT-rich domains, respectively. This differs from the 50% and >90% domain purities in structures greater than one micron that was previously observed in similar blend films [1], likely because of the much lower molecular weight of the polymer materials.

Figure 2 shows the modelled and observed SoXBIC current images and figure 3 presents the spatially averaged currents, normalised against the photon flux. While there is considerable noise in the experimental images, it is clear that the PFB:F8BT segregated domain pattern is present in the 320 eV and 284.7 eV current images and is just observable in the 280 eV and 287.7 eV images as well. The fact that the pattern observed in the current images does not reverse contrast, even at 284.7 eV, where F8BT absorbs more strongly than PFB, suggests that the PFB-rich phase may have better charge transport properties than the F8BT-rich phase. Previous current mapping experiments at visible wavelengths have also shown enhanced current from the PFB-rich domains [4].

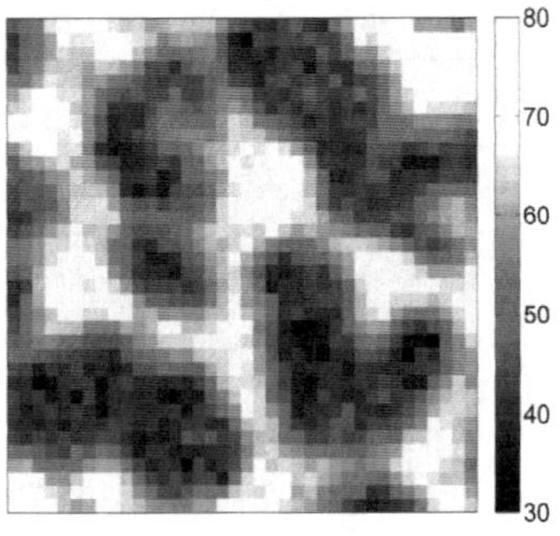

Figure 1. Percentage F8BT composition map. Image size is 1.9 μm by 1.8 μm.

Of the model variants shown in figures 2 and 3, the "full collection" model is the simplest and involves a straightforward calculation of the number of photons absorbed between the device electrodes, multiplied by the photon energy (assumed proportional to the number of charges induced per photon). A further consideration of beam damage was added to account for lower than expected current signal at higher photon energies by subtracting a scaled cumulative sum of the total charges induced by the preceding image scans. Note that the SoXBIC charge generation applies equally to all device layers that lie between the electrodes and hence the model also includes contributions due to charges induced in the PEDOT:PSS layer. The other, "boundary", model assumes that the PFB:F8BT layer is a poor conductor of holes and the PEDOT:PSS is a poor conductor of electrons such that only charges close to the polymer/polymer interface (within 10 nm was the condition utilised here) are contributing to the device current. The effect of PFB:F8BT phase-dependent charge transport on the variation in total current with photon energy was also investigated, but found to not differ significantly from the corresponding models (figure 3) that ignore PFB:F8BT phase effects beyond that of the initial soft X-ray absorption. The "boundary" model appears to be a slightly better fit to the experimental current signal data, both in terms of total current and current image contrast, however the data presented here is not of sufficient quality to draw further conclusions on the device operation.

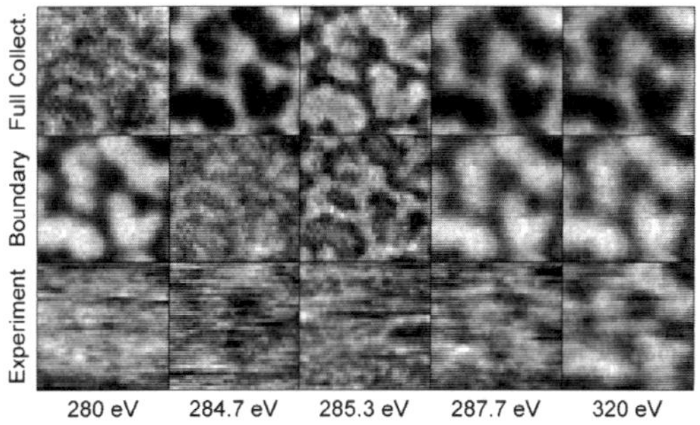

Figure 2. SoXBIC current images. High (low) current values are represented by bright (dark) pixels. Image size is 1.9 μm by 1.8 μm.

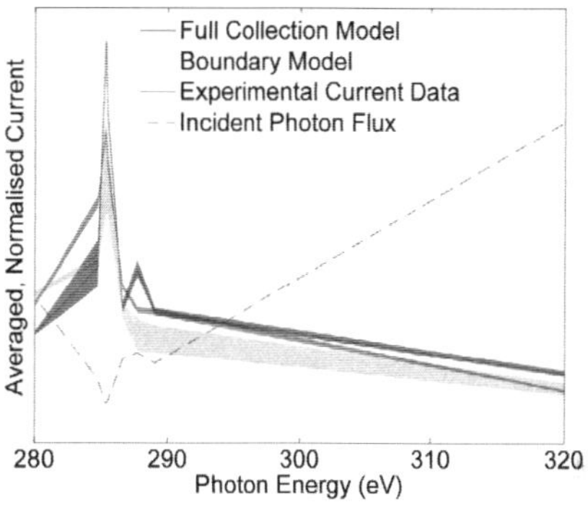

Figure 3. Photon flux normalised and spatially averaged, modelled and experimental currents. Spatial variations in current (image contrast), are shown via line width.

5. Conclusions

We have demonstrated the experimental measurement of charge transport maps with nanoscale structures that correspond to the observed PFB:F8BT phase domain structures within the active polymer layer of a polymer solar cell device. While the lack of contrast reversal in the current signal data suggests that PFB-rich domains have better charge transport properties than F8BT-rich domains, the data presented here is not of sufficient quality to draw further conclusions on the device operation.

Acknowledgements

This work was supported by the U. S. Department of Energy under Contract Nos. DE-FG02-98ER45737 and DE-AC02-05CH11231.

References

[1] McNeill C R, Watts B, Thomsen L, Ade H, Greenham N C and Dastoor P C 2007 *Macromolecules* **40** 3263
[2] Vyvenko O F, Buonassisi T, Isatratov A A, Hieslmair H, Thompson A C, Schindler R and Weber E R 2002 *J. Appl. Phys.* **91** 3614
[3] Jablonski A and Powell C J 1999 *J. Electron Spectrosc. Relat. Phenom.* **100** 137
[4] McNeill C R, Frohne H, Holdswoth J L and Dastoor P C 2004 *Nanoletters* **4** 2503

Journal of Physics: Conference Series **186** (2009) 012024 doi:10.1088/1742-6596/186/1/012024

High power soft x-ray source based on a discharge plasma

Markus Benk[1], David Schäfer[2], Thomas Wilhein[2] and Klaus Bergmann[1]

[1] Fraunhofer Institute for Laser Technology, Steinbachstraße 15, 52074 Aachen, Germany
[2] Institute for X-Optics, RheinAhrCampus Remagen, University of Applied Sciences Koblenz, Südallee 2, 53424 Remagen, Germany

E-mail: markus.benk@ilt.fraunhofer.de

Abstract. We report on a gas discharge plasma as a light source for laboratory based soft x-ray microscopy in the water-window spectral range. At 2.88 nm wavelength the source produces a radiant intensity of up to $4 \cdot 10^{13}$ photons/(sr pulse) narrowband line emission. Employing a matched grazing incidence collector, a photon flux at the specimen of $\sim 10^7$ photons/(μm^2 s) can be obtained at 1000 Hz repetition rate. This allows for imaging of thick hydrated biological samples at an exposure time below 10 s. Images are presented using this source in a first laboratory scale microscope. Image quality and exposure times are promising in view of commercializing laboratory transmission x-ray microscopy.

1. Introduction

Full-field transmission x-ray microscopy is a striking methode for investigating thick hydrated biological samples. To achieve an adequate signal-to-noise ratio, a radiation dose of 10^8 photons/μm^2 at the sample is required [1]. Tomographic imaging enables to fully tap the potential of x-ray microscopy. In order to aquire the desired amount of images within an appropriate periode of time, exposure time for a single image has to amount below ten seconds. This demands a photon flux of at least $\sim 10^7$ photons/(μm^2 s) at the sample. To realize a commercial synchrotron-independent microscope, compact sources capable to deliver that flux are needed. In this paper a gas discharge plasma is presented as an alternative light source for laboratory microscopy in the water-window spectral range.

2. Pinch plasma as a compact source for x-ray microscopic imaging

Here the hollow cathode triggered pinch plasma is under investigation, which has also been discussed as a source for EUV-lithography [2]. At present this concept is advanced to efficiently emit 2.88 nm line radiation at high average power. The gas is heated in a pulsed high current discharge of 12 to 20 J electric energy. Part of this energy is reemitted by the $1s^2$-1s2p transition of helium-like ions in the nitrogen plasma. The source produces a radiant intensity of $2 \cdot 10^{13}$ photons/(sr pulse line) at 12 J and $4 \cdot 10^{13}$ photons/(sr pulse line) at 20 J. Radiant intensity was measured using transmission grating spectrometers calibrated at synchrotron facilities. The lateral diameter of the plasma column is 1175 μm. Spectral bandwidth of the emission line is estimated to be less than $\Delta\lambda/\lambda = 1/550$ [3], which is sufficient for high resolution zone plate imaging. Efficient operation of the source at a repetition rate of 1000 Hz is demonstrated [3] leading to a power input of 12 kW to 20 kW.

3. Setup

The source-collector-module consists of the source peripherals, the electrode system where the radiation is generated, a polycapillary as presented in [4] providing vacuum separation between source and collector chamber, an ellipsoidal grazing incidence collector with central stop and a 200 nm titanium filter virtually limiting the spectrum to the 2.88 nm emission line. Distance between source and collector focus is 600 mm.

The polycapillary consists of divergent 50 μm capillaries with a length of about 20 mm covering a solid angle of $15 \cdot 10^{-3}$ sr from the source. The transmission for the 2.88 nm line amounts 0.33. A transmission of 0.7 as theoretical limit is aimed for. Conductance of the structure can be estimated to be ~0.01 l/s. In figure 1 vacuum level ratio between source and collector chamber calculated for differential pumping at 300 l/s and a microscopic image of the capillaries are shown. At the source side of the polycapillary pressure amounts about 10^{-2} mbar, which allows to obtain a vacuum level of ~10^{-6} mbar in the collector chamber for this pumping speed. The titanium filter is placed between collector and intermediate focus and can also be used to separate the vacuums of collector chamber and microscope.

At present a non matched gold coated grazing incidence collector, designed for different tasks is empoyed. Collection solid angle amounts $1.56 \cdot 10^{-3}$ sr and theoretical magnification is 0.67.

4. Results

Without polycapillary and spectral filter a central photon flux of $5.1 \cdot 10^3$ photons/(μm^2 pulse) collectible in-band radiation is detected. In figure 2 an image of the intermediate focus and a profile plot across the focus are shown. The image is acquired using a Ce:YAG converter crystal at the focal plane and a CCD-camera. The focus is large in comparison to the focus in an LPP microscope due to the relatively large source volume. Only the very center of the spot is used for sample illumination. A good level of illumination homogeneity is achieved.

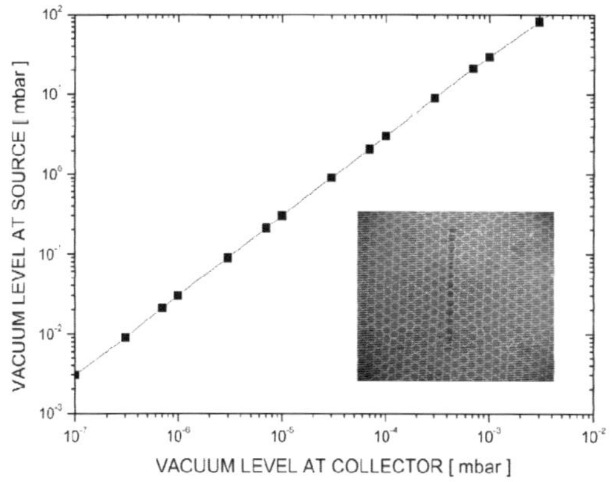

Figure 1. Vacuum level at source and in the collector chamber and microscopic image of capillaries

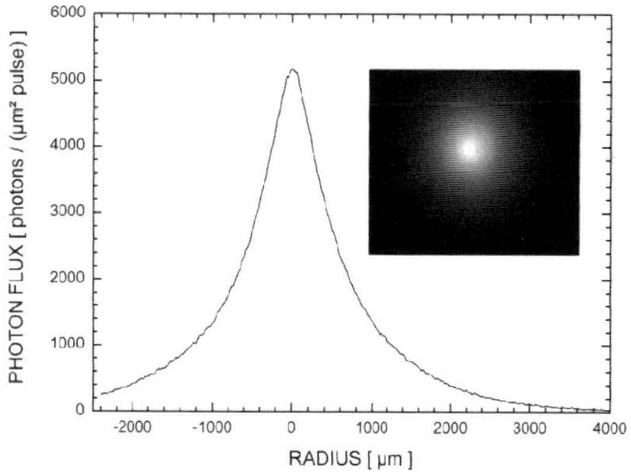

Figure 2. CCD-image of the intermediate focus and profile plot across the focus

In figure 3 microscopic images of dried rat embryonic fibroblast cells are shown. The images are taken using the presented source-collector-module and a microscope with zone plate optic and back thinned CCD camera, described in [5]. A resolution below 40 nm is achieved with the setup [5], which corresponds to the diffraction limit of the zone plate having an outer zone width of 30 nm.

Both images are exposed at 12 Joule pulse energy with polycapillary and titanium filter installed. The left image is exposed 15 seconds at repetition rate of 600 Hz in order to demonstrate exposure times for biological samples attainable with the non-matched collector. The right image is taken at low repetition rate. It is rich in contrast and resolves small features.

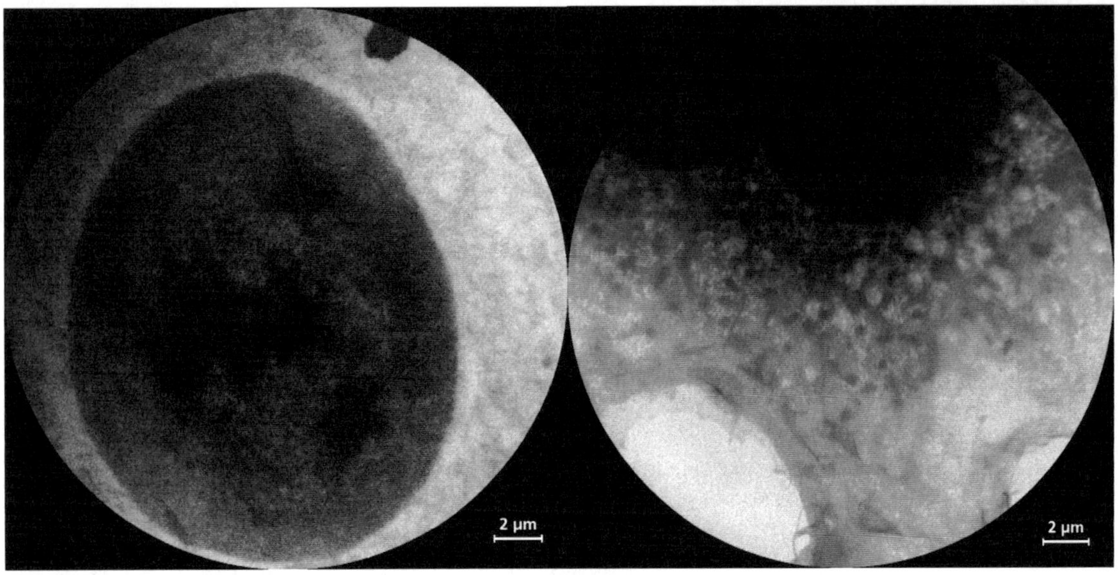

Figure 3. Microscopic images of dried rat embryonic fibroblast cells.

5. Outlook

Ray tracing simulations of source and collector are carried out in order to estimate possible photon flux at the specimen. The collector is modelled as a nickel coated grazing incidence ellipsoidal mirror with a collection solid angle of $1.48 \cdot 10^{-3}$ sr and a magnification of 0.37. The apertures of collector and zone plate are matched. The simulation shows that applying this collector, at a repetition rate of 1000 Hz a flux of $2 \cdot 10^{7}$ photons/(μm^2 s) collectible in-band radiation can be achieved. A factor 0.5 for the transmission of polycapillary and spectral filter is taken into account. Future activities aim for applying this matched collector and for improving the brilliance of the source in order to finally achieve a photon flux of $\sim 10^{8}$ photons/(μm^2 s) at the sample.

Acknowledgements
The polycapillary is provided by Aniouar Bjeoumikhov from the Institute for Scientific Instruments.
The zone plate employed in the microscope setup is provided by Gerd Schneider from BESSY.
This project is supported by the German Federal Ministry of Education and Research under contract number 13N8914.

References

[1] Gölz P 1990 *X-Ray Microscopy III* (Berlin: Springer) *Springer Series in opt. Science* **67**
[2] Pankert J et al 2005 Hollow-Cathode-Triggered Plasma Pinch Discharge *EUV Sources for Lithography* ed V Bakshi (Bellingham: SPIE) pp 395-411
[3] Bergmann K, Küpper F and Benk M 2008 Soft x-ray emission from a pulsed gas discharge in a pseudospark like electrode geometry *J. Appl. Phys.* **103** 123304
[4] Arkadiev V A, Beloglazov V I, Bzhaumikhov A, A Gorny H E, Langhoff N, Margushev Z, Schmalz J, Wedell R and Wolff H 1998 *Proceedings of SPIE* **3444** pp 436-442
[5] Schäfer D, Benk M, Bergmann K and Wiesemann U, Nisius T and Wilhein T 2009 *Journal of Physics: Conference Serie (IOP), Proceedings of the 9th International Conference on X-Ray Microscopy*, submitted for publication

First application experiments with the Stockholm compact soft x-ray microscope

M Bertilson[1], O von Hofsten[1], J Thieme[2], M Lindblom[1], A Holmberg[1], P Takman[1], U Vogt[1], H Hertz[1]

[1] Biomedical and X-Ray Physics, Royal Institute of Technology/Albanova, SE-10691, Stockholm, Sweden

[2] Institute for X-Ray Physics, University of Göttingen, Friedrich-Hund-Platz 1, D-37077 Göttingen, Germany

michael.bertilson@biox.kth.se

Abstract. Most soft x-ray microscopes operating in the water window (λ = 2.3 - 4.4 nm) rely on synchrotron radiation sources. In the future we believe scientists will use soft x-ray microscopes as one imaging tool among others in their own laboratory. For this purpose we have developed a full field soft x-ray microscope with a laser-plasma source compact enough to fit on an optical table. In this contribution we describe the current status of this microscope now featuring stable operation at λ = 3.37 nm or λ = 2.48 nm. In-house fabricated single element zone plates offering the possibility to perform phase contrast imaging have been implemented. We also report on the first application experiments for compact soft x-ray microscopy, including results from studies of clay minerals and colloids existing in nature and results from phase optics experiments. Planned upgrades of the microscope include increasing the source brightness, implementing more efficient condenser optics, and installing a cryo sample stage for tomography. These improvements will open up for further applications, especially in the field of biological imaging.

1. Introduction

Soft x-ray microscopy is now a successful imaging technique. Soft x-rays in the water-window (λ = 2.3 – 4.4 nm) offer a natural contrast mechanism, which opens up for applications in, e.g., bio imaging and soil science. Since the early days of soft x-ray microscopy synchrotron radiation sources have been used. The limited accessibility of these large facilities prevents this imaging technique from being one of the standard tools among scientists. For this reason we developed the first full-field soft x-ray microscope based a compact liquid-jet laser-plasma source making the entire system fit on an optical table [1, 2]. Brighter compact sources together with more efficient condenser and objective optics have made compact soft x-ray microscopes able to capture images with a quality comparable to synchrotron-based microscopes within reasonable exposure times. With these improvements the commercial interest has increased and a few companies are now developing microscopes based on compact sources.

In this paper we report on the current status, recent applications and future plans of the Stockholm compact microscope.

2. The microscope

The microscope has demonstrated stable operation at two wavelengths within the water-window, $\lambda = 3.37$ nm [1] or $\lambda = 2.48$ nm [2]. The microscope consists of a high-brightness liquid-jet-target laser-plasma source, condenser optics, the sample environment, an objective zone plate and a soft x-ray sensitive CCD detector. The arrangements for the two operation modes are illustrated in figure 1.

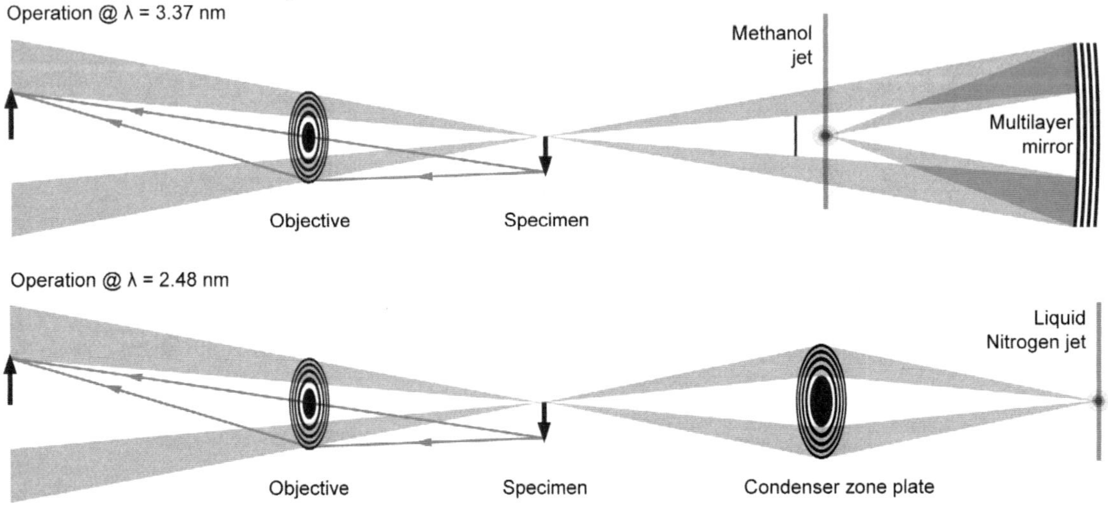

Figure 1. Illustrations showing the basic components of the microscope for operation at $\lambda = 3.37$ nm (above) and $\lambda = 2.48$ nm (below)

The soft x-ray emitting plasma is generated by focusing a pulsed high power laser (~3 ns, 100 Hz, 100 mJ) onto a liquid jet target. For operation at $\lambda = 3.37$ nm a methanol is used and for operation at $\lambda = 2.48$ nm liquid nitrogen is used. The generated photon flux is 10^{11} - 10^{12} ph./(pulse × sr × line) with a line monochromacy of $\lambda/\Delta\lambda > 500$. A 300 nm chromium filter blocks the green light scattered from the source. For details on the sources see refs. [3, 4].

The condenser optics images the source directly onto the sample (critical illumination). A Cr/Sc multilayer mirror, 58 mm in diameter with an average reflectivity of 3.14%, is used for operation at $\lambda = 3.37$ nm [5]. When operating at $\lambda = 2.48$ nm, a 49 nm outer zone width condenser zone plate, 4.53 mm in diameter, is used [6].

Samples are kept in a helium atmosphere at normal room pressure. Two types of sample holders can be used. The first is a ±8° tiltable sample stage for silicon nitride membranes which offers the possibility to do stereo imaging of dry and wet samples. Here, the wet samples are sandwiched between two membranes. The other holder is a rotatable capillary holder, shown in figure 3, which makes the collection of tomographic data of wet and dry samples possible.

In-house fabricated high-resolution nickel zone plates [6] are employed as objectives. So far zone plates with outer zone widths down to 25 nm have been used resulting in a measureable resolution better than 30 nm half-period [2]. Single element differential interference contrast (DIC) zone plates [7] and Zernike-like phase contrast [8] zone plates have been developed and offer contrast enhanced imaging.

The magnified image of the sample is detected by a cooled, back-illuminated and thinned x-ray sensitive CCD with 2048 × 2048 pixels, each pixel 13.5 μm × 13.5 μm in size.

3. Applications

Soft x-ray microscopy is the only full-field microscopy technique capable of imaging several μm thick wet samples with spatial resolutions down to tens of nanometer. This makes it an interesting technique for studying soil particles in their aqueous environment. Figure 2 shows one example from the first experiment where the compact soft x-ray microscope was used for imaging colloidal particles in an aqueous soil suspension [9, 10].

Figure 3 shows one projection image of a diatom mounted on a capillary. The stereo and capillary holders open up for new applications where more depth information is needed. A projection series of 50 images for later tomographic reconstruction could be recorded in ~5 h.

The microscope also serves as a test bench for in-house fabricated zone plates. For example, DIC [7] and Zernike phase contrast [8] zone plates have been tested and characterized in the microscope. The microscope can easily be rebuilt to a compact instrument for diffraction efficiency measurements of zone plates, which gives important feedback to the fabrication process [11].

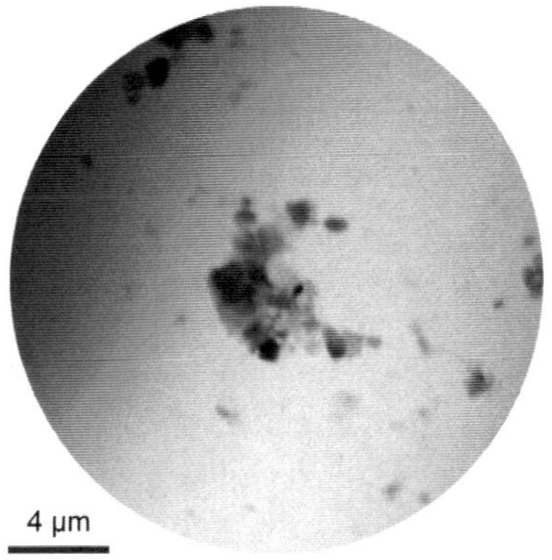

Figure 2. Colloidal Chernozem soil particles in a 5-8 μm thick aqueous suspension between two Si_3N_4 membranes. The image was obtained in a 600 s exposure @ λ = 3.37 nm with a 25 nm outer zone width zone plate.

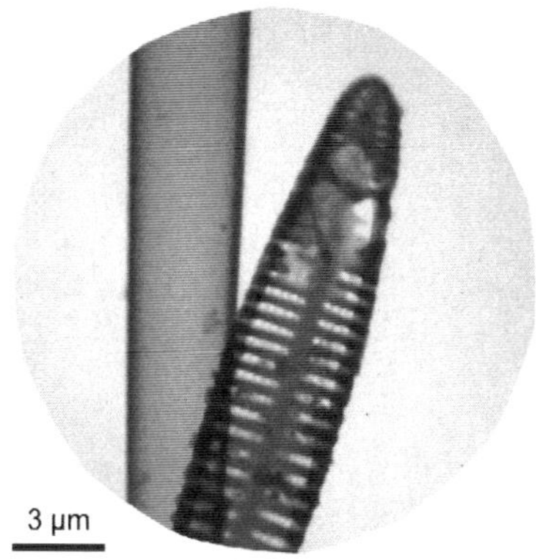

Figure 3. A diatom mounted on a glas capillary (capillary provided by S. C. Gleber and J. Sedlmair). The image was obtained in a 120 s exposure @ λ = 3.37 nm with a 30 nm outer zone width zone plate. The rotatable capillary holder enables stereo microscopy and tomography.

4. Outlook

Future upgrades of the microscope include a new laser (Diode-pumped Nd:YAG, 2 kHz, 150 mJ, <1 ns pulses) for plasma generation, a new multilayer mirror condenser for the λ = 2.48 nm operation mode and a cryo sample holder with a motorized goniometer. These upgrades will reduce the exposure time. This is important for thick wet samples, for dynamic studies of soils, for recoding tomographic data and for reducing problems with drift during exposures. Furthermore, the cryo sample stage will make it possible to record projection series of bio-samples without loss of detail due to radiation damage, which is important for high-resolution tomographic reconstructions.

References

[1] Berglund M, Rymell L, Peuker M, Wilhein T, Hertz H M 2000 *J. Microsc.* **197** 268
[2] Takman P, Stollberg H, Johansson G A, Holmberg A, Lindblom M, Hertz H M 2007 *J. Microsc.* **226** 175
[3] de Groot J, Hemberg O, Holmberg A, Hertz H M 2003 *Appl. Phys.* **94** 3717
[4] Jansson P A C, Vogt U, Hertz H M 2005 *Rev. Sci. Instrum.* **76** 043503
[5] Stollberg H, Yulin S, Takman P A C, Hertz H M 2006 *Rev. Sci. Instrum.* **77** 123101
[6] Holmberg A, Rehbein S, Hertz H M 2004 *Microel. Engin.* **73-74** 639
[7] Bertilson M, von Hofsten O, Lindblom M, Wilhein T, Hertz H M, Vogt U 2008 *Appl. Phys. Lett.* **92** 064104
[8] von Hofsten O, Bertilson M, Lindblom M, Holmberg A, Vogt U 2008 *Opt. Lett.* **33** 932
[9] Thieme J, Sedlmair J, Gleber S C, Bertilson M, von Hofsten O, Takman P, Hertz H 2008 *XRM conf. proc.*
[10] Gleber S C, Sedlmair J, Bertilson M, von Hofsten O, Heim S, Guttman P, Hertz H, Fischer P, Thieme J 2008 *XRM conf. proc.*
[11] Bertilson M, Takman P, Holmberg A, Vogt U, Hertz H M 2007 *Rev. Sci. Instrum.* **78** 026103

High resolution full-field imaging of nanostructures using compact extreme ultraviolet lasers

F Brizuela[1], C Brewer[1], S Fernandez[1], D Martz[1], M Marconi[1], W Chao[2], E H Anderson[2], A V Vinogradov[3], I A Artyukov[3], A G Ponomareko[4], V V Kondratenko[4], D T Attwood[2], K A Bertness[5], N A Sanford[5], J J Rocca[1] and C S Menoni[1]

[1] NSF Engineering Research Center for Extreme Ultraviolet Science & Technology and Department of Electrical and Computer Engineering, Colorado State University, USA
[2] NSF Engineering Research Center for Extreme Ultraviolet Science & Technology and Center for X-Ray Optics. Lawrence Berkeley National Laboratory, USA
[3] P. N. Lebedev Physical Institute, Russia
[4] National Technical University "KhPI", Ukraine
[5] National Institute of Standards and Technology, Boulder, USA

E-mail: brizuela@engr.colostate.edu

Abstract. Recent advances in the development of high peak brightness table-top extreme ultraviolet (EUV) and soft x-ray (SRX) lasers have opened new opportunities for the demonstration of compact full-field EUV/SXR microscopes capable of capturing images with short exposures down to a single laser shot. We demonstrate the practical application of table-top zone plate EUV microscopes that can image nanostructures with a spatial resolution of 54 nm and below and exposure times as short as 1.2 ns, the duration of a single laser shot.

The development of compact full-field extreme ultraviolet (EUV) and soft x-ray (SXR) microscopes relies on the availability of compact illumination sources with sufficient brightness in this region of the electromagnetic spectrum. Several table-top EUV/SXR sources have been successfully used to demonstrate full-field zone plate microscopy including plasma sources, high-harmonic generation (HHG) and lasers [1-3].

We have demonstrated wavelength-scalable high resolution full-field microscopes based on two table-top EUV lasers developed at Colorado State University [4-6]. One microscope uses a compact desk-top capillary discharge laser that emits at 46.9 nm wavelength [7]. The laser emission is created in a highly ionized Argon plasma column generated by fast electrical discharge excitation. The laser output consists of highly monochromatic ($\Delta\lambda/\lambda \sim 5\text{x}10^{-5}$) pulses with ~10 μJ of energy (2.4×10^{12} photons/ pulse) at a repetition rate of up to 12 Hz. The degree of spatial coherence of the illumination can be varied with the length of the plasma column that constitutes the gain medium. The 22 cm long capillary selected for these experiments produces optimized, partially coherent illumination that limits speckle effects that can degrade image quality, while still providing sufficient photon flux for single shot imaging. This system can operate in transmission and reflection modes rendering images with spatial resolution as high as 54 nm with exposure times ranging from ~ 20 sec down to a single laser shot, equivalent to ~1 ns [4, 5].

The second microscope is based on a 13.2 nm wavelength table-top laser. The laser pulses are generated by the amplification of spontaneous emission in a transient population inversion produced by electron impact excitation in a transition of nickel-like cadmium ions [8, 9]. A 4 mm long plasma is created by heating a 4 mm wide Cd slab target with a sequence of pulses from a chirped-pulse-amplification Ti:Sapphire laser system. Operating at typically 5 Hz repetition rate, the EUV laser produces highly monochromatic ($\Delta\lambda/\lambda < 1x10^{-4}$) pulses of 13.2 nm wavelength, 5 ps duration and ~200 nJ energy (~10^{12} photons/pulse) resulting in an average power of ~1 μW. This transmission microscope can obtain images with spatial resolution better than 38 nm with acquisition times of less than 20 seconds [6].

The optics in the microscopes is a combination of reflective and diffractive elements. The condenser in the 46.9 nm microscope is a Sc/Si multilayer coated Schwarzschild. The 13.2 nm wavelength microscope uses a Fresnel zone plate (FZP) condenser. The objectives in both systems are FZP with a typical efficiency of 5-10%. Specifically for the 46.9 nm microscope, a new freestanding zone plate technology that connects the circular zones with pseudo-random bridges was developed to reduce absorption and increase their efficiency [10]. The FZPs of the 46.9 nm microscope have outer zone widths of Δr=200 nm, 120 nm, 70 nm. The 13.2 nm microscope is equipped with zone plate objectives with Δr=80 nm and 50 nm. In both microscopes, the images are captured using a EUV sensitive back-illuminated CCD detector.

The spatial resolution of the microscopes was assessed by evaluating the intensity modulation of transmission gratings. EUV images of periodic gratings with half-periods down to 38 nm were analyzed to build the modulation transfer function (MTF) for every objective zone plate available. Figure 1 shows the MTF of the highest numerical aperture objective for each microscope along with grating images used to obtain the data. A spatial resolution of 54 nm was determined for the 46.9 nm wavelength microscope. With the 13.2 nm wavelength microscope we were able to clearly resolve the smallest patterns available on the sample with a half-period of 38 nm.

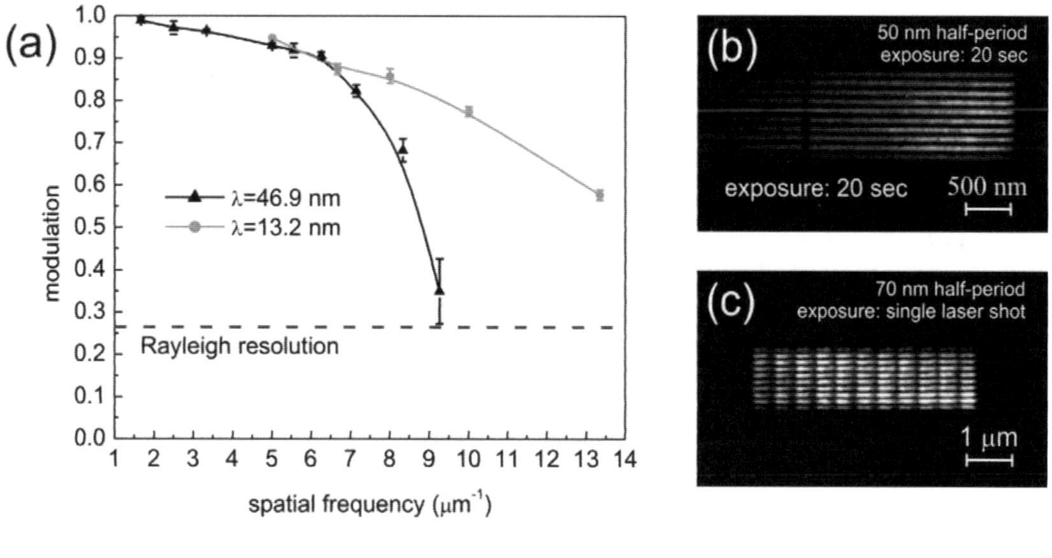

Figure 1. (a) Modulation Transfer Functions (MTFs) for the 46.9 nm microscope using the FZP objective with Δr=70 nm (triangles) and for the 13.2 nm microscope with the FZP objective with Δr=50 nm (circles). EUV images of gratings with a half-period of: (b) 50 nm and (c) 70 nm used to build the MTF shown in (a).

We have exploited the versatility of the 46.9 nm microscope to image a variety of nanostructures. Figure 2.a) is a reflection mode EUV image of a GaN nanowire connecting two Ti contacts grown on a Si wafer [11]. Figure 2 b) and c) are EUV transmission images of a diatom and a carbon nanotube, respectively that were deposited onto Si membranes. The EUV image of the diatom obtained with the Δr=120 nm FZP objective and an exposure of 5 seconds reveals a periodic pattern of 200 nm half-period. The 50 nm diameter carbon nanotube EUV image was captured with a single laser shot using the Δr=70 nm FZP. The flash imaging capability coupled with the ease in sample preparation make the microscope suited for tracing dynamics of such nanoscale systems.

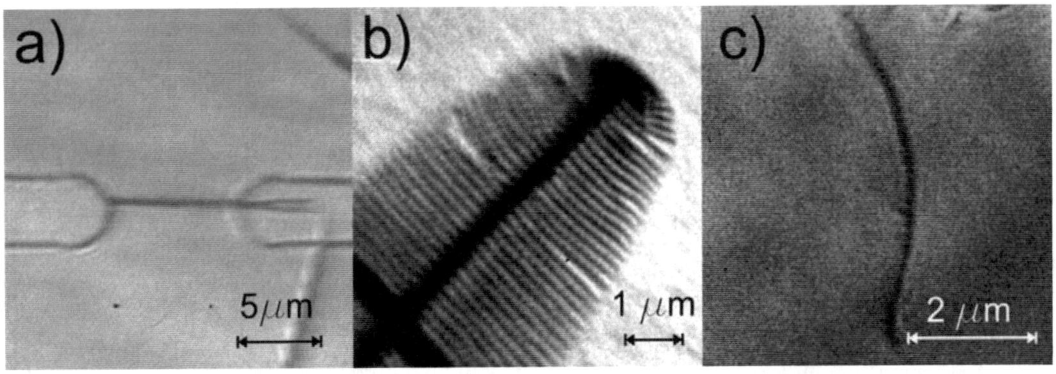

Figure 2. a) Reflection mode image of a GaN nanowire connecting two Ti contacts on a Si wafer. Transmission images of: b) a diatom and c) a carbon nanotube on a Si membrane. The latter was obtained with a single laser shot.

In summary we have demonstrated the practical use of EUV microscopes based on table-top EUV lasers to capture good quality images of nanoscale objects with exposures as short as a single laser shot. We anticipate that using recently developed shorter wavelength, high brightness, table-top lasers it will be possible for compact full-field microscopes to reach a spatial resolution approaching 15 nm with picosecond temporal resolution for a wide range of nanoscience and nanotechnology applications.

This work was supported by the Engineering Research Centers Program of the National Science Foundation under NSF Award Number EEC-0310717, and by the Cooperative Gran program of the U.S. Civilian Research and Development Foundation under Award Number 2845-MO-06. We acknowledge Prof. R. Solanki for the diatom samples.

References

[1] Takman P A C *et al.* 2007 *J. Microsc.* **226**(2) 175–81
[2] Weiland M *et al.* 2005 *Ultramicroscopy* **102** 93–100
[3] Chao W *et al.* 2005 *Nature* **435** 1210–13
[4] Brewer C A *et al.* 2008 *Opt. Lett.* **33**(5) 518–20
[5] Brizuela F *et al.* 2005 *Opt. Express* **13**(11) 3983–88
[6] Vaschenko G *et al.* 2006 *Opt. Lett.* **31**(9) 1214–16
[7] Heinbuch S *et al.* 2005 *Opt. Express* **13** 4050–55
[8] Rocca J J *et al.* 2005 *Opt. Lett.* **30**(19) 2581-83
[9] Wang Y *et al.* 2005 *Phys. Rev.* A **72** 053807
[10] Anderson E H 2006 *IEEE JQE* **42**(1) 27–35
[11] Tanner S M *et al.* 2007 *Appl. Phys. Lett.* **91** 203117

Laboratory x-ray micro imaging: Sources, optics, systems and applications

H M Hertz, M Bertilson, E Chubarova, J Ewald, S-C Gleber[+], O Hemberg, M Henriksson[*], O v Hofsten, A Holmberg, M Lindblom, E Mudry, M Otendal, J Reinspach, M Schlie, P Skoglund, P Takman, J Thieme[+], J Sedlmair[+], R Tjörnhammar, T Tuohimaa, M Vita[*], and U Vogt

Biomedical and X-Ray Physics, Department of Applied Physics,
Royal Inst. of Technol./Albanova, SE-106 91 Stockholm, Sweden

[*)]MTC, Karolinska Inst., SE-171 77 Stockholm, Sweden
[+)]Inst. for X-Ray Physics, Göttingen, Germany

E-mail: hertz@biox.kth.se

Abstract. We summarize the recent progress in laboratory-scale soft and hard x-ray micro imaging in Stockholm. Our soft x-ray work is based on liquid-jet laser-plasma sources which are combined with diffractive and multilayer optics to form laboratory x-ray microscopes. In the hard x-ray regime the imaging is based on a liquid-metal-jet electron-impact source which provides the necessary coherence to allow phase-contrast imaging with high fidelity.

1. Introduction

The Stockholm group is focused on laboratory-scale micro imaging in both the soft x-ray and hard x-ray regimes. The work includes sources, optics, imaging systems, and applications. In the soft x-ray regime the target area is laboratory microscopy. The microscopes described in section 2 are now approaching a maturity and reliability that allows real-world applications. Here thin samples in an aqueous environment have been imaged with high resolution and contrast. In the hard x-ray regime work is focused on a new small spot/high flux source and applications of the source. This microfocus source is based on a liquid-metal-jet-anode, which shows promise to significantly increase brightness compared to existing sources. It enables high-resolution imaging of thick samples with high resolution and adequate exposure times. A high spatial coherence allows significant contrast enhancement by phase imaging. The hard x-ray and soft x-ray projects share the goal to provide unique laboratory imaging solutions. Both rely on our optics fabrication capacity.

2. Soft x-ray sources, optics and microscopy

Here we briefly summarize our work in the soft x-ray regime. It is described in more detail in several contributions in this volume. We demonstrated the first sub-visible-resolution laser-plasma laboratory water-window x-ray microscope [1]. The present version operates at λ=2.48 nm or 3.37 nm and is based on a nitrogen or methanol liquid-jet laser-plasma source, respectively [2]. High-resolution 20 nm and 25-nm Ni zone plates are fabricated in-house, cf. figure 1 [3]. Here

Figure 1. 20 nm Ni zone plate

high process control in, e.g., the electroplating step has been found to be of outmost importance for producing high-efficiency narrow-line width zone plates with decent yield. Recent experimental and theoretical development also involves the fabrication and application of novel diffractive optical elements for DIC and Zernike phase-contrast microscopy [4]. The condenser is either a normal-incidence multilayer optic or diffractive optics.

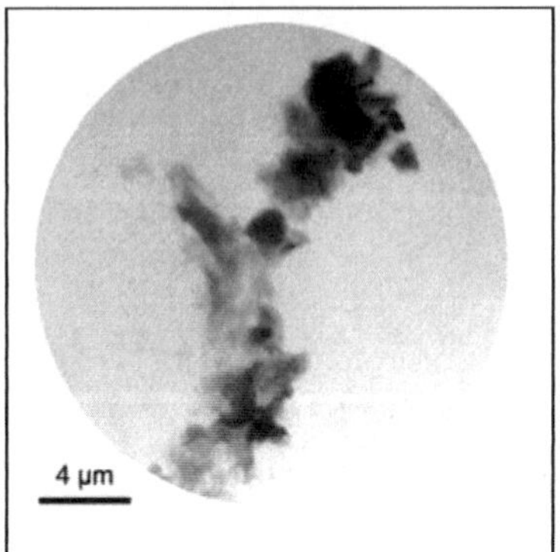

Figure 2. Example of a soil science sample imaged with the compact XRM.

The microscopes now have the reliability and user-friendliness that allows serious work on real-world applications. An early application is soil science, where the instrument demonstrates synchrotron-quality imaging and user-friendliness by taking hundred pictures per day [5]. Figure 2 shows an example. The same type of samples have been used to demonstrate stereo-imaging [6]. Finally, the first tomography on a laboratory water-window microscope has been demonstrated [7]. In the near future the microscope will be equipped with a cryo stage to allow biological imaging. Furthermore, a new high-power laser will result in significantly reduced exposure times, hopefully below 10 s for wet samples.

3. Hard x-ray sources and imaging

3.1. The liquid-jet-anode electron-impact x-ray source

We have developed a microfocus source based on a novel type of anode, the liquid-metal jet. [8]. This electron-impact source concept shows potential for a 100-1000x increase in brightness compared to present systems. The concept has been demonstrated with a tin-jet anode [8] (for 25 keV line emission), with a gallium-jet anode [9] (for 9 keV line emission) and also on non-metal jets such as methanol [10]. As an example, figure 3 shows the source emission with a tin anode. We typically operate with a 5-10 µm x-ray spot and up to a few 100 W of 50-100 keV electron beams.

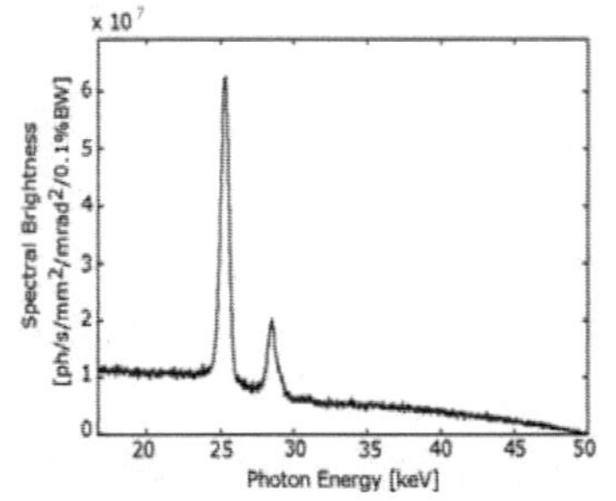

Figure 3. Source spectrum from electron-impact microfocus source based on a liquid-tin-jet anode.

Recent progress includes demonstration of long-term operation of a high-brightness microfocus source based on this concept. The tube uses a Ga/In/Sn alloy with strong line emission at 9.25 keV (comparable to conventional copper anode sources) as well as significant continuum radiation up 100 keV. The spot size is 5-20 µm with an electron-beam power up to 200 W and a power density of ~10 W/µm. This is >10x more than for present state-of-the-art microfocus tubes. This second-generation source incorporates a continuous recycling system for uninterrupted operation as well as reduced footprint and enhanced reliability. This source shows promise to enable many laboratory-scale x-ray applications, today limited by source brightness. Also, for novel imaging modalities, like phase-contrast imaging, source brightness is crucial. The system is now mature enough for scientific and industrial applications. We are therefore making the source commercially available [11].

3.2. High-resolution phase-contrast imaging

Phase-contrast methods increase contrast, detail and selectivity in medical and non-medical x-ray imaging. Such imaging requires a small source spot (to provide high resolution and sufficient spatial coherence) and a high flux (to give reasonable exposure times). To date phase-contrast imaging has therefore primarily been performed at high-brightness synchrotron radiation sources. However, the vast majority of x-ray imaging is performed with compact electron-impact sources in clinics or laboratories. Unfortunately, present compact x-ray sources do not provide the necessary spatial coherence with sufficient power to allow laboratory-scale high-resolution phase-contrast imaging with adequate exposure times.

The liquid-jet anode source is suitable for x-ray imaging, both in conventional absorption and with phase. In both cases the potential for small-spot/high flux operation (high brightness) is the key to provide high-resolution imaging with adequate exposure times. Figure 4 shows absorption imaging of a 20 lp/mm mammography test chart. The contrast is excellent. The observed contrast compares well to theoretical simulations. The source was operated with a tin anode at 50 keV with a 5 μm spot.

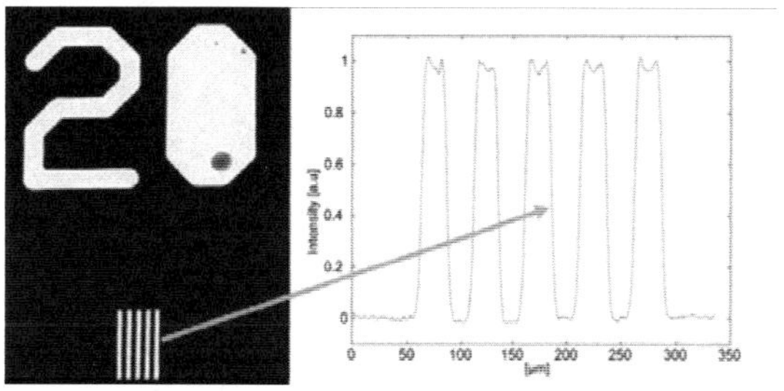

Figure 4. High-resolution absorption imaging of 20 lp/mm mammography test charts.

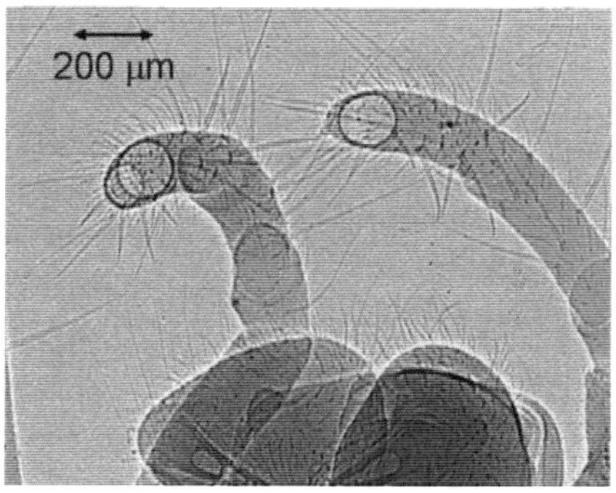

Figure 5. High-resolution phase-contrast image of a spider´s head and pedipalps.

The small-spot source provides sufficient coherence to allow high-quality high-resolution phase-contrast imaging [11]. Figure 5 shows a phase image of a spider. The sub-5 μm hair on the head and pedipalps are clearly visible. We note that these hair are not observable in absorption imaging. Also for the phase imaging we have excellent comparison between modeling and experiments. The source was operated with a tin anode at 50 keV with a 5 μm spot.

We are presently investigating the source's applicability for small-animal imaging experiments, especially for tumour detection. Here mice with defined human neuroblastoma tumours are imaged with phase-contrast. Preliminary results look encouraging. Finally, we have shown that the source can be scaled to high power, thereby paving the way for laboratory-scale high-contrast and high-resolution x-ray imaging with exposure times comparable with present rotating-anode-based systems.

4. Acknowledgements

The authors gratefully acknowledge the financial support from the Swedish Science Research Council, the Foundation for Strategic Research, the Swedish Agency for Innovation Systems, the EU 6th Framework, the Wallenberg Foundation, the Gustafsson Foundation and KTH-KI Life Science Imaging Consortium.

References

[1] M Berglund et. al., J. Microsc. **197**, 268 (2000); P Takman et. al, J. Microsc. **226**, 175 (2007).
[2] M Bertilson et. al., in this volume.
[3] A Holmberg et. al., in this volume.
[4] O v Hofsten et. al., in this volume.
[5] J Thieme et. al., in this volume.
[6] S-C Gleber et. al., in this volume.
[7] M Bertilson et. al., in preparation
[8] O Hemberg et. al., Appl. Phys. Lett. **83**, 1483 (2003).
[9] M Otendal et. al., Rev. Sci. Instr. **79**, 016102 (2008)
[10] T Tuohimaa et. al. Appl. Phys. Lett. **92**, 233509 (2008).
[11] www.excillum.com
[12] T Tuohimaa et al, Appl. Phys. Lett. **91**, 074104 (2007).

A Compact Soft X-Ray Microscope using an Electrode-less Z-Pinch Source

S F Horne, J Silterra, W Holber

Energetiq Technology, Inc; Woburn MA, 01801 USA

Email: shorne@energetiq.com

Abstract: Soft X-rays (< 1Kev) are of medical interest both for imaging and microdosimetry applications. X-ray sources at this low energy present a technological challenge. Synchrotrons, while very powerful and flexible, are enormously expensive national research facilities. Conventional X-ray sources based on electron bombardment can be compact and inexpensive, but low x-ray production efficiencies at low electron energies restrict this approach to very low power applications. Laser-based sources tend to be expensive and unreliable. Energetiq Technology, Inc. (Woburn, MA, USA) markets a 92 eV, 10W(2pi sr) electrode-less Z-pinch source developed for advanced semiconductor lithography. A modified version of this commercial product has produced 400 mW at 430 eV (2pi sr), appropriate for water window soft X-ray microscopy. The US NIH has funded Energetiq to design and construct a demonstration microscope using this source, coupled to a condenser optic, as the illumination system. The design of the condenser optic matches the unique characteristics of the source to the illumination requirements of the microscope, which is otherwise a conventional design. A separate program is underway to develop a microbeam system, in conjunction with the RARAF facility at Columbia University, NY, USA. The objective is to develop a focused, sub-micron beam capable of delivering > 1 Gy/second to the nucleus of a living cell. While most facilities of this type are coupled to a large and expensive particle accelerator, the Z-pinch X-ray source enables a compact, stand-alone design suitable to a small laboratory. The major technical issues in this system involve development of suitable focusing X-ray optics. Current status of these programs will be reported. (Supported by NIH grants 5R44RR022488-03 and 5R44RR023753-03)

1. Introduction:

Soft x-ray microscopy shows enormous promise for imaging cellular structures at resolutions beyond what can be achieved in optical microscopy, and with much simpler sample preparation than is required for electron microscopy[1]. In addition, the lower radiation dose required (compared to electron microscopy) allows tomographic investigation of subcellular structures in three dimensions. Existing synchrotron based microscopes have shown the potential of the soft x-ray microscope as a research tool, but they have the disadvantage of being tied to a massive light source at only a few national laboratories. A suitably bright, compact, low cost light source is the enabling innovation needed to realize a commercial biological soft x-ray microscope.

Energetiq has developed a unique light source based on an electrodeless, inductively coupled plasma Z-pinch. A commercial version optimized for EUV semiconductor lithography[2] has been modified to produce light in the soft x-ray range, and a program is under way to develop a demonstration microscope based on use of this source.

The overall design of the microscope is relatively conventional [3,4]. An illumination system composed of the source and condenser optics generates a hollow cone of light with a focus at the sample. An objective lens with short focal length produces a magnified image. As in other microscopes of this type, we use a zone plate objective, and match the illuminator numerical aperture to that of the zone plate. A cooled, back-thinned CCD camera detects the image.

The chromatic aberration inherent in a diffractive optic imposes spectral purity requirements on the illumination system. If the illumination is not nearly monochromatic, resolution will suffer. This requirement leads one to consider rather simple radiating systems. Helium-like nitrogen (N VI) is nearly ideal, with the $1S^2$-1S2P transition at about 430 eV.

2. Source Performance

The spectral performance of the source was measured using a graded multi-layer mirror as a spectrometer, as shown in Figure 1. The mirror is treated as a Bragg diffractor. The grazing angle is varied by rotating the mirror; given the grazing angle calculated to the brightest pixel in the CCD, the photon energy can be calculated. The spectrum produced in this manner is shown in Figure 2.

We plot versus photon energy. The peaks at 430 and 426 eV are the primary resonance lines of helium-like nitrogen. The widths of the peaks are entirely instrumental. We estimate the power at 426 ev to be down by a factor of roughly 10 from that at 430 eV; we expect negligible chromatic aberration at the image due to this component.

By rotating the mirror to throw the beam onto the X-ray diode, in-band power can be measured as well. We estimate the power radiated into 2π to be about 0.2 – 0.4 W, depending on source operating conditions.

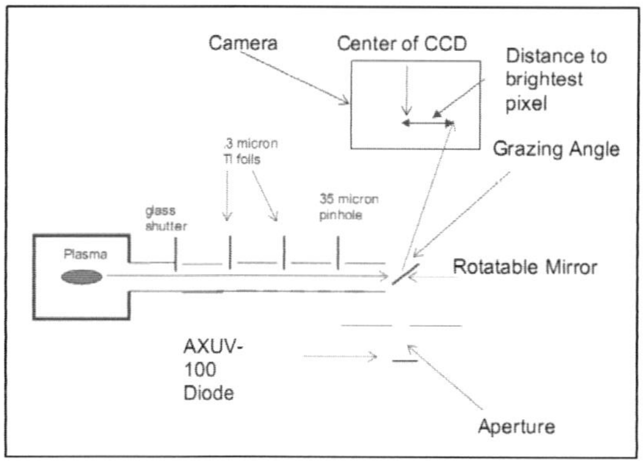

Figure 1: Spectral and Power Diagnostic

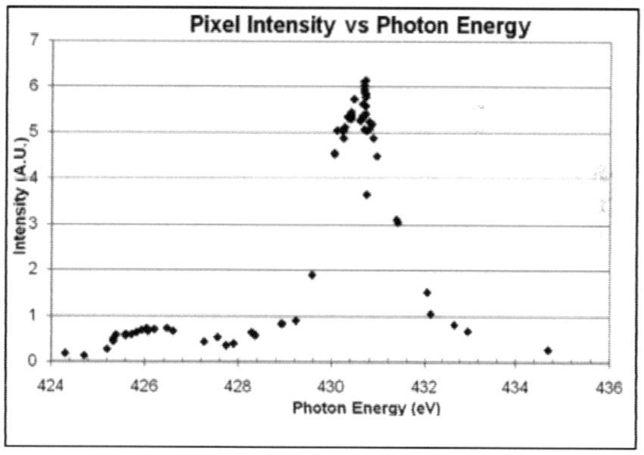

Figure 2: Measured Spectrum

3. Condenser Optic

The condenser optic is an elliptical grazing incidence mirror. The focal points are 1 M apart, and the output numerical aperture is matched to that of the objective zone plate. A copper-nickel alloy was selected both for its resistance to oxidation and corrosion, and for good machining properties. A surface finish of 4 nm RMS was achieved using conventional diamond turning. Testing of the condenser is still under way; figure 3 shows a scan with a 25 micron pinhole/diode across the focal spot. The design demagnification of the prototype is about 1 / 4 , which, with an 800 micron FWHM plasma should yield a spot size of 200 microns FWHM. In fact, the observed FWHM is about twice the ideal size. We believe this defect is due primarily to imperfections in both the figure and mounting geometry of the condenser, and will be corrected.

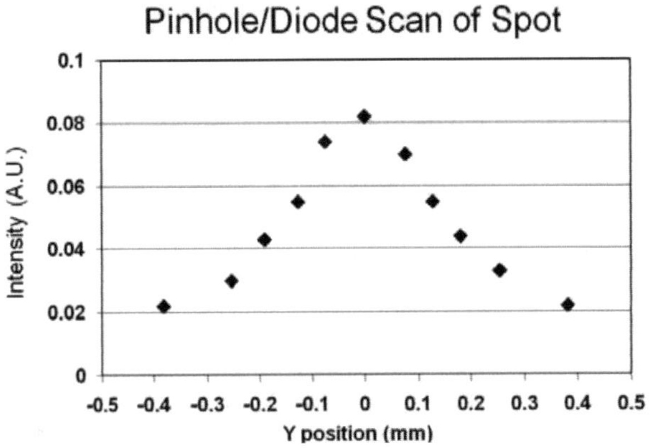

Figure 1: Profile of illumination field

4. Objective lens and camera beamline

The objective portion of the microscope was designed to use zone plates of 120 microns diameter, in conjunction with a standard 1K x 1K, 13 micron pixel size back-thinned CCD X-ray camera.
We have acquired zone plates with outer zone widths (dr) of 40, 30, and 25 nm; the condenser numerical aperture was designed to match the .057 NA of the 25 nm zone plate. We have simulated the optical behaviour of the hollow-cone illumination (simulations courtesy of Patrick Naulleau, LBNL). The basic issue is that for illumination within the NA of the objective, the sample is observed in transmission. However, if the NA of the objective is overfilled, photons are diffracted into the NA of the objective by the smaller sample features – increasing (theoretical) resolution, but at the cost of potentially introducing halos and artefacts into the image. The simulation results imply that we will have acceptable contrast, without significant diffraction-produced artefacts.

In parallel with the microscope program, we are also exploiting many of the same technologies to produce a microbeam system, in conjunction with Columbia University (RARAF). The goal is to produce a sub-micron spot capable of delivering ~ 1 Gy/second to a cell nucleus. Motivated by x-ray proximity printing technology, we plan to use an optic similar to the condenser described above, in combination with an ion-milled aperture. The optics are currently in design.

5. References

[1] Sayre D, Kirz J, Feder R , Kim DM, and Spiller E, *Science* **196** 1339 (1977).
[2] Blackborow P, Gustafson D, Smith D K, Besen, M, Horne S, D'Agostino R, Minami Y, Denbeaux, G, 2007 Application of the Energetiq EQ-10 electrodeless Z-Pinch EUV light source in outgassing and exposure of EUV photoresist *Emerging Lithographic Technologies XI.: Proc SPIE, (San Jose, USA Feb 25-Mar 2* Vol 6517 ed M J Lercel pp. 65171
[3] Attwood D 2000 *Soft X-rays and Extreme Ultraviolet Radiation:Principles and Applications* (Cambridge University Press)
[4] Kim, K W et al. 2006 *Phys. Med. Biol.* **51** N99

Journal of Physics: Conference Series **186** (2009) 012029 doi:10.1088/1742-6596/186/1/012029

Three-dimensional observation of biological specimens using a laser plasma soft x-ray microscope

M Hoshino[1, 2] **and S Aoki**[1]

[1]Graduate School of Pure and Applied Sciences, University of Tsukuba, 1-1-1 Ten-noudai, Tsukuba, Ibaraki 305-8573, Japan

E-mail: hoshino@spring8.or.jp

Abstract. Laser plasma soft x-ray microscope using Wolter mirror optics was applied to three-dimensional observation of a biological specimen. To obtain a high-resolution image with an x-ray charge-coupled device camera, a high-magnification imaging system with a relay mirror was used. To obtain the three-dimensional image of a biological specimen, imaging tomography was used. The three-dimensional image of a diatom was obtained with a spatial resolution of 250nm.

1. Introduction

A soft x-ray microscope is a powerful tool for the observation of hydrated biological cells [1]. By the recent progresses of a synchrotron light source and zone plate optics, a spatial resolution of the microscope better than 15nm has been obtained [2]. On the other hand, the development of a laboratory scale compact soft x-ray microscope has been also performed for the practical uses as an *in-situ* high-resolution microscope which has possibility to become a complementary microscope between electron microscopes and optical microscopes [3]. In our previous study, high-resolution observation of biological specimens using a laboratory-scale laser plasma soft x-ray microscope with a Wolter mirror was demonstrated [4]. In that case, however, a high-resolution photographic plate was used as a detector instead of an x-ray charge-coupled device (CCD). After that, a fundamental improvement of a system so that a high-resolution and magnification x-ray image could be obtained with a CCD camera was carried out by using a relay mirror system [5]. Accordingly, high-resolution imaging with a CCD camera enabled us to perform three-dimensional (3D) imaging using x-ray tomography. In this paper, a result of 3D imaging of a biological test sample using a laser plasma soft x-ray microscope with Wolter mirror optics and its discussion are shown.

2. Optical system for soft x-ray imaging tomography

A schematic diagram of an optical setup for soft x-ray imaging tomography is shown in figure 1. Quasi-monochromatic x-rays of 3.2nm in wavelength (water window x-rays) were generated by laser produced plasma. To obtain a high-resolution and magnification image, a relay mirror system with an inclined optical axis (-155mrad) was used. The details of the microscope such as mirror parameters are shown in reference 4 and 5. Due to the inclined optical axis of a relay mirror, the image distortion was generated at the off-axis region of a final image. Fortunately, this image distortion could be

[2] *Present address*: Japan Synchrotron Radiation Research Institute (JASRI/SPring-8), 1-1-1 Kouto, Sayo-cho, Sayo-gun, Hyogo 679-5198, Japan

analyzed and corrected by using a geometrical ray-trace calculation [6]. The x-ray images of a copper mesh edge are shown in figure 2(a). A left side image represents a raw image (before correction) and a right side one is a corrected image. The prominent distortion of the image was effectively corrected. The edge-response of a corrected image and a line spread function (LSF) is shown in figure 2(b) and 2(c), respectively. The full-width half maximum (FWHM) of a LSF was estimated to be 90nm. Therefore, x-ray imaging with a spatial resolution better than 100nm is possible even though a digital image correction is performed.

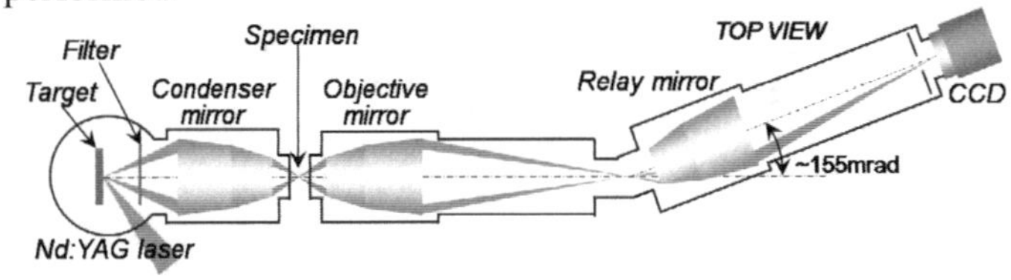

Figure 1. Schematic diagram of a laser plasma soft x-ray microscope using Wolter mirror optics. X-ray wavelength: 3.2nm quasi-monochromatic x-rays.

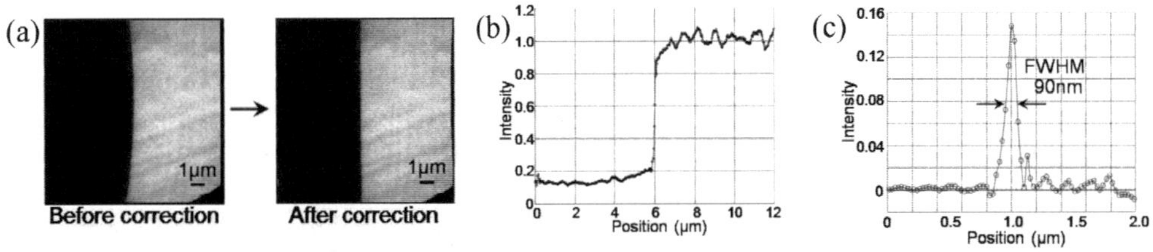

Figure 2. (a) X-ray images of a copper mesh (#400) edge before correction (left) and after correction (right). (b) Line intensity profile obtained from a corrected image. (c) Line spread function obtained from the edge-response.

3. Laser plasma soft x-ray imaging tomography

In the imaging microscope using an objective lens, the 3D image based on x-ray tomography can be obtained as well as projection type x-ray tomography. In the case of imaging tomography, the x-ray image in focus can be regarded as an approximate projection image and a conventional reconstruction method can be applied [7]. In the theoretical resolution imaging, the depth of focus (DOF) of an objective Wolter mirror is calculated by using a following equation,

$$|\Delta z| \leq \frac{1}{2(1-\varepsilon^2)} \cdot \frac{\lambda}{NA^2} \tag{1}$$

where ε is the annular ratio and NA is the numerical aperture. This equation can be derived from the diffraction integral [8]. From this equation, the theoretical DOF of an objective mirror (ε: 0.71, NA: 0.2) used in the microscope was estimated to be $|\Delta z| \leq 83$nm (λ: 3.2nm). In the actual imaging system, however, the apparent DOF becomes large because the spatial resolution obtained is lower than a theoretical value. Since a spatial resolution of the objective Wolter mirror is represented as $\delta = 0.44\lambda/NA$, the apparent NA when a spatial resolution is 90nm (see figure 2(c)) is estimated to be 0.0156. Then, the apparent DOF is estimated to be $|\Delta z| \leq 13\mu$m (λ: 3.2nm) from equation (1). This implies that 3D imaging of a biological specimen whose size is several microns is possible with this spatial resolution.

In this imaging tomography system, samples were set at the tip of a tapered glass capillary to avoid the excess absorption of soft x-rays by the glass capillary. A convolution back-projection method with a Shepp-Logan filter was used for the reconstruction. As a test sample, a fragment of a diatom which produced relatively high image contrast compared with typical biological cells was observed. 50 projection images were obtained through 180° rotation. A reconstructed 3D image of a diatom is shown in figure 3(a). A vessel like structure which is called as *frustule* can be observed. A vertical sectional image obtained from a reconstructed dataset is shown in figure 3(b). The intensity profiles at line A and B are shown in figure 3(c) and 3(d), respectively. From the edge-response of a sectional image of a glass capillary, a spatial resolution of a tomogram was estimated to be 250nm. In this case, a vertical axis of the profiles represents a reconstructed value which corresponds to a linear absorption coefficient (LAC) of a sample. From the experimental value, a LAC of a diatom shell is approximately half of that of a glass capillary whereas a main component of a diatom shell is SiO_2. Therefore, the density of a diatom shell is not as high as that of a glass capillary. On the other hand, a theoretical LAC of a glass capillary (Pyrex) is estimated to be 2.2μm^{-1}(λ: 3.2nm). In the experimental value, however, only one third of a theoretical one is obtained. This may be due to the increase of a background level at the image plane, which is uniformly overlaid with scattering x-ray generated by figure errors on Wolter mirror surfaces [9]. Then, the apparent high transmission was obtained even if it was an almost opaque object. The increase of a background level can be also observed in the edge-response shown in figure 2(b). Although 3.2nm soft x-ray transmission for copper mesh was equal to zero, the apparent transmission could be seen. For the quantitative analysis of the density of a sample, the influence of the background scattering must be considered.

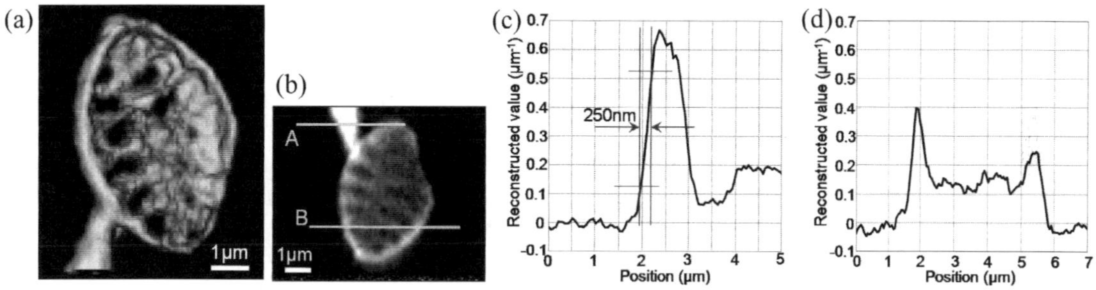

Figure 3. (a) 3D image of a fragment of a diatom reconstructed from 50 projection images. Exposure: laser 40shots per projection. 3D image was developed by MicroAVS. (b) Vertical sectional image. (c) Intensity profile at line A. (d) Intensity profile at line B.

4. Conclusion

A laser plasma soft x-ray microscope using Wolter mirror optics was applied to 3D imaging of a biological specimen. Although a diatom used as a test sample was naturally dried condition, the 3D image based on imaging tomography could be obtained. In the future, a laboratory-scale cryo-microscope and microtomography will be possible. The total acquisition time to obtain all the projection images (50 projection images) was less than 20min. By using a high-intense laser for the excitation of plasma, single pulse exposure imaging will be possible. Then, acquisition time to obtain all the projection images will be drastically shortened, which is within a few minutes or less, as well as the improvement of a spatial resolution resulted from the vibration free imaging.

References

[1] Kirz J, Jacobsen C and Howells M 1995 *Q. Rev. BioPhys.* **28** 33
[2] Chao W, Harteneck B D, Liddle A J, Anderson E H and Attwood D T 2005 *Nature* **435** 1210
[3] Takman P A C, Stollberg H, Johansson G A, Holmberg A, Lindblom M and Hertz H M 2007 *J. Microscopy* **226** 175
[4] Hoshino M and Aoki S 2006 *Jpn. J. Appl. Phys.* **45** 989
[5] Hoshino M and Aoki S 2006 *Jpn. J. Appl. Phys.* **45** L654
[6] Hoshino M and Aoki S 2008 *Appl. Phys. Express* **1** 067005
[7] Weiβ D, Schneider G, Niemann B, Guttmann P, Rudolph D and Schmahl G 2000 *Ultramicroscopy* **84** 158
[8] Born M and Wolf E 1997 *Principles of Optics 6th ed.* (Cambridge: Cambridge University press) p 437
[9] Sugisaki K, Takahashi S, Yoshidomi Y, Shinada K, Mitomi O, Uchishiba E, Hamada R, Kato T and Aoki S 1998 *Appl. Opt.* **37** 8057.

EUV microscopy for defect inspection by dark-field mapping and zone plate zooming

Larissa Juschkin[1,2], Ralf Freiberger[1,2] and Klaus Bergmann[3]

[1]RWTH Aachen University, Chair for Technology of Optical Systems, Germany
[2]JARA – Fundamentals of Future Information Technology
[3]Fraunhofer Institute for Laser Technology, Aachen, Germany

larissa.juschkin@ilt.fhg.de, ralf.freiberger@ilt.fhg.de, klaus.bergmann@ilt.fhg.de

Abstract. A laboratory scale EUV microscope is presented to be used, e.g., in future inspection of EUV masks and mask blanks in extreme ultraviolet lithography. The system can be operated in bright and dark field mode. For defect inspection purpose the dark field mode is preferred, with increased contrast and sensitivity of the system to small structures. The characteristics of the used Schwarzschild objective as imaging component such as large object field and moderate magnification become advantageous for high process speeds, whereas the detector pixel size (13 μm) does not give the spatial resolution in principle possible with the imaging optics. The presence of a defect causes a spot (with otherwise dark background) to appear on the detector. As necessary it can be zoomed in with the help of a second magnification step. For this purpose we suggest to employ a zone plate adapted to the system. The method's feasibility is demonstrated by means of experiments on test structures and the apparatus is characterized with regard to design parameters for commercial systems.

1. Introduction

Microscopes using extreme ultraviolet light are capable of detecting all types of defects on surfaces on a length scale down to several tens of nanometers. Many applications, such as the mapping of mask blanks for EUV lithography, require an ability to scan large surfaces for the presence of small printable defects as rapidly as possible favorably by scattered-light measurements in the EUV region.

Since the 90's mostly works were published on microscopy with extreme ultraviolet and soft x-ray radiation, which use synchrotron sources and Schwarzschild objectives (SSO) [1] - [5]. The achieved spatial resolutions with photon energies of approximately 100 eV amounted to < 100 nm. With the availability of wavelength-adapted EUV multilayer coatings and compact, efficient sources (gas discharge based or laser produced) for this spectral region the EUV technology is used for setting up EUV laboratory microscopes with reflecting optics [6] - [8].

For the demonstration of the application potential of laboratory scale devices a first synchrotron-independent EUV transmission microscope was developed at the Fraunhofer Institute for Laser Technology (ILT) together with the Fraunhofer IOF and IWS in Jena and Dresden, Germany. The concept is based on the combination of a xenon based gas discharge EUV source of AIXUV GmbH with a multilayer mirror based optical system and a CCD detector (see figure 1).

The pulsed EUV radiation (pulse duration ~ 100 ns, repetition rate 50 Hz) is focused in the vacuum onto the sample using a grazing incidence collector with a highly reflecting inner coating of Mo/C manufactured by IWS. The collector is a ring ellipsoid with an object-side numeric aperture of $NA_{coll,obj} = 0.16$ and a source-side numeric aperture of $NA_{coll,src} = 0.04$ and has focal lengths of $f_{coll,src}$ = 885 mm and $f_{coll,obj}$ = 200 mm. For imaging a Schwarzschild objective (manufacturer IOF) is used which consists of two spherical multilayer mirrors (60 bilayers Mo/Si) with a spectral bandwidth of about 1 nm centered at 13.5 nm. The SSO has a magnification factor of 21.34, an object-side numeric aperture of $NA_{SSO,obj} = 0.22$ and a focal length of $f_{SSO,eff}$ = 26.95 mm. The theoretical resolution of the SSO is determined by the optical setup and the mirror quality to be about 70 – 100 nm. As detector a back illuminated CCD camera is used, which limits the resolution of the microscope by its pixel size of 13 μm to approximately 600 nm. The shift into the sub-100 nm range is possible by employing a detector with smaller pixels or using a second magnification step, as described below.

2. Defect inspection by dark-field mapping

The EUV microscope is adapted for the operation in the dark field mode by using a zero order beam stop (figure 1). This beam stop in front of the objective blocks the direct light in the optical path and ensures that only the light scattered by surface defects (particles on a thin foil) is collected and imaged. This increases the contrast and the sensitivity of the system to small defects. Note the differences in the numerical apertures of the illumination and observation beam paths which enable our technical realization of this scheme ($NA_{coll,obj} = 0.16$ smaller as $NA_{SSO,obj} = 0.22$).

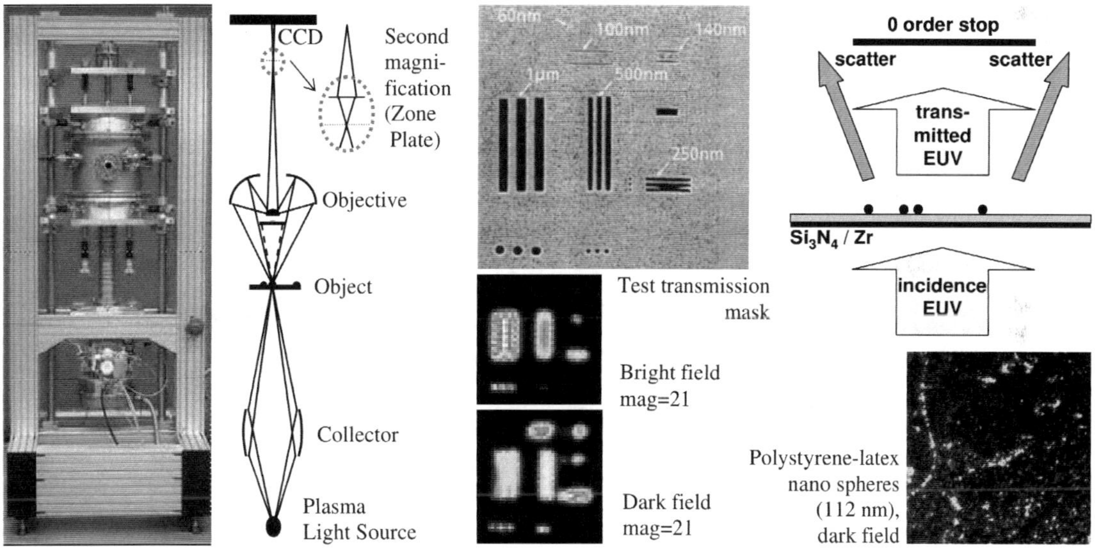

Figure 1. EUV microscope and examples of CCD-images.

The measurements of different test-structures are also shown in figure 1: right-bottom shows the dark field image of a cluster of polystyrene latex spheres with a diameter of 112 nm on a 150 nm thick silicon nitride foil, middle-bottom the direct comparison of bright and dark field measurements of a structured transmission mask (FIB-image at the middle-top for comparison). Small structures below 250 nm exhibit clearly better contrast with EUV dark field images compared with 1 μm and 500 nm structures. The smallest measurable structure size is limited with dark field measurements by the solid angle of the objective ($NA_{SSO,obj} = 0.22$) and the wavelength to approximately 60 nm. We estimated the required dose for detection of the latex spheres at the object level to be about 0.1 mJ / cm^2, which corresponds to only few EUV pulses of the source with the present setup. Hence the characteristics of Schwarzschild objectives such as large object field and moderate magnification become advantageous for high process speeds. The scan speed is then only limited by the read out time and size of the CCD. For applications such as mask blank or mask inspection the results suggest achievable scan speed of up to 3 cm^2 / hour, which is close to the specification of the semiconductor industry.

3. Defect inspection by zone plate zooming

Here the possibility of a zooming procedure in the EUV is examined for resolved imaging and localization of small defects and/or structures. This is realized by using a zone plate adapted to the system as a second magnification step, which requires a magnification of about 10 to 20 and a resolution of only ~1 μm. We designed such a zone plate and realized this magnification principle. The measurements have been accomplished in bright field transmission mode of the microscope.

The NA_{ZP} of the ZP is aligned to the CCD-side NA of the SSO ($NA_{SSO,det} = 0.01$). Already by limiting the number of zones to 30, which also matches the bandwidth of our SSO well, it is possible to get an adequate $NA_{ZP} = 0.011$. The use of a ZP with a focal length of 3.2 mm at a distance of about 4 cm before the CCD with proper sample refocus enables an increase of the EUV microscope's magnification by approximately factor 10 to an overall system magnification of 210. The spatial resolution also rises, so that structures can be resolved down to 100 nm.

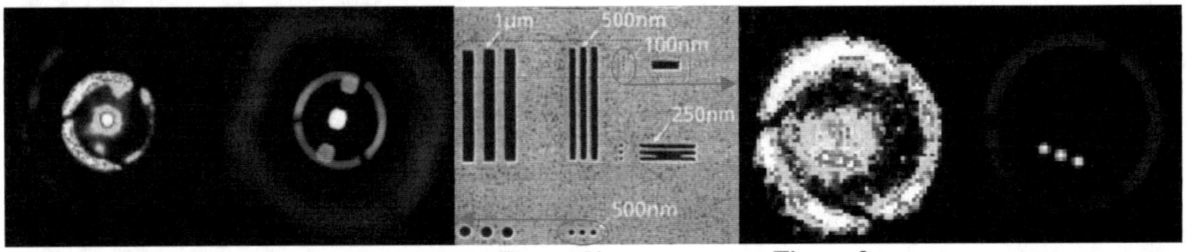

Figure 2a.
CCD-image and simulation. Three points Ø = 500 nm, dist = 500 nm.

Figure 2b.
Transmission-mask (FIB-image).

Figure 2c.
CCD-image and simulation. Three points Ø = 100 nm, dist = 100 nm.

Figure 2a-c shows a comparison between CCD-images and simulations by Zemax®. A ZP is a diffractive optical element, so that higher orders of the diffractive light will appear. With an order selecting aperture light of an unwanted order can be faded out. An additional central stop in the middle of the ZP blocks light from the zero order. In our experimental setup a bright annulus of light that derives from the zero order of the illumination beam path remains. In the center of the annulus only light of the (focused) first order appears. In the CCD-images and the simulations displayed above, three circular points can be seen. At the object-plane (a transmission mask) the distance of the three points on the left CCD-Image is 500 nm, and on the right CCD-Image there are other three points with distance of 100 nm, which are clearly resolved. The magnification with this setup is 224.

4. Summary and outlook

The dark field images exhibit high sensitivity to structures smaller than 250 nm although the resolution of the microscope is limited by the CCD pixel size and magnification of the Schwarzschild objective to approximately 600 nm. An optional second magnification step behind the SSO by means of a zone plate opens the possibility to scan a region on the object very quickly and if necessary to zoom in on the region afterwards by inserting a ZP into the beam path and refocusing sample or detector. This concept is of particular interest for applications where a preferably high throughput is required such as mask blank defect inspection for EUV lithography.

References

[1] Bjorkholm J E, Bokor J, Eichner L, et.al. 1990 *J. Vac. Sci. Technol.* B **8** 6
[2] DiCicco D S, Kim D, Rosser R, Suckewer S 1992 *Optics Letters* 17
[3] Hoover R B, Baker P C, Shealy D, et.al. 1992 *Proc. SPIE* 1742, pp 660-673
[4] Artioukov I A, Vinogradov A V, Asadchikov V E, et.al. 1995 *Optics Letters* Vol. 20 (24)
[5] Toyoda M, Shitani Y, Yanagihara M, et.al. 2000 *Jpn. J. Appl. Phys.* 39, pp 1926-1929
[6] Kinoshita H, Hamamoto K, Sakaya N, et.al. 2007 *Jpn. J. Appl. Phys.* 46, pp 6113-6117
[7] Barty A, Taylor J S, Hudyma R, et.al. 2002 *Proc. SPIE* 1889, 1073
[8] Brunton A, et.al. 2005 *SPIE* 5751-06

Development of a soft x-ray plasma camera with a Fresnel zone plate to image laser produced plasmas

M. Kado, M. Mori, M. Nishiuchi, M. Ishino, and T. Kawachi

8-1, Umemidai, Kizugawa, Kyoto, 619-0215, Japan

E-mail: kado.masataka@jaea.go.jp

Abstract. A soft x-ray plasma camera operated at 3.35nm in the water window x-ray region is developed and demonstrated imaging gas jet plasmas of several spices produced with a 10TW Ti: sapphire laser. The plasma camera consists of a 300nm thick $Ag/Ti/Si_3N_4$ x-ray band pass filter with bandwidth of 1.43nm to cut visible light and also to reduce colour aberration of the Fresnel zone plate, a Fresnel zone plate with diameter of 1mm and outermost zone width of 300nm, and a soft x-ray CCD camera. The magnification of the plasma camera is 10. The soft x-ray plasma camera powered by a Fresnel zone plate is a very powerful tool to observe laser produced plasmas since it is 1000 times brighter and has 5 times higher spatial resolution comparing ordinary x-ray pinhole camera. The soft x-ray images of helium, nitrogen, argon, krypton, and xenon gas jet plasmas are obtained changing gas pressure from 0.01MPa to 1MPa.

1. Introduction

Soft x-ray microscopes using x-ray wavelength from 2.3 to 4.4nm, which is so called "water window", are very powerful tools to image wet biological specimens with high spatial resolution of nm level without any artifacts such as drying, fixing, and staining. [1][2][3]. It will make possible to image a flash photograph at a certain moment of moving biological specimens in natural environment [4] and will help to solve functional role of the specimens.

In order to develop x-ray microscopes it is very important to produce bright and uniform soft x-ray source in water window region and also to examine the x-ray source in detail. X-ray pinhole camera is the general diagnostic to image laser produced plasmas [5]. The spatial resolution and the acceptance angle of the x-ray pinhole camera are both limited to the size of the pinhole. In order to obtain higher spatial resolution one the acceptance angle has to be sacrificed.

Fresnel zone plates have long been used and are very powerful tools to image laser-produced plasma x-ray sources with high spatial resolution [6][7]. We have proposed to use a Fresnel zone plate replacing the pinhole and named it as a soft x-ray plasma camera. The advantages of the soft x-ray plasma camera comparing the ordinary x-ray pinhole camera are its large acceptance angle and the high spatial resolution. Both of large acceptance angle and higher spatial resolution are achievable at same time. The disadvantage of using a Fresnel zone plate is that the chromatic aberration of the Fresnel zone plate degrades the spatial resolution. In order to reduce the degradation of spatial resolution a band path filter to select x-rays in water window region is installed and reduces the spectral bandwidth of the x-ray source.

2. Configuration of the soft x-ray plasma camera

Journal of Physics: Conference Series **186** (2009) 012031 doi:10.1088/1742-6596/186/1/012031

The soft x-ray plasma camera is consisted of a Fresnel zone plate, a band path filter, and a back illuminated soft x-ray CCD camera. The magnification of the plasma camera is 10. The focal length and the diameter of the Fresnel zone plate are 90.9mm and 1mm and the outermost zone width is about 300nm. The band path filter is composed of three thin layers of titanium, silver, and silicon nitride and thicknesses of each layer are 100nm. The central wavelength of the band path filter is 3.35nm and bandwidth is 1.43nm in FWHM. The transmission of the filter is about 28% at the peak. As the x-ray detector a back illuminated soft x-ray CCD camera is used and the pixel number and the size are 2048x2048 and 13.5μm, respectively.

Comparing to the ordinary x-ray pinhole camera with a 10μm pinhole the diameter of the Fresnel zone plate is 100 times larger and that gives 10000 times higher sensitivity. The efficiency of the Fresnel zone plate is about 10% although the efficiency of the pinhole is usually 100%. Total brightness of the soft x-ray plasma camera is 1000 times higher than that of the x-ray pinhole camera and still the spatial resolution of the plasma camera is 5 times better than that of the x-ray pinhole camera. In order to obtain the same spatial resolution by the soft x-ray plasma camera the brightness is reduced to 4%.

3. Characterization of laser plasma x-ray source

The soft x-ray plasma camera is used to characterize water window x-ray plasma sources produced with pulsed laser irradiation onto gas jet plasmas. Several kinds of gas jets such as argon, krypton, nitrogen, and helium are irradiated with 10TW Ti:Sapphire laser and the images of the gas jet plasmas are obtained with the soft x-ray plasma camera. The laser energy and pulse duration on target are 165mJ and 36fs, respectively. The gas pressure is varied from 0.1 to 0.73MPa. Shown in figure 1 is the typical x-ray image of the argon gas jet plasma. There is a large and dim x-ray emission covering

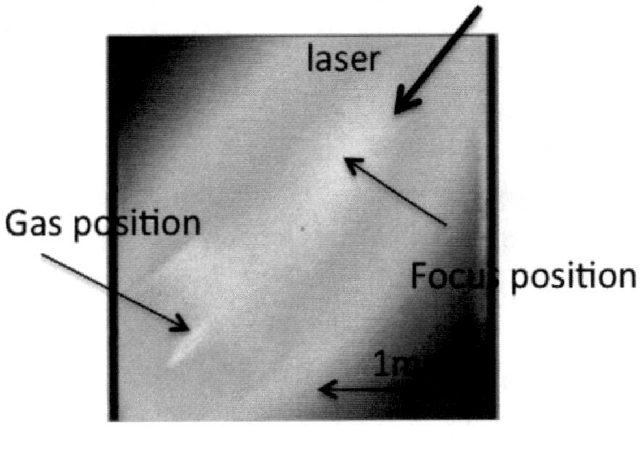

Figure 1. Typical x-ray image of the argon gas jet plasma is shown. The laser propagates from top and right to down and left. The gas jet is placed from 1.5mm away from laser focal spot.

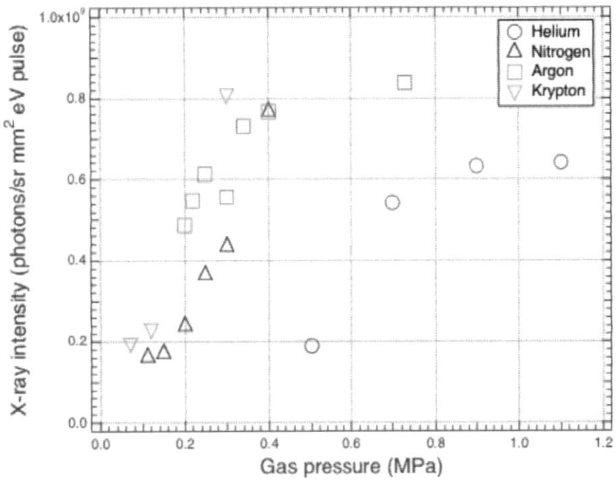

Figure 2. Gas pressure dependence of the x-ray intensities of helium, nitrogen, argon, and krypton gas jet plasmas is shown. X-ray intensity increases as the atomic number of gas species increases and also gas pressure increases.

from laser focal position to the gas jet position and there is also a sharp strong emission at the gas jet position.

The separation of laser focal spot and gas jet position is varied moving gas jet from 0.5 to 2mm and the position of the sharp strong x-ray emissions moves corresponding to the position of gas jet. It is apparent that the sharp strong x-ray emissions are from gas jet plasmas. The size of the gas jet plasma x-ray source is estimated and the lateral size of the gas jet plasma x-ray source is about 80μm and longitudinal size is about 300μm. X-ray intensity is also estimated taking account of the transmission efficiency of the band path filter, diffraction efficiency and the solid angle of the Fresnel zone plate, and quantum efficiency of the soft x-ray CCD camera. The x-ray intensity of the argon gas jet plasma with gas pressure of 0.73MPa is 8.37×10^{8} (photons/sr mm^2 eV pulse) and is converted to 3.00×10^{-6}J in the water window x-ray region. The conversion efficiency from incident laser energy to the water window x-ray region is calculated to be about 0.0022%.

The x-ray intensities of the gas jet plasma x-ray sources are obtained from various species of gas jet such as argon, krypton, nitrogen, and helium and gas pressure. Shown in figure 2 is the gas pressure dependence of the x-ray intensity of various gas jet plasmas.

4. Summary

In summary a soft x-ray plasma camera with a Fresnel zone plate to achieve high spatial resolution and high sensitivity is developed and used to characterize gas water window x-ray source using gas jet plasmas. The soft x-ray plasma camera is expected 1000 times higher sensitivity and 5 times higher spatial resolution. The soft x-ray plasma camera is operated at the wavelength of 3.35nm and the magnification is 10. The several species of gas jet plasmas such as helium, nitrogen, argon, and krypton are irradiated with a Ti:Sapphire pulsed laser with 136mJ in 36fs pulse duration.

Dependences of the x-ray intensities of the gas jet plasmas on gas species and gas pressure are obtained. X-ray intensity increases as the atomic number of gas species increases and also gas pressure increases. The x-ray energy of the argon gas jet plasma with gas pressure of 0.73MPa is 3.00×10^{-6}J in the water window x-ray region and the conversion efficiency from incident laser energy to the water window x-ray region is calculated to be about 0.0022%.

Acknowledgment

This research was partially supported by the Ministry of Education, Science, Sports and Culture, Grant-in-Aid for Scientific Research (C), 18540398, 2006. The Fresnel zone plate is fabricated and supplied by Applied Nanotools, Inc. in Canada.

References

[1] M. Kado, M. C. Richardson, J. M. Rajyaguru, M. J. Muszynski, Y. Yamamoto, Proc. of Experimental Biology and Medicine **220**, 21, 1999.
[2] J. M. Rajyaguru, M. Kado, M. C. Richardson, and M. J. Muszynski, Biophysical Journal Vol. **72**, pp. 1521-1526, 1997.
[3] J. M. Rajyaguru, M. Kado, K. Nekula, M. C. Richardson, and M. J. Muszynski, Microbiology **143**, pp. 733-738, 1997.
[4] M. Kado, H. Daido, Y. Yamamoto, K. Shinohara, and M. C. Richardson, Proc. of 8th Int. Conf. X-ray Microscopy, pp. 41-43, 2006.
[5] M. Nakai, J. Plasma Fusion Res., **81** (81-92), 2005.
[6] H.G. Ahlstrom, Physics of Laser Fusion II, LLNL UCRL-53106 report, 1982.
[7] H. Azechi and Y. Tamari, J. Plasma Fusion Res., **79** (398-401), 2003.

Near-Edge X-Ray Absorption Fine Structure Measurements Using a Laser Plasma XUV Source

C. Peth[1], F. Barkusky[1], J. Sedlmair[2], S.-C. Gleber[2], E. Novakova[2], J. Niemeyer[3], J. Thieme[2], T. Salditt[2], K. Mann[1]

[1] Laser-Laboratorium Goettingen, Hans-Adolf-Krebs-Weg 1, D-37077 Goettingen, Germany

[2] Institute for X-Ray Physics, Georg-August-University, Friedrich-Hund-Platz 1, D-37077 Goettingen, Germany

[3]Institut für Angewandte Biotechnolgie, Marie-Curie-Strasse 7, 37079 Goettingen, Germany

kmann@llg-ev.de

Abstract. We present a compact setup for near-edge x-ray absorption spectroscopy at the carbon K-edge based on a laser-driven plasma source. Thin polymer films were investigated, showing good agreement with corresponding synchrotron data. Furthermore we have examined the carbon near-edge structure of phospholipids and fulvic acids, providing detailed information on intermolecular binding states.

1. Introduction

The progress in development of laboratory-scale soft x-ray sources in recent years has enabled experimental techniques that could be performed before almost exclusively at synchrotron sources. Table-top soft x-ray sources of high brilliance are now used for various applications, e.g. x-ray microscopy [1], lensless diffractive imaging [2], photoelectron spectroscopy [3] or absorption spectroscopy [4]. The latter, among other techniques, accomplishes the investigation of the near edge x-ray absorption fine structure (NEXAFS).

NEXAFS is a well established method for elemental and compositional analysis of a sample yielding also surface sensitive information [5]. In particular, NEXAFS is used to study the structure of intermolecular bonds of polymers by probing the electronic transition from the core level to unoccupied states. Since each element has a characteristic core binding energy NEXAFS spectra contain element specific information. The analysis of these unique spectroscopic fingerprints allows the identification and distinction of different polymers [6].

In this paper we present NEXAFS measurements that were obtained by using a flexible laboratory scale setup based on a laser-driven plasma source. The table-top setup can be used for NEXAFS experiments in transmission as well as reflection under grazing incidence conditions. For transmission measurements thin films have to be used due to the low penetration depth of soft x-rays into matter. In contrast, NEXAFS experiments performed in reflection mode (ReflEXAFS) offer the advantage that thin film preparation is not necessary. Moreover, the surface sensitivity is strongly increased.

2. Experimental Setup

The experimental setup for the generation and characterization of soft x-ray radiation emitted from laser plasmas in the "water window" is shown schematically in Figure 1. A Nd:YAG laser beam (Innolas, 1064 nm, 1 Hz, 800 mJ, 7 ns) is focused into a pulsed gas puff target centred in a vacuum chamber, as described in detail elsewhere [7]. The laser focus has a diameter of about 60 μm, yielding power densities of up to 4 x 10^{12} W/cm² that are sufficient to ignite a hot and dense plasma. Krypton is employed as target gas (backing pressure 25 bar), accomplishing broadband radiation (Kr XXV – Kr XXXVI) in the spectral range of the "water window". Due to the small mean free path of the soft x-ray radiation at atmospheric pressure the target vacuum chamber is evacuated to approx. 10^{-4} mbar.

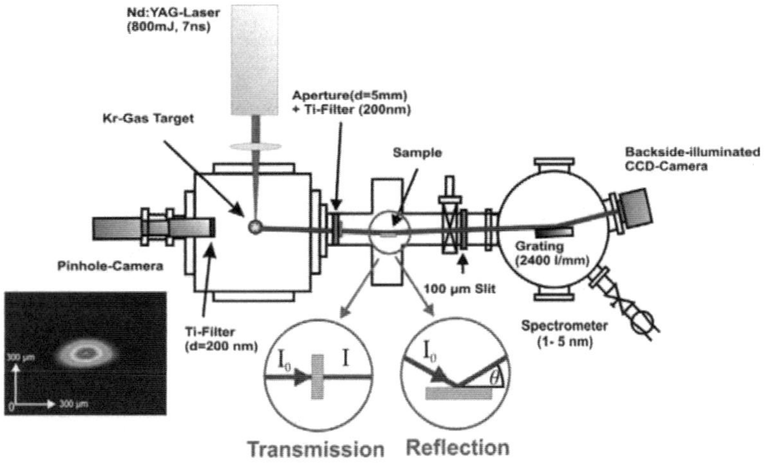

Figure 1. Experimental setup of the laser-plasma XUV source used for NEXAFS experiments. The insert on the left shows a picture of the krypton plasma taken by the pinhole camera.

An XUV spectrometer (1 – 5 nm) was used both for the spectral investigation of the plasma source and the NEXAFS experiments. The spectrometer consists of a 100 μm entrance slit, an aberration corrected flat-field grating (Hitachi, 2400 lines/mm) and a back-side illuminated CCD camera (Roper Scientific, pixel size 13 μm). The resolution of this spectrometer was experimentally determined to be $\lambda/\Delta\lambda \approx 200$ @ 2.87 nm. It was mounted 90° to the laser beam and opposite to the pinhole camera (cf. Fig. 1). To block visible radiation from the plasma and scattered laser light a titanium foil (200 nm thickness) was positioned between plasma source and sample.

To calibrate the spectrometer nitrogen was used as target gas. The spectrum of the nitrogen plasma, ignited under the same experimental conditions as krypton, consists of several lines (e.g. $1s^2$ - 1s2p transition of N VI at 2.8787 nm) that were used for spectral calibration.

3. Results

For thin samples it is possible to determine the x-ray absorption fine structure by measuring the transmitted flux through the sample. The optical density can be evaluated according to Lambert-Beer's law: $\mu(E) \cdot d = -\ln(I/I_0)$. $\mu(E)$ is the linear energy dependent absorption coefficient, d the sample thickness, I the transmitted and I_0 the reference intensity, respectively.

Figure 2 shows NEXAFS spectra that were recorded in transmission for a polyimide film (d=200 nm), a dried phospholipid layer (1,2-Dioleoyl-sn-Glycero-3-Phosphatidilserine (DOPS)) and a fulvic acid from aquatic origin. The lipid multilayers and the fulvic acid were prepared on thin Si_3N_4 membranes (d=100nm). The spectra show several sharp peaks below the carbon K-absorption edge which can be attributed to C 1s – π^* transitions and broader features above the edge that belong to C 1s - σ^* transitions. To evaluate the data multi-Gaussian/Lorentzian fits were performed using the software SpecFit. In Table 1 the identified features are summarized.

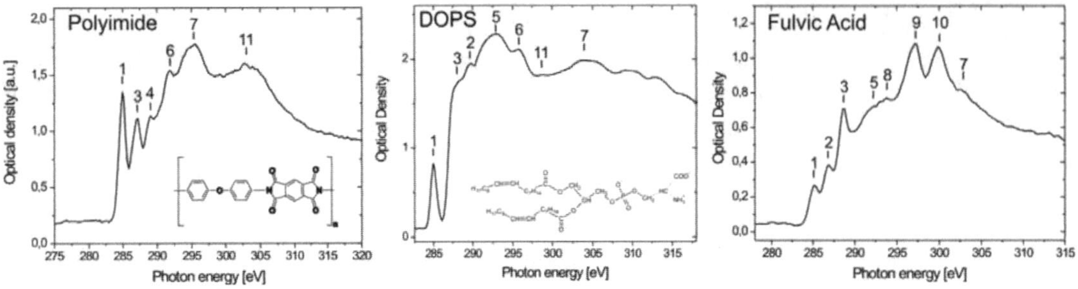

Figure 2. NEXAFS spectra of a polyimide film (d=200 nm), a phospholipid layer (DOPS) and a fulvic acid.

Polyimide films were used as a first test sample and the comparison with corresponding synchrotron data shows a good agreement. The peak positions deviate by less than 0.4 eV from reference data [8]. The lipid spectrum exhibits a sharp peak that can clearly be assigned to C=C bond of the unsaturated fatty acid chains. In the NEXAFS spectrum of the fulvic acid also potassium resonances can be identified.

Table 1. Energies and assignments of features in the NEXAFS spectra of Poyimide (PI), a phospholipid layer (DOPC) and a fulvic acid.

Feature	Energy [eV]			Assignment
	PI	DOPS	Fulvic Acid	
1	285.2	285.0	285.2	1s → π* (C=C)
2			286.3	1s → 3s /Ryd.
3	287.3	288.0	288.6	1s → π* (C=O)
4	289.3			1s → π*
5		292.8	291.8	1s → σ*(C-C)
6	291.6	295.8		1s → σ* (C-O, C-N)
7	295.0	302.9	302.4	1s → σ*(C=C)
8			293.7	1s → 4p (COOH)
9 / 10			296.9 / 299.8	K L_{III} / K L_{II}
11	303.0	298.7		1s → σ*(C=O)

4. Conclusion

The results demonstrate that it is possible to investigate the near edge absorption fine structure by utilizing broad-band radiation from a laboratory-scale laser driven plasma source. In the future the spectral resolution will be improved by using higher resolution gratings.

The financial support by SFB755 "Nanoscale Photonic Imaging" is gratefully acknowledged.

References

[1] Takman P, Stollberg H and Hertz H 2007 *Journal of Microscopy* **226** 175
[2] Sandberg R, Paul A, Raymondson D et al. 2007 *Phys. Rev. Lett.* **99** 098103
[3] Kondo H, Tomie T and Shimizu H 1996 *Appl. Phys. Lett.* **69**, 182
[4] Vogt U, Wilhein T, Stiel H and Legall H 2004 *Rev. Sci. Instr.* **75** 4606
[5] Stöhr J *NEXAFS Spectroscopy* 2003 Springer Series in Surface Science
[6] Dhez O, Ade H and Urquhart S 2003 *J. El. Spectr. Rel. Phen.* **128** 85
[7] Peth C, Barkusky F, Mann K 2008 *J. Phys. D.***41** 105202
[8] Jordan-Sweet J, Kovac C, Goldberg Met al. 1988 *J. Chem. Phys.* **89** 2482

Journal of Physics: Conference Series **186** (2009) 012033 doi:10.1088/1742-6596/186/1/012033

Optical setup for tabletop soft X-ray microscopy using electrical discharge sources

David Schäfer[1], Markus Benk[2], Klaus Bergmann[2], Thomas Nisius[1], Urs Wiesemann[3] and Thomas Wilhein[1]

[1]Institute for X-Optics, RheinAhrCampus Remagen, University of Applied Sciences Koblenz, Südallee 2, 53424 Remagen, Germany
[2]Fraunhofer Institute for Laser Technology ILT, Steinbachstr. 15, 52074 Aachen, Germany
[3]ACCEL Instruments GmbH, Friedrich-Ebert-Str. 1, 51429 Bergisch Gladbach, Germany

E-mail: schaefer@rheinahrcampus.de

Abstract. We report on an optical setup for compact full-field transmission microscopy developed for imaging with laboratory scale X-ray sources such as gas discharge plasmas. The optical setup consists of a reflective condenser and a zone plate objective imaging in the water window region (wavelength 2.3 - 4.4 nm). The microscope is able to visualize object structures in the sub-40 nm region. The radiation for the microscope is generated by a gas discharge in nitrogen with an input power of 10 kW and few Watt/(2π·sr) narrowband line emission. A gold coated axially symmetrical elliptical mirror is used to collect the radiation and forms a focused spot on the sample. Having a diameter of several 100 μm the spot is larger than the utilized imaging field, which leads to uniform illumination of the object. The mirror can be adjusted by a 5-axes piezo driven translation stage to optimize the flux in the object plane. The He-α line emission (2.88 nm) is selected of the nitrogen emission spectrum by means of a titanium thin film filter providing monochromatic sample illumination. A micro zone plate generates a magnified image detected by a back illuminated TE-cooled CCD camera. At high repetition rates the source is suitable for imaging test objects like Siemens stars, latex spheres and diatoms with magnifications of 1000 at exposure times in the range of a few seconds.

1. Introduction

Nowadays soft X-ray microscopy is a promising technique providing information of samples not easily obtainable with visible light or electron microscopy [1]. Especially biological specimen in their aqueous environment show a high natural contrast in the water window spectral region (λ = 2.3 nm to 4.4 nm) due to the high absorption differences between carbon (protein) and water. With the availability of table-top X-ray sources such as laser induced plasma [2] or gas discharge plasma [3], X-ray microscopy becomes independent of large facilities like storage rings. However, reducing exposure times to a few seconds for thick biological samples is still a task.

In this publication we present a compact full field transmission X-ray microscope based on a gas discharge light source using a grazing incidence collector for sample illumination and a micro zone plate as imaging optic.

2. Experimental setup

The schematic of the compact soft X-ray microscope is shown in figure 1. A gas discharge with pseudo spark like electrode geometry and 10 – 20 Joule pulse energy is used as X-ray source [4, 5, 6]. It is operated at repetition rates of up to 1 kHz and has a lateral source diameter of 1175 μm (FWHM). The resonant $1s^2$-1s2p transition of Helium like nitrogen at 2.88 nm wavelength is used for imaging which is realized by means of a titanium metal foil (thickness 200 nm) to provide monochromatic sample illumination. With calibrated grating spectrographs a photon number for the He-α line emission of $2\cdot10^{13}$ photons/(pulse·sr) is measured at 10 J pulse energy.

To mitigate debris and separate the vacuum between the source ($\sim1\cdot10^{-2}$ mbar) and optic chambers ($\sim1\cdot10^{-6}$ mbar) a capillary consisting of several glass channels with 50 μm diameter is mounted in front of the optical setup. The structure of the capillary is matched to the aperture of the collector and has a measured transmission of 33% (theoretical value 80%). The transmitted light is focused on the sample by a grazing incidence gold coated elliptical collector (ZEISS, axes: 600.27 mm and 18 mm). The collector has a gain of about 100 and demagnifies the source to a focal spot of about 800 μm diameter providing uniform sample illumination.

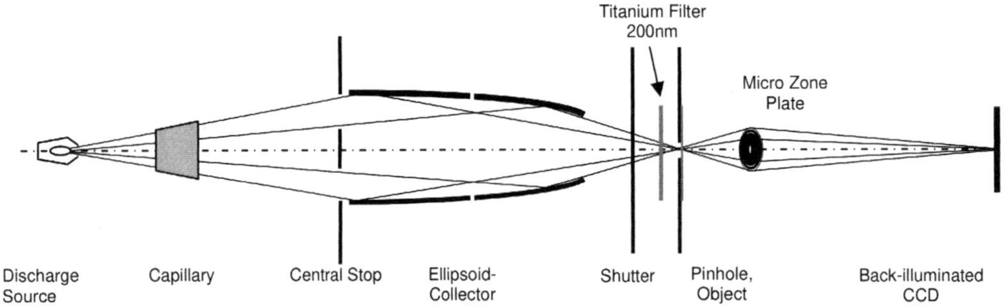

Figure 1. Schematic drawing of the optical setup.

A central stop is mounted in front of the collector to avoid direct light from the source passing through the sample reaching the detector. The sample is dissected on a Si_3N_4-membrane closely behind an aperture. A micro zone plate generates a magnified image of the sample on a TE-cooled, back-thinned CCD-detector with 1340 x 1300 pixels and a pixel size of 20 μm squared (Princeton Instruments). The zone plate has 468 zones and a diameter of 56 μm with an outermost zone width of 30 nm. The focal length at 2.88 nm wavelength is 585 μm. The diffraction limited spatial resolution of the zone plate is 37 nm.

All optical components are mounted on motorized stages and can be positioned independently. The collector and the focusing direction stage of the zone plate are piezo driven for fine adjustment.

3. Experimental results

First experiments were done with lithographic test structures to show the performance of the microscope. In figure 2 (left) a microscopic image of a Siemens star is shown. It was exposed for 60 s at 300 Hz repetition rate with a magnification of 1057x. The inner structures with 100 nm feature size are clearly resolved. To quantify the achieved resolution a high contrast region in the image has been examined. On the right of figure 2 the intensity plot of the marked area in the image of the Siemens star is illustrated. The plot is averaged over a width of 5 pixel and demonstrates an intensity rise from 10% to 90% within 37 nm, which corresponds to the theoretical diffraction limit of the zone plate.

To present the applicability of the microscope for biological research dried cells have been examined. Figure 3 shows rat embryonic fibroblast cells (left 2 min, 400 Hz; middle 5 min, 200 Hz; both 1057x). Inside the left image filament structures of about 50 nm size are visible. The picture on the right illustrates a HeLa TK-cell (5 min, 300 Hz, 1057x), a human epithelial cell line derived from a cervix carcinoma stopped during cell division.

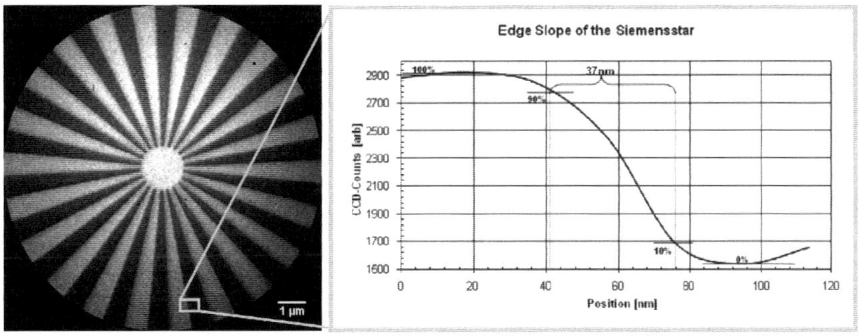

Figure 2. Microscopic image of a Siemens star (left) and a plot of the marked area demonstrating an edge slope (10% to 90%) below 40 nm (right).

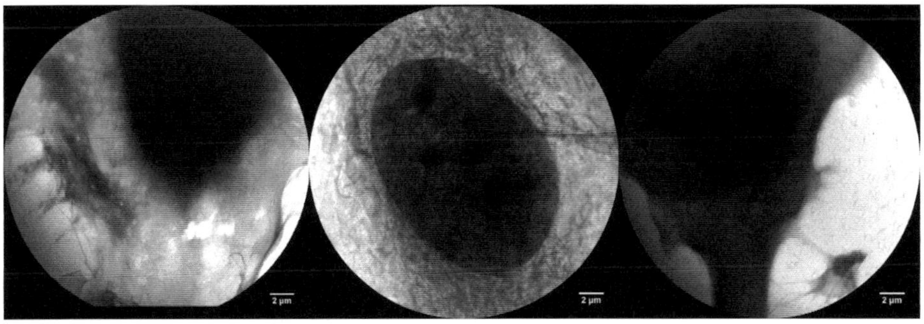

Figure 3. Dried biological cells imaged with a magnification of 1057x. Details see text.

4. Conclusion and Outlook

The developed optical setup is capable of resolving structures below 40 nm at exposure times of a few ten seconds with a magnification of 1000x. Better resolution is attainable by using a zone plate with smaller outermost zone width. The results we achieved show that using this concept laboratory X-ray microscopy of biological samples is within reach. To reduce the exposure time to the range of a few seconds several ideas will be realized in the near future. First, the source will be improved for more brilliance and stability at repetition rates above 1 kHz. Second, a new collector with matched geometry to the micro zone plate and a smaller focal spot will be integrated.

Acknowledgments

The authors gratefully acknowledge Ruth Barth from the Institute of Physical Chemistry of the University of Heidelberg for providing the biological specimen used in the experiments and Gerd Schneider from BESSY for kindly providing the micro zone plate. This work has been supported by the German Federal Ministry of Education and Research under contract No. FKZ 13N8914

References

[1] Jacobson C 1999 *Trends Cell Biol.* **9** 44-47
[2] Rymell L, Hertz H M and Engström L 1994 *X-ray Microscopy IV*, Aristov V V and Erko A I (Chernogolovka: Bogorodskii Pechatnik Publishers) 365-368
[3] Schmahl G et al. 1992 *X-ray Microscopy* **III**, Michette A G, Morrison G R and Buckley C J (Berlin Heidelberg: Springer) 66-69
[4] Bergmann K, Küpper F, Benk M 2008 *J. Appl. Phy.* **103**(12) 123304
[5] Benk M, Schäfer D, Wilhein T and Bermann K 2008 High power soft x-ray source based on a discharge plasma *This volume*
[6] Benk M, Bergmann K, Schäfer D and Wilhein T 2008 *Opt. Lett.* **33** 2359-2361

NEXAFS spectroscopy with a laser plasma x-ray source on soil samples

J Sedlmair[1], S-C Geber[1], C Peth[2], K Mann[2] and J Thieme[1]

[1]Institute for X-Ray Physics, Georg-August-University Göttingen, Friedrich-Hund-Platz 1, D-37077

[2]Laser-Laboratorium Göttingen e.V., Hans-Adolf-Krebs-Weg 1, D-37077

E-mail: `jsedlma@gwdg.de`

Abstract. Humic substances are post-mortal organic substances without an exact chemical structure. Their large specific surface is important for transport processes, especially in soils. We analyzed the NEXAFS spectra of humic substances, from which the amount of certain chemical compounds such as aromatic and aliphatic groups can be verified by the resonances of their binding energy. For the experiments, a compact table-top setup working with a laser plasma source was used. NEXAFS makes it possible to distinguish between samples, even if they contain the same composits, because information about the specific functional groups in the sample is supplied.
The evaluation was carried out using the program SpecFit. It was developed on IDL within our group and allows to fit the NEXAFS-data with a combination of arctangent, Gaussian and Lorentzian curves.

1. Introduction

Humic substances make up a major part of the natural organic matter (NOM), not only in soils, but also in water (rivers, lakes, oceans and their sediments) and organic deposits (peat, brown coals and shale). They can be divided into humins, humic acids and fulvic acids, according to their color or to their solubility depending on the pH of the solution [1].
Humic substances are also the reason for the characteristic brown color of decaying plants. They have an amorphous macro-molecular structure of high molecular weight. Thus, since humic substances do not have an exactly defined chemical structure, it is of key importance to find reliable ways to measure their properties.

2. NEXAFS-spectroscopy and SpecFit

With near edge x-ray absorption fine structure spectroscopy (NEXAFS) it is not only possible to characterize humic substances from different origins, even if they contain the same composites, but also to easily distinguish between the operationally defined fractions fulvic acid, humic acid and humins [3]. This stems from the fact, that NEXAFS reveals the binding forms within a sample by measuring the sum over all possible final states of the electrons and so yields information about the specific functional groups [2]. Thus, NEXAFS-spectroscopy is a powerful means to especially characterize environmental samples.
Because of the lack of an exact structure, it is more suitable (compared to linear combination methods) to fit the spectra with a combination of arctangent, Gaussian and Lorentzian curves

using the evaluation program SpecFit, which was developed within our group using Interactive Data Language (IDL) [4]. Figure 1 shows an example of the evaluation with SpecFit on the basis of a data set we gained from a fulvic acid. The arctangent function simulates the absorption edge, whereas the Gaussian curve denotes the resonance peak [5] [2]. The Lorentz curve takes the broadening of the resonance into account, which is an intrinsic feature of the experimental apparatus.

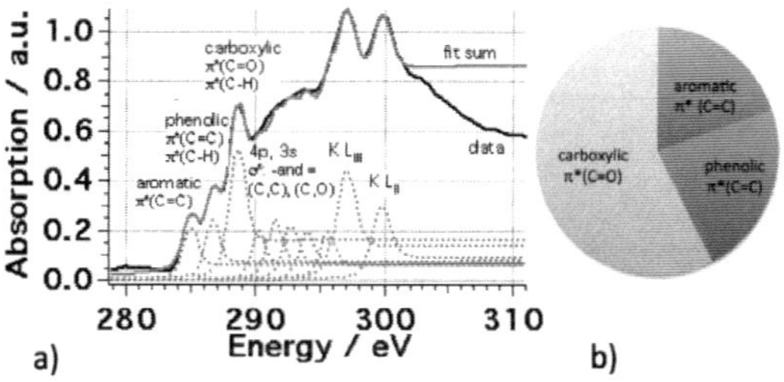

Figure 1. **a)** NEXAFS-spectrum of fulvic acid evaluated with a combination of Gauss-, Lorentz- and arctan- curves using the program SpecFit. **b)** Comparison of the areas under the first three Gauss-peaks.

The assigned peaks can then be related to the values of the resonances of certain binding energies, for instance single or double bond between two C-atoms.
We decided to assign a peak and an absorption edge to every resonance over the method of defining one general arctangent function to approximate the absorption edge and several Lorentz-Gaussian curves to fit the peaks. That way, the real peak heights are acquired, especially for the important peaks before and right at the absorption edge. The parameters were chosen in a way to find the optimal fit for the data curve.

3. Materials and Methods

We analyzed two different sample groups. The first group consisted of four samples from the area around Göttingen, representing the different fractions of humic substances. The second group was one sample of humic substances (purchased from Sigma Aldrich), which was extracted using three different solvents. All humic substances have been mixed with Millipore-water. We used 1% -standard solutions, which would be diluted further, if necessary. The basic procedure was the following: The glass vessels (2ml, Eppendorf) containing the solutions of humic substances and water were treated with ultrasound for several minutes and afterwards 1-2µl were dropped on a Si_3N_4-foil. The water evaporated and the foil with the dry sample was mounted on the sample holder and installed in the table-top NEXAFS-spectrometer (Laser Laboratorium Göttingen), which is explained in reference [6]. The Si_3N_4-foils (100nm thick, frame $5{\times}5mm^2$, window $1.5{\times}1.5mm^2$) were purchased from Silson Ltd., Northampton, England.

4. Results and Discussion

First, we compared four samples containing different amounts of humic substances with each other, see figure 2. The three parts, which humic substances can be divided into, can clearly be differentiated. Humins in the very humus-rich chernozem contains the highest amount of aromatic carbon. This sample was from a chernozem-soil [7]. Humic acid, here in the form of its salt (Ka-humate) is a frequently used fertilizer. The fulvic acid was of aquatic origin. Both acids contain significantly less carbon than the chernozem. The smallest content of humus could be found in the luvisol-sample. Luvisol is a soil, which contains clay minerals and calcite and normally is poor in nutrients (humus). As the aromatic content decreases, the carboxylic groups become more and more observable in the spectrum. Another feature, which is also growing, is the amount of potassium, observable by the increasing peaks of the L_{III} and L_{II} peaks.
The second series of humic substances samples characterized the influence of the extraction solvent on the humic fraction. We analyzed one reference sample of humic acid and the same sample after three different extraction methods (figure 3). The measurements reveal, that the

choice of the solvent influences the chemical composition of the sample. This should be taken into account, when analyzing soil samples or other specimens that need to be extracted.

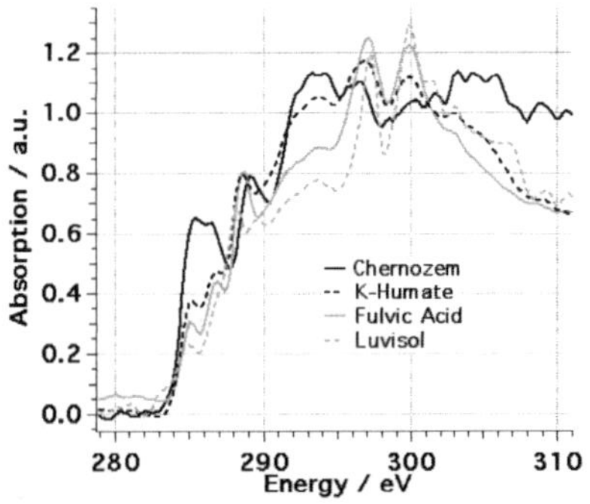

Figure 2. Normalized NEXAFS-spectra of the four different soil-samples. The evaluation has been done with SpecFit. However, the difference in peak heights of the aromatic, carboxylic and potassium peaks is clearly visible.

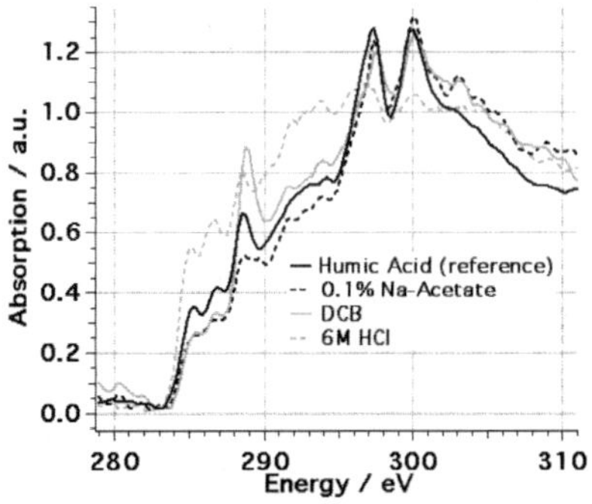

Figure 3. Normalized NEXAFS-spectra of the reference sample and the result of the three different extraction methods. The evaluation has been done with SpecFit. Again, the different peak heights of the aromatic, carboxylic and potassium peaks are clearly visible.

5. Conclusion

We analyzed two groups of soil samples regarding their carbon-content. One group consisted of different fractions of humic substances, whereas the other group was made up of one humic substances sample (reference) and the result of the extraction of this sample using three different solvents.The experiments show, that the spectrometer operating with a laser-driven plasma source is a good alternative to synchrotron-based spectro-microscopes, if spatial resolution is not necessary.

All of the samples were clearly distinguishable concerning the three major carbon-binding groups (aromatic, phenolic and carboxylic) and more; where the information was derived from the fitting method. This evaluation, using a combination of Lorentzian, Gaussian and arctangent curves, is an improvement to methods using only one general arctangent curve to approach the absorption edge, especially when working with samples without exactly specified content such as biological and environmental specimens.

Acknowledgement

This work has been supported by the DFG within the Collaboratve Research Center SFB 755 "Nanoscale Photonic Imaging".

We would like to thank Andreas Kappler and Iris Bauer from the Eberhard-Karls-University Tübingen for the extracted humic substances and Jürgen Niemeyer from the Institute for Applied Biotechnology (IBT) Göttingen for the humic substances samples from the Göttingen area.

References

[1] Stevenson F J 1982 *Humus Chemistry: Genesis, Composition, Reactions* (John Wiley & Sons Inc)
[2] Stöhr J 1992 *NEXAFS Spectroscopy* (Berlin: Springer Verlag)
[3] Thieme J, Schmidt C and Niemeyer J 2000 *X-Ray Microscopy: Proceedings of the Sixth International Conference (Berkeley, CA 1999)*, ed Meyer-Ilse W, Warwick T and Attwood D T (Berkeley: AIP Conference Proceedings) p 30-306
[4] Gleber G 2003 Diploma thesis (Göttingen university)
[5] Thieme J, McNulty I, Vogt S and Paterson D 2007 *Environ Sci Technol.* **41**(20) p 6885-6889
[6] Peth C, Barkusky F and Mann K 2007 *J. Phys. D: Appl. Phys.* **41**(105202)
[7] Ahl C, Meyer B, Jörgensen R and Hugenroth P 1985 *Mitteilgn. Dtsch. Bodenkundl. Gesellsch.* **42** p 437-453

Journal of Physics: Conference Series **186** (2009) 012035 doi:10.1088/1742-6596/186/1/012035

Gain in spatial resolution of X-ray laboratory microtomographs with enlarging X-ray optical elements.

R A Senin[1], A V Buzmakov[1], A V Konovko[2], I S Smirnov[3], A S Geranin[3] and V E Asadchikov[1]

[1]A V Shubnikov Institute of Crystallography, Leninsky pr, 59 Moscow, Russia
[2] M V Lomonosov Moscow state university, Leninskie gori, 1, Moscow, Russia
[3] Moscow institution of electronics and mathematics, Moscow, Russia

E-mail: senin@crys.ras.ru

Abstract. Two types of x-ray optical elements are discussed – capillary air bubble lens and asymmetric cut single crystal. These elements are applied to laboratory microtomography. Several reconstructions using laboratory x-ray source (2 kW, 1x1 mm^2 size) and the optical elements are obtained. Best voxel size of 1 μm is achieved using two crossed asymmetric cut single crystal (Bragg magnifier) and 14x14 μm pixel CCD camera. Two times magnifying x-ray microtomograph with 7.5 μm resolution was built using capillary air bubble lens.

X-ray microtomograpy is a powerful tool for the structure investigations. Unfortunately most of laboratory setups have two main disadvantages. First, as a "white" beam is used it makes difficult to reconstruct the true map of linear absorption coefficients (μ) [1]. Second, the spatial resolution of the tool is limited by the resolution of CCD camera used or by the source size for projection magnification [2].

The only known setup having another construction is a laboratory microtomograph by XRadia [3]. In this case the resolution is determined by a zone plate and monochromaticity provided by order sorting aperture. Why do we need monochromatic irradiation and optics? Narrow energy range simplifies reconstruction procedure, and the use of X-ray optics provides a better spatial resolution.

In the previous experiments it was demonstrated [4,5,6] the air bubble lens and Bragg magnifier can be applied to improve laboratory X-ray imaging. Two microtomographs based on these elements are constructed and represented in this article

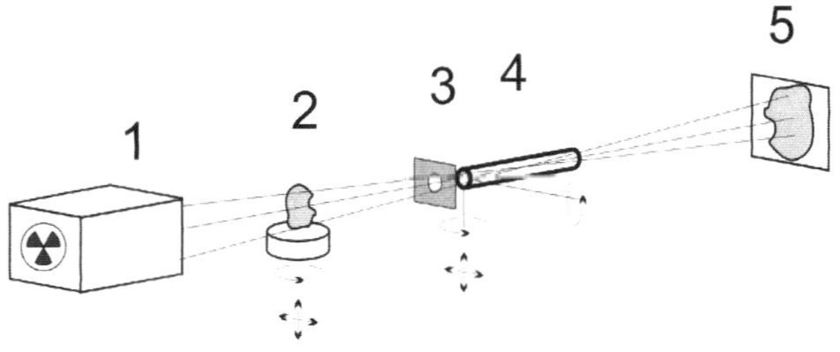

Figure 1. Sketch of a microtomograph with x-ray refracting lens. (1) X-ray source, (2) the specimen, (3) diaphragm, (4) refracting lens, (5) 2D detector.

Journal of Physics: Conference Series **186** (2009) 012035 doi:10.1088/1742-6596/186/1/012035

At first, capillary air bubble lens (67 air bubbles, 200 microns diameter of every bubble, 15 microns bridge between bubbles) [7] is used as x-ray optical element behind a investigated object. As a source X-ray Cr tube with 2 kW power and 1x1 mm2 spot size was applied in the experiments. Such a big source illuminates the object and allows us to collect the x-rays from the object into 200 microns aperture. At 5.4 keV (Cr Kα) the lens has nearly 1 mm field of view and more than 2 mm field depth at 2 times magnification. Thus, the image stays always sharp during the rotation. According to our calculations the spatial resolution of the lens is better than 1 μm. The scheme of this setup is given in Fig.1.

The largest magnification we obtain with this optical element is 14, but low intensity of the conventional X-ray tube limits the magnification in laboratory microtomographs. The result of reconstruction of a gecko tooth inner structure with two times optical magnification is shown in Fig. 3a. Spatial resolution in this reconstruction is two times better than that one without optics and the voxel size is ~7.5 μm

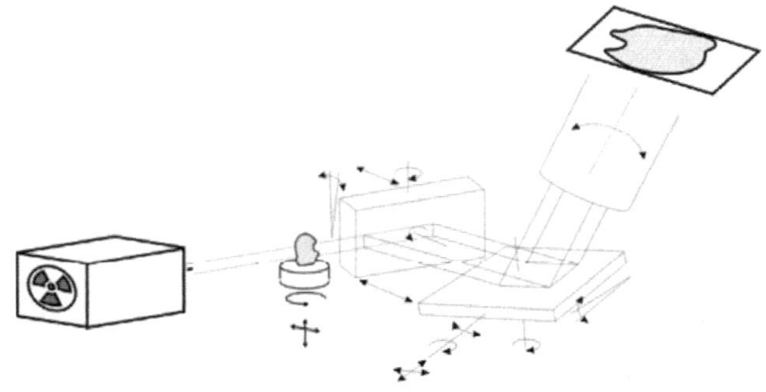

Figure 2. The image formation for 2D magnification (two crystals are used).

Second element is a pair of crossed asymmetric cut single crystals (Si (2 2 0)) [8,9]. Each crystal magnify only in one direction. Crystals were cut with asymmetry coefficient 15 for 8 keV (Cu Kα). The scheme of this setup is given in Fig.2. The same tooth of a gecko was used as a test object. In Fig. 3b one can see reconstructed image of its structure with ~1 μm voxel size. The image is computer sectioned to show its inner structure.

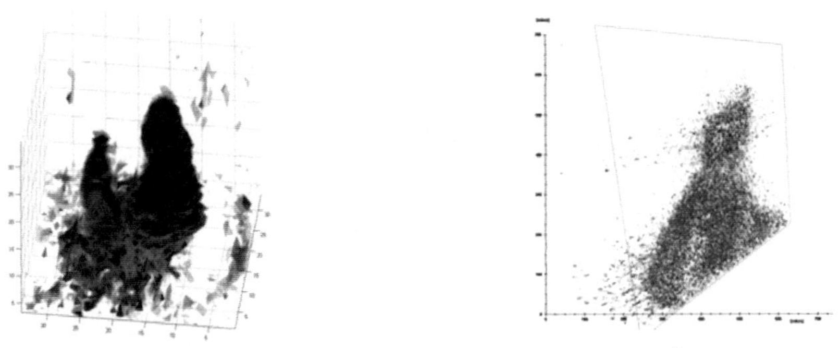

a b

Figure 3. The result of reconstruction of teeth of a gecko obtained using microtomograph with (a) lens, X-ray magnification x2 and (b) crystals, X-ray magnification x14.

In this work we demonstrated the possibility to construct two new types of laboratory monochromatic microtomographs which allowing to achieve 1 μm voxel size with 14x14 μm pixel CCD camera. Quite better resolution can be obtained using crossed single crystals but capillary air bubble lens has as a big advantage in much easier alignment.

Acknowledgments The work of R.A.S. was partially supported by Grant of the President of Russian Federation No. MK- 4434.2007.2

References

[1] W.Kalender Computed Tomography: Fundamentals, System Technology, Image Quality, Applications, 2nd edition // Med. Phys. Volume 33, Issue 8, pp. 3076-3076 (August 2006)
[2] www.skyscan.be
[3] www.xradia.com
[4] V.E.Asadchikov, B.V.Mchedlishvili, Yu.V.Ponomarev, A.A.Postnov. R.A.Senin, T.V.Tsiganova. X-ray microscopy with the usage of asymmetric reflection from single crystal. *JETP Letters*, 2001, **73**. I.4 pp 205-209.
[5] A.V. Andreev, V. E. Asadchikov, A. V. Buzmakov, A.A. Konovko, S.V. Kuzin, A.A. Perzov, Y.V. Ponomarev, R.A. Senin, I.S. Smirnov, S.V. Shestov, V.N. Shkurko Two-dimensional image magnification by X-ray microscope of asymmetric reflection. JETP Letters, vol 85, №. 1, pp. 106-108
[6] V.E. Asadchikov, A.V. Buzmakov, A.V. Popov, R.A. Senin ,I.V. Suloev, A.V.Vinogradov, «X-ray magnification with the use of the microcapillary lens: simulations and experiments», *Proceedings of SPIE* -- **6163**. Saratov Fall Meeting 2005: Optical Technologies in Biophysics and Medicine VII, Valery V. Tuchin, Editor, 61631G (Jul. 24, 2006)
[7] V.E.Asadchikov, Yu.I.Dudchik, N.N.Kolchevsky, F.F.Komarov, A.V.Popov, R.A.Senin, I.V.Suloev, A.V.Vinogradov. The usage of multielement refractive lens for imaging at5.4 kev. Seventh international conference on X-ray microscopy. Book of abstracts. P.54.
[8] W.J. Bøttinger, R.C.Dobbyn, H.E. Burdette and M.Kuriyama. Real Time Topography with X-Ray Image Magnification. Nucl. Inst. and Meth. vol. 195. (1982) pp. 355-361.
[9] M.Stampanoni, G.L. Borchert, R. Abela, P. Ruegsegger, ,2003a. *Appl. Phys. Lett.* 82 (17), 2922.

2 Dimensional Position Sensitive XAFS by Using In-house X-ray Spectrometer

Kozo Shinoda[1], Shigeru Suzuki[1], Masaru Kuribayashi[2] and Takeyoshi Taguchi[2]

[1] Institute of Multidisciplinary Research for Advanced Materials, Tohoku University, 1-1, Katahira, 2-chome, Aoba-ku, Sendai 980-8577, Japan.

[2] X-ray Diffraction Division, Rigaku Corporation, 3-9-12, Matsubara, Akishima, Tokyo 196-8666, Japan.

shinoda@tagen.tohoku.ac.jp

Abstract. Position sensitive XAFS measurements by using the in-house X-ray spectrometer and the position sensitive X-ray detector were attempted. The in-house spectrometer produces monochromized divergent X-ray beam. Therefore, an extended direct-beam image projected on the detector can be taken, and its extension rate is depending on the arrangement of the sample and the detector. A position sensitive XAFS measurement was demonstrated by using Ni metal foil and NiO powder as a model sample. Ge(220) or Si(400) Johansson-type bent single-crystal was used as monochromator, and Mo and LaB_6 were used as the target and filament, respectively. Tube voltage and current were operated at 16 kV and 100 mA (1.6 kW). XAFS spectra were measured by transmission-mode with sample set/reset method and required time of each experiment is about 6 hours in total. It was confirmed that metal and oxide spectra with enough quality for structural analysis were clearly separated by each position in the sample.

1. Introduction

X-ray absorption spectroscopy (XAS) was applied for analyzing the local structure in various material systems by using the Fourier calculation method in the early 70's and plays an important role in structural chemistry and physics. Since synchrotron radiation facilities were provided for XAS experiments, this technique has been applied to various field of engineering, material science, environmental science, catalytic science, and so on. In-house x-ray absorption spectrometers have been developed since 80's [1].

In the early 00's, a new in-house spectrometer apparatus of which size is smaller due to using a compact x-ray source was developed [2]. This spectrometer, Rigaku R-XAS looper, showed good performance at experimental energy range of 5 to 25 keV, and was available to apply to measurements at S K absorption edge (~2.5 keV) [3]. On the other hand, the pixel-array type position sensitive area detector namely D/teX-25 has recently developed for high-speed and high-sensitivity x-ray diffractometry by Rigaku corporation, and its detecting window size is 12.8 mm x 12.8 mm including 512 pixel x 512 pixel with about 24 micrometer of pixel size [4]. The in-house spectrometer is employed a focused optical system with Johansson-type bent monochromator and the output beam is divergent in vertical direction within 4 degree in divergence angle as shown in figure 1.

The window size of the position sensitive detector is suitable to detect the direct beam radiated and monochromized by the in-house spectrometer. Achievement of a position sensitive XAS is expected by using the in-house spectrometer with the 2-dimensional detector.

2. Experimental

In this study, all experiments were carried out at around Ni K absorption edge (8332 eV) using Ge(220) and Si(400) monochromator for the measurements with standard and high energy resolution, respectively. Mo and LaB_6 was respectively used as x-ray source target and anode filament, and operating tube voltage and current were 16 kV and 100 mA (1.6 kW).

To evaluate the quality of monochromized x-ray radiation, the image of the direct incident beam was detected at the x-ray beam energy of 8400 eV corresponding to just above the Ni K absorption edge. The spectrometer has two slits of beam-width limiting slit placed just before the focusing point and divergence slit placed at 60 mm from beam source point. A model sample of a tape stuck metal nickel foil and pasted NiO powder was used for the experiments of the position sensitive XAS.

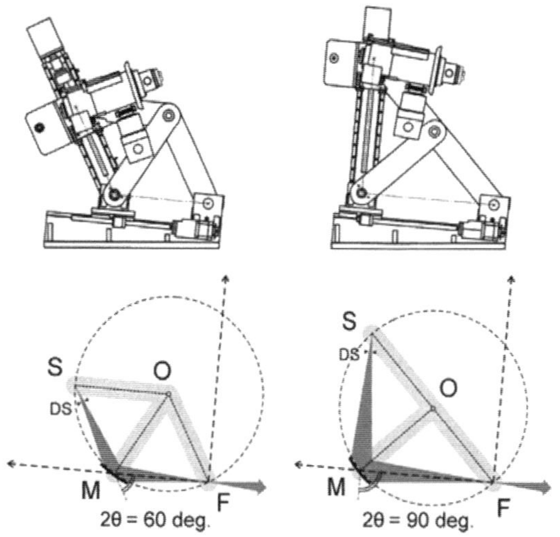

Figure 1. Schematic diagram of the in-house spectrometer and its optical system. S : x-ray source, M : monochromator, F : focus point, O : center of Rowland circle, DS : divergence slit.

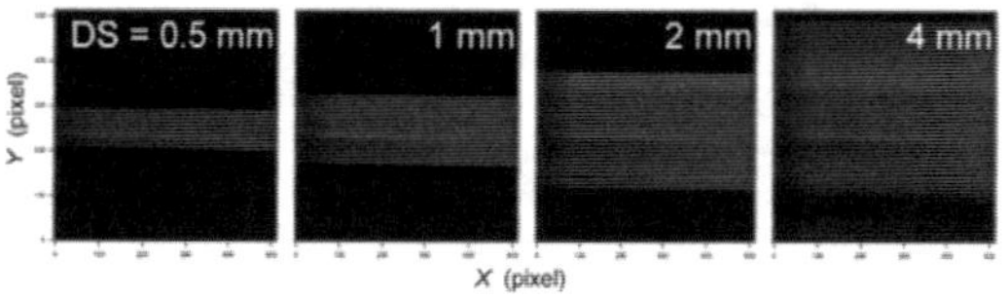

Figure 2. Direct beam images with different DS size using Si(400) monochromator at 8400 eV.

3. Results

3.1. Direct image of incident beam

Spectrometer was positioned at the energy of 8400 eV using Ge(220) or Si(400) monochromator, and direct beam images were obtained by the detector D/teX-25 located at 220 mm behind the focus point F in figure 1. Figure 2 shows the direct beam images with various condition of divergence slit (DS) using Si(400) as monochromator. The vertical divergence of direct beam is 4 degree when the DS size is 8 mm. Figure 3 shows the direct beam images with different size of the beam width slit and fixed size of DS. The output beam from the spectrometer has divergence vertically by Bragg reflection from bent monochromator and horizontally due to beam width. Therefore, the vertical width of beam in the window of the detector is same irrespective of the kind of monochromator crystal. On the other hand, the

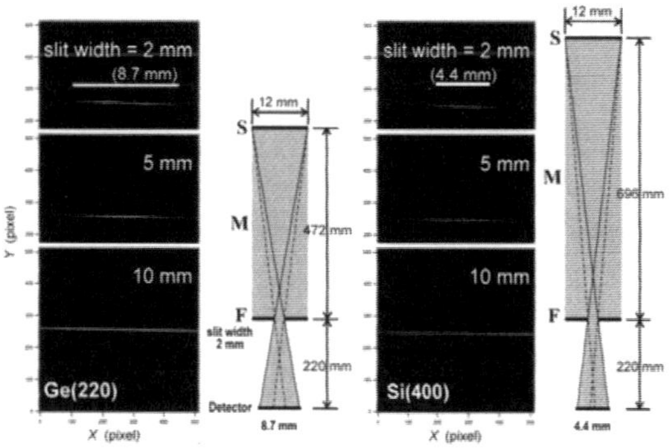

Figure 3. Direct beam images with different size of the beam width slit using Ge(220) (left) and Si(400) (right).

horizontal width depends on monochromator crystal at the same energy of x-ray because of difference of the path length between x-ray source point S and focus point F. The value of lengths are shown in the inset figures.

3.2. Position sensitive XAFS measurement
When an analysis of local structure or chemical state is applied to a multiphase material with position dependence, there are two way of approach; one is a method with scanning the irradiating position using microsized beam, the other is the method in this study. The position sensitive XAS was demonstrated by using a model sample of Ni metal foil and NiO powder were put on a adhesive paper tape. The photograph of the sample is shown in figure 4.

The image of direct beam transmitted through the model sample is shown in figure 5. The intensity map images were measured using Ge(220) monochromator at 231 points of the energy around Ni K absorption edge. The required time for the measurements of the intensity of transmitted through the sample and of blank was totally for 6 hours. The resultant XAFS spectra of areas 1 and 2 in figure 5 are shown in figure 6. The pattern profile of these spectra shows the features of each Ni metal and NiO. In case of using Si(400) as monochromator, measured spectra have an applicable quality to analysis of the chemical state of absorbing element by using the profile of XANES (x-ray absorption near-edge structure).

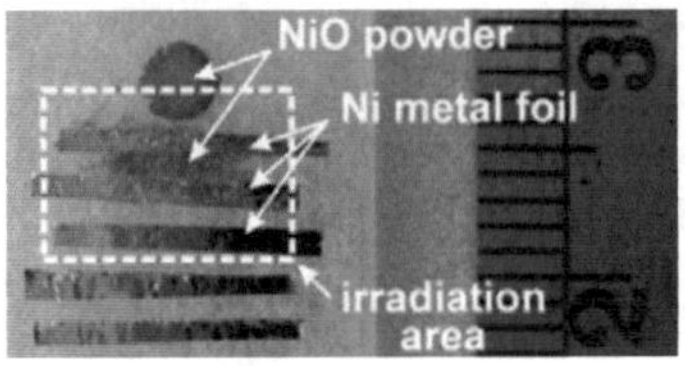

Figure 4. The photograph of the Ni-NiO model sample.

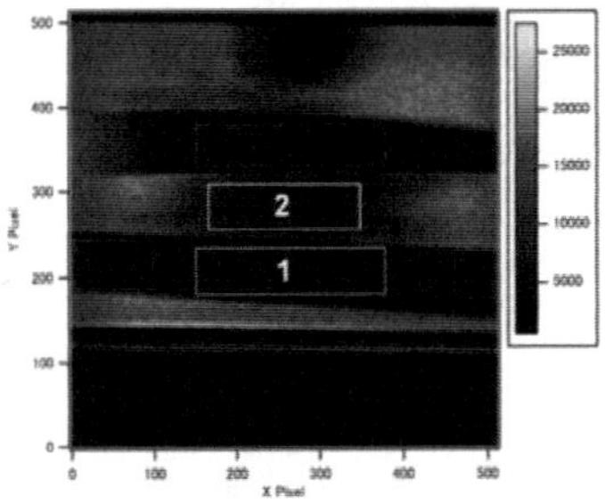

Figure 5. Direct intensity distribution of the beam transmitted through the model sample.

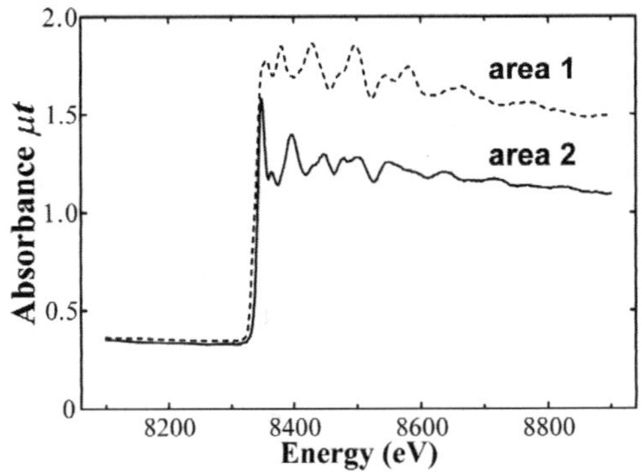

Figure 6. XAFS spectra obtained by extracting the intensity of the area 1 and 2 in fig. 5.

4. Summary
The results of demonstration in this study indicated that the combination of the in-house x-ray spectrometer and the position sensitive array detector is effective in the measurements and analysis of the position sensitive XAS.

References
[1] Tohji K, Udagawa Y, Tanabe S and Ueno A 1984 *J. Am. Chem. Soc.* **106** 612
Tohji K, Udagawa Y, Tanabe S, Ida T and Ueno A 1984 *J. Am. Chem. Soc.* **106** 5172
Tohji K and Udagawa Y 1985 *J. Phys. Chem.* **89** 83
[2] Taguchi T, Harada J, Kiku A, Tohji K and Shinoda K 2001 *J. Synchrotron Rad.* **8** 363
[3] Taguchi T, Shinoda K and Tohji K 2005 *Physica Scripta* **T115** 1017
[4] T. Taguchi, Powder Diffr., 21 (2006), 97-101.

Journal of Physics: Conference Series **186** (2009) 012037 doi:10.1088/1742-6596/186/1/012037

Laboratory X-ray microscopy with reflective – refractive lens system

L. Sveda[1,2], **J. Marsik**[3], **M. Horvath**[3], **L. Pina**[1,3], **Yu. I. Dudchik**[4], **V. Semencova**[3], **R. Havlikova**[1] **and V. Jelinek**[3]

[1]CTU, FNSPE, V Holesovickach 2, CZ-18000 Praha 8, Czech Republic
[2]Elya Solutions s.r.o., Pobrezni 95/74, CZ-18600 Praha 8, Czech Republic
[3]Rigaku Innovative Technologies Europe s.r.o., Novodvorska 994, CZ-14221 Praha 4, Czech Republic
[4]Institute of Applied Physics Problems, Kurchatova 7, 220064 Minsk, Belarus

E-mail: svedal@troja.fjfi.cvut.cz

1. Abstract

Compound refractive lens (CRL), consisting of a number of in-line placed concave microlenses made of low-Z material, is a unique imaging X-ray device. An experiment combining the CRL manufactured by the method described in [1] and working as an objective with the reflective micromirror as ta condenser and X-ray microfocus source is presented. The system was used as an X-ray microscope working at 8 keV X-rays and also for demonstration of demagnified imaging.

2. Introduction

High resolution x-ray microscsopy based on the laboratory x-ray sources has already reached the level of commercial products [2, 3]. Achievable resolution currently breaks the 50 nm level. Such systems, typically, use a table top x-ray source, collecting optics and objective optics. While there are several options for the collecting optics (capillary, zone plate, grazing incidence mirrors, etc.), zone plate is used as the objective optics in the majority of high resolution systems. The objective optics creates a geometrically enlarged image of the sample on the imaging detector.

Still, there are especially industrial and engineering applications, where the resolution of the order of $\sim 1\ \mu$m is sufficient. These include small electronic parts imaging, imaging XRF, etc. The other large application area, which was actually the original reason for our work, are the university teaching activities. All these experiments would largely benefit from much lower price of the components. While the zone plate costs 5 kEUR typically, the price of the micromirror or the CRL especially is much lower.

Compound refractive lens (CRL) is generally a refracting X-ray optics. It consists of a number of individual lenses aligned one behind each other with typically $\sim 100\ \mu$m radius. Originally, CRLs were developed by Snigirev [4], were it focused only in one direction using a set of holes. The CRL developed by Dudchik [1], which is called a Microcapillary lens, is based on the same principle, but consists of a number of precisely aligned gas (air) bubbles inside the plastic filled capillary. The oretical resolution is $\geq 0.2\ \mu$m and demonstrated measurement at synchrotrons shown $\geq 0.3\ \mu$m[5].

Micromirrors are grazing incidence X-ray mirrors developed by Reflex s.r.o. and were used originally as a condenser for Bede Microsource. Ellipsoidal or parabolic gold layered mirrors with diameter $\sim$ 1 mm, length $\sim$ 50 mm, source-detector distances $\sim 250 - 600$ mm. The mirrors are generally manufactured using the replication technology [6]. Different mirror materials were also tested, including Molybdenum instead of Gold and recently also the tests with multilayers inside [7] were undertaken.

3. Experimental setup

X-rays were emitted from the X-ray tube (40 keV, 2 mA, $CuK\alpha$ line, $\sim 50 - 80$ μm spot size). They were further collected the sample by the Micromirror ME 600 (ellipsoidal mirror, gold coated, source–mirror dist. $x_1 \sim 15$ mm, and source–detector dist. $x_d \sim 600$ mm). This mirror is optimized for 8 keV, however is not optimized for the lens itself. The CRL used in the experiment was a 38 mm long capillary filled with 121 bubbles with $R = 100$ μm radius. Focal length, calculated and measured at synchrotrons as well, is $\sim$ 109 mm if measured from the lens center. Two X-ray detectors were used within the experiment. Reflex RX1 Fast X-ray Camera was used for rough measurements and adjustments. It has 620×460 pixels, 23.5×23.5 μm^2 each. Prototype of Reflex high resolution μCRON camera was used for actual measurements. It has 1376×1038 pixels with effective pixel size $\sim 0.645 \times 0.645$ μm^2.

Two gold meshes were used as a sample. One of them had 25 μm period and 5 μm thick wires, while the second gold mesh had 85 μm period.

4. Results

First experiments were done in order to verify the quality of the lens itself. Sample — the mesh with 1000 periods per inch — was placed close to the source. The image created by the lens was then captured by the μCRON camera with magnification $M = 1/5$. The result after subtracting the source image and further image processing is shown in Figure 1.

The mesh inside the rectangular structure is clearly visible. Mesh period after the demagnification is $\sim$ 5 μm, each of the periods can be perfectly distinguished. Thus, it has been verified that the resolution of the lens and the μCRON camera is $\sim$ 1 μm.

Tilted rectangular structure containing the image of the mesh is just an artefact. It is due to the combination of the source profile, pinholes and screens used to guide the radiation and image processing.

A combination of the lens and micromirror was the next experiment. It demonstrated the properties of the combination and was used as a testbed for gaining experience for future modifications (not presented here). The mirror was used as the condenser while the lens was used as the objective.

Ideally, the parameters of the mirror and the lens have to be mutually optimised to match each other, however the one used for the tests was not optimised as this is planned in the future. In order to shorten the source–detector distance, and thus increase the signal, the lens was placed into the mirror focus, while the sample was in the focus of the lens ($\sim$ 200 mm in front of the lens). Thus only a circle of the sample was illuminated. This relatively unusual arrangement is sometimes used in visible light illumination experiments in order to make the illumination homogeneous.

The beamstop was not used in our experiments. This gave the possibility to study the lens from two different point of views as the resulting image was a composition of source image created by the lens working as a pinhole and image of the mesh created by the lens as an objective. The beamstop was applied later on virtually using the image processing methods.

One of the results after applying the virtual beamstop (removing the pinhole effect) is shown in Figure 2. Only a circle is seen, exactly as expected, with mesh bars visible as shadows. The lens setup was chosen to be have a 1:1 magnification. The image was obtained after 5 minute exposure time.

In fact, the image created by the pinhole was much brighter if compared to the one shown in figure. This is due to absorption inside the lens. The rays creating a pinhole image are going through the center of the lens, where the absorbtion is the smallest. On the other hand, the objective image is created by rays passing through the lens at the angle and therefore at larger distances from optical axis and thus encountering much larger absorbtion.

Supposing that the mirror/lens matches properly, imaging properties should be ruled by imaging properties of the CRL [5]. Either a smaller diameter of condenser or longer focal length is thus needed. If enlarged focal length is used, however, absorbtion in the air becomes important and has to be taken into account. An eample of an image which can be expected if everything goes OK is shown at Figure 1. Expected resolution is therefore $\sim 1\ \mu$m.

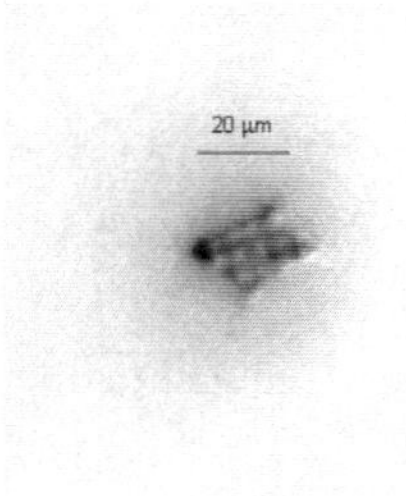

Figure 1. Mesh with 1000 periods per inch imaged by the lens with magnification $M = 1/5$. Exposure time is 15 minutes.

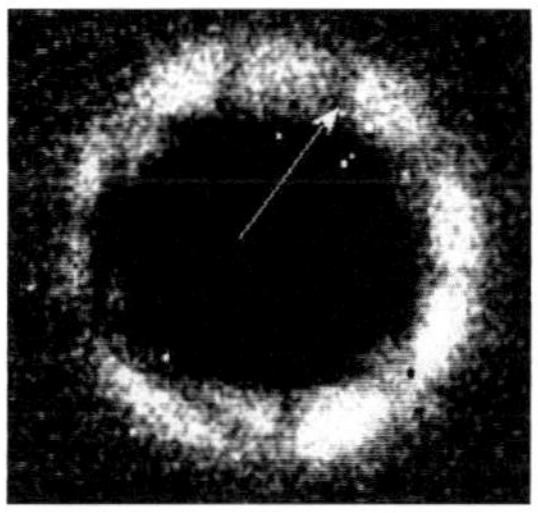

Figure 2. An example of imaged circular section of the mesh after 5 minutes exposure. The mesh is visible as a shadow at the circle (white arrow). The image is $360 \times 260\ \mu m^2$ large.

5. Conclusions

It has been shown that the Microcapillary lens can be used for imaging in combination with the X-ray tube in both magnifying and demagnifying mode. Further, demonstration of combination of the Micromirror and the Microcapillary lens was shown. It has been demonstrated that mutual optimization of mirror and lens is crucial for future imaging experiments. Currently available elements are not usable for efficient x-ray microscopy imaging.

Acknowledgments

We acknowledge the support of the Ministry of Industry and Trade of the Czech republic, grant FT-TA3/112 and Ministry of education, youth and sports of the Czech republic, grant MSM6840770022.

References

[1] Dudchik Y I and Kolchevsky N N 1999 *Nuclear Instruments and Methods in Physics Research A* **421** 361–364

[2] Yin G C, Song Y F, Tang M T, Chen F R, Liang K S, Duewer F W, Feser M, Yun W and Shieh H P D 2006 *Applied Physics Letters* **89** 221122–+

[3] http://www.xradia.com

[4] Snigirev A, Kohn V, Snigireva I and Lengeler B 1996 *Nature* **384** 49–51

[5] Piestrup M A, Gary C K, Park H, Harris J L, Cremer J T, Pantell R H, Dudchik Y I, Kolchevsky N N and Komarov F F 2005 *Applied Physics Letters* **86** 131104–+

[6] Pina L, Inneman A, Hudec R and Arndt U W 1999 *Proc. SPIE Vol. 3766, p. 299-307, X-Ray Optics, Instruments, and Missions II, Richard B. Hoover; Arthur B. Walker; Eds.*

[7] Sveda L, Inneman A, Semencova V, Pina L, Hudec R and Havlikova R 2007 *Advances in X-Ray/EUV Optics and Components II, Khounsary, A. M.; Morawe, Ch.; Goto, S.*

Compact phase-contrast soft X-ray microscopy

O von Hofsten, M Bertilson, M Lindblom, A Holmberg, H M Hertz and U Vogt

Biomedical and X-Ray Physics, Department of Applied Physics, Royal Institute of Technology/Albanova, SE-10691 Stockholm, Sweden

Email: olov.hofsten@biox.kth.se

Abstract. For nearly all elements, the real part, δ, of the complex index of refraction n ($n = 1 - \delta + i\beta$) is larger than the imaginary part, β, in the x-ray region. Since only β is used in absorption contrast, phase-contrast imaging techniques which give access to δ are very important. In this paper we present two different implementations of phase contrast in our compact soft x-ray microscope, differential-interference contrast and Zernike phase contrast.

1. Introduction

Different phase-contrast x-ray microscopy methods have been demonstrated at synchrotron light sources. However, the implementation of these techniques in our compact soft x-ray microscope are not straightforward. The main reason is the difference in sample illumination by the condenser or, more precisely, the degree of spatial coherence in the sample plane, which always will be partially coherent. We have simulated the image formation process in a full-field microscope including partial coherence [1]. With the help of this program, different phase-contrast methods have been simulated [2], and the results of the simulation are the basis for two different methods we have verified experimentally. Both techniques are based on the use of diffractive-optical-element (DOE) objectives, i.e., the zone-plate objective is exchanged by a modified optic. Here, one DOE optic gives differential interference contrast (DIC), while the other shows Zernike phase contrast [3, 4]. Both optics are suited for use in compact X-ray microscopes.

The performance of both DIC and Zernike DOE objectives were evaluated in comparison with a normal zone plate by imaging a nickel Siemens star pattern and linear grating test objects. Images obtained with the phase-contrast optics exhibit typical DIC or Zernike contrast enhancement in addition to the normal absorption contrast.

2. Theory

The idea of a single-element phase-contrast optic was first proposed by Chang et al [5]. A similar optic was then used in phase-contrast microscopy by Di Fabrizio [6], and later by Vogt [7] and Chang [8], with a high degree of coherence in the condenser illumination. The latter experiment was carried out by covering parts of the condenser, in order to achieve circular polarized radiation. Since then, other types of single-element phase-contrast optics have been developed and tested [9, 10]. In compact soft X-ray microscopy, the degree of coherence in the sample plane is low due to the photon economy and thus new types of optics were developed to satisfy this constraint.

The Zernike DOE and the DIC-DOE are similar in the way they manipulate the optical field since both optics make a type of filtering in the Fourier plane. However, the contrast effect is fundamentally different. Whereas the Zernike DOE phase-shifts the undiffracted field relative to the diffracted field by $\pm\pi/2$ yielding contrast for phase-contributing parts of an object, the DIC-DOE creates a point-spread function (PSF) with two spots. These spots have a certain separation, called the shear, and a phase difference, called the bias. The shear and the bias are important parameters in achieving the phase-contrast effect.

2.1. *Single-element Zernike phase-contrast optic*

Zernike phase contrast is an established technique in x-ray microscopy when using an arrangement where a phase ring phase-shifts the undiffracted light [11, 12]. However, in soft x-ray microscopy, focal lengths are short which makes a phase ring more difficult to implement. By moving the phase-shifting ring to the zone-plate itself and instead shifting the appropriate zones, the arrangement is simplified. The result can be viewed in figure 1.c. It is important to emphasize that this is only possible if the Fourier plane and the zone plate plane are comparable. This will affect the available field-of-view depending on how coherent the illumination of the sample is.

With Zernike phase contrast, one may either phase-shift the diffracted light by $\pi/2$ (positive Zernike phase contrast) or $-\pi/2$ (negative phase contrast). For X-rays, the positive Zernike configuration results in an inverted contrast of the sample, as compared to absorption contrast.

2.2. *Differential-interference contrast optic*

In a previous publication, the illumination requirements for single-element DIC imaging were thoroughly investigated [2]. It can be shown that by making a so-called phase cut in the optical element, a PSF with two spots can be created. The shear and the bias of the resulting PSF depend on the position of the phase cut, which phase shifts the zones on one side by π (see figure 1.a). It can also be shown that a useful position of this phase cut is at half the radius of the zone plate, creating the side-cut DIC-DOE. This position yields a shear equal to the resolution limit of the optic, and a bias of $\pi/3$. Due to the small shear produced, this optic does not degrade the resolution.

A necessary requirement for the side-cut DIC-DOE is to have a sufficient degree of coherence in the sample plane. The shear distance in the sample plane needs to be coherently illuminated. If this is fulfilled, two points in the sample plane that are separated by the shear can interfere in the image plane. If the degree of coherence is partially coherent over the shear, the image contrast will decrease.

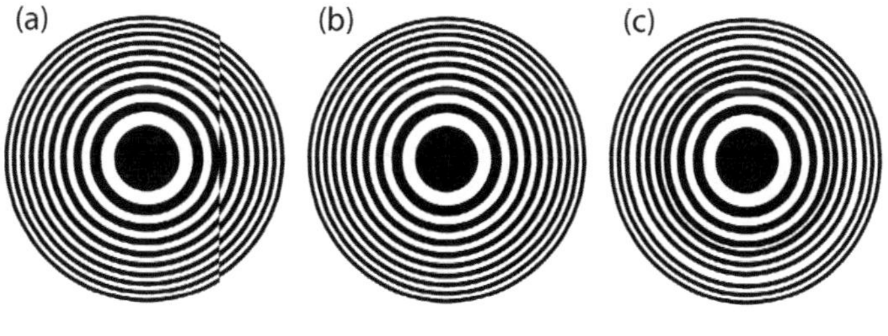

Figure 1. (a) The DIC-DOE, (b) a normal zone plate and (c) the Zernike zone plate.

3. Results

Each type of phase-contrast DOE was manufactured in nickel [13]. The outermost zone was 50 nm and the diameter was 50 or 75 μm. The resulting image for each phase-contrast optic was compared to an equivalent zone plate without any phase shifts. The imaged objects were also manufactured in nickel, with a thickness of approximately 100 nm.

Figure 2 shows images taken with the compact soft X-ray microscope using the two types of phase-contrast DOEs and a corresponding normal zone plate. The wavelength was 2.48 nm and the condenser had an outermost zone width of 50 nm and a diameter of 4.53 mm. The resulting numerical aperture ratio was $m=0.5$ which was enough to give sufficient degree of coherence in the sample plane. When compared to an equivalent zone plate without any phase shifts, both optics yielded a measurable contrast increase for the given object [3, 4]. In addition, the Zernike zone plate showed inverted contrast as expected for the positive Zernike configuration. Although not visible in figure 2, the field-of-view was limited in the Zernike contrast arrangement. This is a consequence of integrating the phase plate with the zone plate, but depends on the divergence of the illumination. The Zernike contrast image also suffered from halo and shade-off effects, which are known artifacts from optical microscopy.

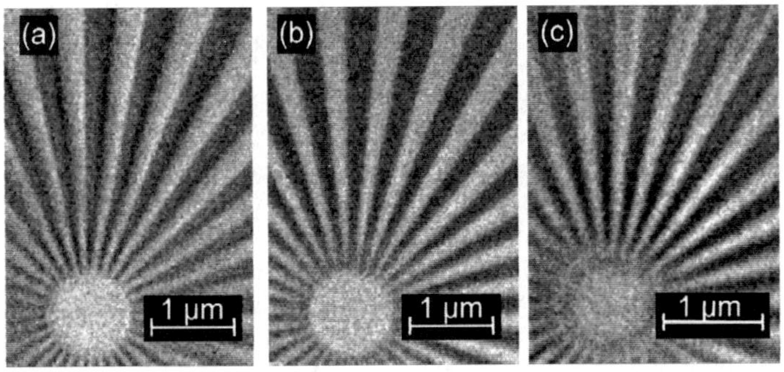

Figure 2. Images taken in the compact soft x-ray microscope using (a) a side-cut DIC-DOE, (b) a normal zone plate, and (c) a Zernike zone plate. Phase-contrast effects are clearly visible in (a) and (c), where contrast reversal is also visible. Note the difference in scale between (b) and (c).

4. Conclusions

Compact soft x-ray microscopy can benefit from single-element phase-contrast optics since implementation and design is straight-forward. Contrast increase is possible, resulting in a lower absorbed dose and/or reduced exposure times. The Zernike zone plate has the advantage of improved contrast in all directions, while it is limited in field-of-view. The DIC-DOE can provide a larger field-of-view but will only show contrast in one direction. Further investigation in the benefits and disadvantages of each optic is necessary before choosing one above the other.

References

[1] von Hofsten O, Takman PAC, Vogt U 2007 *Ultramicroscopy* **107** 604
[2] von Hofsten O, Bertilson M, Vogt U 2008 *Opt Express* **16** 1132
[3] von Hofsten O, Bertilson M, Holmberg A, Vogt U 2008 *Opt Lett* **33** 932
[4] Bertilson M, von Hofsten O, Vogt U 2008 *Appl Phys Lett* **92** 064104
[5] Chang C, Naulleau P, Anderson E, Rosfjord K, Attwood D 2002 *Applied Optics* **41** 7384
[6] Di Fabrizio E, Cojoc D, Cabrini S, Kaulich B, Susini J, Facci P, et al. 2003 *Opt Express* **11** 2278
[7] Vogt U, Lindblom M, Jansson PAC, Tuohimaa TT, Holmberg A, Hertz HM, et al. 2005 *Opt Lett* **30** 2167
[8] Chang C, Sakdinawat A, Fischer P, Anderson EH, Attwood DT 2006 *Opt Lett* **31** 1564
[9] Sakdinawat A, Liu Y 2008 *Opt Lett* **32** 2635
[10] Sakdinawat A, Liu Y 2008 *Opt Express* **16** 1559
[11] Schmahl G, Rudolph D, Schneider G, Guttmann P, Niemann B 1994 *Optik* **97** 181
[12] Neuhausler U, Schneider G, Ludwig W, Meyer MA, Zschech E, Hambach D 2003 *J Phys D* **36** 79
[13] Holmberg A, Rehbein S, Hertz HM 2004 *Microelectron Eng* **73-74** 639

X-ray imaging with laser-Compton scattering X-ray at AIST

M Yasumoto[1], R Kuroda[1], H Ikeura-Sekiguchi[1], H Toyokawa[1], N Sei[1], H Ogawa[1], M Koike[1], K Yamada[1], K Mori[2], F Sakai[3], S Tamura[4]

[1] Research Institute of Instrumentation Frontier, AIST, Umezono 1-1-1, Tsukuba, Ibaraki, 305-8568, Japan
[2] Ibaraki Prefectural University of Health Sciences, Ami 4669-2, Inashiki, Ibaraki 300-0394, Japan
[3] Sumitomo Heavy Industries, Ltd. (SHI), 2-1-1, Yatocho, Nishitokyo, Tokyo 188-8585, Japan
[4] Photonics Research Institute, AIST, Midorigaoka 1-8-31, Ikeda, Osaka, 563-8577, Japan

E-mail: m.yasumoto@aist.go.jp

Abstract. An X-ray facility with a compact electron linear accelerator via laser-Compton scattering (LCS X-ray) is recently installed at AIST. The LCS X-ray facility in approximately 8-m square consists of a 40-MeV linac, a femtosecond terawatt laser and an X-ray beamline. The LCS X-ray facility produces quasi-monochromatic X-rays with energy in the range of 12 ~ 40 keV which are commonly used in many studies currently carried out at synchrotron radiation facilities. The LCS X-ray is suitable for phase-contrast imaging such as in-line method, diffraction-enhanced imaging method, X-ray interferometry method, because of the quasi-monochoromaticity and the small source size. These advantages produce much clearer image of weakly absorbing specimens in the radiography of clinical and industrial application. Here we report on the feasibility study of the in-line phase-contrast imaging of biological samples using the LCS X-ray. Our experimental arrangement is widely used in the in-line method. The weakly absorbing parts of the samples are successfully imaged under the phase-contrast condition.

1. Introduction

An X-ray facility with a compact electron linear accelerator via laser-Compton scattering (LCS X-ray) is recently installed at National Institute for Advanced Industrial Science and technology (AIST) in Japan [1, 2]. The LCS X-ray has comparable qualities such as small source size, tunable energy, high brightness to those provided by other large synchrotron radiation facility. The compact X-ray source is suitable for imaging application (radiography) in medical and industrial applications. Thus, we started new X-ray imaging project in these applications. Here, we report on status of the new X-ray imaging project.

2. Laser -Compton scattering X-ray (LCS X-ray)

Our LCS X-ray facility is placed in approximately 10-m square room. The facility consists of a photo-cathode RF gun, a S-band accelerator tube of 40-MeV, a femto-second terawatt laser (Ti:Sa

Journal of Physics: Conference Series **186** (2009) 012039 doi:10.1088/1742-6596/186/1/012039

laser) and an X-ray beamline. Figure 1 shows the schematic layout of the LCS X-ray facility at AIST. The X-ray beamline has ~2-m square area provided for X-ray application users.

The LCS X-ray has uniqueness of ultra-short, widely tunable energy and small source size compared with other X-ray sources. The LCS X-ray facility produces quasi-monochromatic X-rays with energy in the range of 12 ~ 40 keV which is commonly used in many studies currently carried out at other synchrotron radiation facilities. The LCS X-ray is incoherent beam in terms of both time and space. Table 1 shows the characteristics of the LCS X-ray at AIST. The pulse width is ultara-shorted bunch character of femoto-second order duration. The short bunched X-ray beam is suitable for the time-resolved application. On the other hand, the small source size of the LCS X-ray is capable to realize the phase-contrast imaging (PCI) which becomes a routine technique with synchrotron radiation X-ray [3, 4, 5].

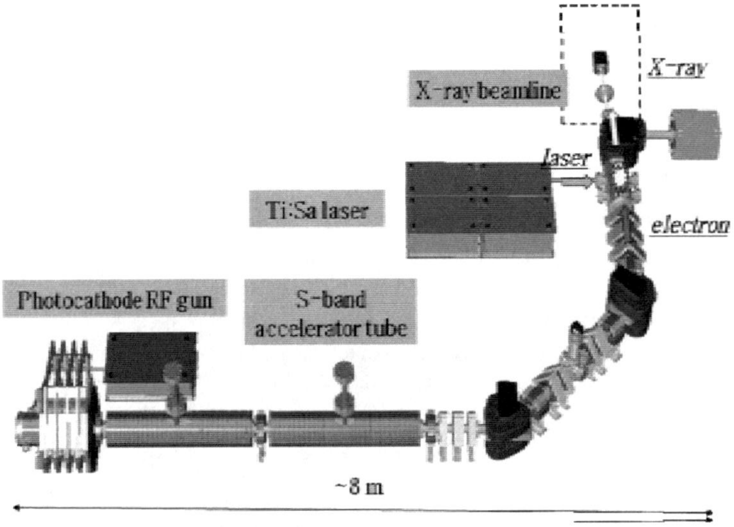

Figure 1 Schematics of LCS X-ray facility.

Table 1. LCS X-ray characteristics at AIST

Energy tunable range	12 ~ 40keV
Pulse width	150 fs ~ 3ps
Energy spread	4% @ 2.5 mrad 10% @ 7.5 mrad
Source diameter	~40 μm
Repetition rate	10 pps
Photon flux	~10^7 s^{-1}

Table 2. Specifications of HARP camera.

Gain	~ 200 times
Lines of resolution	> 700 TV lines
Frame rate	10 (up to 30) fps
Effective pixels	1280 × 1024
Resolution	12 bit

3. X-ray imaging system using LCS X-ray

By using the LCS X-ray we explore multiple imaging which is categorized with real-time imaging, 3-dimenstional (3D) imaging and high-contrast imaging.

In order to realize the real-time imaging we introduced a high sensitive X-ray imaging system that composes of the X-ray image-intensifier and a high sensitivity camera. The camera has a special image pick-up tube with an amorphous selenium (a-Se) photoconductive target and is called HARP (High-gain Avalanche Rushing amorphous Photoconductor). The HARP camera has sensitivities about 100 times greater than CCD cameras. Table 2 indicates the specifications of the HARP camera system. Therefore, the system can realize the real-time imaging synchronized with the X-ray pulse beam (10 pps).

We demonstrated the phase-contrast imaging with the LCS X-ray [6]. The observed specimen was dissected rat lumbar vertebrae fixed in formalin. The distances between the specimen and an image detector (imaging plate) are 750 mm and 40 mm, which are corresponding to the conditions of phase-contrast image and absorption-contrast image, respectively. The former The X-ray image under the former condition is much clearer than the image of the latter condition in the view of small bone structure. This result leads to that the LCS X-ray can be applied with this enhancement technique to the medical clinic.

4. Conclusions

We investigated advantages of the LCS X-ray imaging with showing the unique facility. The LCS X-ray imaging system is relatively compact rather than the imaging system on synchrotron radiation facility. In our project, the LCS X-ray radiography with hard X-ray has been applied to phase-contrast method. Thus, the LCS X-ray source is of considerable interest and will be potentially replaced with other X-ray sources in the near future.

References:

[1] Kuroda R et al. 2007 *Int. J. Mod. Phys.* B **21** 488
[2] Toyokawa H et al. 2007 *Proc. of Particle Accelerator Conference 2007 (PAC07)* 121-123
[3] Mori K et al. J. 2002 *Synchrotron Rad.* **9** 143.
[4] Wilkins S. W. et al. 1996 *Nature* **384** 335.
[5] Shimao D. et al. 2005 *Jpn. J. Appl. Phy.* **44** 450.
[6] Sekiguchi-Ikeura H et al. 2008 *Appl. Phy. Lett.* **92** 131107

Journal of Physics: Conference Series **186** (2009) 012040 doi:10.1088/1742-6596/186/1/012040

Multimodal imaging for the detection of sub-micron particles in the gas-exchange region of the mammalian lung

David Haberthür[1], Manuela Semmler-Behnke[2], Shinji Takenaka[2], Wolfgang G. Kreyling[2], Marco Stampanoni[3,4], Akira Tsuda[5] and Johannes C. Schittny[1]

1 Institute of Anatomy, University of Bern, Switzerland
2 Institute for Inhalation Biology, Helmholtz Zentrum München, Germany
3 Swiss Light Source, Paul Scherrer Institut, Switzerland
4 Institute for Biomedical Engineering, University and ETH Zürich, Switzerland
5 Physiology Program, Harvard School of Public Health, USA

E-mail: schittny@ana.unibe.ch

Abstract. The deposition sites of inhaled aerosols in the gas-exchange region of the lung represent one of the key parameters needed for the understanding of the interaction between these particles and lung tissue. In order to develop a method for three-dimensional imaging of sub-micron particles in lung tissue we applied gold particles (200 and 700 nm) to rat lungs by intratracheal instillation. The samples were scanned at TOMCAT, the beamline for TOmographic Microscopy and Coherent rAdiology experimenTs at the Swiss Light Source. The 200 nm particles were slightly below the detection capabilities of TOMCAT. Therefore, their localization was obtained only by electron microscopy. At a voxel size of 350 nm we observed single and clustered gold particles (700 nm) in alveoli, alveolar ducts, and small bronchioli. The locations of the gold particles were verified by transmission electron microscopical serial sections. We observed a very high correlation between these two imaging modalities. We conclude that a combination of x-ray tomographic microscopy and electron microscopy allows the full unrestricted 3D localization of particles smaller than the resolution of x-ray tomographic microscopy. We are planning to use this method for the verification of the simulation of particle deposition in the airway tree.

1. Objective

To study the deposition of sub-micron particles in the mammalian lung, we used synchrotron radiation based x-ray tomographic microscopy (SRXTM) to record tomographic images of lung samples with high resolution (350 nm). In addition to SRXTM, we used conventional transmission electron microscopy (TEM) to obtain high resolution images (resolution smaller than 1 nm).

2. Materials and Methods

We applied 200 nm and 700 nm gold particles to rat lungs by intratracheal instillation. Thirty minutes after instillation, the lungs were fixed with 2.5 % glutaraldehyde by vascular perfusion, stained according to standard protocols for TEM and embedded in Epon. The samples were shaped to cylinders with a diameter of either 0.6 or 1.2 mm and a length of several millimeters using a watchmakers lathe and then scanned at TOMCAT [1], the beamline for TOmographic Microscopy and Coherent rAdiology experimenTs at the Swiss Light Source at the Paul Scherrer Institute in Villigen, Switzerland. The samples were scanned at a wavelength of 11.5 keV, post processed on a 20-node Linux cluster using filtered backprojection, resulting in an image stack of 2048×2048×2048 pixels with isotropic voxels of 350 nm side length. The samples were visualized in three dimensions using an isosurface computed with Imaris x64 5.72 (Bitplane AG, Switzerland) on an AMD 64-based windows computer [2]. After tomographic image acquisition serial sections of the sample were cut and processed for transmission electron microscopy [3].

3. Results

We have been able to observe single and clustered gold particles in alveoli, alveolar ducts, and small bronchioli while imaging them at a voxel side length of 350 nm with the use of SRXTM (Fig. 1a and 1b). We were able to verify the locations of the observed gold particles by TEM (Fig. 1c).

Particles with a size of 200 nm — a size smaller than the resolution achievable at TOMCAT — were imaged using TEM after selecting a particular alveolus in three dimensional visualizations obtained from SRXTM data. Thirty minutes after instillation, particles were observed in cells — mainly macrophages — as well as in airspace. Most of the particles located in the airspace were in close contact to protein precipitations (Data not shown).

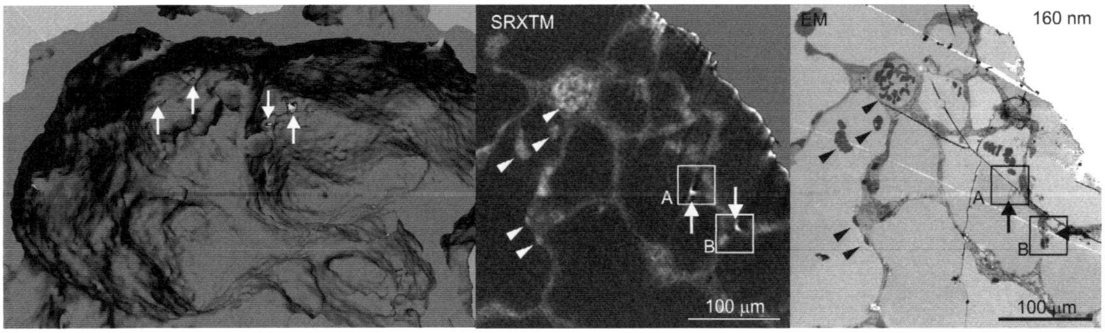

(a) (b) (c)

Figure 1: (a): Three dimensional isosurface visualization of gold particles in the lung obtained from SRXTM data. Gold grains (yellow) are deposited in terminal airspaces. The air-tissue interface was removed on top of the gold grains during the visualization process in order to show the gold particles (arrows). (b): Detailed view of virtual SRXTM section obtained at TOMCAT containing two gold particles (arrows). The arrowheads are pointing to erythrocytes which are lighting up in the SRXTM images due to their high iron content of the hemoglobin. Due to the high contrast of gold the particles appear larger in the SRXTM images than they are in reality and we observed spiked image artifacts going out from the particles. (c): Corresponding TEM section of the virtual SRXTM section. The thin black lines in (c) represent folds in the serial section. The white lines in (c) represent knife marks from the sectioning process.

4. Discussion

A very high correlation between the two imaging modalities was observed. The virtual slices obtained from the SRXTM image stack (Fig. 1b) and the real slices obtained using TEM (Fig. 1c) have only been corrected for rotation and magnification. The correct vertical position of the serial

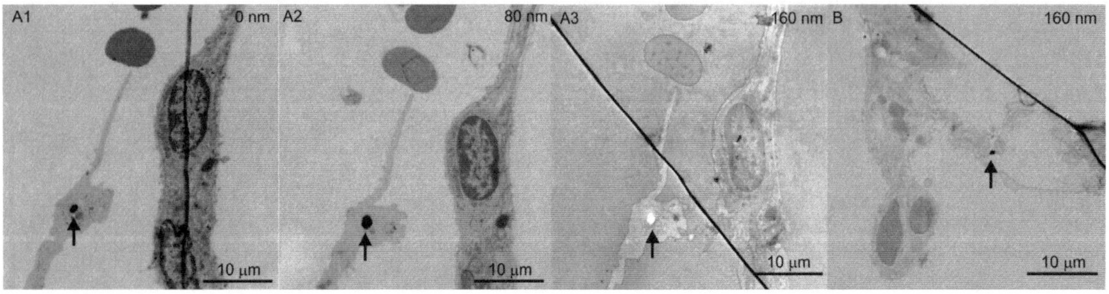

(a) (b) (c) (d)

Figure 2: Transmission electron microscopy images of 700 nm gold particles: (a)–(c): Close-up view of square A of Fig. 1b showing a gold particle (arrows) in consecutive TEM sections which are 80 nm apart. (d): Gold grain observed in square B in Fig. 1b (arrow). Roughly half of the gold grains observed were located inside the cells, e.g. macrophages. The thin black lines visible in subfigures (a), (c) and (d) represent folds in the TEM section.

TEM section in the sample was obtained through rigorous alignment and precise positioning of the sample in the microtome.

We have been able to track the gold particles over consecutive serial sections, even if they sometimes were not directly visible. As can be seen in Fig. 2c, we sometimes only observed a hole where we would have expected to see the gold particle (arrow). Since the gold grains are much harder than Epon and do not stick well to the resin, it is expected that the grains will be pulled out of the Epon block as soon as half of the grain is cut.

5. Conclusions

We conclude that the combination of SRXTM and TEM allows the three dimensional localization of particles in the mammalian lung. In a synergistic way, we used SRXTM to obtain the full unrestricted 3D access and TEM to verify the localization of the particles in the 3D-space with very high resolution.

We are planning to use this method for the detection and localization of inhaled particles and as a mean of providing data for airflow simulation in the mammalian lung.

Acknowledgments

We thank Christoph Hintermüller and Federica Marone for expert help at TOMCAT and Bettina De Breuyn as well as Christoph Lehmann for the preparation of the samples. This work has been supported by Swiss National Science Foundation grant 3100A0-109874 and by US National Heart, Lung, and Blood Institute grant HL-070542.

References

[1] Stampanoni M, Groso A, Isenegger A, Mikuljan G, Chen Q, Meister D, Lange M, Betemps R, Henein S and Abela R 2007 *AIP Conference Proceedings* **879** 848–851 URL `http://link.aip.org/link/?APC/879/848/1`

[2] Tsuda A, Filipovic N, Haberthür D, Dickie R, Matsui Y, Stampanoni M and Schittny J C 2008 *J Appl Physiol* **105** 964–976 URL `http://jap.physiology.org/cgi/content/abstract/105/3/964`

[3] Mund S I, Stampanoni M and Schittny J C 2008 *Developmental Dynamics* **237** 2108–2116 URL `http://dx.doi.org/10.1002/dvdy.21633`

Energy-tunable full-field x-ray microscopy: Cryo-tomography and nano-spectroscopy with the new BESSY TXM

S Heim, P Guttmann, S Rehbein, S Werner and G Schneider
Helmholtz-Zentrum Berlin für Materialien und Energie GmbH, Wilhelm-Conrad-Röntgen-Campus, BESSY II, Albert-Einstein-Strasse 15, 12489 Berlin, Germany

E-mail: sheim@bessy.de

Abstract. We describe the new full-field transmission x-ray microscope (TXM) for automated cryo-tomography and nano-spectroscopy which is installed at the BESSY undulator U41. For the first time in soft x-ray microscopy, an elliptically shaped single-bounce glass capillary is employed as condenser which collects the divergent radiation emerging from the exit slit of a focusing spherical grating monochromator (FSGM). The new TXM overcomes several limitations of conventional TXM setups, it provides an energy resolution up to $E/\Delta E = 10^4$ which is well suited for spectromicroscopy. Furthermore, no pinhole is required close to the sample plane which facilitates flat sample holder geometry for nano-tomography. With its high energy resolution, the TXM also supports imaging in high orders of diffraction of the zone plate objective. Using this approach, a spatial resolution of 14 nm lines and spaces is demonstrated. In addition, we present selected applications from materials and life sciences.

1. The new BESSY TXM

Driven by severe limitations of conventional TXM setups which do not permit spectromicroscopy studies and are restricted to glass capillary holders for tomography applications, the BESSY microscopy group developed a new microscope based on an advanced x-ray optical element [1] and a high precision cryo tomography stage. The separation of monochromatization and object illumination is key to the new optical concept. A conventional grating monochromator provides the high energy resolution while a single-bounce ellipsoidal glass capillary employed as achromatic condenser illuminates the object. This optical concept works ideally in combination with an undulator source providing the required collimated x-ray beam and high photon flux.

The new BESSY TXM also significantly improves the possibilities to study cryogenic or heated samples for life and materials sciences. Taking advantage of the developments over the last decades in cryo electron tomography, the sample stage of the TXM is an adaption of a state-of-the-art TEM stage. Therefore, the new microscope provides the same excellent mechanical accuracy and temperature stability known from cryo TEM, a tilt range of $\pm 80°$ and lateral travel ranges of ± 1 mm with a bidirectional reproducibility of < 15 nm. The translation axes are mounted with respect to the rotation axis in order to make it possible to establish eucentricity for any part of the specimen, i.e. bring any point of the sample onto the spatially fixed tilt axis. Together with the high photon flux and the fully automated operation of the microscope's control software, the aquisition time for a full tomographic tilt series is reduced to less than 20 minutes.

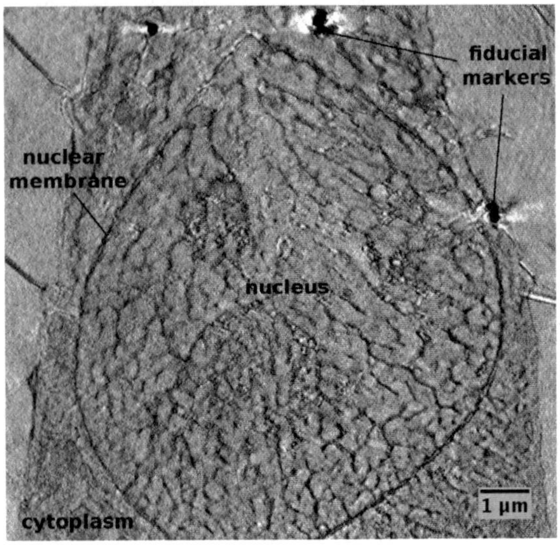

Figure 1. Slice through the reconstructed local linear absorption coefficient of the nucleus of a mouse adenocarcinoma cell; 131 projections recorded at $E = 510\,\text{eV}$.

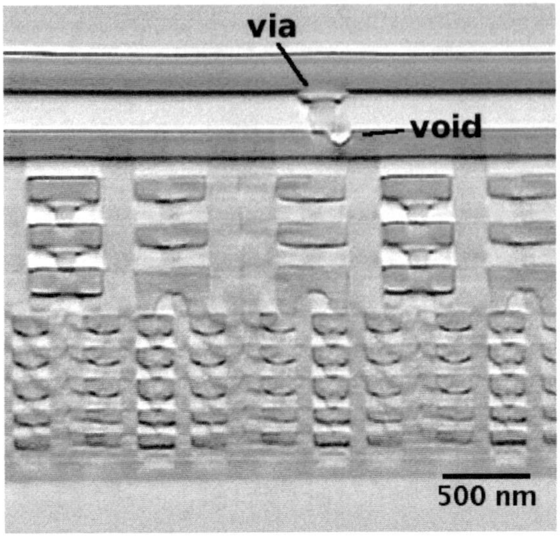

Figure 2. Slice from a tomographic reconstruction of a stress migration sample imaged at $E = 700\,\text{eV}$, visualizing the location of the void close to the via.

2. Applications

2.1. Cryo-tomography of frozen-hydrated mammalian cells

Full-field x-ray microscopy has proven to be an excellent tool for 3-D studies of whole frozen-hydrated biological samples [2, 3]. However, tomography of flat grown mammalian cells was not possible as they do not fit into glass capillary sample holders. Our new TXM overcomes this severe limitation, because its optical concept requires no monochromator pinhole close to the object plane. In the new TXM, it is possible to study adherent cells grown on TEM grids. As an example, figure 1 shows one slice of a tomographic reconstruction of a frozen-hydrated mouse cell imaged at 510 eV. The voxel size is 9.8 nm, which is also the thickness of the slice. The nuclear membrane with its double structure and internal structures of the nucleus are visualized.

2.2. Stress-migration in advanced copper interconnect structures

Besides electromigration in semiconductor devices, stress migration is one of the most important failure mechanisms in advanced copper interconnects. As a result of lattice vacancy migration due to local mechanical stress at the via/plate interface small holes called voids form. However, the void formation mechanism is not fully understood. X-ray microscopy permits to study dynamic void evolution during stress migration. An exemplary micrograph from the stress-migration studies is shown in figure 2. This slice through the reconstructed volume of a 1 μm thick IC lamella shows that the void is located in the copper interconnect line directly below the via. Depending on their size, larger voids lead to unacceptable resistance increase and ultimately cause open circuit or other circuit malfunction.

2.3. Imaging in the third order of diffraction

The resolving power of zone plates scales with the order of diffraction. By employing high orders of diffraction, it is possible to increase the resolution without the need for manufacturing smaller outermost zone widths. In order to demonstrate the resolution achievable with high order imaging, we designed a multi-layer test sample with B_4C/Cr lines and spaces of 5 different

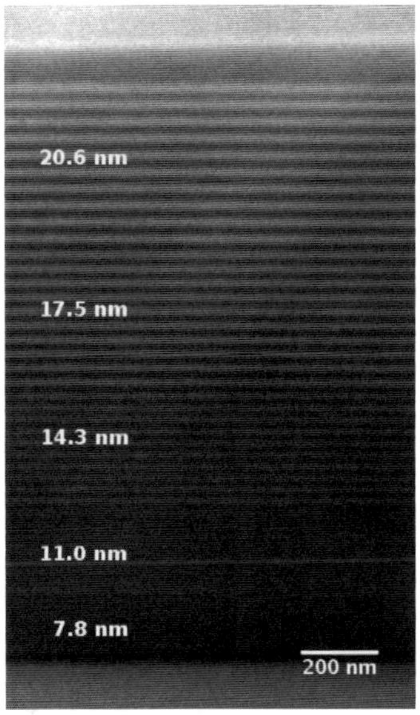

Figure 3. Bi_4C/Cr multi-layer test structure imaged in the 3rd diffraction order using a $dr_N = 25$ nm zone plate, $E = 700$ eV, $E/\Delta E = 1700$, $t_{\text{exp}} = 15.7$ s

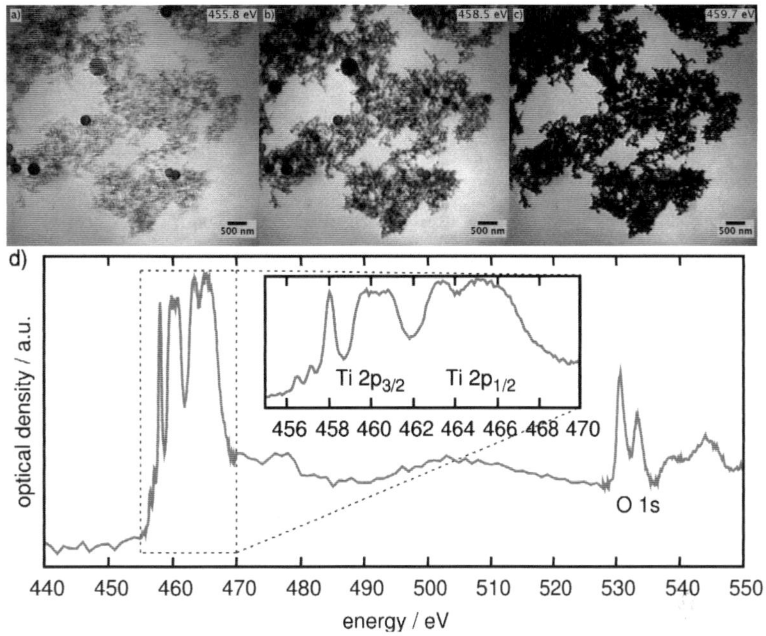

Figure 4. a)-c) Selected micrographs of TiO_2 particles imaged at different photon energies. d) Spectrum for position indicated in a) - c) extracted from the whole energy stack (440-550 eV, $E/\Delta E = 7200$, total 380 images).

widths down to 7.8 nm. Figure 3 shows the result of a resolution test experiment performed with the BESSY TXM at 700 eV photon energy with a 25 nm zone plate operating in the 3rd order. Note that lines and spaces of 14.3 nm are clearly resolved while the 11.0 nm structures still show some modulation.

2.4. Nano-spectroscopy of TiO_2 particles

The spectromicroscopy capabilities of the new TXM are demonstrated with TiO_2 particles. An image stack was taken in the energy range from the Ti L-edge up to the oxygen K-edge with an energy resolution of $E/\Delta E \geq 7200$ (cf. figure 4). Note the combined excellent spectral and spatial resolution in a large field of view.

Acknowledgments

The authors are greatly indebted to S. Braun (IWS, Dresden) and Y. Ritz (AMD, Dresden) for the fabrication of the multi-layer test sample and W. Müller (NIH, Bethesda) for the cell preparation. This work was supported by the EU in the 6th framework program under contract number RII3-CT-2004-506008, the Human Frontier Science Program (HFSP) Research Grant Ref. RGP0053/2005-C and the German Federal Ministry of Education and Research (BMBF) under contract number 05KS4BY1/7.

References

[1] Guttmann P, Zeng X, Feser M, Heim S, Yun W and Schneider G, 2008, *this conference*
[2] Schneider G, *Ultramicroscopy* **75** (1998) 85-104
[3] Weiss D, Schneider G, Niemann B, Guttmann P, Rudolph D, Schmahl G, *Ultramicroscopy* **84** (2000) 185–197
[4] Thieme J, Schmahl G, Rudolph D, Umbach E (Eds.): X-Ray Microscopy and Spectromicroscopy, Proceedings of the Fifth International Conference on X-Ray Microscopy, Springer-Verlag Heidelberg (1998)

Journal of Physics: Conference Series **186** (2009) 012042 doi:10.1088/1742-6596/186/1/012042

X-ray Tomographic Microscopy at TOMCAT

F Marone[1], C Hintermüller[1], S McDonald[1,2], R Abela[1], G Mikuljan[1], A Isenegger[1], M Stampanoni[1,3]

[1] Swiss Light Source, Paul Scherrer Institut, 5232 Villigen, Switzerland
[2] Department of Radiology, University of Lausanne Medical School, Lausanne, Switzerland
[3] Institute for Biomedical Engineering, University and ETH Zurich, 8092 Zurich, Switzerland

federica.marone@psi.ch

Abstract. Synchrotron-based X-ray Tomographic Microscopy is a powerful technique for fast non-destructive, high resolution quantitative volumetric investigations on diverse samples. At the TOMCAT (TOmographic Microscopy and Coherent rAdiology experimenTs) beamline at the Swiss Light Source, synchrotron light is delivered by a 2.9 T superbend. The main optical component, a Double Crystal Multilayer Monochromator, covers an energy range between 8 and 45 keV. The standard TOMCAT detector offers field of views ranging from 0.75x0.75 mm^2 up to 12.1x12.1 mm^2 with a pixel size of 0.37 μm and 5.92 μm, respectively. In addition to routine measurements, which exploit the absorption contrast, the high coherence of the source also enables phase contrast tomography, implemented with two complementary techniques (Modified Transport of Intensity approach and Grating Interferometry). Typical acquisition times for a tomogram are in the order of few minutes, ensuring high throughput and allowing for semi-dynamical investigations. Raw data are automatically post-processed online and full reconstructed volumes are available shortly after a scan with minimal user intervention.

1. Introduction

X-ray tomographic microscopy is a powerful technique for fast non-destructive, high resolution, quantitative 3D investigations on an ample variety of samples. Absorption based and "edge-enhanced" tomographic experiments have been routinely performed at TOMCAT [1] since its kick off in June 2006. At TOMCAT we also perform phase contrast tomography. We offer to the user community two complementary techniques based on a Modified Transport of Intensity approach [2] and a Differential Phase Contrast method (DPC) with a grating interferometer [3]. The first phase contrast approach yields a good approximation of the 3D distribution of the refractive index decrement of a weakly absorbing object from a single tomographic dataset without need of additional hardware. It is particularly suited for small specimens when high resolution is required. The second phase contrast technique, the Differential Phase Contrast (DPC) method [3] is characterized by a higher sensitivity, by moderate resolution and gives the best results with larger samples. During the past year, considerable efforts have been spent on fully integrating this approach at the beamline, both in terms of fast acquisition and reconstruction. TOMCAT is one of the few beamlines in the world offering this new feature to the user community.

2. Beamline overview

Synchrotron light is delivered to the beamline by a 2.9 T superbend, with a critical energy of 11.1 keV. In terms of optics, TOMCAT has been kept simple, to prevent degradation of the beam profile. The only optical elements present are a Chemical Vapor Deposited (CVD) diamond window, separating the UHV sector of the machine from the HV of the beamline and a fixed-exit double crystal multilayer monochromator (DCMM), which covers an energy range from 8 to 45 keV (energy bandwith: 2-3% with multilayers, 0.018% with Si111) and it is located at approximately 7 m from the source in the front end. This unusual location guarantees a beam size in the experimental hutch, at 20 m from the source, of 40 mm horizontally and 7-4 mm vertically (for energies between 10 and 30 keV).

The TOMCAT endstation, in its standard setup features the beam conditioner block, the sample manipulator unit and the microscope. The standard TOMCAT detector offers field of views ranging from 0.75x0.75 mm^2 up to 12.1x12.1mm^2 with a pixel size of 0.37 μm and 5.92 μm, respectively. A more efficient second system, based on a high-aperture tandem 1:1 configuration and accepting a diagonal up to 40 mm, is also available. The CCD camera (PCO2000) has a 2048x2048 pixels chip (7.4x7.4 μm^2 pitch). This standard setup is mostly used for routine absorption experiments and when the Modified Transport of Intensity approach [2] is applied.

For the DPC method, ad-hoc hardware is required (figure 1). The hard X-ray interferometer works with 2 gratings, a phase grating and an absorption grating, positioned at uneven Talbot distances. A phase stepping approach is used to transform the local fringes position of the interference pattern into signal intensity variations, containing information on the phase gradient of the object. The required grid translation for this phase stepping approach is performed with a piezo stage [4], since speed, accuracy and reproducibility are mandatory. The 2 gratings are mounted, at 25 m from the source, on a massive support completely decoupled from the sample stage to avoid vibrations during sample rotation and translation. Phase matching is usually necessary to avoid large phase jumps: the sample needs to be scanned in a wet environment. For this purpose we developed an aquarium [4], consisting of an aluminum tank with entrance and exit Kapton windows along the beam path. The sample holder comes into the aquarium through a hole at its bottom. A gasket system at the bottom of the tank ensures smooth sample rotation, translation and centering.

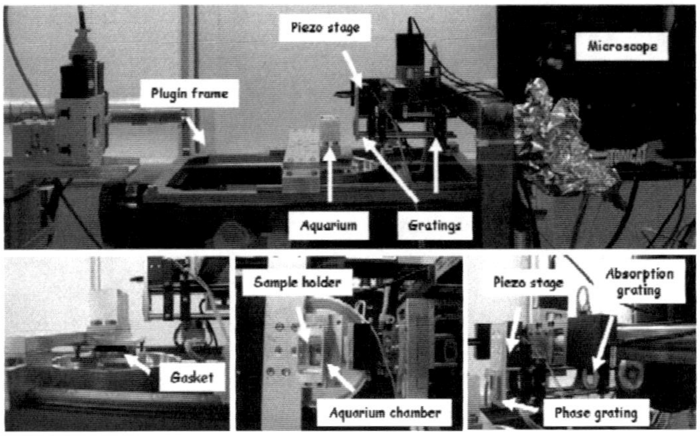

Figure 1. Setup of the TOMCAT endstation for DPC experiments. Top: overview; Bottom: *left*: detail of the gasket system allowing rotation, translation and centering of a specimen in the aquarium chamber - *center*: top view of the aquarium - *right*: detail of the gratings assembly.

3. Software

Fast acquisition of high quality data at TOMCAT is achieved by the combination of a continuous scanning protocol with an optimized acquisition software, where the sample rotates continuously over a 180 degrees range, while images are simultaneously collected by the camera. The optimized acquisition software [5] implemented at TOMCAT exploits the 4 GB internal RAM memory of the detector, where acquired images are transferred at very high rates (up to 14.7 fps at full frame) and its First In First Out (FIFO) buffer acquisition mode. The same approach with continuous sample rotation and FIFO acquisition mode is also used for DPC experiments, where a complete continuous scan is acquired for each phase step. Our new acquisition scheme makes high throughput phase contrast studies possible.

During acquisition, a selection of sinograms is computed on the fly, allowing quality check of the recorded tomogram right after the end of a scan through a user friendly web based application [5]. This application is also used to select and fine tune the reconstruction parameters (e.g. center of rotation and filter).

The complete reconstruction of tomographic datasets, from the raw projections to the 3D volume, occurs on a 5 node Linux cluster (2 dual core Xeon processors clocked at 3.0 GHz per node). Such job can be submitted to the cluster directly from the mentioned web application, with just one mouse click. Each single step of this reconstruction pipeline has been optimized for speed, to be able to fully exploit the advantages of our fast acquisition protocol. A high resolution 2048x2048 (1501 projections) dataset is reconstructed on the cluster in less than 12 minutes.

4. Application

The different sensitivity and resolution achieved with the two phase contrast imaging approaches implemented at TOMCAT is illustrated in figure 2. Two tomograms of the same post-transplant human myocardium biopsy have been acquired with the two techniques and the same coronal slice is shown. The higher sensitivity of the DPC method and the higher resolution achieved with the Modified Transport of Intensity approach are clear. The resolution of the DPC method depends on the Talbot distance and is limited by the grating pitch [3]. In contrast, for the propagation method [2], Fresnel fringes need to be detected and therefore high resolution radiographs are imperative. The best results are obtained with a 10x magnification objective, a 1.5x1.5 mm^2 field of view and a 0.74x0.74 μm^2 pixel size.

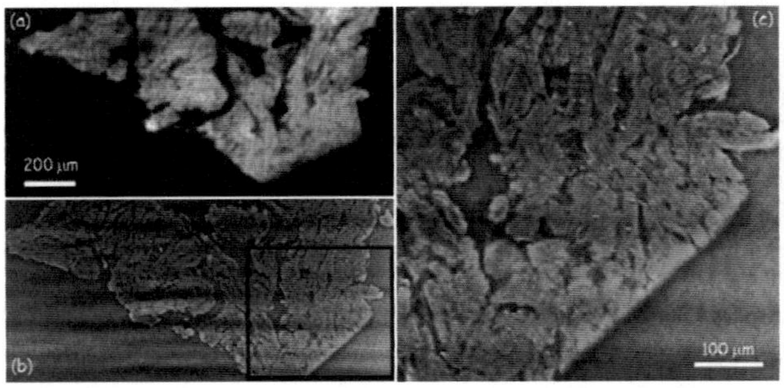

Figure 2. Comparison of two tomograms of a post-transplant human myocardium, embedded in solid paraffin (Sample courtesy of Dr. S. De Marchi, Inselspital, Bern, Switzerland and Dr. M. Gugger, Pathologisches Institut, University of Bern, Switzerland) obtained with the two phase contrast techniques implemented at TOMCAT - (a) DPC tomogram (1.4 µm pixel size, 500 projections, 25 keV, 3rd Talbot distance, 9 phase steps, 3.981 and 2 µm pitch for the phase and absorption gratings, respectively). (b) Tomogram acquired and reconstructed using the Modified Transport of Intensity approach (0.7 µm pixel size, 1500 projections, 20 keV, 32 mm sample-detector distance). (c) Magnification of the region enclosed in the black box in (b).

Acknowledgments

S. McDonald was supported by the Centre d'Imagerie Biomédicale (CIBM) of the UNIL, EPFL, UNIGE, CHUV and HUG. We thank F. Pfeiffer for valuable discussion, C. David and C. Grünzweig for the fabrication of the x-ray optics and O. Bunk for the implementation of the piezo stage.

References

[1] Stampanoni M, Grošo A, Isenegger A, Mikuljan G, Chen Q, Bertrand A, Henein S, Betemps R, Frommherz U, Böhler P, Meister D, Lange M and Abela R 2006 *Developments in X-Ray Tomography V: Proc. SPIE Vol 6318 (San Diego, USA, 13-17 August 2006)* ed. U. Bonse pp. 63180M1-14.
[2] Grošo A, Abela R and Stampanoni M 2006 *Optics Express* **14** 8103-8110.
[3] Weitkamp T, Diaz A, David C, Pfeiffer F, Stampanoni M, Cloetens P and Ziegler E 2005 *Optics Express* **12** 6296-6304.
[4] McDonald S, Marone F, Hintermüller C, Mikuljan G, David C, Pfeiffer F and Stampanoni M in preparation.
[5] Hintermüller C, Marone F, Isenegger A and Stampanoni M in preparation.

High-throughput, high-resolution X-ray phase contrast tomographic microscopy for visualisation of soft tissue

S A McDonald[1,2], F Marone[1], C Hintermüller[1], J-C Bensadoun[3], P Aebischer[3] and M Stampanoni[1,4]

[1]Swiss Light Source, Paul Scherrer Institut, 5232 Villigen PSI, Switzerland.
[2]Department of Radiology, University of Lausanne Medical School, Lausanne, Switzerland.
[3]EPFL, School of Life Sciences, Station 15, 1015 Lausanne, Switzerland.
[4]Institute for Biomedical Engineering, University and ETH Zurich, 8092 Zurich, Switzerland.

E-mail: samuel.mcdonald@psi.ch

Abstract. The use of conventional absorption based X-ray microtomography can become limited for samples showing only very weak absorption contrast. However, a wide range of samples studied in biology and materials science can produce significant phase shifts of the X-ray beam, and thus the use of the phase signal can provide substantially increased contrast and therefore new and otherwise inaccessible information. The application of two approaches for high-throughput, high-resolution X-ray phase contrast tomography, both available on the TOMCAT beamline of the SLS, is illustrated. Differential Phase Contrast (DPC) imaging uses a grating interferometer and a phase-stepping technique. It has been integrated into the beamline environment on TOMCAT in terms of the fast acquisition and reconstruction of data and the availability to scan samples within an aqueous environment. The second phase contrast approach is a modified transfer of intensity approach that can yield the 3D distribution of the phase (refractive index) of a weakly absorbing object from a single tomographic dataset. These methods are being used for the evaluation of cell integrity in 3D, with the specific aim of following and analyzing progressive cell degeneration to increase knowledge of the mechanistic events of neurodegenerative disorders such as Parkinson's disease.

1. Introduction

Phase sensitive X-ray imaging methods can provide substantially increased contrast over conventional absorption based imaging, in particular for biological samples, and therefore exploiting the coherent nature of synchrotron radiation has evident advantages. Phase contrast tomography gives fast access to 3D structural information in near-native biological samples without need for time consuming tissue alteration/modification for contrast enhancement, such as casting and staining as currently required by most high-resolution imaging techniques. An added advantage is that phase signals are produced with much lower dose deposition, which can be very important when radiation damage becomes an issue. Various phase-sensitive X-ray imaging methods have been developed, including interferometric methods [1], propagation methods [2], and techniques using a crystal analyzer, such as diffraction enhanced imaging [3]. These methods differ significantly in terms of the required experimental setup.

In this paper the application of two complementary techniques for high-throughput, high-resolution X-ray phase contrast tomography is illustrated. A Modified Transport of Intensity approach [4] yields a good approximation of the 3D phase distribution of a weakly absorbing object from a single tomographic dataset without the need of additional hardware. It is particularly suited for small specimens when high resolution is required. The Differential Phase Contrast (DPC) method [5], based on grating interferometry, is characterized by a higher sensitivity and suited to larger samples. These methods are being used towards the evaluation of soft tissue and cell integrity in 3D, with the specific aim of following and analysing progressive cell degeneration to increase knowledge of the mechanistic events of neurodegenerative disorders such as Parkinson's disease.

2. Cell Visualisation

The possibility to visualise the integrity of encapsulated cell populations will allow a more robust screening of encapsulated clones for clinical application. Unstained human fibroblast cells were encapsulated within a polymeric hollow fibre containing PVA foam. X-ray microtomography measurements of the capsule were carried out both in absorption and using a phase contrast approach based on Transport of Intensity, which can yield the 3D distribution of the phase (refractive index) of a weakly absorbing object from a single tomographic dataset [4]. Figure 1(a) compares reconstructed slices from datasets scanned using both methods.

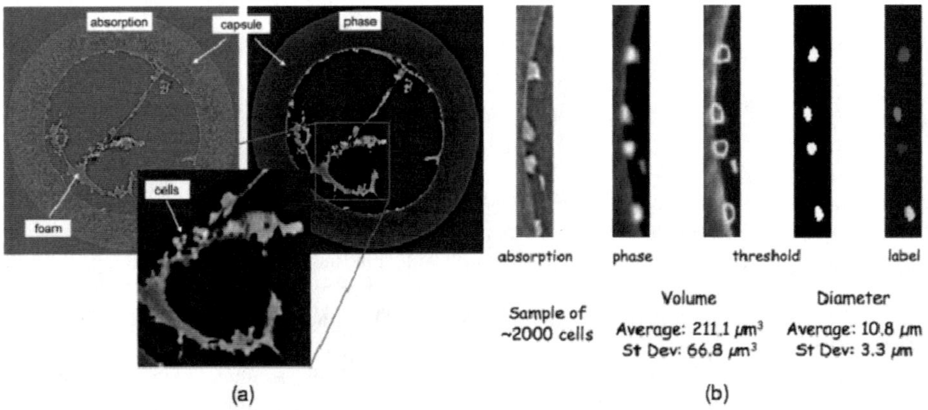

Figure 1. (a) Showing slices through the polymer capsule as scanned using conventional absorption and a Transport of Intensity phase algorithm. The diameter of the capsule is 0.7 mm. Inset: section of foam with cells attached. (b) Labelling of cells based on a nearest neighbourhood method to give isolated objects in 3D.

Use of the Transport of Intensity phase algorithm has enabled easier segmentation of the cells from the PVA foam. Steps have been taken to make some quantitative analysis of the cells. Figure 1(b) illustrates the method taken to label the cells for a small section attached to the wall of the capsule. Clustering of cells has been observed.

3. Visualisation of the Substantia Nigra of a Rat Brain

Differential Phase Contrast (DPC) imaging uses a grating interferometer, consisting of a phase grating and an analyser absorption grating, and a phase stepping technique. Figure 2 shows the experimental setup installed at the TOMCAT beamline of the Swiss Light Source.

A focus has been placed on the fast acquisition and reconstruction of data. A full DPC scan over 9 phase steps and 721 projections can be performed in 20 minutes (energy 17.5 keV, 5th Talbot distance). The technique has been used to visualise the internal soft tissue features of a rat brain. Figure 3(a) shows an *overview* slice through the reconstructed 3D volume of the brain. Density differences can be observed at a spatial resolution of 10 microns.

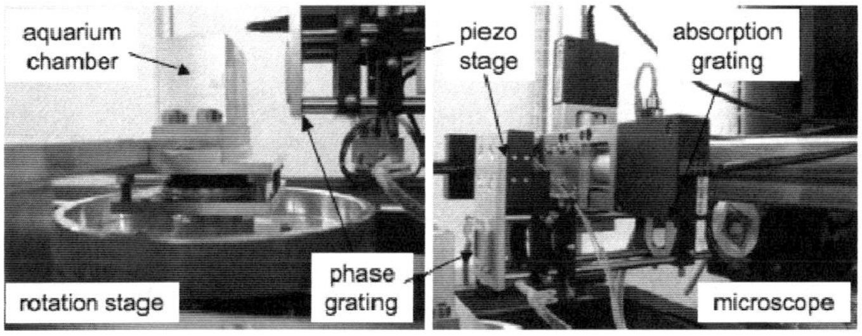

Figure 2. Showing the setup of the interferometer with aquarium at the TOMCAT beamline (SLS). The brain was scanned within liquid paraffin.

Figure 3(b) shows a reconstructed slice of a *local* region of interest of the brain. The tissue structure is observed at a spatial resolution of 3.6 microns. The aim now is to focus around the region of the substantia nigra to follow the progressive degeneration of the dopaminergic neurons.

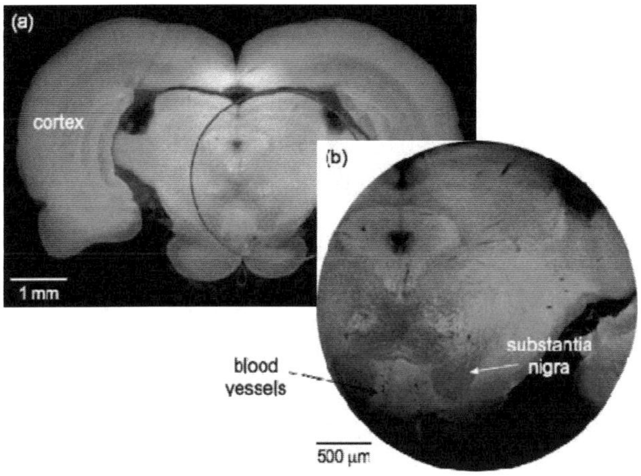

Figure 3. Showing (a) a reconstructed overview slice through the 3D tomographic volume of a rat brain, and (b) a slice of the local region of interest indicated.

4. Summary

Differential Phase Contrast microtomography with grating interferometry has enabled the visualisation of density differences in the soft tissue of a rat brain, with fast scan times. The Transport of Intensity phase contrast algorithm has facilitated the segmentation of encapsulated cells compared to conventional absorption contrast.

Acknowledgements

The authors would like to thank Franz Pfeiffer for fruitful discussion and precious hints, Christian David and Christian Grünzweig for the grating fabrications. S A McDonald was supported by the Centre d'Imagerie Biomédicale (CIBM) of the Université de Lausanne (UNIL), École Polytechnique Fédérale de Lausanne (EPFL), Université de Genève (UNIGE), Centre Hospitalier Universitaire Vaudois Lausanne (CHUV) and Hôpitaux Universitaires de Genève (HUG).

References

[1] Momose A 2003 *Optics Express* **11** 2303
[2] Cloetens P, Ludwig W, Baruchel J, Dyck D V, Landuyt J V, Guigay J P and Schlenker M 1999 *Appl. Phys. Lett.* **75** 2912
[3] Chapman D, Thomlinson W, Johnston R E, Washburn D, Pisano E, Gmür N, Zhong Z, Menk R, Arfelli F and Sayers D 1997 *Phys. Med. Biol.* **42** 2015
[4] Groso A, Abela R and Stampanoni M 2006 *Optics Express* **14** 8103
[5] Weitkamp T, Diaz A, David C, Pfeiffer F, Stampanoni M, Cloetens P and Ziegler E 2005 *Optics Express* **13** 6296

X-ray phase tomography with a Talbot interferometer in combination with an X-ray imaging microscope

Atsushi Momose[1], Yoshihiro Takeda[1], Wataru Yashiro[1], Akihisa Takeuchi[2], Yoshio Suzuki[2]

[1]Department of Advanced Materials Science, Graduate School of Frontier Sciences, The University of Tokyo, 5-1-5 Kashiwanoha, Kashiwa, Chiba 277-8561, Japan.
[2]JASRI/SPring-8, 1-1-1 Kouto, Mikazuki, Hyogo 679-5198, Japan.

E-mail: momose@mml.k.u-tokyo.ac.jp

Abstract. An X-ray Talbot interferometer was combined with an X-ray imaging microscope to generate differential phase contrast. X-ray phase tomography was attained based on the measurement of the differential phase shift using the fringe scanning method. Polymer blend samples were observed and phase separation structures were depicted. The spatial resolution was 1.3 μm in this test experiment, which will be easily improved by increasing the magnification.

1. Introduction

An X-ray Talbot interferometer (XTI) consisting of two transmission gratings is attractive as a device for X-ray phase imaging due to various advantages in its optics [1, 2, 3, 4]. One of these advantages is the availability of spherical-wave X-rays, which cannot be used for X-ray phase imaging techniques using crystals. This advantage allows us to combine the XTI with an X-ray imaging microscope. Subsequently, a phase-sensitive mode can be appended to a conventional X-ray imaging microscope that functions under an amplitude-contrast mode.

X-ray phase imaging with the XTI was demonstrated using plane-wave X-rays and the spatial resolution was subsequently limited by the pitch of the gratings. To overcome this limitation, however, it is not easy to shorten the pitch because one of the gratings, which should be of the amplitude type, must have a structure of several tens of microns in height. Possible approaches are the combinations with a projection microscope [5] and an imaging microscope [6, 7]. We consider the latter to be more attractive because the problem of blur due to Fresnel diffraction, which is equivalent to defocusing in an imaging microscope, is eliminated.

We report an attempt at phase tomography with an X-ray imaging microscope using a Fresnel zone plate in combination with the XTI located in front of the imaging plane. The observation of the phase-separation structure in polymer blends is demonstrated.

2. Experimental setup

Experiments were performed using 9 keV synchrotron radiation X-rays at the undulator beamline 20XU of SPring-8, Japan. Figure 1(a) shows the schematics of the constructed X-ray imaging microscope, which consisted of a Fresnel zone plate (FZP), a CCD-based X-ray image detector,

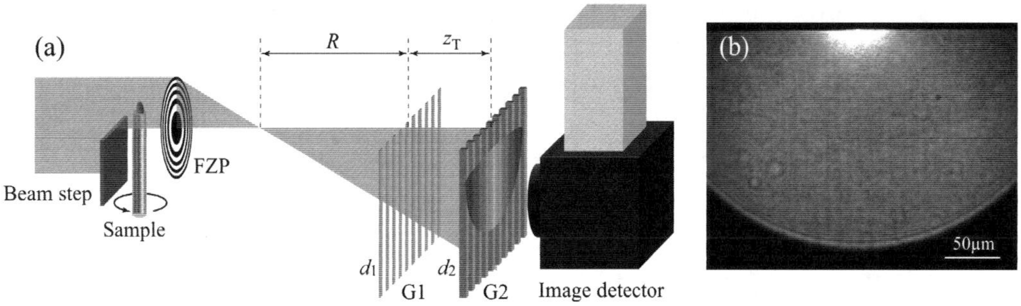

Figure 1. (a) Setup of phase-sensitive X-ray microscope consisting of the FZP and XTI, and (b) field of view.

and an XTI. A sample was placed at the object plane of the FZP, and the image detector was set at the imaging plane. An XTI, consisting of a $\pi/2$ phase grating G1 and an amplitude grating G2, was placed immediately in front of the image detector. The deflection of X-rays, or the slope of the X-ray wave front, caused by passing thought an object is measured by the XTI.

When the pitches of G1 (d_1) and G2 (d_2) satisfy $d_2/d_1 = (R + z_T)/R$, where R and z_T are the distance from the focus of the microscope to G1 and the spacing between G1 and G2, respectively, a flat field is generated. In addition, $z_T = pd_1d_2/\lambda$ should be satisfied in order to operate the XTI under spherical-wave illumination [8], where p is an positive half integer when a $\pi/2$ phase grating is used for G1. λ is the wavelength.

Note that the wave-front slope becomes shallow at the image plane when the microscope magnification is increased. This implies that the sensitivity to the X-ray deflection is inversely proportional to the magnification. The sensitivity and magnification should be balanced, and the magnification was set to 20.2 in this study; the outermost zone width and the diameter of the FZP (NTT-AT, ATN/FZP-S86/416) were 86 nm and 416 μm, respectively, and the image detector was located about 5.5 m downstream of the FZP. d_1 and d_2 were designed to be 8.0 μm and 8.3 μm, respectively, and z_T was 24 cm. The effective pixel size of the image detector was 4.34 μm, which corresponded to 0.215 μm on the object plane.

Figure 1(b) is the field of view of the constructed microscope. The vertical stripes were caused by the pitch non-uniformity of the gratings. Ring patterns were speckle noise due to the dust on the beam line window and FZP. These features in the field of view were, however, not crucial for phase tomography because their effect could be reduced by subtracting the data without any sample from that with a sample.

The rotation axis of the sample was aligned perpendicularly to the diffraction vector of the gratings; therefore, the derivative of the X-ray phase shift (differential phase image) in the direction parallel to the tomography reconstruction plane was measured. Phase tomograms that map the refractive index were reconstructed from the differential phase images measured at plural angular positions of the sample rotation.

3. Experimental result

The differential phase images were measured using the five-step fringe-scanning method [4] at every angular position of the sample rotation with a 0.5° step over 180° in order to reconstruct a phase tomogram. Samples were a binary blend of polystyren (PS) and polymethylmethacrylate (PMMA) and a ternary blend of PS, PMMA, and polybutadiene (PB). Figure 2(a) shows a tomogram of the PS/PMMA binary blend. The PMMA-rich region with a bright gray scale was differentiated from the PS-rich region (dark gray level). In the case of the PS/PMMA/PB

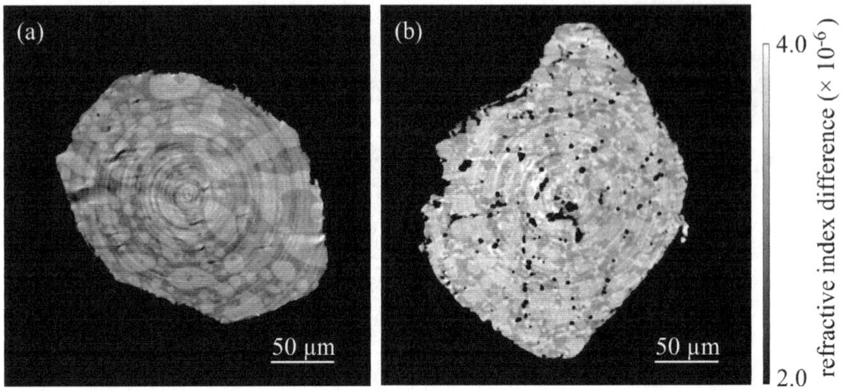

Figure 2. Phase tomograms of PS/PMMA binary polymer blend (a) and PS/PMMA/PB ternary blend (b).

ternary blend, three regions of characteristic gray levels are seen in Fig. 2(b), where the black area in the sample corresponds to voids. The spatial resolution was evaluated as 1.3 μm from the profile across the sample surface in the tomogram. Although the resolution is still considered to be determined by the mechanism of the XTI (and probably by the drift of the sample during the scan in this measurement), a value much smaller than the grating pitch was attained.

4. Conclusion

A novel differential phase-contrast X-ray microscope was presented. An XTI can be considered to be a phase-sensitive X-ray image detector. The idea was to replace a normal image detector in the conventional X-ray imaging microscope with the XTI in order to generate the differential phase contrast. The quantitative measurement of the differential phase shift using the fringe scanning method enabled X-ray phase tomography. The reconstructed microscope was used for the tomographic observation of polymer blends. The phase separation structures could be depicted with a spatial resolution of 1.3 μm.

Acknowledgments

The experiment was performed under the approval of SPring-8 committee 2007B1130. This study was financially supported by Grant-in-Aid for Scientific Research (B) of The Ministry of Education, Culture, Sports, Science and Technology.

References

[1] Momose A, Kawamoto S, Koyama I, Hamaishi Y, Takai K and Suzuki Y 2003 *Jpn. J. Appl. Phys.* **42** L866
[2] Weitkamp T, Nöhammer B, Diaz A and David C 2005 *Appl. Phys. Lett.* **86** 054101
[3] Weitkamp T, Daiz A and David C 2005 *Opt. Express* **13** 6296
[4] Momose A, Yashiro W, Takeda Y, Suzuki Y and Hattori T 2006 *Jpn. J. Appl. Phys.* **45** 5254
[5] Engelhardt M, Baumann J, Schuster M, Kottler C, Pfeiffer F, Bunk O and David C 2007 *Appl. Phys. Lett.* **90** 224101
[6] Momose A, Yashiro W, Moritake M, Takeda Y, Uesugi K, Takeuchi A, Suzuki Y, Tanaka M and Hattori T 2006 *SPIE Proc.* 6318 63180T.
[7] Takeda Y, Yashiro W, Hattori T, Takeuchi A, Suzuki Y and Momose A 2008 *Appl. Phys. Express* submitted.
[8] Yashiro W, Takeda Y and Momose A 2008 *J. Opt. Soc. Am. A* **25** 2025

3-d chemical imaging with STXM tomography

Martin Obst, Jian Wang and **Adam P. Hitchcock**

BIMR, McMaster University, Hamilton, ON L8S 4M1 Canada

E-mail: aph@mcmaster.ca

Abstract. Three dimensional (3-d) chemical mapping using angle scan tomography in a soft X-ray scanning transmission X-ray microscope (STXM) has been applied to quantitative chemical mapping in three dimensions of calcium carbonate biomineralization by planktonic freshwater cyanobacteria of the strain *Synechococcus leopoliensis* PCC 7942. Aspects of making and mounting grid sections, acquiring multi-energy, multi-angle sequences, and data analysis are discussed.

1. Introduction

Previous tomography studies using full field soft X-ray microscopy [1,2] or scanning transmission X-ray microscopy on plunge frozen specimens (cryo-STXM) [3] have been performed at only a single photon energy, which provides mainly density/thickness contrast with limited and potentially ambiguous chemical information. In contrast, when tomograms are acquired with images at multiple photon energies at each angle [4-7], it is possible to provide spectrally verifiable, quantitative 3-d chemical mapping through analysis of the species-specific, near edge X-ray absorption spectrum (NEXAFS) of each voxel. We have recently extended the types of sample mounting that we are able to use to include ~120 μm wide strips of 400 mesh Cu transmission electron microscopy grids. Although these supports are delicate and challenging to mount and align, since the sample is in a dry state, they do allow access to the full photon energy range of STXM, in contrast to pulled glass capillaries which, to date, have only been used successfully in the region just below the O 1s threshold [4-7]. Here we describe recent studies with C 1s and Ca 2p NEXAFS of calcium carbonate biomineralization on the surface of planktonic freshwater cyanobacteria of the strain *Synechococcus leopoliensis* PCC 7942. This work is an extension to 3-d mapping of an extensive study of the same system using STXM without tomography [8] and should be viewed as a preliminary report rather than the final exposition of the data.

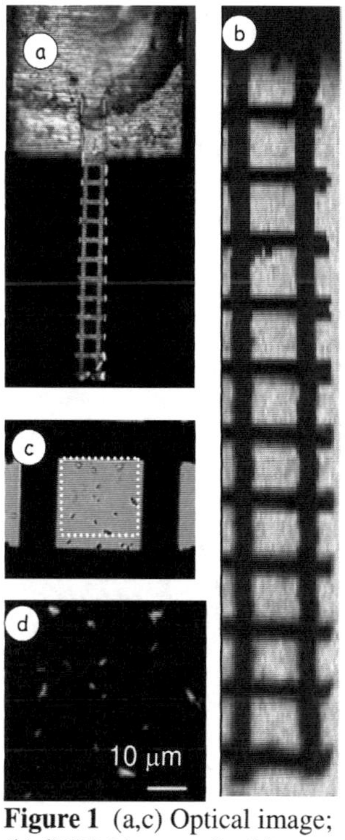

Figure 1 (a,c) Optical image; (b) STXM image of grid-section attached to the tomo-rotator. (d) STXM OD image (352.6 eV) of one grid square.

2. Experimental

The miniature stepping motor, alignment system, as well as the acquisition and analysis procedures have been described in detail elsewhere [6,9]. Figure 1 displays optical and STXM images of a grid section mounted on the rotation holder. In the STXM image Fig. 1b, the lower part of the grid section is slightly out of focus since the section was not perfectly flat after sample mounting. The width is sufficiently narrow that full 360° rotation is possible, even with the standard 290 μm sample-OSA spacing used at the C 1s edge. However the combination of excessive absorption, foreshortening, and ultimately masking of the view by the grid bars limit data collection to -70 and +70° (where 0° corresponds to having the sample plane orthogonal to the X-ray propagation axis) even for samples near the centre of the grid square. For these initial studies we emphasized collection of more data in the energy domain and thus the angle spacing was only every 3 or 5°, which, along with the incomplete coverage of the angle domain and thus missing wedge effect [10], might lead to artifacts in the tomographic reconstructions.

The cyanobacteria were cultured in the nutrient-poor medium Z/10 and incubated 12 hours in a solution of $NaHCO_3/CaCl_2$ which was ~6 times saturated with respect to calcite [8]. Lacey carbon coated 400 mesh Cu grids were dipped into the cell suspension and the water was sucked away using filter paper. The grid was air dried under an optical microscope. A stripe of grid squares containing multiple cells was chosen and sectioned using a fresh scalpel blade. The grid stripe was then aligned and glued to the rod of the rotation stage (Fig. 1a) using a micromanipulator.

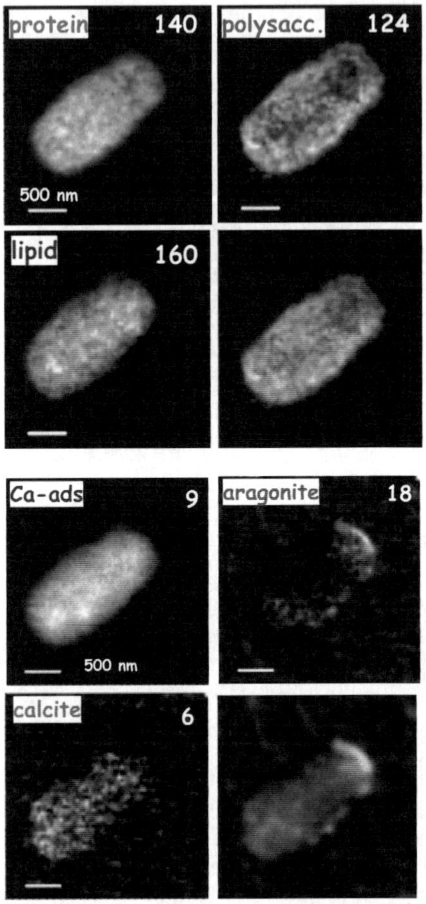

Figure 2. Component chemical maps and color-coded composite map for a single projection of one *Synechococcus leopoliensis PCC 7942* bacterium. (upper) **C 1s** results highlighting the biological components, and (lower) **Ca 2p** results for the Ca-adsorbed, aragonite and calcite components. Bold numbers indicate maximum thickness in nm for pure white (black = 0).

3. Results

Figure 2 presents the results of quantitative chemical mapping of a single *S. leopoliensis* cell at the C 1s (upper) and Ca 2p (lower) edges. The results are in agreement with other work [8]. While these results are based on measurements with an extensive sampling of the spectral space (88 C 1s energies and 89 Ca 2p energies), when a sub-set of these images is carefully chosen so as to select energies with good contrast among the species, then similar results can be obtained. Based on such explorations a reduced set of 5 C 1s and 6 Ca 2p

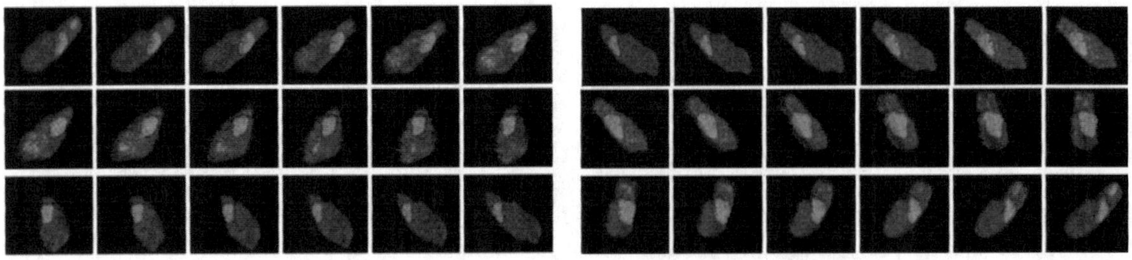

Figure 3 Projection views (after alignment) of the net protein ($OD_{288.2}$- $OD_{280.0}$, green) and Ca 2p signal ($OD_{352.6}$ – $OD_{350.2}$). The Ca 2p signal is displayed in 2 ranges with segmentation based on optical density per thickness (blue ~ $2.5{\cdot}10^{-3}$OD / nm (~12%) Ca adsorbed to extracellular polymers, red ~ $4.4{\cdot}10^{-3}$OD / nm (~ 22%) aragonite, assuming that the red volume contains solely aragonite and organic carbon, and the blue volume contains solely the adsorbed Ca species and organic carbon.)

energies was chosen, and then simultaneous C 1s and Ca 2p 11-energy image sequences were measured at 46 angles between -70 and +70°.

Figure 3 presents some of the results in the form of 36 projections in which the Ca signal ($OD_{352.6} - OD_{350.2}$, total Ca, shown in blue) is combined with the 'biological signal' ($OD_{288.2} - OD_{280.0}$, which is dominated by protein, shown in green). Further analysis to extract detailed speciation (protein, polysaccharide and lipid from C1s; adsorbed-Ca, aragonite and calcite from Ca 2p) is in progress and will be reported elsewhere.

Figure 4 is a rendering of the net Ca signal determined as the difference of the signal at 350.2 eV (no Ca), and 352.6 eV (max of the $Ca_{3/2} \rightarrow 3d$ signal, which is common to all species). There is a rather uniform layer of Ca, which the detailed speciation studies indicate is dominated by the adsorbed-Ca component. In addition there is a 'hot spot' (red) which is most likely the start of an aragonite deposit.

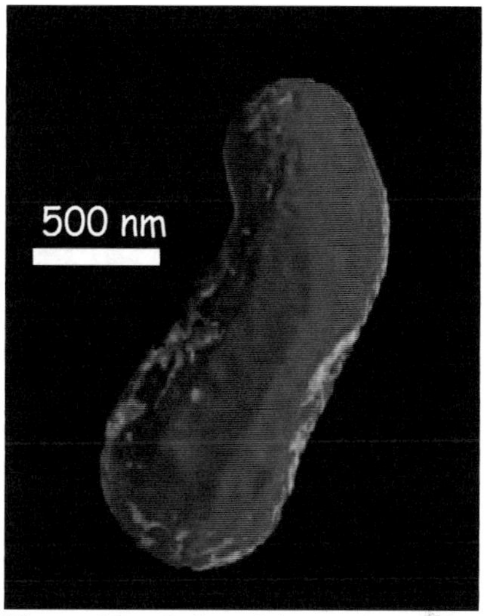

Figure 4 Rendering of the Ca distribution from a single *S. leopoliensis* bacterium, derived from the $OD_{352.6} - OD_{350.2}$ signal. A region of higher Ca signal level is identified by the red coloured voxels.

S. leopoliensis cells are known to facilitate the precipitation of calcite on their surface. Other Ca 2p spectromicroscopic studies [8] have identified $CaCO_3$ with aragonite-like short range order to be a precursor phase of calcite precipitation. 3d-chemical mapping will now allow us to further investigate and possibly determine if the $CaCO_3$ phases originate from the cell membrane, or if extracellular polymers act as a template for mineral nucleation.

Acknowledgements
We thank Göran Johansson for excellent contributions to initial establishment of STXM tomography; Tolek Tyliszczak and David Kilcoyne for providing essential support for the microscopes. Research supported by NSERC, Canadian Foundation for Innovation, and the Canada Research Chair program, STXM data was measured at ALS STXM532 (ALS is supported by BES, U.S. Department of Energy).

References
[1] Weiß D, Schneider G, Niemann B, Guttmann P, Rudolph D, and Schmahl G 2000 *Ultramicroscopy*, **84** 185
[2] Larabell C A and Le Gros M A 2004 *Molecular Biology of the Cell*, **15** 957
[3] Wang Y, Jacobsen C, Maser J and Osanna A 2000 *J. Microscopy*, **197** 8
[4] Johansson G A, Dynes J J, Hitchcock A P Tyliszczak T, Swerhone G D W and Lawrence J R 2006 *Proc. SPIE* **6318** 1-I
[5] Johansson G A, Dynes J J, Hitchcock A P Tyliszczak T, Swerhone G D W and Lawrence J R 2006 *Microscopy and Microanalysis*, **12** S02 1412
[6] Johansson G A, Tyliszczak T, Mitchell G E, Keefe M and Hitchcock A P 2007 *J. Synchrotron Radiation* **14** 395
[7] Hitchcock A P, Johansson G A, Mitchell G E, Keefe M H and Tyliszcak T 2008 *Appl. Phys. A* **92** 447
[8] Obst M, Dynes J J, Lawrence J R, Swerhone G D W, Karunakaran C, Kaznatcheev K V, D. Bertwistle D, Benzerara K, Kilcoyne A L D, Tyliszczak T and Hitchcock A P 2009 *Geochimica et Cosmochimica Acta* under review.
[9] Ress D B, Harlow M L, Marshall R M, McMahan U J 2004 *Structure* **12** 1763
[10] Baba N, Katayama E 2008 *Ultramicroscopy* **108** 239

9th International Conference on X-Ray Microscopy IOP Publishing
Journal of Physics: Conference Series **186** (2009) 012046 doi:10.1088/1742-6596/186/1/012046

High-sensitivity phase-contrast tomography of rat brain in phosphate buffered saline

Franz Pfeiffer[1], Christian David[2], Oliver Bunk[2], Carole Poitry-Yamate[3], Rolf Grütter[3], Bert Müller[4], Timm Weitkamp[5]

[1]Department of Physics, Technical University Munich, 85748 Garching, Germany
[2]Paul Scherrer Institut, 5232 Villigen PSI, Switzerland
[3]Ecole Polytechnique Fédérale de Lausanne, 1015 Lausanne, Switzerland
[4]Biomaterials Science Center, University of Basel, 4031 Basel, Switzerland
[5]European Synchrotron Radiation Facility, B.P. 220, 38043 Grenoble Cedex, France

E-mail: `franz.pfeiffer@ph.tum.de`

Abstract. We report advances and complementary results concerning a recently developed method for high-sensitivity grating-based x-ray phase-contrast tomography. In particular we demonstrate how the soft tissue sensitivity of the technique can be used to obtain in-vitro tomographic images of rat brain specimens. Contrary to our previous experiments with fixated specimen (chemically modified or formalin fixed), the present results on the rat's brain are closer to the in-vivo situation. The findings are particularly important from a clinical point of view, since a similar approach using three gratings can be implemented with more readily available x-ray sources, such as standard x-ray tubes.

Phase-sensitive x-ray imaging, which uses the phase shift rather than the absorption as the imaging signal has the potential of substantially increased contrast in soft biological tissue [1, 2]. While various phase-sensitive x-ray imaging methods were developed in the past years, we have particularly focused on the development of grating-based differential phase-contrast imaging [3, 4]. In previous work we have demonstrated how phase-contrast tomography with a grating interferometer setup can yield images of biomedical specimens with unprecedented soft tissue sensitivity at synchrotron x-ray sources [4, 5, 6]. In particular we have shown that this x-ray method can be used to discern between subtle details in the tissue structure of animal brains [4, 6] and other organs [5], an application field which until now has almost exclusively been reserved for other, *i.e.*, MRI techniques.

Contrary to the above mentioned previous studies, where the samples were fixated using formalin, the results presented in this contribution were obtained on rat brain specimens prepared in phosphate buffered saline (PBS). This is particularly important to evaluate the real potential of the method for future pre-clinical or clinical in-vivo applications. In the latter context, experiments based on formalin-fixed specimens could be misleading, since formalin-fixation (like most other fixation methods) significantly alters the chemical conditions of the sample, thus prohibiting an assessment of the performance of phase-contrast tomography under real in-vivo conditions.[1]

[1] To prepare the brain samples prior to extraction from the animal, we used microwave irradiation to the head of anesthetized rats (4kW at 2450 MHz for 2s). This procedure preserves the neurochemical environment of the entire brain in situ.

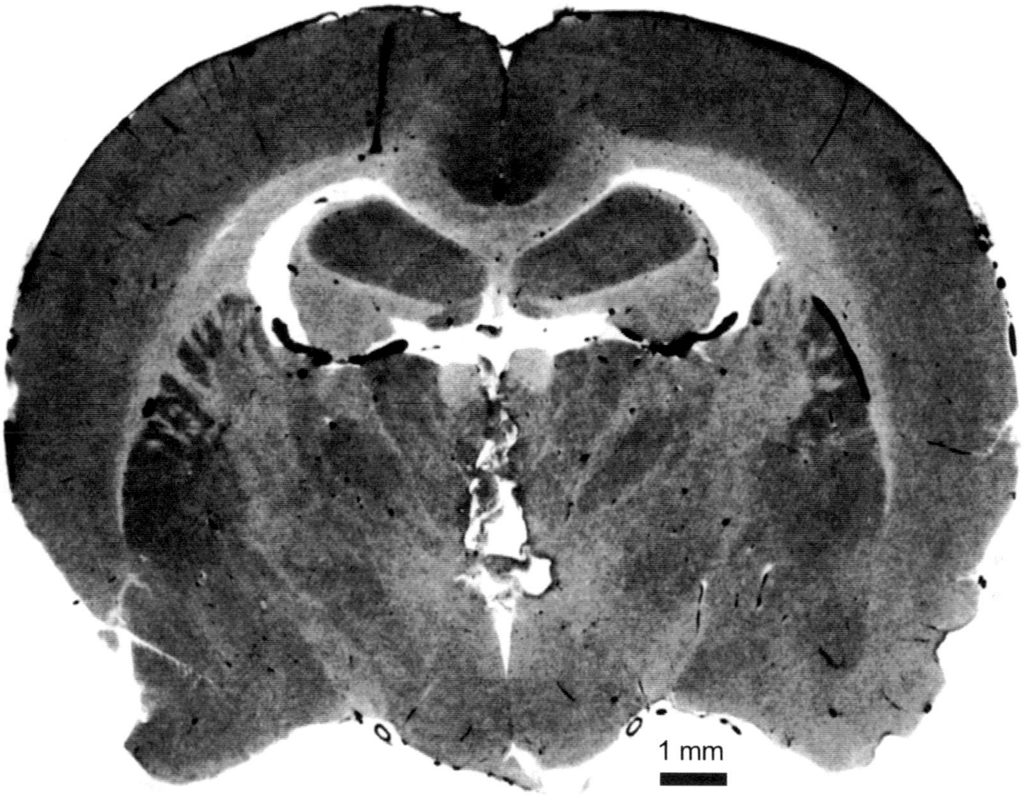

Figure 1. One out of several hundred reconstructed tomographic slices of a rat brain specimen measured in phosphate buffered saline (PBS). The image is displayed on a linear gray scale.

The principles of differential phase-contrast imaging using a grating interferometer have been described in detail in [3, 4], and are only briefly reviewed here. The experimental arrangement consists of the specimen on a high-precision rotation stage, a phase grating G1, and an analyzer absorption grating G2. The first grating (G1) acts as phase mask, and imprints periodic phase modulations onto the x-ray wave front. The second grating (G2) with absorbing lines and the same periodicity and orientation as the fringes created by G1 is placed in the detection plane, immediately in front of the detector. When one of the gratings is scanned along the transverse direction ('phase stepping'), the intensity signal in each pixel in the detector plane oscillates as a function of the grating position. The fundamental idea of the method is to analyze, for each pixel, the changes of these oscillations, when an object is placed into the x-ray beam. Three dimensional (3D) information of the specimen using computed tomography (CT) can be obtained, when the specimen is rotated around the rotation axis and a phase stepping series is recorded for each angle. The differential phase-contrast projection tomography data set can be reconstructed by using a modified filter kernel (Hilbert transform) in combination with standard filtered back-projection algorithms [9, 10].

The imaging experiments were carried out at the beamline ID19 of the European Synchrotron Radiation Facility (ESRF, Grenoble). A monochromatic x-ray beam of 26 keV was used for the measurements. The interferometer was placed at a distance of 145 m from the wiggler source. The gratings were fabricated by a process involving photolithography, deep etching into silicon and electroplating of gold. The first grating (Si phase grating, G1) had a period of: $p_1 = 3.99\ \mu$m and a height of $h_1 = 31.7$ μm. The corresponding values for the second grating (gold absorber grating, G2) were $p_1 = 2.00$ μm and $h_2 = 30$ μm. To achieve a very high angular, and thus phase sensitivity, the distance between G1 and G2 was chosen to be as large as 376 mm (ninth fractional Talbot distance). The images were recorded using a 15 μm micron thick polycrystalline gadolinium oxysulphide scintillation screen with a magnifying optical lens system and a cooled charge coupled device (CCD). We used the FReLoN 2000 (a Fast-Readout, Low-Noise CCD developed at the ESRF) with 2048×2048 pixels, with a final effective pixel size of 15.0 × 15.0 μm^2.

Figure 1 shows one out of several hundred reconstructed virtual tomographic slices of the rat brain specimen measured in phosphate buffered solution (PBS). 1001 projections, each consisting of four images, within a total exposure time of approx. 33 minutes (0.5 seconds exposure time per individual image) were used for this reconstruction. As for formalin fixation [4], one clearly observes numerous anatomical features well above the noise level. Although such differentiation is usually hardly possible based on x-ray CT scans, our method clearly resolves these small density differences in the brain tissue structure. Based on the standard deviation of the gray values in background (PBS) regions of the reconstructed tomographic slices we deduce a measurement sensitivity for the real part of the refractive index of 0.7×10^{-10}. This corresponds to an electron density sensitivity of 0.04 e/nm^3 and a mass density sensitivity of approximately 0.1 mg/cm^3 for aqueous specimens.

In summary we have demonstrated how an improved grating based phase-contrast tomography setup yields images of biomedical specimens with unprecedented soft tissue sensitivity at synchrotron x-ray sources. In particular we have shown that this x-ray method can be used to discern between subtle details of the tissue structure of animal brains, an application field which until now has almost exclusively been reserved for other, *e.g.*, MRI techniques. While the results can be applied immediately for biomedical studies at synchrotron x-ray sources, they are potentially interesting from a clinical point of view, since a similar approach can be implemented with more readily available x-ray sources, such as standard x-ray tubes [11, 12].

We gratefully acknowledge J. Bruder and C. Grünzweig for the grating fabrication and Sabrina Lang, Georg Schulz, Marco Germann, and Hans Deyle for their help during the experiments. This work was supported by the ESRF (project MD-328) by allocation of beam time.

References

[1] Fitzgerald R 2000 *Physics Today* **53** 23-27
[2] Momose A 2005 *Jap. J. Appl. Phys.* **44** 6355-6359
[3] Weitkamp T, Diaz A, David C, Pfeiffer F, Stampanoni M, Cloetens P and Ziegler E 2005 *Optics Express* **13** 6296-6304
[4] Pfeiffer F, Bunk O, David C, Bech M, Le Duc G, Bravin A and Cloetens P 2007 *Phys. Med. Biol.* **52** 6923-6930
[5] Weitkamp T, David C, Cloetens P, Bunk O and Pfeiffer F 2008 *Eur. J. Radiology* in press, doi:10.1016/j.ejrad.2008.04.031
[6] F. Pfeiffer, Bunk O, David C, Bech M, Le Duc G, Bravin A and Cloetens P 2008 *Proc. SPIE* **7078**, 707816
[7] Login GR, Leonard JB and Dvorak AM 1998 *Methods* **15** 107
[8] Poitry-Yamate C L, Lei H and Gruetter R 2008 submitted
[9] Faris GW and Byer RL 1988 *Appl. Opt.* **27** 5202-5206
[10] Pfeiffer F, Kottler C, Bunk O and David C 2007 *Phys. Rev. Lett.* **98** 108105
[11] Pfeiffer F, Weitkamp T, Bunk O and David C 2006 *Nature Physics* **2** 258-261
[12] Pfeiffer F, Bech M, Bunk O, Kraft P, Eikenberry E, Brönnimann C, Grünzweig C and David C 2008 *Nature Materials* **7** 134-137

The synchrotron-based imaging station for micro-radiography and-tomography at the BAMline (BESSY)

H. Riesemeier[1] A. Rack[2], S. Zabler[3], J. Goebbels[1], B. R. Müller[1], J. Banhart[4]
[1]Federal Institute for Materials Research and Testing (BAM), Berlin Germany
[2]Forschungszentrum Karlsruhe / K.I.T. – ANKA, Germany
[3]Max Planck Institute of Colloids and Interfaces, Golm, Germany
[4]Hahn-Meitner-Institute / Helmholtz Centre Berlin, Germany
E-mail: heinrich.riesemeier@bam.de

Abstract. Third generation synchrotron light sources allow one to image micro-structured, multi-component specimens with different contrast modalities and resolutions up to the submicrometer range, thanks to their high brilliance and partial coherence. An imaging setup for micro-tomography and -radiography installed at BESSY as part of its first hard X-ray beamline (BAMline) can now be used for absorption, refraction as well as phase contrast. The experimental station is dedicated to in-house research and applications by external users. Monochromatic synchrotron light between 6 keV and 80 keV is attained via a fully automated double multilayer monochromator. We show the suitability of the setup for several applications from materials research.

1. Introduction

In this article we shall describe the imaging instrumentation of the BAMline at the German storage ring BESSY in Berlin (Berliner Elektronenspeicherring — Gesellschaft für Synchrotronstrahlung m.b.H.). BESSY is a third generation synchrotron light source operating since 1998 (ring energy of 1.7 GeV) [1]. The BAMline is BESSY's first hard X-ray beamline equipped with a superconducting 7 T wavelength shifter insertion device (operated by the Bundesanstalt für Materialforschung und -prüfung (Federal Institute for Materials Research and Testing, BAM), Germany) [2, 3]. The beamline is dedicated to host several experimental methods which include an imaging set-up since 2001 [3].

2. Experimental setup

In Fig. 1 the optical section of the BAMline is displayed. A double multilayer monochromator (DMM) and/or a double crystal monochromator (DCM) can be used to monochromatize the synchrotron light coming from the superconducting 7 T wavelength shifter (7T-WLS) insertion device. A vertical focusing option is available for the DMM and a horizontal focusing option for the DCM. For further beam condition, slits are positioned upstream (aperture/primary – in-vacuum) and downstream of the monochromators (in-vacuum) as well as in the experimental hutch upstream of a differential pumped double foil Kapton window. Several polished metal foils (Al, Be, Cu) placed in front of the DMM are used to absorb low energy X-rays which otherwise would be transmitted by the DMM by total reflection and consequently deteriorate the DMM's monochromaticity. The filters also lower the heat load on the first multilayer mirror. For monochromaticity reasons the layer-structure of the DMM has been optimized to suppress higher harmonic wavelengths (the multilayer coating consists of 150 double layers W/Si leading to an energy bandwidth of 1.7% in the range between 5 keV and 50 keV — experimentally verified at 20 keV with an DCM energy scan downstream of the DMM). Additionally, the beamline can be operated in white beam mode as well.

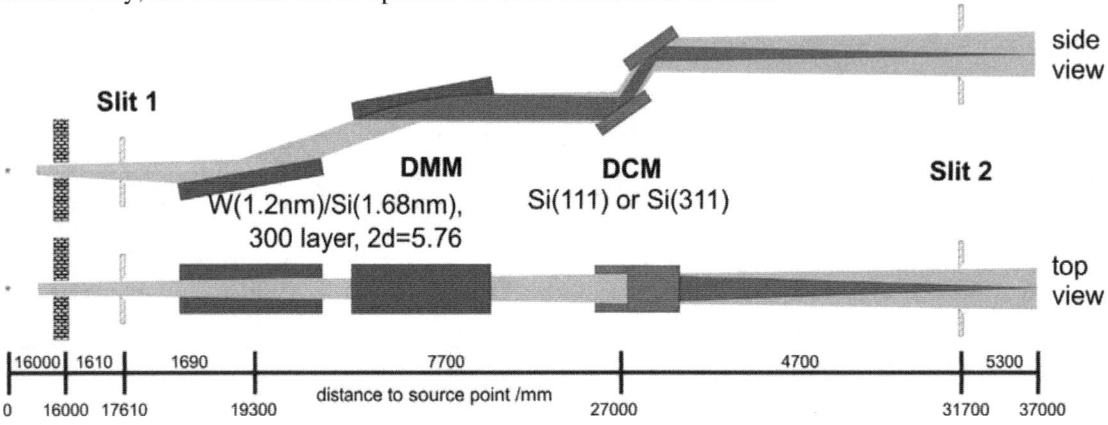

Figure 1. Principle sketch of the BAMline's X-ray beampath (top: side view, bottom: top view). The radiation originating from the insertion device (7T-WLS) can be monochromatized using a double multilayer monochromator (DMM) and/or a double crystal monochromator (DCM) which includes a vertical focusing option for the DMM and a horizontal for the DCM. The position of the tomography stage is in the experimental hutch at a distance of 35 m from the source [1, 2].

The photon flux measured downstream of the DMM is plotted for different photon energies in Fig. 2. Due to the limited amount of photons available only the DMM is used for micro-tomography. The bending optic of the second multilayer was successfully applied to collimate the beam, resulting in a larger photon flux density and higher acquisition speeds in absorption contrast modes [3].

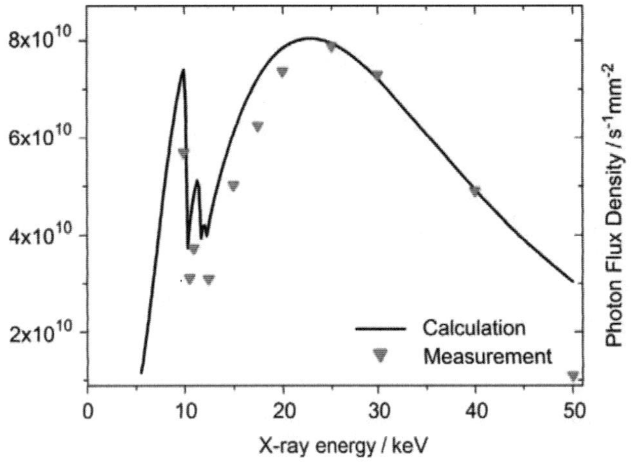

Figure 2. Photon flux density behind the double multilayer monochromator at the BAMline; measured and calculated (using Reflec) with a ring energy of 1.7 GeV, a ring current of 200 mA, a magnetic flux density of 7 T at the wavelength shifter and a distance from the source of 35 m. At 7 keV a flux density of was measured with a focused beam. Around 11 keV typical flux reduction due to the tungsten L absorption edges can be seen (the multilayer coating consists of 150 double layers W/Si) [1, 2].

Indirect X-ray pixel detectors containing thin transparent high-Z scintillators are used in order to record spatially resolved radiographs [4]: an inhouse designed system optimized for larger fields of view ("macroscope") combining Rodenstock objectives with classical visible light tele-objectives (tandem design). Additionally, a commercially available microscope designed by the French Company OptiquePeter (Lyon) is used for high resolution imaging [3, 5]. As visible light camera a PCO.4000 (Interline Transfer CCD, PCO AG, Germany) is used for micro-tomography and a PCO.1200hs (CMOS) for fast radiography.

3. Applications

In Fig. 3 an example application of the magnesium-aluminum alloy AZ91 is shown. Imaging the real part yields an absorption tomogram with insufficient material contrast (cf. Fig. 3, left). Increasing the sample-to-detector distance results in Fig. 3 (middle) with enhancement of the structural interfaces. Finally we used a method based on contrast transfer functions in order to retrieve the phase maps from a set of four tomographic scans, then compute the holo-tomogram which is shown in Fig. 3 – right. The X-ray energy used for these images was E = 23 keV, 900 radiographic projections were recorded at each propagation distance and with an effective pixel size of 1.6μm (resolution <4 μm,).

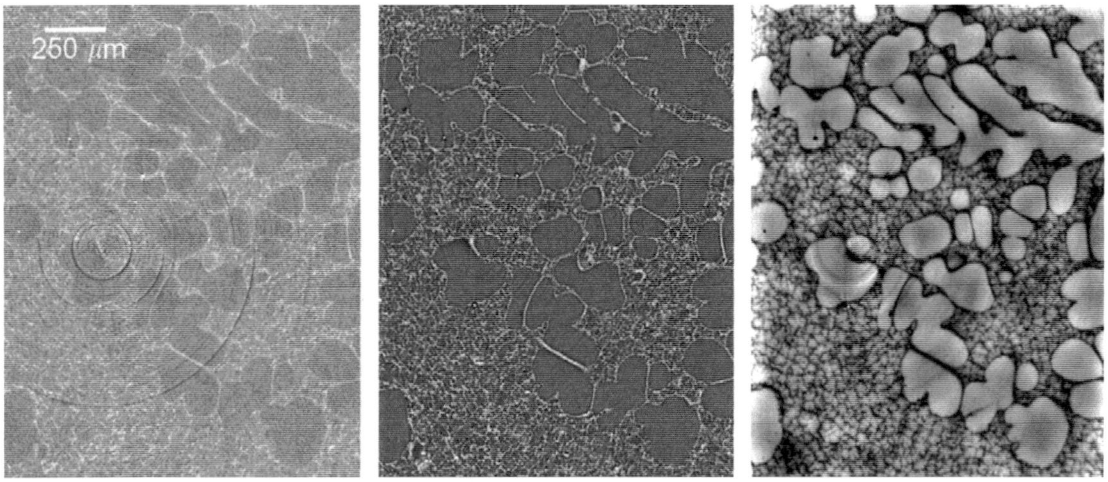

Figure 3. Left: absorption tomogram from a specimen consisting of the alloy AZ91 (magnesium and aluminium, similar attenuation coefficients), middle: same slice but as phase contrast tomography calculated from Fresnel-propagated projection images (edge enhancement), right: holotomography of this slice combining different propagation distances [3].

4. Summery

In this article we have briefly introduced the high resolution synchrotron-based radiography and tomography setup at the BAMline (BESSY), suitable to investigate specimens, in a non-destructive manner, with resolutions up to approximately 1 μm (as verified with an Xradia test pattern) and contrast modes measuring the local attenuation coefficient (synchrotron microtomography), the local electron density (holotomography) and inner surfaces and interfaces (refraction enhanced tomography). The high potential of this device is illustrated by investigations of specimens originating from materials research (commercial magnesium–aluminum thixo-alloy AZ91). Optimization approaches such as using a dedicated scintillator concept have been tested successfully, resulting in a higher performance of our setup [3]. Future further improvements include new scintillator development with materials such as LSO:Tb. Other forms of imaging, including white beam and methods with dual energy imaging have shown promising results as well.

5. References

[1] W. Görner, M.P. Hentschel, B.R. Müller, H. Riesemeier, M. Krumrey, G. Ulm, W. Diete, U. Klein and R. Frahm, Nucl. Instr. and Meth. A 467-468 (2001), 703-706

[2] H. Riesemeier, K. Ecker, W. Görner, B.R. Müller, M. Radtke and M. Krumrey, X-ray Spectr., 34 (160 - 163) 2005

[3] A. Rack, S. Zabler B.R. Müller, H. Riesemeier, G. Weidemann, A. Lange, J. Goebbels, M. Hentschel and W. Görner, Nucl. Instr. and Meth. A 586 (2008), 327-344

[4] U. Bonse, F. Busch, Prog. Phys. Molec. Biolog. 65 (1996) 133-169

[5] A. N. Danilewsky, A. Rack, J. Wittge, T. Weitkamp, R. Simon, H. Riesemeier, T. Baumbach, Nucl. Instr. and Meth. B 266 (2008), 2035-2040

Hard x-ray imaging holography and tomography

Yoshio Suzuki, Akihisa Takeuchi, Kentaro Uesugi

JASRI /Spring-8, Koto 1-1-1 Sayo, Hyogo 679-5198, Japan

yoshio@spring8.or.jp

Abstract. Phase-contrast microtomography with imaging holographic microscope has been developed. The imaging holography optics consists of X-ray microscope with Fresnel zone plate objective and wavefront-dividing interferometer with total-reflection mirror. One half of incident beam is used to illuminate the object, and the other half is used as the reference wave. Hologram of the magnified image is generated at the image plane of microscope by the interference with reference wave. Phase-contrast tomography can be carried out, by combining holographic wave-front reconstruction with conventional computed-tomography technique. Preliminary results with test specimens are shown.

1. Introduction

Phase-contrast imaging is an attractive method for X-ray microscopy, since the phase-contrast is much higher than the absorption contrast in the hard X-ray region, and the phase-contrast is essentially proportional to the electron density of object, while the absorption contrast strongly depends on the atomic number of component elements. Recently, some methods of interferometric microscopy have been proposed in the hard X-ray region: use of specially designed Fresnel zone plate (FZP) objective [1-2], a common-path interferometer with two FZPs [3], and use of wavefront-dividing interferometer with prism optics [4-5]. Here, instrumentation for imaging holography with a total-reflection mirror interferometer and its application to phase-contrast micro-tomography is described. The optical system for imaging holography is shown in Fig. 1. A half of illuminating beam is used as a reference wave, and the other half is object wave. The reference beam is deflected by a total reflection mirror placed behind the back-focal plane of FZP objective, and is superimposed onto the magnified image of object in order to generate interference fringes. The recorded fringe image is called hologram or interferogram that includes the whole information of wavefront through the object.

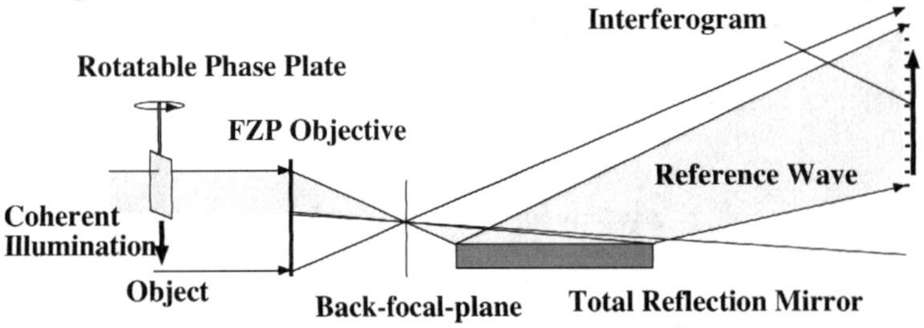

Figure 1. Optical system of imaging holography with a total-reflection-mirror interferometer.

2. Experimental setup

A schematic diagram of the entire experimental setup is shown in Fig. 2. The undulator radiation is monochromatized by a double-crystal monochromator (DCM). The interferometer requires only the one-dimensional coherence, i.e. only the vertical coherence in this case. Then, a slit with a vertical opening of 10 μm is placed just behind the monochromator to define the spatial coherence in the vertical direction, while the horizontal coherence is determined by the source size of undulator radiation [5].

The diameter of objective FZP is 155 μm, and outermost zone width is 100 nm. The focal length at 1.0 Å is 155 mm. The total-reflection mirror is an uncoated fused-quart plane mirror with surface figure error less than 30 nm. A rotatable phase plate made of 60 μm-thick-quartz is inserted at the object plane for fringe scanning. The interferogram at the image plane is taken with a cooled CCD camera (C4742-98-24A, Hamamatsu Photonics, Japan) coupled with a relay lens and a scintillator (P43) screen. The magnification of the imaging microscope was 42 for 1.0 Å X-rays, and the conversion pixel size of imaging detector was 0.103 μm. The specimen is mounted on a high precision rotating-stage that has a wobbling error less than ±0.1 μm for CT scan.

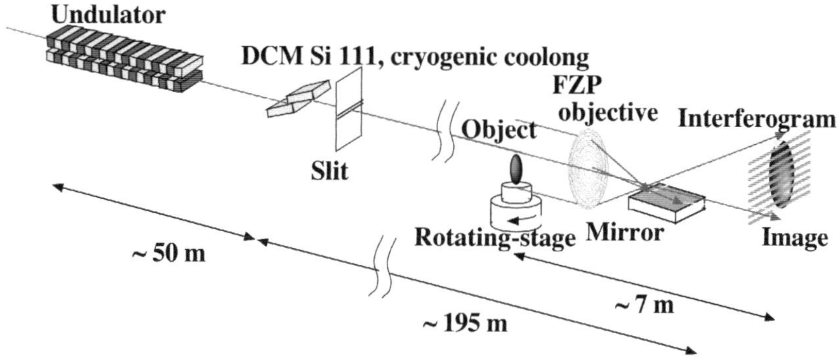

Figure 2. Experimental setup at BL20XU of SPring-8.

3. Holographic Image Reconstruction

An example of measured hologram is shown in Fig. 3. There are two methods for numerical image reconstruction from the measured hologram. One is filtered-Fourier/back-Fourier transform (FFBF) method (optical reconstruction) that is equivalent to the Leith-Upatnieks off-axis holography. The other is fringe-scan method. Image reconstruction from single hologram can be done by the FFBF method, while the fringe scan method requires three or more interferograms. However, the spatial resolution of FFBF reconstruction perpendicular to the carrier fringe is limited by the interference fringe pitch, while the spatial resolution of the fringe scan method does not depend on the career fringe.

Figure 3. Measured interferogram. Sample is resolution test pattern as shown in Fig. 4. Exposure time: 6 s. X-ray wavelength: 1.0 Å. The observed fringe period is about 50μm.

Figure 4 shows the results of image reconstruction. The spatial resolution of the reconstructed image is estimated about 0.2 μm in the fringe-scan method. The resolution of FFBF reconstruction is about 1.2 μm (0.6 μm line and 0.6 μm space) in the vertical direction that is nearly equal to the career fringe pitch (50 μm period in the image plane and magnification of 42), while the spatial resolution in the horizontal direction is almost equivalent to that of the fringe scan method.

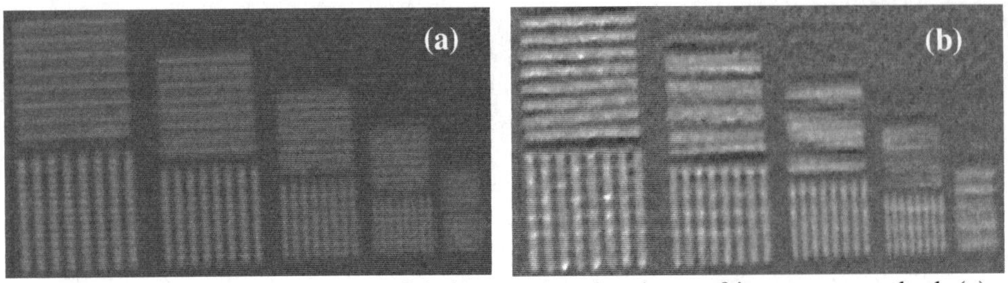

Figure 4 Reconstructed images of hologram by the 4-step fringe-scan method (a), and by the Fourier transform method (b). The test pattern is periodic line and space made of 0.5 μm-thick tantalum, from left, 0.6 μm line and 0.6 μm space, 0.5 μm line/space, 0.4μm, 0.3 μm, and 0.2 μm, respectively. Flat-field-correction was done by subtracting the reconstructed image without specimen.

4. Phase-contrast microtomography

MicroCT results with a test specimen are shown in Fig. 5. The object is composed of heavy element material (phosphor P43: Gd_2O_2S grain, ρ = 7.3), light element material (glycol-phthalate: 85% phthalic anhydride-15% ethylene glycol), and intermediate density material (SiO_2 fiber, ~5 μm in diameter, ρ = 2.2). The transmission of 5 μm-thick samples is calculated to be 0.6 for Gd_2O_2S, 0.99 for SiO_2, and 0.999 for glycol-phthalate at 1.0 Å, respectively. Therefore, only the P43 grain is visible in the absorption contrast image. In the phase-contrast CT, the entire specimen structures are observable, as shown in Fig. 5. The light element materials would be visible by absorption contrast, if the optimized x-ray energy, soft x-rays in this case, is used. However, for lower energy x-rays, radiation damage is much serious than that in the hard x-ray region, and an important problem is that the light element materials hidden behind the heavy element structure cannot be seen, because almost all the x-rays are stopped by the heavy element materials. It is considered that the phase-contrast imaging is useful especially for these complex structure samples composed of light and heavy element materials.

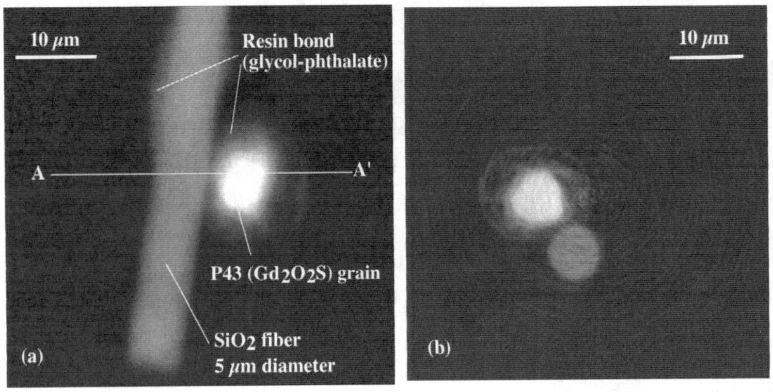

Figuire 5. Phase-contrast CT image of test object. (a) Reprojection image of the 3D-reconstructed object. (b) Section image at A-A' plane. Image reconstruction of hologram by 4-step fringe-scan method. 360 projections, 0.5° step for CT scan. Exposure time: 6 s/hologram. X-ray wavelength: 1.0 Å.

The experiments were performed under the approval of the SPring-8 Proposal Review Committee (Proposal Numbers: 2005A0070, 2006A1563, 2006B1443)

References

[1] Koyama T, Kagoshima Y, Wada I, Saikubo A, Shimose K, Hayashi K, Tsusaka Y, Matsui J 2004 Jpn. J. Appl. Phys. **43**, L421-423

[2] Koyama T, Tsuji T, Yoshida K, Takano H, Tsusaka Y, Kagoshima Y 2006 Jpn. J. Appl. Phys. **45** L1159-1161

[3] Watanabe N, Hoshino M, Sato M, Takeda Y, Namiki T, Aoki S, Takeuchi A, Suzuki Y 2006 Proc. of 8th Int. Conf. X-ray Microscopy, IPAP Conf. Ser. **7** 372-374

[4] Suzuki Y, Takeuchi A 2005 Rev. Sci. Instrum. **76** 093702

[5] Suzuki Y, Takeuchi A 2007 AIP Conference Proceedings **879** 1305-1307.

Journal of Physics: Conference Series **186** (2009) 012049 doi:10.1088/1742-6596/186/1/012049

Fast X-ray micro-CT for real-time 4D observation

H Takano, K Yoshida, T Tsuji, T Koyama, Y Tsusaka and Y Kagoshima
Graduate School of Material Science, University of Hyogo, 3-2-1 Kouto, Kamigori, Ako, Hyogo 678-1297, Japan

E-mail: htakano@sci.u-hyogo.ac.jp

Abstract. Fast X-ray computed tomography (CT) system with sub-second order measurement for single CT acquisition has been developed. The system, consisting of a high-speed sample rotation stage and a high-speed X-ray camera, is constructed at synchrotron radiation beamline in order to utilize fully intense X-rays. A time-resolving CT movie (i.e. 4D CT) can be available by operating the fast CT system continuously. Real-time observation of water absorbing process of super-absorbent polymer (SAP) has been successfully performed with the 4D CT operation.

1. Introduction

X-ray CT technique is essential application utilizing X-ray's transparent characteristics. The technique can reveal inner structures of objects and has been widely applied for not only medical diagnosis but material science and so on. According to recent development of X-ray sources and X-ray optics, micro-CT techniques with high-spatial resolution have been developed and the systems that can resolve below 100 nm order are not special. Otherwise, new approach for x-ray CT, improvement of temporal resolution for observing dynamics of blind inner structures is very promising. This approach has been recently developed for medical diagnosis CT system and excellent multi-slice CT system with sub-second resolution is commercially available [1]. For micro-CT based on synchrotron radiation source, some works of fast CT techniques have been reported [2, 3]. Although they successfully captured high-speed jet or in-situ heart beating, their techniques are available only for repetitive phenomena. In order to realize real-time fast CT, a high-speed X-ray camera for micro-imaging has been developed. The camera system has ability to acquire images with hundreds to a thousand frames per second (fps) and real-time fast micro-CT system can be configured by using high-speed sample rotation system. The fast micro-CT system with the spatial resolution of 30 μm and the fastest acquisition time for single CT of 0.16 s has been successfully developed. In this paper, system components and performance of the fast micro-CT are reported. And as an excellent application of the system for real-time 4D observation, water absorbing process of a piece of SAP set in glass capillary is introduced.

2. System components and experimental set-up

The key components of the fast micro-CT system are a sample rotation stage and an imaging detector. The rotating stage is driven by DC motor with the fastest rotating rate of 3.125 rotations per second (rps). The stage is based on sample rotator generally used for powder diffractometer and the wobbling of rotation centre is tuned within 4 μm. The fast X-ray camera system we developed is visible-light conversion type and consists of P-43 scintillator (10 μm thickness), relay lens optics, and high-speed

CMOS sensor. The relay lens optics has two C-mount camera lens with F = 0.95 and a 2x extension lens. The optical path is bent with a plane mirror to avoid direct incidence of X-ray to image sensor. The high-speed CMOS sensor (Hamamatsu, C8201) having 256 x 256 arrays of 20 μm-sized square pixel can take images with 250 fps (1000 fps is also possible with 128 x 128 arrays readout), and continuous image acquisition for over 40 s is possible on 512MB RAM memory. Therefore, effective pixel size of the fast X-ray camera system is 10 μm and the view field is 2.56 mm.
The fast micro-CT system is constructed at Hyogo-ID beamline (BL24XU) of synchrotron radiation facility SPring-8. The beamline has a figure-8 type hard X-ray undulator, strong emission with relatively low heat load is available, as the light source and a silicon double-crystal monochromator (DCM). Monochromatized X-ray beam with the energy of 10 keV, the intensity of 5 x 10^{12} photons/s, and the size of about 1.5 mm squares is obtained at experimental hutch about 70 m apart from the undulator source. The overview of the beamline and components of the fast micro-CT system are shown in Fig.1. The X-ray imaging optics we adopted is the simplest projection mode for utilizing the strong intensity, and fully statistic image can be obtained for single exposure (4 ms) with this system. The distance between sample and image detector is set as close as possible in order to reduce edge-enhancement by refraction, and a rotating beam diffuser consisting of sandpapers is installed to reduce unevenness of view field caused by speckle noise from coherent undulator illumination. The control system is very simple to avoid dead-time during measurements. There is no synchronized signal between sample rotation and image acquisition. Just the projection images are started to be acquired with 250 fps after the rotation rate of the sample stage is fully stabilized. Time series dataset of projection images over 40 s (i.e. over 10000 images!) is obtained by the continuous measurement at once. The number of projection images for single CT reconstruction depends on the sample rotation rate. 40 projection images covering 180 degree can be obtained in 0.16 s where the rotation is driven with the highest rate of 3.125 rps. Time series CT data, we call this as 4D CT, can also be obtained by extracting single CT at every 90 degrees from the dataset. Then, 4D CT observation with the fastest interval of 0.08 s is possible as a real-time movie.
In the case of fast CT, number of projection images seems too small to reconstruct fine structure though the effective pixel size of the detector is as small as 10 μm. The spatial resolution of fast CT is evaluated by reconstructed wall images of thin glass capillary with the wall thickness of 10 μm. Blur of the reconstructed walls, corresponding to the spatial resolution, is investigated for various projection numbers. The spatial resolution around 30 μm is obtained in any view field with the projection numbers down to 90 (0.36 s for 1CT). For the fastest projection number of 40, only the centre view field within 0.8 mm satisfies 30 μm resolution. The evaluated spatial resolution is always inferior to effective pixel size of 10 μm. The degradation is caused from not CT system but image detector because an edge-response of projection image is evaluated to 30 μm.

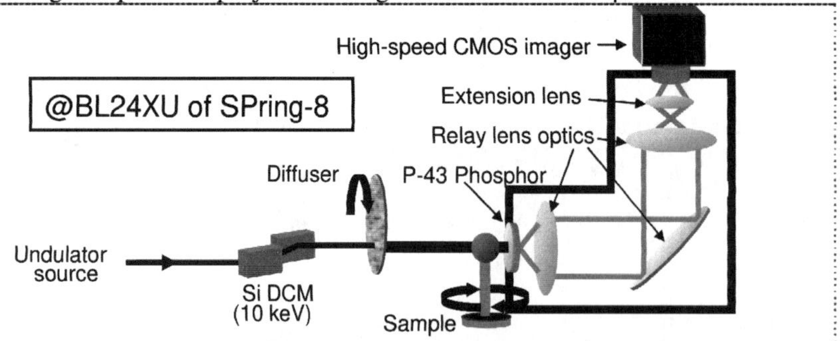

Figure 1. Schematics of fast X-ray micro-CT system at BL24XU of SPring-8.

3. Real-time 4D observation of water absorbing process of SAP

The SAP, consisting of sodium polyacrylate ([-CH_2-CH(COO-Na^+)-]$_n$), can absorb water of the hundreds times weight for itself. The material is widely applied for various products. We attempt to

Journal of Physics: Conference Series **186** (2009) 012049 doi:10.1088/1742-6596/186/1/012049

observe the water absorbing process of the SAP as 4D CT application. A piece of SAP taken from a disposal diaper for baby is put into glass capillary tube with the diameter of 1 mm. A water injection needle is set above the upper edge of the glass capillary. A water drop is injected to SAP just after starting a measurement with the sample rotation rate of 1.2 rps (0.36 s for single CT acquisition). A 4D CT movie consisting of 226 3D CT reconstructions with the interval of 0.18 s is obtained. Figure 2 shows extracted typical images from the movie data. Water absorbing process and expansion of SAP can be clearly observed. Moreover, time dependant change of absorption coefficient of SAP can be observed by analysing histograms of the reconstructed sectional images. Figure 3 shows the time dependant changes of peak value of the obtained absorption coefficient (corresponds to density). The density rapidly increases just after water injection and slowly decreases close to that of water. This type of information is important for understanding mechanism of the process, and is essential application utilizing advantage of X-ray CT.

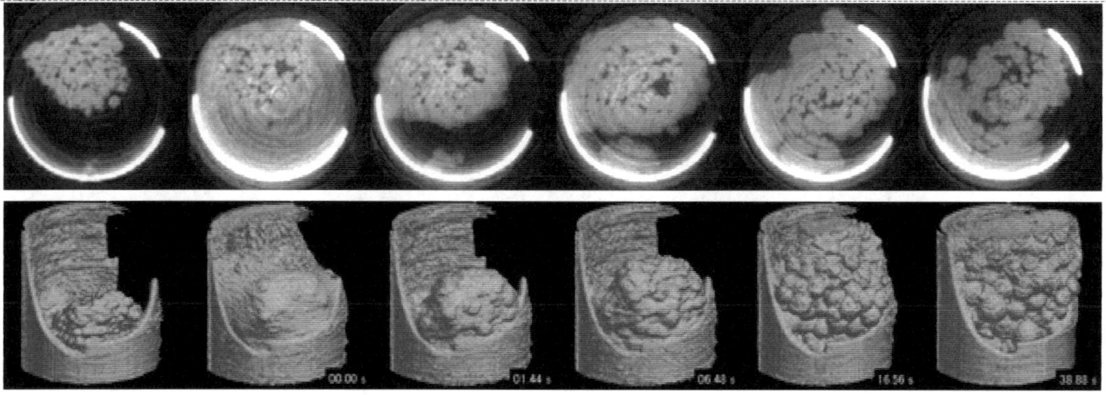

(a) Before (b) Just after (c) 1.5 s later (d) 6.5 s later (e) 16.6 s later (f) 38.9 s later

Figure 2. Extracted images from 4D CT movie of water absorbing SAP. The lower images are volume rendering images and the upper images are the corresponding sectional image of CT reconstruction. The images show typical phase of dynamics. (a) is taken before water injection (dry state), and (b) – (f) are time series images taken from just after water injection.

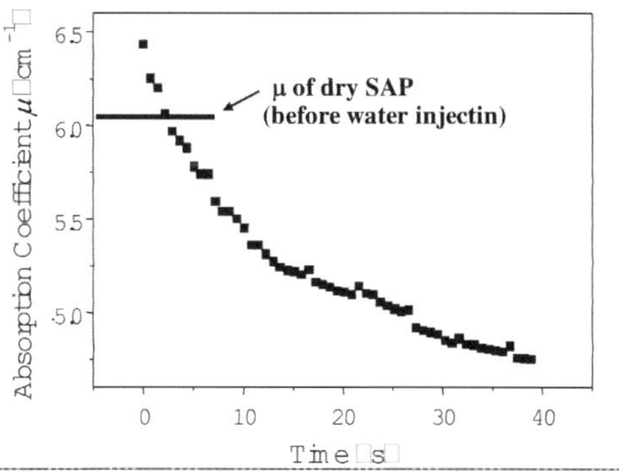

Figure 3. Time dependant changes of absorption coefficient of SAP. The absorption coefficients are obtained by analyzing histograms of sectional images. The value rapidly increases just after water injection and slowly decreases close to that of water.

4D CT technique is quite promising because it gives direct understanding for dynamical phenomena. Various applications will be expected by improving the temporal resolution and the spatial resolution. Although these experiments require the stronger illumination to sample, we can perform further improvement by using raw undulator radiation (without monochromator).

4. References

[1] Rybicki F J *et al.* 2008 *Int. J. Cardiovasc Imaging* **24(5)** 535
[2] Sera T, Yokota H, Fujisaki K, Fukasaku K, Tachibana H, Uesugi K, Yagi N and Himeno R 2008 *Phys. Med. Biol.* **53** 4285
[3] Wang J 2007 *AIP Conf. Proc.* **879** 1535

Journal of Physics: Conference Series **186** (2009) 012050 doi:10.1088/1742-6596/186/1/012050

High-definition high-throughput micro-tomography at SPring-8

Kentaro Uesugi, Akihisa Takeuchi and Yoshio Suzuki
JASRI / SPring-8, Kouto 1-1-1, Sayo, Hyogo 679-5198, Japan

E-mail: ueken@spring8.or.jp

Abstract. In SPring-8, the number of micro-tomography experiments is increasing. We adopted a continuous acquisition method and high speed CCD camera for increasing a throughput. Those techniques reduced the measurement time and increased the throughput. The total measurement time for 1800 projections with 2000×1311 pixels projection images was reduced to 20 minutes or less while that of previous system using "step-and-shoot" method was a few hours. These enable us to image soft materials (tissue) and to do time resolved measurements in three-dimension. The spatial resolution of those systems is around 1μm and the field of view 1 mm in horizontal direction. Using those systems, more than 30 measurements can be done during one day. It also means that the hundreds gigabytes of data per day is obtained.

1. Introduction

X-ray computed tomography is one of the methods for non-destructive investigation of materials. It can obtain the internal structures of materials as a distribution of X-ray linear absorption coefficients (LACs). Usually they are called as CT images. LACs are defined as a function of X-ray energy, chemical composition and density of the materials [1]. Therefore the obtained LACs are unique for materials when monochromatic X-rays are used. That is, the materials are distinguished each other using LACs in an appropriate measurement condition.

Recently there are some investigations about finite element analysis which use the CT images as a beginning condition [2]. In such case, high definition 3-D image is required for increasing data points in the simulations. The high definition tomography system also satisfies the requirement of users. They are "larger field of view and higher resolution image".

Here, high definition micro-tomography system is described with a spatial resolution of about 1μm at BL20XU and BL47XU in SPring-8.

2. Micro-tomography system

The system consists of light source, double crystal monochromator (Si(111), liquid nitrogen cooled), high speed X-ray shutter, beam diffuser, high precision sample stages and high resolution X-ray image detector (Fig. 1). The X-ray energy ranges are 8 keV to 37 keV at BL20XU and 6 keV to 37 keV at BL47XU, respectively. The distances from light source to experimental set up are 240m and 48m for BL20XU and BL47XU, respectively. At BL47XU a pair of vertical deflecting mirrors can be used to suppress the higher harmonics. Beam diffuser is a spinning piece of copy paper placed 160m and 4m upstream from sample at BL20XU and BL47XU, respectively. And it is used for increasing the stability of apparent X-ray and decreasing the speckle noise.

High resolution X-ray image detector (Hamamatsu Photonics, AA50) consists of single crystal phosphor screen ($Lu_2SiO_5:Ce^+$, 10μm-thick), objective lens (Nikon, CF Plan 20X EPI ELWD), mirror and high definition CCD camera (Tab. 1). The field of view of the detector is 1mm × 0.65mm. The CCD camera is usually used with 2×2 binning mode and the effective pixel size is 0.5 μm × 0.5 μm.

Table 1. Characteristics of cooled CCD camera, Hamamatsu Photonics, C4880-41S.

CCD type	Interline transfer
Format	4000(H) × 2624(V)
Pixel size	5.9 μm × 5.9 μm
Active area	23.6 mm × 15.5 mm
Full well	16000 electrons
Frame rate	1.7Hz
Read-out noise	5 electrons r.m.s
A/D converter	12 bits
Cooling temp.	-50°C

The spatial resolution of the detector is about 1μm restricted by the thickness of phosphor screen [3,4]. And high precision rotation stage has the accuracy better than 0.2 μm. Therefore the spatial resolution of the micro-tomography system also achieves 1μm in three-dimension.

The exposure time for each projection is from 200 msec to 300 msec with X-ray flux density about 1.0E+12 photons/mm^2/s. Total scan time is 20 minutes for 1800 projections with "continuously rotated (on the fly scan)". Direct beam images (without sample) are taken before and after obtaining sample images (30 images for each).

Convolution back projection method is used for CT reconstruction. The reconstruction software uses a multi thread technique. The reconstruction time is about 10 sec for a slice (2000 pixels, 1800 projections) using the software on Intel Core2 Quad (Q6600) processor.

3. Results

CT images of Cu-Al concentric test patterns are shown in Fig. 2. The X-ray energy was 15keV and the number of projections was 900. The total scan time was 15 minutes. The 2 μm and 4 μm pitches are resolved with the system.

A ductile iron including graphite spheres was imaged with this system and obtained the size and distribution of graphite spheres. The 3-D analysis showed that the shapes were almost spherical rather than ellipsoidal and the radius distributed between 15 μm to 25 μm. And the sphere's distribution was ordered with the radial distance of 50 μm.

These kinds of information can be obtained only by X-ray micro-tomography because of its three-dimensional nature. Especially the monochromatic X-ray provides the quantitative information about contrast of the image, namely LACs. This will helpful for investigation of materials.

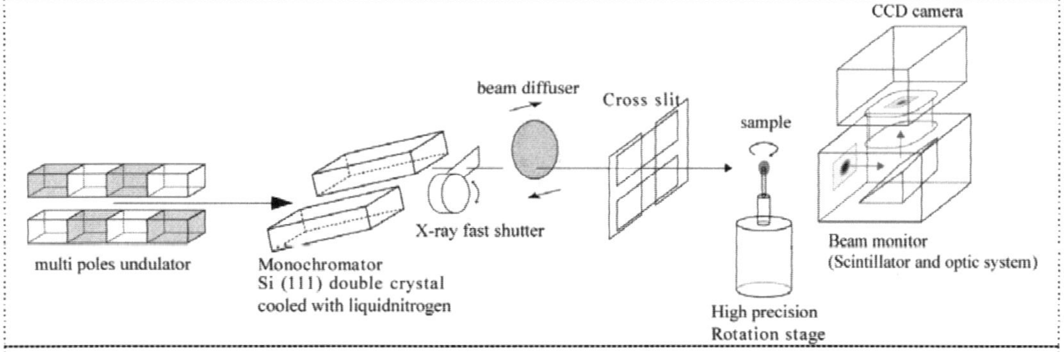

Figure 1. Schematic illustration of parallel beam microtomography at BL20XU and BL47XU.

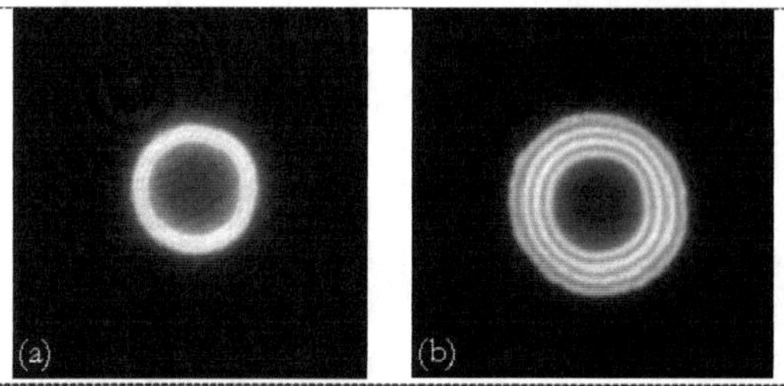

Figure 2. CT images of Al-Cu concentric test patterns. Bright area shows Cu region. The field of view is 94μm x 94μm. (a) Pitch is 2μm. (b) Pitch is 4μm.

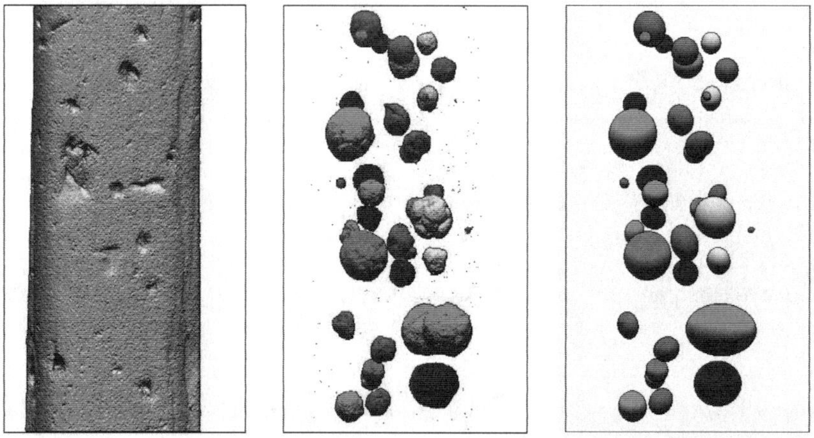

Figure 3. A ductile iron including graphite spheres was imaged with micro-tomography and obtained the size and distribution of graphite spheres. During the image processing, the graphite spheres are treated as triaxial ellipsoids. The width and height of each image are 0.3mm and 1.0mm, respectively.

Acknowledgement
The synchrotron radiation experiments were performed at SPring-8 with the approval of the Japan Synchrotron Radiation Research Institute (JASRI) (proposal nos. 2003B0108-NM-np, R03B47XU-0018N, R03B47XU-0056N, R04A47XU-0033N and R04A47XU-0044N). Professor Yasuda (Osaka Univ.) and Dr. Nakano (GSJ/AIST) kindly provided a ductile iron and its 3-D analysis results.

References
[1] Chantler, C.T., Olsen, K., Dragoset, R.A., Chang, J., Kishore, A.R., Kotochigova, S.A., and Zucker, D.S. 2005, X-Ray Form Factor, Attenuation and Scattering Tables (version 2.1): http://physics.nist.gov/ffast
[2] H. Toda, M. Takata, T. Ohgaki, M.Kobayashi, T. Kobayashi, K. Uesugi, K. Makii and Y. Aruga 2006 Advanced engineering and Materials **8**(6), 1-9
[3] K. Uesugi, Y. Suzuki, N. Yagi, A. Tsuchiyama, T. Nakano, 2001 Nuclear Instruments and Methods in Physical Research A, **467-468**,853-856
[4] H. Takano, Y. Suzuki, K. Uesugi, A. Takeuchi and N. Yagi 2001 Proc. SPIE, **4499**, 126-133.

Coherent Imaging at FLASH

H.N. Chapman[1], S. Bajt[2], A. Barty[3], W.H. Benner[3], M.J. Bogan[3], S. Boutet[4], A. Cavalleri[5], S. Duesterer[2], M. Frank[3], J. Hajdu[6], S.P. Hau-Riege[3], B. Iwan[6], S. Marchesini[7], A. Sakdinawat[7], K. Sokolowski-Tinten[8], M.M. Seibert[6], N. Timneanu[6], R. Treusch[2], B.W. Woods[3].

1. Centre for Free-Electron Laser Science, DESY and Universität Hamburg, Notkestrasse 85, 22607 Hamburg, Germany.
2. HASYLAB, DESY, Notkestrasse 85, 22607 Hamburg, Germany.
3. Lawrence Livermore National Laboratory, Livermore CA 94550, USA.
4. Stanford Linear Accelerator Laboratory, Menlo Park, CA 94025, USA.
5. Department of Physics, Clarendon Laboratory, University of Oxford, Parks Road, Oxford OX1 3PU, UK.
6. Laboratory of Molecular Biophysics, Department of Cell and Molecular Biology, Uppsala University, Husargatan 3, Box 596, SE-75124 Uppsala, Sweden.
7. Lawrence Berkeley National Laboratory, Berkeley, CA 94720, USA.
8. Institut für Experimentelle Physik, Universität Duisburg-Essen, Lotharstraße 1, 47048 Duisburg, Germany.

Henry.Chapman@desy.de

Abstract. We have carried out high-resolution single-pulse coherent diffractive imaging at the FLASH free-electron laser. The intense focused FEL pulse gives a high-resolution low-noise coherent diffraction pattern of an object before that object turns into a plasma and explodes. In particular we are developing imaging of biological specimens beyond conventional radiation damage resolution limits, developing imaging of ultrafast processes, and testing methods to characterize and perform single-particle imaging.

1. Imaging at the FLASH Free-Electron Laser

The FLASH FEL, located at the Deutsches Elektronen-Synchrotron (DESY) laboratory in Hamburg, generates short, high-power soft-X-ray pulses by the principle of self-amplification of spontaneous emission [1]. This facility currently operates at fundamental wavelengths between 6 nm and above 50 nm, with pulse energies above 50 μJ and pulse durations less than 30 fs. Such short pulses permit the high-resolution and ultrafast imaging of a sample before the pulse destroys it [2].

In our coherent diffractive imaging experiments, we focus the FEL pulse to a 25-μm diameter spot to illuminate an isolated sample, which scatters the light to give a coherent and continuous diffraction pattern. We record this forward-directed diffraction pattern on a bare CCD after reflection from a multilayer-coated plane mirror angled by 45° to the incident beam [3]. The optical path length from the sample to the CCD is 53 mm. For a pixel size of 20 μm and a wavelength of 7 nm, for example, this camera length allows us to adequately sample (above the Nyquist rate) the intensity fringe pattern due to interference of waves scattered from two object points separated by 9.2 μm. These conditions

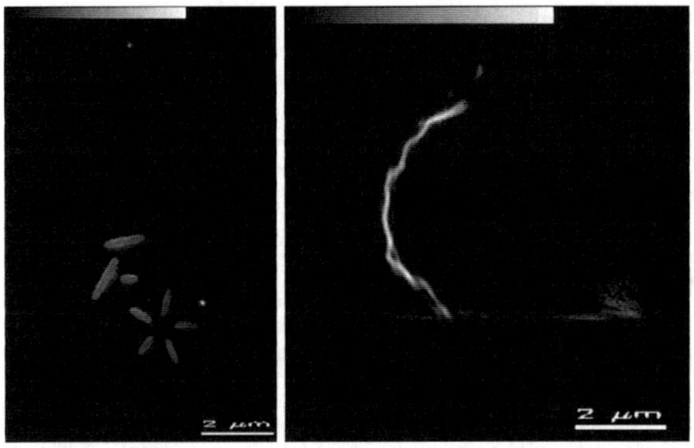

Figure 1. (left) Reconstructed image of a test object with two 145-nm diameter spheres, reconstructed to 40-nm resolution (adapted from [7]). (right) Reconstructed image of a *Spiroplasma melliferum* cell. Both images were reconstructed from single-pulse FEL coherent diffraction patterns at 13.5 nm wavelength.

allow the diffraction pattern to be phased and then inverted to give a high-resolution image of the sample. There are advantages to this mode of microscopy when using FEL pulses: the retrieved image is complex-valued, giving image contrast both by absorption and phase-shifting or refractive properties of the sample; the diffraction pattern is insensitive to the transverse sample position; and "focusing" of the image is performed as part of the computational reconstruction process. These properties make diffractive imaging ideal for studying objects that are injected into vacuum on the fly, and for high-resolution single-shot imaging where the sample does not survive the beam.

2. High-Resolution Imaging

Our apparatus can be used to position fixed samples on silicon nitride window arrays into the beam, or to shoot particles across the beam. Examples are shown in Fig. 1. The test object was etched with a focused ion beam (FIB) after a few latex spheres of 145 nm were deposited on the silicon nitride membrane. Both the FIB object and the two latex spheres were reconstructed to a resolution of 40 nm. We also performed the first ever reconstruction of a single cell using a single FEL pulse. The single cell organism is *Spiroplasma melliferum*, an elongated helical bacteria, and the image was reconstructed to 70 nm resolution. We additionally developed nanofabricated coded apertures to increase the efficiency for single shot ultrafast holographic imaging of weakly scattering objects [4].

We successfully developed and demonstrated a particle injection system and diagnostics [5]. The injector operates as a continuously refreshed beam of free nanoparticles delivered through a differentially pumped aerodynamic lens, in a "shotgun" mode (unsynchronized to the FLASH pulses). We operated with pulse trains of 10 μs spacing. The particles, traveling faster than 100 m/s, were only hit by a single pulse. A summary of some results is shown in Fig. 2. During one run we operated the injector continuously for 18.7 hours, performing 26 sample changes with 14 different samples (Fig. 2 C). The CCD collected images continuously and 1873 of 16639 patterns contained particle scattering information, giving an average hit rate of 0.05 Hz (including all sample change down time). Samples included size-selected latex spheres, Ag cubes, DNA nanocomplexes, picoplankton, and lipid micelles. We have reconstructed images of injected whole hydrated picoplankton cells and latex dimers. Our results confirm the feasibility of single-particle diffractive imaging.

3. Ultrafast Imaging

The ultrafast duration of FLASH pulses allows the study of processes with a temporal resolution of tens of femtoseconds. The soft-X-ray FEL source gives significantly higher spatial resolution than achievable with optical lasers and a higher penetration and larger signal to noise than short-pulse electron microscopy. These capabilities open up the detailed study of non-repetitive processes, such as crack formation, phase separation and nucleation, plasma formation, and spallation and ablation. We demonstrated ultrafast coherent X-ray diffraction of ablating foils pumped with a visible laser synchronized to the FEL pulses.

We have phased time-resolved coherent diffraction patterns of laser-ablating microstructures [6]. Many identical structures were manufactured by focused ion-beam milling. We see that, from the

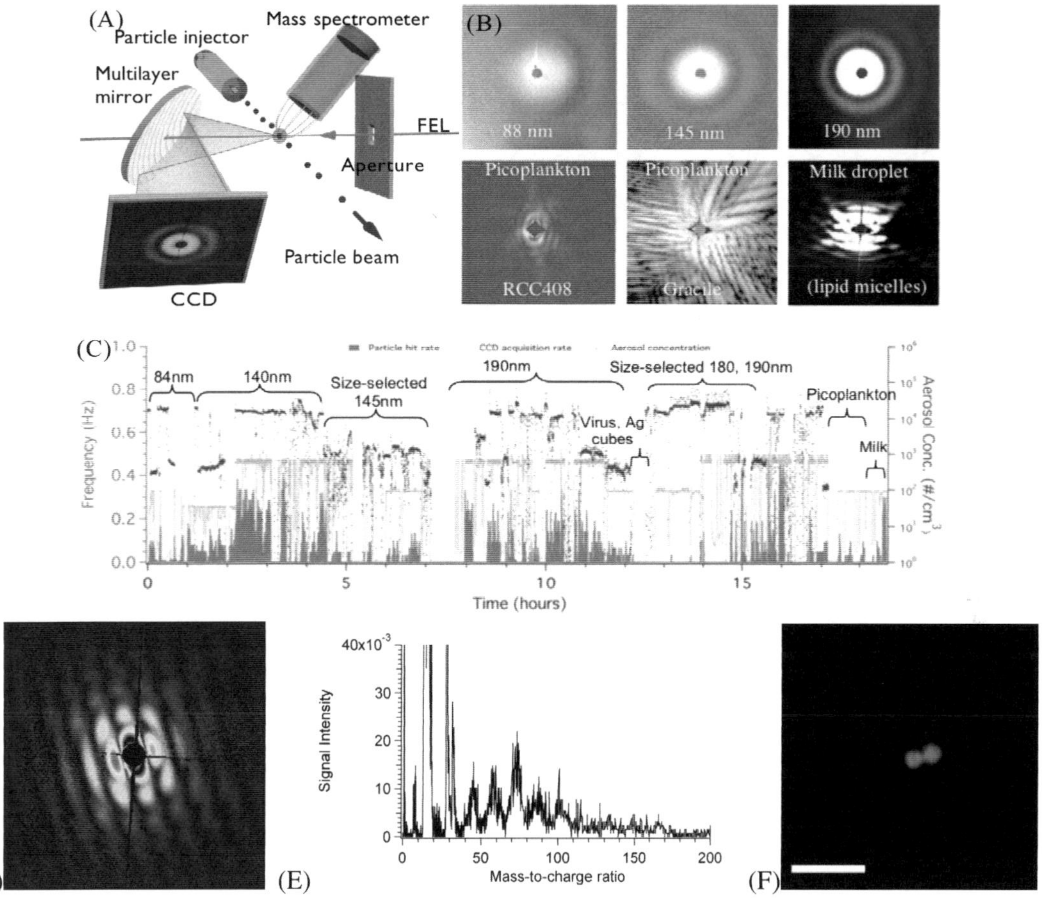

Figure 2. (A) Schematic of the diffraction camera for FLASH set up in particle injection mode, (adapted from [5]) (B) Example patterns of spherical nanoparticles and biological samples such as single cells and lipid micelles. (C) Summary of a particle injection run at FLASH showing particle hit rate (diffraction pattern acquisition rate), CCD readout rate and the aerosol concentration as measured by differential mobility. (D) Diffraction image collected during injection of megadalton scaffolded DNA complexes with a tamper of sucrose. (E) Time-of-flight mass spectrometer detected ions specific to this type of particle. (F) The image reconstruction showed that two particles were irradiated simultaneously by a single FLASH pulse (scalebar is 1 micron). (A) & (D-F) adapted from [5].

reconstructed images, and correlations of patterns at various time delays to the pattern of the undisturbed object, order in the structure is progressively lost starting from short length scales. This structural rearrangement progresses at close to the sound speed in the material [6]. Our initial experiments show the ability to capture single-shot images of nanoscale dynamics at timescales of individual atomic motion

References

[1] V. Ayvazyan *et al.*, Eur. Phys. J. D **37**, 297–303 (2006)
[2] H. N. Chapman *et al.*, Nat. Phys. **2**, 839–843 (2006).
[3] S. Bajt *et al.*, Appl. Opt., **47**, 1673–1683 (2008).
[4] S. Marchesini *et al.*, Nat. Photon. **2**, 560 - 563 (2008).
[5] M. J. Bogan *et al.*, Nano Letters **8**, 310–316 (2008).
[6] A. Barty *et al.*, Nat. Photon. **2**, 415–419 (2008).
[7] S. Boutet *et al.*, http://dx.doi.org/10.1016/j.elspec.2008.06.00

Journal of Physics: Conference Series **186** (2009) 012052 doi:10.1088/1742-6596/186/1/012052

Coherent laser scanning diffraction microscopy

Martin Dierolf[1,2], Pierre Thibault[1], Cameron M Kewish[1], Andreas Menzel[1], Oliver Bunk[1] and Franz Pfeiffer[1,2]

[1] Paul Scherrer Institut, 5232 Villigen PSI, Switzerland

[2] École Polytechnique Fédérale de Lausanne, 1015 Lausanne, Switzerland

E-mail: martin.dierolf@psi.ch

Abstract. Coherent diffractive imaging (CDI) is a promising approach to high-resolution x-ray microscopy. While CDI typically has a rather limited field of view, this problem can be solved by ptychography, a technique for which an extended object is raster scanned by a compact coherent illumination probe. Significant overlap of illumination for adjacent scan points allows then a self-consistent reconstruction from the entirety of collected coherent diffraction patterns. However, current reconstruction schemes require accurate *a priori* knowledge of the probe. Our recently developed new algorithm for ptychographic data sets allows us to simultaneously reconstruct both object and illuminating probe. We demonstrate the application of the new method in a test experiment with visible laser light showing that intricate illumination functions can be retrieved reliably.

1. Introduction

In coherent diffractive imaging (CDI) the specimen's complex-valued transmission function is reconstructed from far-field diffraction intensities by means of phase retrieval algorithms. The achievable resolution only depends on the angular range over which one can detect the diffraction signal. Reconstruction is usually based on an iterative projection algorithm, for instance, the widely used hybrid input-output (HIO) algorithm by Fienup [1] and its variants. To ensure sufficient sampling of the diffraction pattern, either the sample has to be isolated or the illuminating beam has to be confined to a limited region. Although frequently found to be experimentally challenging, the latter approach allows one to obtain images of extended, non-compact specimens by raster scanning. One possibility is to individually reconstruct the diffraction patterns and combine the resulting images afterwards [2]. In contrast ptychography allows a self-consistent, simultaneous reconstruction of the full scanned area from multiple diffraction patterns, given sufficient overlap of adjacent illuminations. Originally developed in the 1970s for electron crystallography [3], ptychography has also been succesfully applied to non-crystalline specimens [4–6], recently benefiting from the combination with iterative phase retrieval in the so-called "ptychographical iterative engine" (PIE) [5–7]. However, accurate knowledge of the illumination probe is required *a priori* or has to be obtained experimentally.

We recently developed a new reconstruction scheme which allows simultaneous reconstruction of both object and illuminating probe [8] from a ptychographic data set. In this article this is demonstrated with visible laser light data collected in a focused-beam geometry.

2. Reconstruction algorithm

The phase retrieval problem for ptychographic data sets can be reformulated as finding the intersection of two sets of constraints. The "Fourier constraint" enforces the Fourier modulus of the object's exit wave to be consistent with the measured intensities in the diffraction pattern,

$$I(\mathbf{q}) = |\mathcal{F}\{\psi_j(\mathbf{r})\}|^2 . \tag{1}$$

The "overlap constraint" states that at each point of the scan the exit wave can be factorized as a illuminating probe P and a complex-valued object transmission function O,

$$\psi_j(\mathbf{r}) = P(\mathbf{r} - \mathbf{r}_j)O(\mathbf{r}). \tag{2}$$

Exit waves from adjacent scan positions are coupled for non-zero overlap of the corresponding illuminations, which leads to a high redundancy and gives the method its efficiency. The solution of the ptychographic phase retrieval problem, *i.e.* the two functions O and P for which both constraints are fulfilled at each point j of the ptychography scan, is determined iteratively with the difference map algorithm [9], which uses a combination of projections of the state vector $\Psi(\mathbf{r}) = (\psi_1(\mathbf{r}), \psi_2(\mathbf{r}), \psi_3(\mathbf{r}), \ldots, \psi_N(\mathbf{r}))$ onto the two constraint sets. For the overlap projection the distance $\|\Psi - \Psi^0\|^2 = \sum_j \sum_{\mathbf{r}} |\psi_j(\mathbf{r}) - \hat{P}(\mathbf{r} - \mathbf{r}_j)\hat{O}(\mathbf{r})|^2$ has to be minimized, leading to coupled equations for probe and object:

$$\hat{O}(\mathbf{r}) = \frac{\sum_j \hat{P}^*(\mathbf{r} - \mathbf{r}_j)\psi_j(\mathbf{r})}{\sum_j |\hat{P}(\mathbf{r} - \mathbf{r}_j)|^2}, \tag{3}$$

$$\hat{P}(\mathbf{r}) = \frac{\sum_j \hat{O}^*(\mathbf{r} + \mathbf{r}_j)\psi_j(\mathbf{r} + \mathbf{r}_j)}{\sum_j |\hat{O}(\mathbf{r} + \mathbf{r}_j)|^2}. \tag{4}$$

For the reconstruction of experimental data, one usually starts with a rough model for the probe $\hat{P}$ and a random guess of the object function $\hat{O}$. While the updated object function is calculated from equation (3), probe retrieval is achieved by iterating the probe in a nested loop according to equation (4). Updating the state vector $\Psi(\mathbf{r})$ using the difference map formalism completes one iteration.

3. Experimental

The experimental setup is illustrated in Figure 1: The sample, a cross-section of a pumpkin stem (see Figure 2 for visible-light micrographs), was placed on a scanning stage close to the focal plane of a lens illuminated by a HeNe laser beam (632.8 nm wavelength). A cooled charge-coupled device (CCD) with 1024×1024 pixels of 24×24 μm^2 size was placed in the focal plane of a second lens (not shown in Figure 1) to collect far-field diffraction patterns. To compensate for the limited dynamic range of the CCD, five frames with different exposure times were combined for each point of the raster grid. The total scan consisted of 11×11 points with 50 μm steps.

The reconstructed images are shown in Figure 3, both (a) the complex transmission function of the object and (b) the probe in the specimen plane. Taking advantage of the chaotic behavior of the difference map algorithm, the images were obtained as averages of 20 reconstruction estimates taken at intervals of five iterations from the steady-state regime of a single run. The pixel size in the reconstruction is 2.6 μm, which is about 45 times smaller than the size of the illuminating probe. The latter corresponds to the intrinsic resolution of this scanning microscope in absence of CDI and shows the potential of the method to overcome intrumental limitations.

The retrieval of the complete illuminating wave field allows us to numerically propagate it to different planes along the optical axis. The longitudinal cut of the propagated probe wave field in Figure 3(c) shows, that the sample was actually placed 1.3 mm upstream from the focus. Although the reconstruction was started with just a disk, resembling a pinhole, this rather complicated illumination function was reliably retrieved.

Journal of Physics: Conference Series **186** (2009) 012052 doi:10.1088/1742-6596/186/1/012052

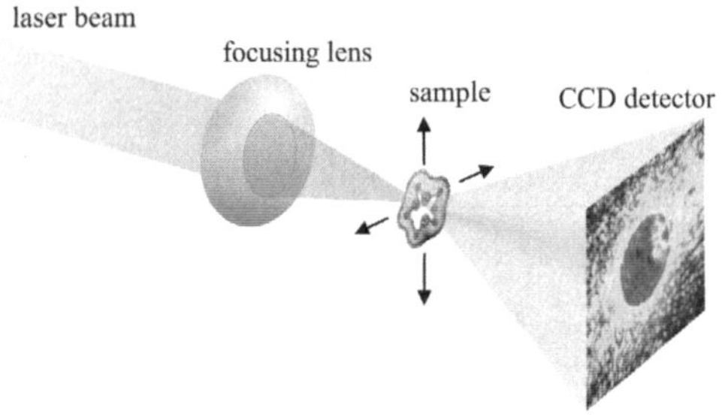

Figure 1. Experimental setup: The incoming beam is focused onto the specimen forming a small probe. The sample (Figure 2) is scanned perpendicular to the beam. At each point a diffraction pattern is collected with the CCD.

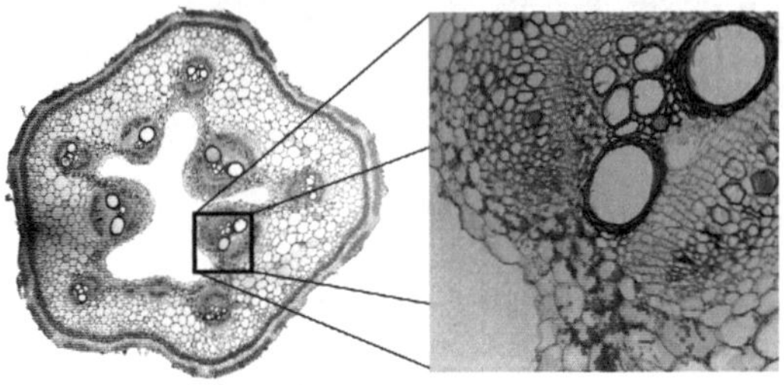

Figure 2. Visible-light micrograph of the specimen, a pumpkin stem cross-section. The magnified square (sidelength 640 μm) corresponds to the field of view in the reconstruction in Figure 3(a).

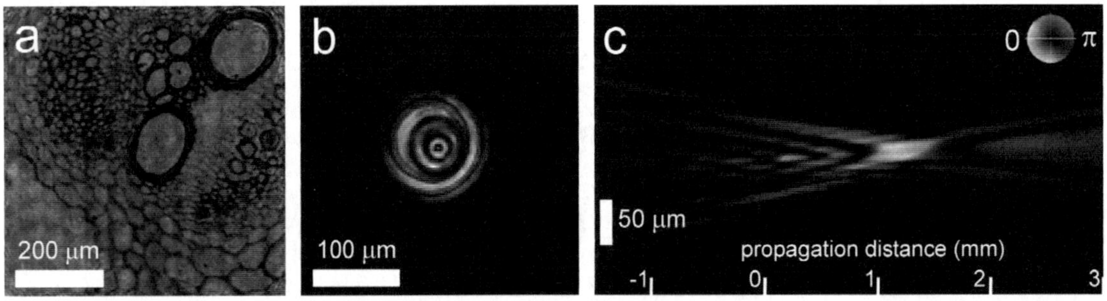

Figure 3. Reconstruction results, with amplitudes mapped to brightness and phases mapped to hue, see color wheel in (c). (a) Complex optical transmission function of object. (b) Reconstructed probe in the specimen plane. (c) Longitudinal cut of numerically propagated probe wave field.

4. Conclusions

Our new reconstruction approach overcomes the requirement of previous ptychography schemes of accurate knowledge of the illumination, which is hard to model and challenging to characterize experimentally. Complicated illumination functions can be retrieved simultaneously with the object, making the technique perfectly suited for, *e.g.*, combination with scanning x-ray transmission microscopy [8] and use in wavefront-sensing applications.

References

[1] Fienup J R 1982 *Appl. Opt.* **21** 2758–2769
[2] Abbey B *et al.* 2008 *Nature Phys.* **4** 394–398
[3] Hegerl R and Hoppe W 1970 *Berichte der Bunsen-Gesellschaft* **74** 1148–1154
[4] Chapman H N 1996 *Ultramicroscopy* **66** 153–172
[5] Rodenburg J M, Hurst A C and Cullis A G 2006 *Ultramicroscopy* **107** 227–231
[6] Rodenburg J M, Hurst A C, Cullis A G, Dobson B R, Pfeiffer F, Bunk O, David C, Jefimovs K and Johnson I 2007 *Phys. Rev. Lett.* **98** 034801
[7] Faulkner H M L and Rodenburg J M 2004 *Phys. Rev. Lett.* **93** 023903
[8] Thibault P, Dierolf M, Menzel A, Bunk O, David C and Pfeiffer F 2008 *Science* **321** 379–382
[9] Elser V 2003 *J. Opt. Soc. Am.* A **20** 40–55

Defocus as an ineffective means of changing spot size for fluctuation microscopy

J. M. Gibson and M.M.J. Treacy*

Argonne National Laboratory, Advanced Photon Source, Argonne, IL 60439
*permanent address: Arizona State University, Department of Physics, P.O. Box 871504, Tempe, AZ 85287 (author e-mail: jmgibson@aps.anl.gov)

Abstract. We examine the use of defocus for changing the effective spot size in fluctuation microscopy. Our conclusions relate to other coherent techniques. While it is possible to increase the illuminated spot size by defocusing, the number of phase oscillations inside the illuminated spot increases. As a result the stationary phase region increases with defocus at a slower rate than the spot size. Therefore, the effective spot size for fluctuation microscopy is not the beam size, and the phase interference is increasingly diluted by incoherent contributions, rendering defocus an ineffective means of changing the spot size for fluctuation microscopy.

In many applications of coherent illumination, it seems desirable to defocus the illumination to control the illuminated spot size. Several authors have used this technique, but little analysis has been done of the effects of defocus on coherent scattering. In a recent paper [1], we identified that transit of a small object over a defocused probe, as might be envisaged in coherent diffraction imaging, can lead to distortions in object reconstruction. In this paper we study the specific effect of defocusing on the speckle patterns from a small sample volume, and show that the effect is more complicated than had previously been assumed [2, 3]. In particular, we explain how the use of defocusing to change the probe size for fluctuation microscopy is ineffective, despite the appearance that it might work [3].

Fluctuation microscopy (FM) [4, 5] examines medium-range order in disordered materials. The method relies on the measurement of statistical fluctuations in the diffraction by a coherent probe from an ensemble of sample volumes, and can be implemented using coherent nanodiffraction patterns obtained in a scanning transmission microscope. Modifying the illuminated spot size is key to the technique. The method has been used with light [6] and x-rays [7], although electrons have most widely been used[4]. The fluctuations, often measured simply by examining the variance of the scattering with Q, have been shown to be very sensitive to medium range order [4]. A particularly simple way to measure a correlation length has been proposed in which the illuminated probe size is varied and the correlation length plotted as a function of the probe size [8].

1. Theory for a Gaussian probe

The conditions for FM require a coherently illuminated region centered over an object. The object is usually much larger than the illuminated region, and data is recorded by moving the object and statistically analyzing the fluctuations in the diffraction patterns. To simplify our discussion we assume that the illumination wavefunction intensity is a Gaussian with standard deviation σ, with a constant phase in the illumination plane where the thin object sits. The amplitude is $\psi = \psi_0 \exp\left(-r^2/4\sigma^2\right)$ where the radial coordinate r is relative to the beam center. Within the Fresnel approximation, defocusing is equivalent to replacing the standard deviation with a

Journal of Physics: Conference Series **186** (2009) 012053 doi:10.1088/1742-6596/186/1/012053

complex $\sigma' = \sqrt{\sigma^2 - i\lambda\,\Delta f/4\pi}$ where λ is the radiation wavelength and Δf the defocus. The resulting amplitude of the illumination is still a Gaussian, with an increased standard deviation $\sigma_d = \sigma\sqrt{1+\Delta^2}$, where the "generalized" defocus is $\Delta = \lambda\,\Delta f/4\pi\sigma^2$. The phase distribution in the object plane in the reduced coordinate $u = r/\sigma$ is given by

$$\phi(u) = \tan^{-1}(\Delta) - \frac{\Delta u^2}{2(1+\Delta^2)} \tag{1}$$

The parabolic phase distribution in equation (1) is not desirable for FM, because the theory requires that the phase be constant (or linearly varying) across the illumination [5]. Since the technique works by summing the scattered components from the illuminated atoms and examining the interference between them, the change in phase across the illumination scrambles the interference. Phase oscillations within the Gaussian envelope are demonstrated in fig. 1 for Δ=2.

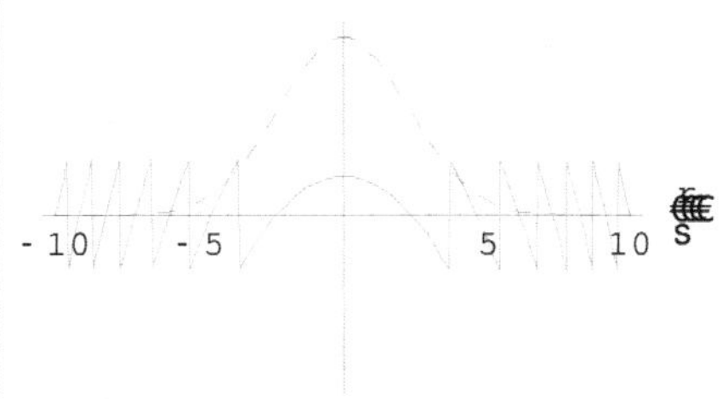

Figure 1: the amplitude (dotted) and phase (solid) distribution of a defocused Gaussian beam (with Δ=2) the phase oscillations represent a shift of π, and the amplitude is arbitrary.

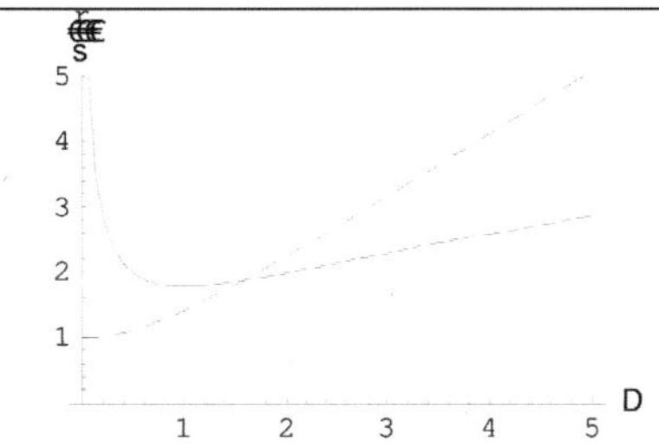

Figure 2: The standard deviation for the amplitude (dashed) and the critical radius of the stationary phase region (solid) as a function of generalized defocus for a Gaussian beam.

There are two serious problems caused by this oscillatory phase: the region over which coherent interference could occur as though the phase was stationary is limited to a smaller region in the center of the probe, and the speckle amplitude is diluted by the "incoherent" scattering from the oscillatory regions. We define the "stationary" phase region as a circular region of radius u_c at the center of the Gaussian where the phase change is limited to less than π/4. (Note that the absolute value of the phase at the origin, $\tan^{-1}(\Delta)$, does not affect any measurements). From equation (1)

$$u_c = \sqrt{\frac{\pi(1+\Delta^2)}{2\Delta}} \tag{2}$$

The critical radius u_c grows with Δ but more slowly than the intensity standard deviation does. Figure 2 shows the functional forms of u_c and σ_d/σ. Asymptotically the critical radius goes like $\sqrt{\Delta}$ whereas the amplitude half-width grows like Δ.

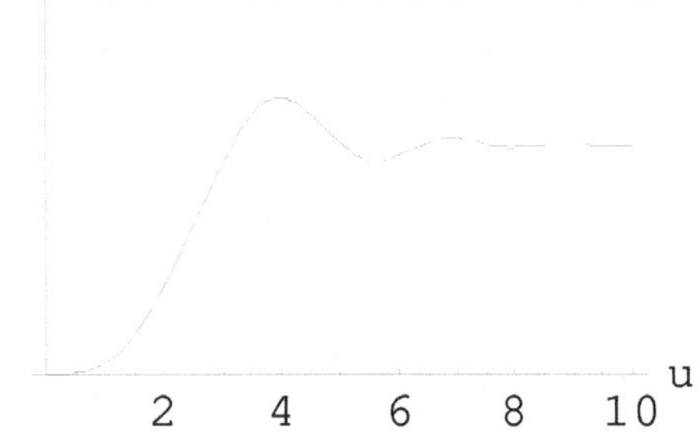

Figure 3: The integrated contributions with increasing distance u to the coherent signal from a chain of correlated atoms at Δ=2, where u_c=1.99.

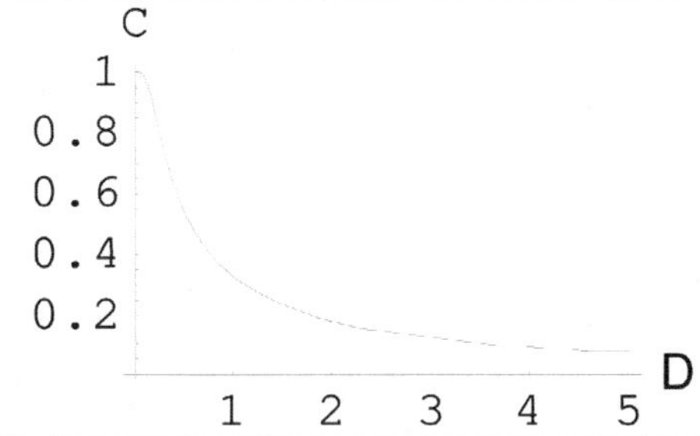

Figure 4: The coherent fraction of the illumination as a function of the generalized defocus.

Fig. 3 shows that only the region up to about the critical radius contributes to the coherent speckle from an ordered region, and that the oscillatory contributions beyond average to zero. So the maximum correlation length which is being probed is $u_c\sigma$ and not σ_d.

The situation is made worse by the fact that the oscillatory parts of the phase add to the incoherent signal, which reduces the signal to noise of the coherent signal that is sought in FM. The coherent fraction can be estimated by integrating up to u_c and normalizing to the total integrated intensity to give the function shown in figure 4

$$C(\Delta) = 1 - \exp\left(-\pi/8\Delta\right) \qquad (3)$$

So now we can see the problem. For example, to get the effective coherent illumination size to increase to $r = 2\sigma$, it is necessary to use a generalized defocus of 2.06 where the coherent fraction is only 17%. Attempts to usefully increase the critical radius above σ are thus severely hampered by the reduced visibility of fringes, and the actual coherent spot size is NOT the same as the defocused spot size (but can in principle be calculated from it). Since FM often relies on the height of speckle peaks as well as just their existence [5], this systematic change in visibility must be at very least accounted for, and most likely renders the signal-to-noise too low for useful application.

While these results were calculated specifically for Gaussian illumination, we expect them to apply to other illumination functions, since the parabolic nature of the emerging wave front, while perturbed by the detailed amplitude distribution, is general, especially at large defocus $\Delta \gg 1$.

2. Comparison to experiment and discussion

Despite the theory above, there appears to be experimental evidence that the coherent illuminated region can be varied by defocusing, both for x-rays [2, 9] and for electrons [3]. However, we can explain this paradox by the simple assumption that the structures that were examined were dominated by random correlations, which is often the case. With random structures, the existence of the phase curvature in the wavefront is just equivalent to an additional random phase factor. In other words no matter what the phase structure for the probe, provided it is well-defined, there will be a combination of scatterers in a random structure which will create speckle coming from atoms across the entire illuminated region. Just by observing the reciprocal space size of speckle from a random sample, one cannot identify the presence of phase curvature. On the other hand, in FM, where you are looking for a signal from medium-range ordering, the phase curvature will scramble the correlations outside the critical radius as described above. While it may be possible to account for this effect and still use the technique, on simple signal-to-noise grounds we conclude that the defocusing technique is not effective for FM. Phase curvature is also likely to affect many other applications of coherent diffraction, such as Coherent Diffraction Imaging which we have considered elsewhere [2].

We acknowledge fruitful discussions with L. Fan, D. Kumar and I. McNulty, and thank the US Department of Energy, Office of Science, Office of Basic Energy Sciences for support to the Advanced Photon Source through contract DE-AC02-06CH11357.

References

1. Gibson, J.M. and M.M.J. Treacy, Phys. Rev. B, **in press** (2008).
2. Retsch, C.C., Y. Wang, S. Frigo, I. McNulty, L. Lurio and G.B. Stephenson. in *Eleventh US National Conference on Synchrotron Radiation Instrumentation*. 1999. Stanford, CA: AIP.
3. Cowley, J.M., Ultramicroscopy, **90**, 197 (2002).
4. Gibson, J.M. and M.M.J. Treacy, Physical Review Letters, **78**, 1074 (1997).
5. Treacy, M.M.J., D.J. Paterson, I. McNulty, J.M. Gibson and L. Fan, Reports on Progress in Physics, **68**, 2899 (2005).
6. Kumar, D, *PhD thesis in preparation, Arizona State University*.
7. Fan, L., D.J. Paterson, I. McNulty, M.M.J. Treacy and J.M. Gibson, Journal of Microscopy, **225**, 41 (2007).
8. Gibson, J.M., M.M.J. Treacy and P.M. Voyles, Ultramicroscopy, **83**, 169 (2000).
9. Fan, L., *private communication*.

Hard X-ray scanning transmission microscopy with a 2*D* pixel array detector

A Menzel[1], C M Kewish[1], M Dierolf[1,2], P Thibault[1], P Kraft[1], O Bunk[1], K Jefimovs[1,*], C David[1] and F Pfeiffer[1,2]

[1] Paul Scherrer Institut, 5232 Villigen PSI, Switzerland
[2] École Polytechnique Fédérale de Lausanne, 1015 Lausanne, Switzerland
[*] current affiliation: EMPA, 8600 Dübendorf, Switzerland

E-mail: andreas.menzel@psi.ch

Abstract. X-ray microscopy is an invaluable tool for the characterization of life and materials science samples on the submicron length scale. Particularly in case of hard X-rays, it combines high penetration power with elemental, chemical, and magnetic specificity. However, contrast decreases drastically at higher photon energies, and imaging modalities beyond mere absorption measurements become increasingly attractive. We discuss the use of a fully pixelated detector for hard X-ray scanning microscopy of a biological specimen in absorption, phase contrast, and dark-field imaging mode.

1. Introduction

In scanning transmission X-ray microscopy (STXM) the photon flux past the specimen is recorded in order to determine the specimen's transmission. Spatial resolution is due to the limited extent of the tightly focused illuminating X-rays. More advanced detection schemes that offer complementary image information exist [1–8]. However, their every-day use is often hindered by the need of dedicated segmented detectors [1–5] or typically slow and noisy readout of fully pixelated 2*D* detectors such as charge-coupled devices (CCDs) [6–8].

Yet detector development has seen significant progress in recent years. For instance, 2*D* arrays of single-photon counters are now available. The PSI-developed PILATUS is such a detector [9]. A PILATUS module has $\sim 100,000$ pixels, each of which is $172 \times 172\,\mu\text{m}^2$ in size and is operated in single-photon counting mode. The PILATUS allows frame rates beyond 100 Hz without readout noise. Its quantum efficiency in the energy range of $4-10$ keV is $> 90\%$.

2. Experimental

Figure 1 shows a STXM scan of a single unstained red blood cell on a Si membrane. The measurements were taken at the cSAXS beamline at the Swiss Light Source (SLS), Paul Scherrer Institut (PSI), Villigen, Switzerland. The photon energy was 6.2 keV corresponding to the wavelength $\lambda = 2$Å. We used a Fresnel zone plate with 100 μm diameter and 100 nm finest zone width as focusing device. Data were collected "on the fly," *i.e.*, while the sample stage was moving along the fast horizontal axis of the scan. Acquisition time per pixel was 20 ms, followed by 8 ms dedicated for readout. Image pixels are $100 \times 100\,\text{nm}^2$. The sample–detector distance was 7.19 m bridged by a flight tube, which was flushed with He gas at atmospheric pressure.

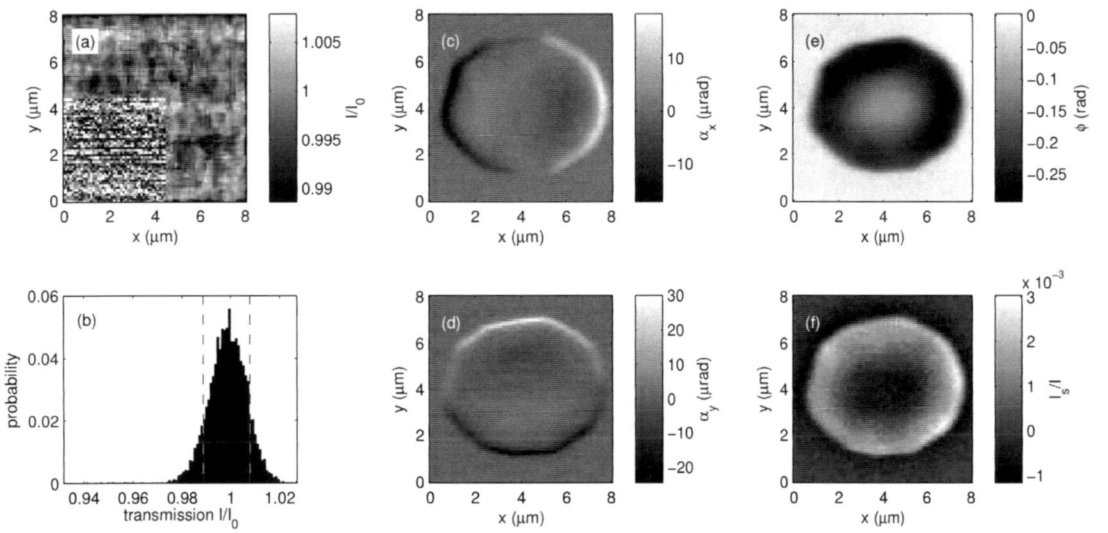

Figure 1. STXM measurements at 6.2 keV of a red blood cell. (a) median-filtered transmission image. The lower left corner shows raw data. The gray scale excludes the extreme 10% of both the dark and the bright side, shown also by dashed lines in the histogram (b). (c) and (d) horizontal and vertical differential phase contrast, respectively. (e) absolute phase contrast, determined using Eq. 1. (f) dark field image, in which scattered light yields contrast.

As demonstrated in the transmission image (a) and its histogram (b), there is essentially no discernible absorption contrast. The image has been median-filtered in order to decrease noise distortions. Nevertheless, statistics are not sufficient to reliably "see" the specimen. The signal-to-noise ratio (SNR) of the transmission measurement is $\sim 1:8$, which was determined after having located the specimen using phase contrast, discussed below, Fig. 1 (e).

In the absence of a specimen the detector is illuminated by an annulus corresponding to the light diffracted by the zone plate and the central stop. A gradient in the phase shift, ϕ, caused by a specimen gives rise to a refraction angle $\alpha_x = \frac{\lambda}{2\pi}\frac{\partial\phi(x,y)}{\partial x}$. Thus, measuring the shift of the illumination on the detector yields differential phase contast (DPC). Figures 1 (c) and (d) show such measurements in the horizontal (x) and vertical (y) directions, respectively. No filtering or image enhancement has been performed. We note that virtually the entire image information is due to shifts by less than a single detector pixel. The statistical noise is in the order of 0.5 μrad in x and 0.7 μrad in y, *i.e.*, even in case of this weakly interacting specimen the SNR in DPC measurements is $\sim 30:1$.

Such quantitative DPC measurements can reliably be integrated [10–12]. We use the real part of Eq. 4 of Ref. 11,

$$\phi = \mathrm{i}\mathcal{F}^{-1}\left\{\frac{q_x\mathcal{F}\phi^x + q_y\mathcal{F}\phi^y}{q_x^2+q_y^2}\right\}, \qquad (1)$$

and note that its imaginary part can be used to signify inconsistencies between the measured DPC signals ϕ^x and ϕ^y. q_x, q_y denote the momentum transfer, and $\mathcal{F}$ and $\mathcal{F}^{-1}$ are the forward and inverse Fourier transform operators. Zero padding the DPC signals and applying Eq. 1 yields the absolute phase shown in Fig. 1 (e), where the gray scale is in radians. Thus we can gain complete knowledge of the specimen's complex-valued optical X-ray transmission function, a fact which allows essentially arbitrary imaging modes or "omni-microscopy" [13].

Finally, since the detector we used for these studies is fully pixelated it allows virtually arbitrary detector responses. Figure 1 (f) shows a dark-field image, where the ratio of light scattered beyond the radius of the illuminated annulus to total detected intensity is plotted. A constant background has been subtracted. This imaging mode probes small-length-scale density fluctuations, which in case of the red blood cell apparently do not reveal additional structural information. It has been remarked earlier [3, 14] that, despite the low count rates used for this measurement, it often yields excellent SNR. Here, SNR $\sim 20:1$.

3. Conclusions

The advent of a new generation of pixel detectors has made possible imaging measurements in unprecendented flexibility and efficiency. Here, we used such a detector for scanning X-ray transmission microscopy. Using imaging modes, such as phase contrast and dark-field imaging, we have demonstrated the viability of such hard X-ray measurements even in case of weakly absorbing biological specimens. The excellent SNRs of these modes enable the microscopist to drastically reduce the dose imparted to the specimen.

More advanced imaging modalities are possible. Instead of merely integrating the scattered light, as we have done here for our dark-field measurement, one can quantitatively analyze the small-angle X-ray scattering pattern for each point of the raster scan. Such studies, including the correlation of scattering behaviour and tissue types and properties are ongoing and will be published elsewhere. One can also use the area detector to "oversample" a coherent diffraction pattern at each point. This allows ptychographic image reconstruction methods, which improve the resolution towards the diffraction-limited case [15–17].

Acknowledgements

We thank J. Als-Nielsen, Niels Bohr Institute, Copenhagen, Denmark, for the sample and X. Donath, PSI, for technical support at the cSAXS beamline. P.T. acknowledges financial support from the Fonds Québécois de la recherche sur la nature et les technologies (FQRNT).

References

[1] Dekkers N and de Lang H 1977 *Philips Technical Review* **37** 1–9
[2] Hawkes P W 1978 *J. Opt.-Nouv. Rev. Opt.* **9** 235–241
[3] Morrison G R and Browne M T 1992 *Rev. Sci. Instrum.* **63** 611–614
[4] Feser M, Jacobsen C, Rehak P and De Geronimo G 2003 *J. Phys. IV* **104** 529–534
[5] Feser M, Hornberger B, Jacobsen C, De Geronimo G, Rehak P, Holl P and Struder L 2006 *Nucl. Instrum. Methods Phys. Res. Sect. A-Accel. Spectrom. Dect. Assoc. Equip.* **565** 841–854
[6] Morrison G and Niemann B 1998 *X-Ray Microscopy and Spectromicroscopy* ed Thiema J, Schmahl G, Rudolph D and Umbach E (Berlin: Springer Verlag)
[7] Morrison G, Eaton W J, Barrett R and Charalambous P 2003 *J. Phys. IV* **104** 547–550
[8] Kaulich B *et al.* 2006 *Proc. 8th Int. Conf. X-ray Microscopy* vol 7 pp 22–25
[9] Broennimann C *et al.* 2006 *J. Synchrotron Radiat.* **13** 120–130
[10] Arnison M R, Larkin K G, Sheppard C J R, Smith N I and Cogswell C J 2004 *J. Microsc.-Oxf.* **214** 7–12
[11] Kottler C, David C, Pfeiffer F and Bunk O 2007 *Opt. Express* **15** 1175–1181
[12] de Jonge M D, Hornberger B, Holzner C, Legnini D, Paterson D, McNulty I, Jacobsen C and Vogt S 2008 *Phys. Rev. Lett.* **100** 163902
[13] Paganin D, Gureyev T E, Mayo S C, Stevenson A W, Nesterets Y I and Wilkins S W 2004 *J. Microsc.-Oxf.* **214** 315–327
[14] Chapman H N, Jacobsen C and Williams S 1996 *Ultramicroscopy* **62** 191–213
[15] Rodenburg J M and Faulkner H M L 2004 *Appl. Phys. Lett.* **85** 4795–4797
[16] Rodenburg J M, Hurst A C, Cullis A G, Dobson B R, Pfeiffer F, Bunk O, David C, Jefimovs K and Johnson I 2007 *Phys. Rev. Lett.* **98** 034801
[17] Thibault P, Dierolf M, Menzel A, Bunk O, David C and Pfeiffer F 2008 *Science* **321** 379–382

Cryo diffraction microscopy: ice conditions and finite supports

H Miao[1], K Downing[3], X Huang[1], J Kirz[2], S Marchesini[2], J Nelson[1], D Shapiro[2], J Steinbrener[1], A Stewart[1,4], and C Jacobsen[1]

[1] Department of Physics and Astronomy, Stony Brook University, Stony Brook, NY 11794, USA

[2] Advanced Light Source, Lawrence Berkeley National Laboratory, 1 Cyclotron Road, Berkeley, CA 94720, USA

[3] Donner Laboratory, Lawrence Berkeley National Laboratory, 1 Cyclotron Road, Berkeley, CA 94720, USA

E-mail: `chris.jacobsen@stonybrook.edu`

Abstract. Cryo specimen conditions greatly alleviate radiation damage in x-ray microscopy. In diffraction microscopy with plane-wave illumination, a finite support constraint is required in order to obtain a reconstructed image, but this demands that there be an area outside the specimen which is free of optical response. We report here observations based on our early attempts in cryo diffraction microscopy, where ice scatter from outside the cell and ice deposition during data taking have been problems. We outline solutions that are being implemented for future experiments.

1. Introduction

X-ray diffraction microscopy offers an approach to 2D and 3D cell imaging with no lens-imposed limits on spatial resolution or efficiency of data recording [1, 2], and our group has used the approach to image freeze-dried yeast cells [3]. However, when using plane wave illumination and far-field data recording, one must use a finite support constraint in order to obtain a reconstructed image [4]. This finite support constraint demands that part of the real-space image plane sampled in the data recording be optically "empty," with no significant variations in absorption or phase shift. While detailed studies are required to understand the phase tolerance of the "empty" background requirement, one would expect that phase shifts that are very small compared to the Rayleigh quarter wave criterion (*i.e.*, ice thickness variations that are very small compared to $\lambda/(4\delta) = 1260$ nm) are required in the "empty" area.

While radiation damage is a potential limitation in any microscopy with ionizing radiation, working with the specimen at cryogenic temperatures can greatly reduce this limitation [5, 6]. In x-ray microscopy, this allows specimens to tolerate radiation doses of up to about 10^{10} Gray before structural changes are observed at the 30–100 nm resolution level [7, 8] (though latent changes can be detected spectroscopically at lower doses [9]). A common method for preparing frozen hydrated specimens is to place cells on a thin membrane (such as ~50 nm thick formvar), blot off the excess liquid, and rapidly plunge into liquid ethane or propane in order to rapidly

[4] Present address: Department of Materials, University of Oxford, Oxford OX1 3PH, UK

freeze the sample and avoid ice crystal formation. With sufficiently thin water layers, one can obtain a vitreous or amorphous ice layer with very uniform transmission as observed in cryo electron microscopy [6], and the use of solutions such as glucose can increase the thickness that can be prepared in an amorphous state. We have already used this approach to obtain some preliminary reconstructions of frozen-hydrated yeast cells [10]. Smooth ice layers are important for cryo x-ray diffraction microscopy, where one wishes to obtain an isolated cell with a uniform, structureless ice layer outside the cell so that a finite support constraint can be applied and a real-space reconstruction can be achieved.

2. Observations of ice buildup

We describe here some initial observations on working with frozen hydrated specimens in x-ray diffraction microscopy. Our experiments were carried out using a homebuilt vacuum chamber system employing a standard transmission electron microscope type goniometer stage (JEOL) and side-entry cryo transfer holder (Gatan) [11]. This chamber is also equipped with a liquid-nitrogen-cooled copper plate of about 80 cm^2 total surface area at a temperature of about -190°C, although the nearest corner of this plate is about 8 cm distant from the specimen. This plate was cooled down about two hours before each experiment reported here. In one test, we loaded formvar-coated but otherwise empty specimen grids into the cryo holder while at room temperature, inserted them into the specimen chamber, and then cooled the grid while in the microscope chamber. After the temperature of the grid stabilized at around -160°C, we took a diffraction pattern of an initially blank area every minute. Surprisingly, a doughnut-shaped ring clearly appeared in the pattern after some time (see Fig. 1). In repeated experiments, the time for this ring to form was highly dependent on the vacuum pressure in the chamber; Fig. 1 shows the azimuthally averaged diffraction signal obtained after 60 minutes when the base pressure was at 3×10^{-6} Torr only a few hours after the chamber had been open to room air, as well as the signal when the chamber had been under high vacuum for more than a day so that the base pressure was at about 1×10^{-7} Torr. We have reasons to believe that the "doughnut" is due to the scattering from either the ice crystals formed in the chamber, or a "wavy" surface of an amorphous ice layer (see [12] for further discussion of possible explanations for this "doughnut" shape). Given that it takes more than one hour to collect the data (especially when going to 3D imaging, where at ALS beamline 9.0.1 the data collection time could be well over several hours), this scattering from ice formed during data collection could prevent us from reconstructing diffraction patterns

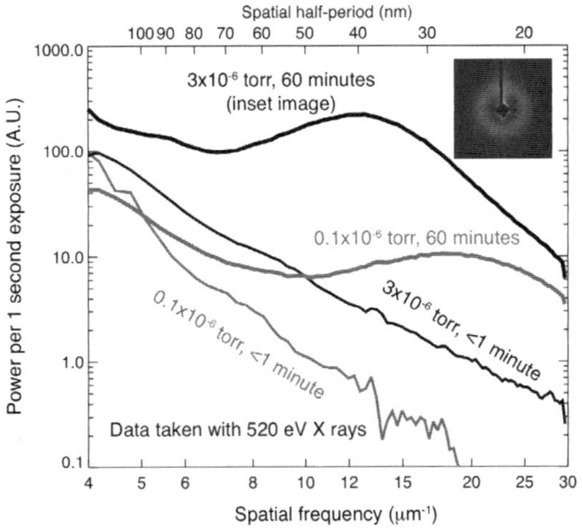

Figure 1. Azimuthal average of coherent 520 eV soft x-ray diffraction patterns of empty formvar coated grids following insertion into our diffraction chamber and in-chamber cooling to $-160°$ C. Shown are data for chamber base pressures of 3×10^{-6} torr and 1×10^{-7} torr, and time durations of either less than one minute or 60 minutes. As can be seen, "rings" of scattering are observed, indicating scattering dominated by a particular range of randomly oriented spatial frequencies (not connected to the 0.366 nm cubic ice ring seen in electron or x-ray diffraction). As can be seen, low base pressure is essential for minimizing the contribution of ice scattering during data recording of cryo specimens.

using a finite support constraint, since the "empty" ice area outside the cell might not be so optically empty to the x-ray beam.

3. Discussion

The above experiments were carried out with a very simple anticontamination plate. When cooling down this anticontaminator plate, the base pressure in the chamber would often drop by a factor of ten, so it certainly was effective in capturing residual vapors in the system. However, ice contamination is well recognized as a potential limit in electron microscopy, where the cryo specimen is often surrounded by plates in very close proximity to the sample—like a few millimeters away. In one example, Cheng *et al.* have achieved a contamination rate of only 0.13 nanometers/hour [13]. While rebuilding our chamber to improve a number of its capabilities, we are replacing our first version of a poorly-adjustable "cryo box" of this type with one that fits both more tightly around the specimen and is also easily adjustable using motorized positioners. When this upgrade is completed, we hope to move forward in the imaging of frozen hydrated specimens using cryo x-ray diffraction microscopy.

Acknowledgments

We wish to thank Eva Nogales at UC Berkeley, and Huilin Li at Brookhaven National Laboratory, for access to their labs and for advice during our early efforts at working with frozen hydrated specimens. We wish to thank the staff at the Advanced Light Source in Berkeley for their excellent assistance, and we gratefully acknowledge support from NIH for the application of x-ray diffraction microscopy to cell imaging under grant GM064846-05.

References

[1] Sayre D 1980 *Imaging Processes and Coherence in Physics* ed Schlenker M (Berlin: Springer-Verlag) pp 229–235

[2] Miao J, Charalambous P, Kirz J and Sayre D 1999 *Nature* **400** 342–344

[3] Shapiro D, Thibault P, Beetz T, Elser V, Howells M, Jacobsen C, Kirz J, Lima E, Miao H, Neiman A M and Sayre D 2005 *Proceedings of the National Academy of Science* **102** 15343–15346

[4] Fienup J 1978 *Optics Letters* **3** 27–29

[5] Taylor K and Glaeser R 1974 *Science* **106** 1036–1037

[6] Steinbrecht R A and Zierold K (eds) 1987 *Cryotechniques in Biological Electron Microscopy* (Berlin: Springer-Verlag)

[7] Schneider G 1998 *Ultramicroscopy* **75** 85–104

[8] Maser J, Osanna A, Wang Y, Jacobsen C, Kirz J, Spector S, Winn B and Tennant D 2000 *Journal of Microscopy* **197** 68–79

[9] Beetz T and Jacobsen C 2003 *Journal of Synchrotron Radiation* **10** 280–283

[10] Lima E 2006 *The Advancement of Biological Imaging Through X-ray Diffraction Microscopy* Ph.D. thesis Department of Physics and Astronomy, Stony Brook University

[11] Beetz T, Howells M, Jacobsen C, Kao C C, Kirz J, Lima E, Mentes T, Miao H, Sanchez-Hanke C, Sayre D and Shapiro D 2005 *Nuclear Instruments and Methods in Physics Research A* **545** 459–468

[12] Miao H 2008 *Studies on X-ray Diffraction Microscopy* Ph.D. thesis Department of Physics and Astronomy, Stony Brook University

[13] Cheng A, Fellmann D, Pulokas J, Potter C and Carragher B 2006 *Journal of Structural Biology* **154** 303–311

Nanostructure analysis by coherent hard X-ray diffraction

Yoshinori Nishino[1], Yukio Takahashi[2], Hideto Kubo[2], Hayato Furukawa[2], Kazuto Yamauchi[2], Kazuhiro Maeshima[3], Naoko Imamoto[3], Ryuta Hirohata[4], Eiichiro Matsubara[4], and Tetsuya Ishikawa[1]

[1] RIKEN SPring-8 Center, 1-1-1 Kouto, Sayo-cho, Sayo-gun, Hyogo 679-5148, Japan
[2] Graduate School of Engineering, Osaka University, 2-1 Yamada-oka, Suita, Osaka 565-0871, Japan
[3] Cellular Dynamics Laboratory, RIKEN, 2-1 Hirosawa, Wako, Saitama 351-0198, Japan
[4] Department of Materials Science and Engineering, Kyoto University, Yoshida, Sakyo, Kyoto 606-8501, Japan

Email: nishino@spring8.or.jp

Abstract. We report our recent activities on x-ray diffraction microscopy. We have been developing hardware and software instruments for efficient recording of high-quality coherent diffraction data. By using the microscope system developed, we carried out coherent diffraction measurement for various samples in materials science and biology, and succeeded in three-dimensional reconstruction revealing the internal structure.

1. Introduction

X-ray diffraction microscopy is an innovative structural analysis method in nanoscience. In the method, the sample structure is reconstructed directly from the oversampled coherent diffraction data by using iterative phase retrieval methods. The reconstruction does not require *a priori* knowledge on the sample structure nor supplemental experimental data. Lenses are not necessary for the microscope, whence diffraction-limited resolution can be achieved in principle. The lensless feature is significant especially for hard x-rays, where it is difficult to fabricate a lens with a high numerical aperture. High penetration power of x-rays offers the unique possibility to achieve high-spatial-resolution imaging for relatively thick samples in a non-destructive manner, and can provide complementary information to other conventional microscopic methods, such as transmission electron microscopy and fluorescence microscopy.

We have been performing diffraction microscopy experiments by using hard x-rays from SPring-8 [1]. In this paper, we describe recent development of our microscope instruments and experimental results of three-dimensional reconstruction.

2. X-ray diffraction microscopy

2.1. Microscope instruments

Figure 1 shows the coherent diffraction microscope we developed. Synchrotron undulator radiation illuminates a sample particle, and the coherent diffraction patterns are recorded using a CCD detector covering a small-angle region around the forward scattering direction. In our measurement, the sample size is typically around 1 μm and is smaller than the transverse coherence area of the incident x-ray wave. Since coherent diffraction signals from a single micron-sized sample particle are weak, all optical components are put in a high vacuum in order to avoid the air scattering noise. A pinhole is installed about 1 m upstream of the sample to illuminate the targeted sample particle only even at the highest incident angle [2]. Guard slits are used to suppress parasitic scatterings from the pinhole and thus to allow high-quality data acquisition.

One of newly introduced features of our instruments is a CCD manipulator. The CCD manipulator can change the distance from the sample to the detector, L, by 700 mm. It helps efficient measurement of samples of different sizes at different incident x-ray energies. The change of the sample-to-detector distance is required to achieve a certain oversampling ratio, the number of the detector pixels in a single speckle, because the speckle size is approximately given by aL/λ. Here, a is the sample size, and λ is the x-ray wavelength, and we assumed that the sample size is smaller than the illuminating coherent beam size. This new feature is beneficial for *e.g.*, element specific imaging using the anomalous scattering, where the incident x-ray energy is tuned near the absorption edge energy of a specific atom [3,4]. The CCD manipulator also helps to move the CCD detector to higher scattering angles to achieve higher spatial resolution.

We also developed semi-automated data acquisition software for efficient three-dimensional measurement at different incident angles. After each change of the incident angle, the software can automatically optimize the sample position to maximize the coherent diffraction intensity, and then start acquiring data. Although the sample is placed approximately on the rotation axis, the repeated sample position alignment helps to compensate the drifts of optics and the x-ray source. Careful alignment of optics is critical in recording high-quality data and in reducing the size of missing data near the forward scattering direction behind the beamstop. During the beamtime, we frequently check the quality of the measured data by manually reconstructing two-dimensional images.

Figure 1. X-ray diffraction microscope installed at BL29XUL in SPring-8.

2.2. Three-dimensional observation of nano-scale internal structures

Our microscope system allows us the acquisition of high-quality coherent diffraction data with a small missing-data size near the forward scattering direction [5], which is essential for faithful reconstruction of the sample structure. Three-dimensional structure can be reconstructed form a series of coherent diffraction data at different incident angles. In the first Born approximation, the three-dimensional reconstruction provides us an electron density map. In the following, we briefly describe two successful examples of our recent three-dimensional reconstructions of samples in materials science and biology.

In materials science application, we observed an aluminum alloy particle [6]. The sample 2014 aluminum alloy is known to contain submicron-sized S-phase (Al_2CuMg) precipitates, which play a key role in controlling the hardness of the alloy. The values of the electron densities in literature show that the precipitate has 25% higher electron density than the aluminum matrix does. In our three-dimensional reconstruction, we observed an internal high-electron-density region inside a ~1 μm sized sample particle, and the high density indicates the region is rich in precipitates.

In biological application, we observed an unstained human chromosome [7]. Figure 2 shows the coherent diffraction pattern from a single chromosome. We obtained high-quality coherent diffraction data at 38 incident angles, and succeeded in reconstructing its three-dimensional structure. This is, to our knowledge, the first successful three-dimensional electron density mapping of a cell organelle by using hard x-rays. The two- and three-dimensional reconstruction shows that high electron density structure extends around the axes of the chromatids. The sample in the experiment was dried in air and is not in biologically ideal condition. We are now developing instruments to observe frozen-hydrated biological samples which are closer to the natural living state.

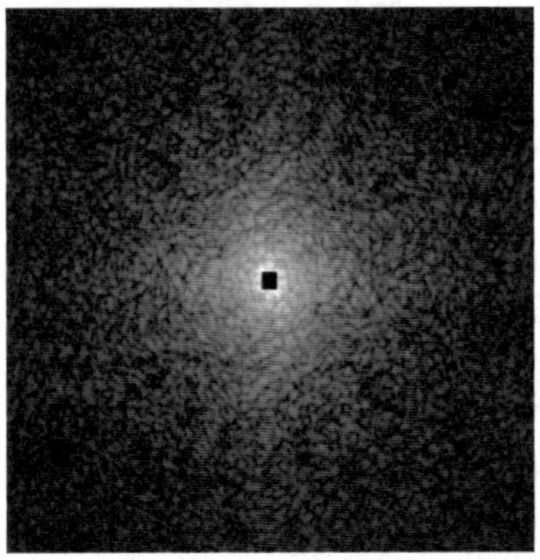

Figure 2. Coherent diffraction pattern from a single unstained human chromosome.

3. Summary

We developed coherent x-ray diffraction microscope system that allows us to record high-quality coherent diffraction data with a small size of the missing data near the forward scattering direction. The recently introduced CCD manipulator is effective especially in the measurement of samples of various sizes at various energies. With the X-ray diffraction microscope hardware together with the semi-automated control software, it became possible to perform three-dimensional data acquisition efficiently. As examples of three-dimensional observation using the microscope, we described successful results for an aluminum alloy and a human chromosome.

Acknowledgments

This work was supported by KAKENHI (19310084 and 18360304) and "Promotion of X-ray Free Electron Laser Research" of MEXT, Japan.

References

[1] Y. Nishino, J. Miao, Y. Kohmura, Y. Takahashi, C. Song, B. Johnson, M. Yamamoto, K. Koike, T. Ebisuzaki, and T. Ishikawa, *IPAP Conference Series* **7**, 386-388 (2006).

[2] Y. Kohmura, Y. Nishino, T. Ishikawa, and J. Miao, *J. Appl. Phys.* **98**, 123105 (2005).

[3] C. Song, *et al.*, *Phys. Rev. Lett.* **100**, 025504 (2008).

[4] Y. Takahashi, H. Kubo, H. Furukawa, K. Yamauchi, E. Matsubara, T. Ishikawa, and Y. Nishino, *Phys. Rev.* B **78**, 092105 (2008).

[5] Y. Nishino, J. Miao, and T. Ishikawa, *Phys. Rev.* B **68**, 220101(R) (2003).

[6] Y. Takahashi, Y. Nishino, T. Ishikawa, and E. Matsubara, *Appl. Phys. Lett.* **90**, 184105 (2007).

[7] Y. Nishino, Y. Takahashi, N. Imamoto, T. Ishikawa, and K. Maeshima, *Phys. Rev. Lett.* **102**, 018101 (2009).

Journal of Physics: Conference Series **186** (2009) 012057 doi:10.1088/1742-6596/186/1/012057

Advances in X-ray microscopy at FLASH using transmissive optics

Thomas Nisius[1], Rolf Früke[1], David Schäfer[1], Marek Wieland[2], Marion Kuhlmann[3], Elke Plönjes[3] and Thomas Wilhein[1]

[1]Institute for X-Optics, RheinAhrCampus Remagen, University of Applied Sciences Koblenz, Südallee 2, 53424 Remagen, Germany

[2]University of Hamburg, Institut für Experimentalphysik, Luruper Chaussee 149, 22761 Hamburg, Germany

[3]DESY, HASYLAB, Notkestrasse 85, 22607 Hamburg, Germany

E-mail: nisius@rheinahrcampus.de

Abstract. In order to successfully plan and perform experiments at high flux EUV- and X-ray sources, e.g. free electron lasers, it is necessary to know which kind of optics, exposed to the full beam, can be used. Due to the high intensities, it is not clear, whether transmissive diffractive optics are applicable, because these optics are usually fabricated on thin membranes which are therefore subject to additional absorption in the desired energy range. Since diffractive optics, especially zone plates, offer the possibility to achieve small spots when used as a focussing element and can also achieve good image quality in microscopic setups, their usage would facilitate many experiments, especially for their easy handling. As a proof of concept, we set up a zone plate based scanning transmission microscope at the unfocussed beamline BL3 at FLASH (DESY/Hamburg). The operating wavelength was 32 nm and 13.8 nm, respectively. While the first attempt, utilizing a zone plate composed of PMMA on silicon substrate failed due to ablation of the PMMA, a second zone plate (chromium on silicon nitride) was successfully used to focus the beam onto different samples (e.g. nickel-mesh, a silicon nitride structured sample and nickel structures on silicon nitride). The resulting focal spot size was estimated from the acquired images to be in the range of 1 µm - 3 µm in diameter. Beside the scanning approach, a full-field transmission microscopy setup was tested. A preliminary image in full field mode was taken with a second zone plate, showing features of a Siemens-star pattern down to approximately 120 nm. After several hours of exposure, no damage was visible to the optics. Beside the optics, different filters (Silicon/Zirconium, Zirconium and Aluminum) have been placed in the beam to evaluate possibilities to further reduce intensity which may be necessary if sensitive detectors are involved. All of the filters withstood the irradiation during the whole experiment.

1. Introduction

Going for shorter wavelengths with the objective to get smaller and smaller structures, e.g. to produce nano-structures by lithography or to aim for higher resolution in imaging setups, EUV and soft X-ray sources have gained a significant importance in the past years. With its promised spatially coherent beam, the free electron laser FLASH at DESY, Hamburg is a new source providing laser like beam properties in the desired spectral region. Additionally, with its short pulse structure, an ideal source for pump-probe experiments is available. In order to successfully perform experiments at FLASH it is first necessary to know the beam characteristics like coherence properties and beam stability. Secondly, it has to be evaluated which kinds of optics, exposed to the full FEL beam, can be used, in terms of optical properties, e.g. imaging quality and also in terms of mechanical and thermal stability. The experiments show that up to now transmission optics are well suited for their use at the FEL, especially for their easy handling and alignment.

2. Experimental setup

In order to evaluate the performance of transmissive optics, a compact scanning microscope was set up. This microscope mainly consists of a zone plate focussing optic with central stop to block direct light transmitted through the optic and an order sorting aperture to get rid of higher diffraction orders, a piezo driven sample scanning stage (1 nm resolution) and an EUV sensitive photodiode detector which can be replaced by a CCD camera during the alignment procedure. For the first experiment ($\lambda = 32$ nm, 10 µJ pulse energy) a zone plate (N = 500, $r_n = 60$ µm, $dr_n = 60$ nm) made from PMMA on silicon substrate was used. The second experiment ($\lambda = 13.8$ nm, 20 µJ pulse energy) was performed with a chromium zone plate (N = 4080, $r_n = 800$ µm, $dr_n = 98$ nm) on silicon nitride membrane. The schematic setup of the scanning transmission microscope is shown in figure 1. To acquire fullfield microscope images, this setup was modified by inserting a second zone plate between object and detector.

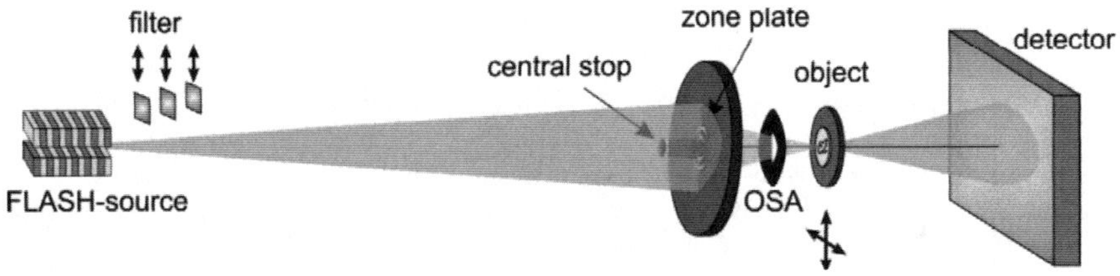

Figure 1: Schematic setup of the compact scanning transmission microscope [1].

3. Experimental results

The first zone plate made from PMMA on silicon substrate, which was choosen due to the good diffraction efficiency and the low absorption of silicon at 32 nm, was ablated by the FEL beam,

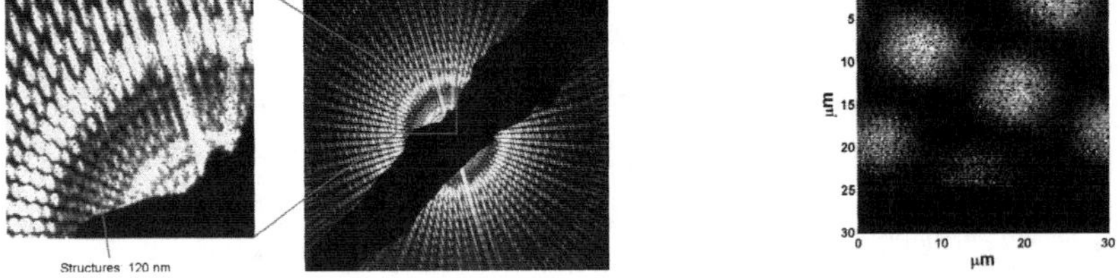

Figure 2: Fullfield image of a siemens star, showing structures down to 120 nm (a) and scanned image of Ni-mesh, 12.6 µm grating constant, 5 µm bars.

but the supporting membrane did not show any defects. The second zone plate was successfully used for imaging experiments and withstood the FEL radiation for several hours despite the high absorption of the silicon nitride membrane (thickness: 100 nm) of about 45% at a wavelength of 13.8 nm [2]. After the experiment, no damage to the optic was visible. As stated earlier, the photodiode detector was replaced by a CCD camera during alignment and a second zone plate was inserted between sample and detector, resulting in a fullfield microscope setup. A Siemens star pattern with structures down to 100 nm was imaged onto the camera, showing that structures down to 120 nm can be resolved (figure 2 (a)).

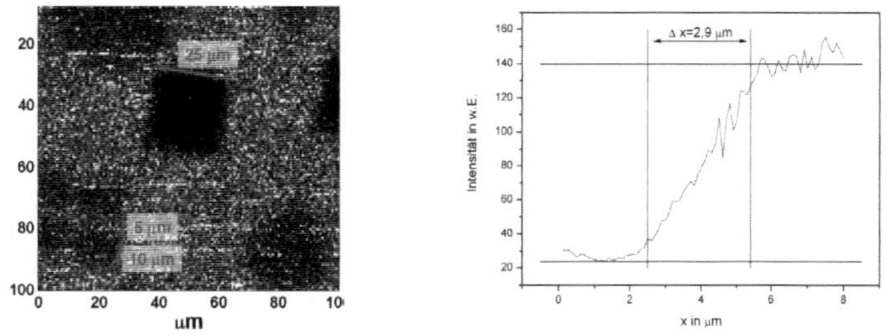

Figure 3: Scanned image of a test square (a) and plot along an edge (b).

Figure 2 (b) shows an image of a nickel mesh with 12.6 µm grating constant obtained with the scanning microscope. In figure 3, it can be seen, that the resulting resolution estimated from a scan across an edge is below 3µm which results in an intensity of about $5{\cdot}10^{10}$ W/cm^2 in the focal spot. In this case, the sample was a nickel square (25 µm x 25 µm) on a silicon nitride membrane. From figure 2 (b) it is obvious, that structures of about 1 µm should still be resolvable as the solid bars of the used mesh have a width of 5 µm. While the scanning microscope setup is able to resolve structures well below 1 µm (as shown in [1]), this resolution could not be reached for different reasons. To some extend, beam instabillities caused by pulse to pulse deviations of the source position may result in a moving focal spot of the zone plate. The image quality is further reduced by beam intensity fluctuations either due to pointing instabilities or due to intensity changes between adjacent pulses. The sample was examined by scanning electron microscopy and x-ray microscopy using a laboratory source [3] after the experiment. Since no evidence for FEL induced damage was found it can be concluded that the observed resolution is not caused by degradation of the sample.

4. Conclusion and Outlook

We have shown, that transmissive diffractive optics are usable in conjunction with the FEL beam. The used filter foils and the optics did not show any degradation, even after several hours of exposure to the full beam. With the scanning microscope aproach we proved a focal spot size of about 1 µm - 3 µm. To improve the achievable resolution, the FEL beam properties have to be investigated further, since we found hints for a slight lateral jumping of the source which in fact decreases the spatial coherence and thus the resolution.

Acknowledgments

The zone plates were produced by P. Charalambous, ZonePlates Ltd., London (UK), C. David, Paul Scherer Institute, Villingen (Switzerland) and D. Weinreich, Institute for X-ray Physics, Göttingen (Germany). The test object with Siemens stars and gratings was produced by S. Rehbein and A. Holmberg, KTH Stockholm (Sweden).

References

[1] Früke R, Kutzner J, Witting T, Zacharias H and Wilhein T 2005 *Europhysics Letters (EPL)* **72** 915–921 URL `http://stacks.iop.org/0295-5075/72/915`
[2] Henke B L, Gullikson E M and Davis J C 1993 *Atomic Data and Nuclear Data Tables* **54** 181–+
[3] Schäfer D, Benk M, Bergmann K, Nisius T, Wiesemann U and Wilhein T 2008 Optical setup for tabletop soft x-ray microscopy using electrical discharge sources 9th International Conference on X-Ray Microscopy

Journal of Physics: Conference Series **186** (2009) 012058 doi:10.1088/1742-6596/186/1/012058

Near diffraction limited coherent diffractive imaging with tabletop soft x-ray sources

Richard L. Sandberg[1], Daisy A. Raymondson[1], William F. Schlotter[2], Kevin Raines[3], Chan La-O-Vorakiat[1], Ariel Paul[1], Margaret M. Murnane[1], Henry C. Kapteyn[1], and Jianwei Miao[3]

[1] Department of Physics and JILA, University of Colorado and NIST, Boulder, Colorado, USA
[2] Stanford Synchrotron Radiation Laboratory, SLAC, Menlo Park, California, USA, currently at current address Institute for Experimental Physics, University of Hamburg, Luruper Chausee 149,22761 Hamburg, Germany
[3] Department of Physics and Astronomy and California NanoSystems Institute, University of California, Los Angeles, California, USA

richard.sandberg@colorado.edu

Abstract. Tabletop coherent x-ray sources hold great promise for practical nanoscale imaging, in particular when coupled with diffractive imaging techniques. In initial work, we demonstrated lensless diffraction imaging using a tabletop high harmonic generation (HHG) source at 29 nm, achieving resolutions ~200 nm. In recent work, we significantly enhanced our diffractive imaging resolution by implementing a new high numerical aperture (up to NA=0.6) scheme and field curvature correction where we achieved sub-100 nm resolution. Here we report the first demonstration of Fourier transform holography (FTH) with a tabletop SXR source, to acquire images with a resolution ≈ 90 nm. The resolution can be refined by applying phase retrieval. Additionally, we show initial results from FTH with 13.5 nm HHG radiation and demonstrate ~180 nm resolution.

1. Introduction

Lensless imaging, also known as x-ray diffractive microscopy, is a relatively new technique that is particularly suited to x-ray imaging, enabling high-resolution imaging of non-crystalline, aperiodic samples such as biological materials [1]. This technique provides a large depth of field, insensitivity to alignment, and near-wavelength-limited resolution without the need for complex optical systems. In past work, we demonstrated lensless diffraction imaging using a tabletop high harmonic (HHG) soft x-ray source at 29 nm, achieving resolutions ~200 nm [2]. By implementing a high numerical aperture (up to NA=0.6) scheme and curvature correction, we achieved resolutions < 100 nm using two approaches. First, we use phase retrieval with the GHIO algorithm to achieve a resolution of 94 nm with a 29 nm HHG source, and a resolution of 71 nm using a 47 nm capillary discharge laser [3].

In this work, we report the first demonstration of Fourier transform holography with a tabletop soft x-ray source, acquiring images with a resolution ≈ 90 nm. Fourier transform holography (FT holography) is a diffractive imaging technique that is complimentary to coherent diffractive imaging (or lensless imaging). Instead of using an iterative phase retrieval algorithm to extract the phase from the measured data, FT holography uses a reference beam that interferes in the far field with the light scattered off of the sample in a hologram. The magnitude of the Fourier transform of the hologram produces the spatial autocorrelation of the sample. For this paper, we used a geometry which produces a reference beam by introducing multiple holes surrounding the sample [4]. In the spatial autocorrelation, each reference hole produces an image of the sample and its complex conjugate. These individual 'sub-images' can be added up to increase the signal to noise ratio of the image [5]. Thus, this technique was chosen as a way to increase the signal to noise of images taken with the high harmonic source that will have decreasing flux at higher photon energies, however, for the photon energies demonstrated here, adding the sub-images was not necessary.

2. Experiment

Figure 1 shows the experimental setup for Fourier Transform Holography with a tabletop soft x-ray source. High harmonic generation occurs when 25 fs pulses from a 4 W Titanium-sapphire laser operating at 3 kHz are focused into a hollow-core capillary gas filled waveguide. The waveguide was filled with 70 Torr of Ar for 29 nm and 400 Torr of He for 13.5 nm generation. Two 200 nm thick filters (Al for 29 nm and Zr for 13.5 nm) remove the near-IR light while transmitting the soft x-ray harmonics. Two Mo-Si multilayer mirrors select either 27th harmonic at 29 nm or the 57th harmonic at 13.5 nm and gently focus it onto a sample a few centimeters from a CCD which records the hologram.

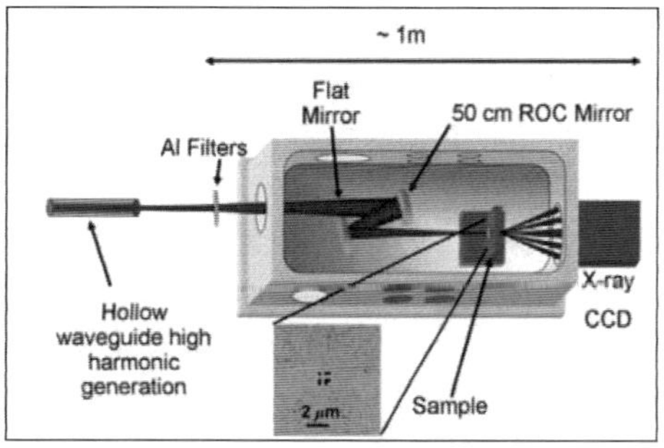

Figure 1. Compact lensless diffractive microscope used with both a 29 nm HHG source and a 13.5 nm HHG source.

2.1. *FT holography with 29 nm illumination.* Figure 2 shows the results with the 29 nm source. A FT holography sample was used with five reference holes each with an approximate diameter of 130 nm. These samples were constructed by milling out holes in 300 nm of gold on a 100 nm thick Si_3N_4 window with a focused ion beam. Figure 2 shows an SEM image of the sample, the collected diffraction pattern from 10 minutes of exposure, the autocorrelation reconstruction, and a lineout showing the reference hole limited resolution of 89 nm. However, with longer exposure times, scattered light can be collected to higher angles that will support higher resolution image. This information can then be extracted using an iterative phase retrieval algorithm [4]. With an exposure time of 80 minutes and refining the image with phase retrieval, we demonstrate a resolution of 53 nm with 29 nm illumination [6].

2.2. *Initial FT holography images with 13.5 nm illumination.* We also present initial images using high harmonic light at the technologically important 13.5 nm wavelength. A Fourier-transform holography sample was used with seven reference holes each with an approximate diameter of 200 nm. Figure 3 shows an SEM image of the sample, the collected diffraction pattern from 40 minutes of exposure, the autocorrelation reconstruction, and a lineout showing the theoretically-limited resolution of ~150 nm based on the reference hole diameter.

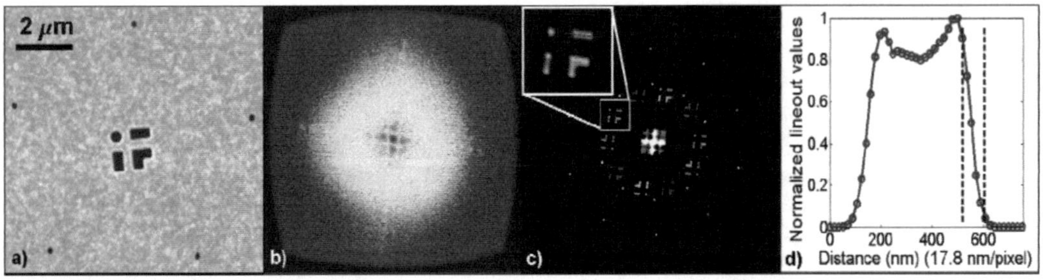

Figure 2. (a) SEM image of sample with five reference holes (b) 29 nm hologram from 10 minutes of exposure after curvature correction (log scale) (c) autocorrelation reconstruction of the sample (d) line cut through one sub-image showing 90 nm resolution.

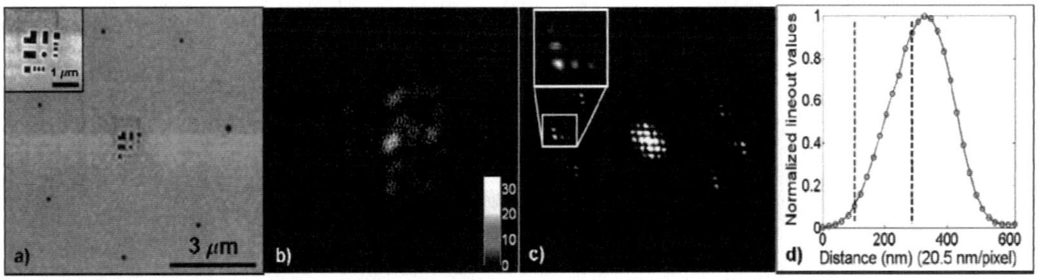

Figure 3. a) SEM image of FTH sample with seven reference holes b) 13.5 nm hologram from 40 minute exposure (log scale) c) autocorrelation reconstruction of the sample d) line cut through one sub-image showing 184 nm resolution.

3. Conclusion

As higher brightness, shorter wavelength, tabletop sources become available, this high numerical aperture scheme – a first for diffractive imaging with x-rays – will make possible near-wavelength resolution below 10 nm. Important applications of this work exist in nanotechnology, lithography, materials science, and biological imaging. Moreover, the ultrafast (fs) duration of the HHG pulses will enable imaging of dynamic systems with high temporal resolution.

References

[1] Miao J, Ishikawa T, Earnest T and Shen Q 2008 *Annu. Rev. Phys. Chem.* **59** 387

[2] Sandberg R L, Paul A, Raymondson D A, Hadrich S, Gaudiosi D M, Holtsnider J, Tobey R I, Cohen O, Murnane M M, Kapteyn H C, Song C, Miao J, Liu Y and Salmassi F 2007 *Phys Rev Lett* **99** 098103

[3] Sandberg R L, Song C, Wachulak P W, Raymondson D A, Paul A, Amirbekian B, Lee E, Sakdinawat A E, La-o-vorakiat C, Marconi M C, Menoni C S, Murnane M M, Rocca J J, Kapteyn H C and Miao J 2008 *Proc. Natl. Acad. Sci. USA* **105** 24-27

[4] Eisebitt S, Lüning J, Schlotter W F, Lörgen M, Hellwig O, Eberhardt W and Stöhr J 2004 *Nature* **432** 885

[5] Schlotter W F, Rick R, Chen K, Scherz A, Stöhr J, Lüning J, Eisebitt S, Gunther Ch, Eberhardt W, Hellwig O and McNulty I 2006 *Appl. Phys. Lett.* **89** 163112

[6] Sandberg R L, Raymondson D A, La-o-vorakiat C, Paul A, Raines K, Miao J, Murnane M M, Kapteyn H C and Schlotter W F 2008 (submitted)

Fresnel Diffraction Correction by Phase-considered Iteration Procedure in Soft X-ray Projection Microscopy

Tatsuo Shiina[1], Tsuyoshi Suzuki[1], Toshio Honda[1], Atsushi Ito[2], Yasuhito Kinjo[3], Hideyuki Yoshimura[4], Keiji Yada[5], Kunio Shinohara[6]

1. Graduate School of Advanced Integration Science, Chiba University
1-33 Yayoi-cho, Inage-ku, Chiba-shi, Chiba 263-8522, Japan
2. School of Engineering, Tokai University
1117 Kitakaname, Hiratsuka-shi, Kanagawa 259-1292 Japan
3. Tokyo Metropolitan Industrial Technology Research Institute
2-11-1 Fukazawa, Setagaya-ku, Tokyo 158-0081 Japan
4. School of Science and Technology, Meiji University
1-1-1 Higashimita, Tama-ku, Kawasaki-shi, Kanagawa 214-8571 Japan
5. Tohken Co., Ltd.
2-27-7 Tamagawa, Chofu-shi, Tokyo 182-0025 Japan
6. Advanced Research Institute for Science and Engineering, Waseda University
3-4-1 Okubo, Shinjuku-ku, Tokyo 169-8555, Japan

shiina@faculty.chiba-u.jp

Abstract. In soft X-ray projection microscopy, it is easy to alter the magnification by changing the distance between the pinhole and the specimen, while the image is blurred because the soft X-rays are diffracted through the propagation from specimen to CCD detector. We corrected the blurred image by the iteration procedure of Fresnel to inverse Fresnel transformation taking phase distribution of the specimen into account. The experiments were conducted at the BL-11A of the Photon Factory, KEK, Japan for the specimens such as glass-capillaries, latex-particles, dried mammalian cells and human chromosomes. Many of those blurred images were corrected adequately by the iteration procedure, though some images such as those which have high-contrast or are overlapped by small cells still remain to be improved.

1. Introduction

Projection microscope has advantages of wide field of view and simple optics in comparison with imaging microscope. It is easy to zoom in a specimen to obtain high-magnification images after capturing it with low magnification. Quality of the obtained image is often impaired at high resolution because it is blurred with fringes caused by Fresnel diffraction. Recently the high-resolution imaging microscope achieves good results [1] with a limited field of view, while various blur-correction methods for the projection microscopy have been developed for the improvement of processing capability [2]. An iterative algorithm for phase retrieval is effective since the phase of the X-rays is not recorded by the CCD detector and therefore a direct, algebraic back propagation is not possible. We also studied the blur-correction methods for the images obtained by the soft X-ray projection microscopy and developed the blur-correction method with an iteration of Fresnel to inverse Fresnel transformation based on physical implication of the blur generation [3]. Here we improved the methods with respect to the restraint condition and the illumination condition in calculation procedure and applied to images of various specimens.

2. Projection microscope and iteration procedure

The system setup is illustrated in Fig.1. The light source was the monochromatic soft X-rays at BL-11A in KEK-PF, Japan. The wavelength was 1.5 to 2.5nm (0.83 to 0.50keV in energy). The detector is a back-illuminated X-ray CCD camera with the pixel pitch of 24.8μm (Hamamatsu C4880-30-26WS). A rotary stage was installed to obtain the CT-reconstruction images. All devices and specimens were in the vacuum chamber. The specimen was observed at x110.7 lateral magnification at the distance of 3mm between the post-pinhole and the specimen. The CCD's FOV became 140μm square at this magnification. In that case, the Fresnel's fringe was observed on the projection image when the diameter of the post-pinhole was less than 5μm. The size of the projection image per pixel was about 0.2μm. Each projection image was integrated in its intensity for 2 - 5 minutes to improve the signal-to-noise ratio. For the elimination of the blur on the projection image, the iteration process was applied. The process repeats the calculation between Fresnel to inverse Fresnel transformations under some restrictions such as addition of the intensity distribution of X-ray illumination in the CCD's FOV, and the convergence condition in the calculation. A fixed sample to detector distance was used in the iteration process. In addition, the phase of a spherical wave was also considered as the radially-propagating beam. The specimens were glass-capillaries, latex-particles, dried mammalian cells (HeLa and MOLT-4) and human chromosomes.

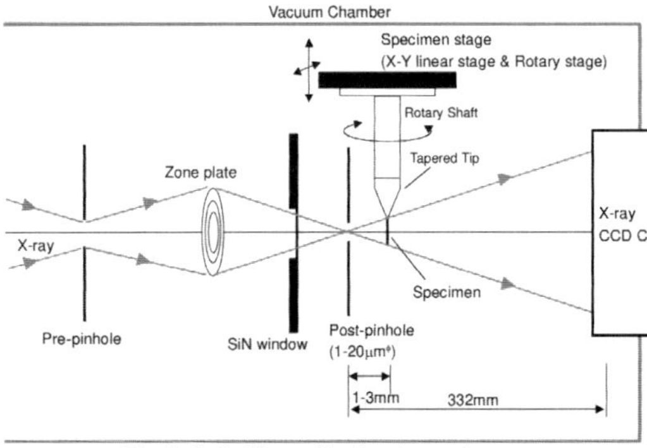

Fig.1 Soft X-ray Projection microscope.

3. Projection images and blur-correction image

Some of the blur-corrected results are shown in Figs.2-4. In these figures, magnification was x166 and the wavelength of the soft X-rays was 1.77nm (0.74keV in energy). In the blurred image of a HeLa cell (Fig.2a), there were bright fringes around the cell and white spots in the cell. They are due to the influence of the Fresnel diffraction. In the corrected image (Fig.2b), they were thoroughly taken out. The nucleus and the cytoplasm in the cell became clear. Figure 3 shows the images of a tapered hollow glass capillary. The apex was less than ϕ1μm. The several Fresnel fringes appeared around the capillary in the blurred image of Fig.3a. It was hard to distinguish the apex and width of the capillary. In view of the pixel-size of the CCD detector and the microscope magnification, the calculated resolution was estimated to be 0.2μm. The thickness of the glass capillary was estimated as about ϕ0.5μm. The hollow structure was reproduced in the corrected image of Fig.3b. Figure 4 shows the observation of the glass capillary with a latex particle on the apex. The magnification was x110.7. In Fig.4a, the several fringes appeared, the feature of the apex of the glass capillary was uncertain. The borders between the capillary and the latex particle were clearly recognized in Fig.4b. Both diameters of the latex particle and the apex of the capillary were 8μmϕ.

We performed the blur-correction for the images of various specimens including MOLT-4 cells and human chromosomes. As a result, it was confirmed that the blur-correction was adequate in many specimens. On the other hand, some of images cannot be corrected the blur properly. In those images, the edge of the object was sharp, or plural small samples were closely distributed. The correction failure of the former image was due to the restraint condition in the calculation. That of the latter was because the fringes from the small samples were overlapped and disappeared by the digitization. The partial coherence of the light source also made difficult to correct the above images because BL-11A is a bending magnet source.

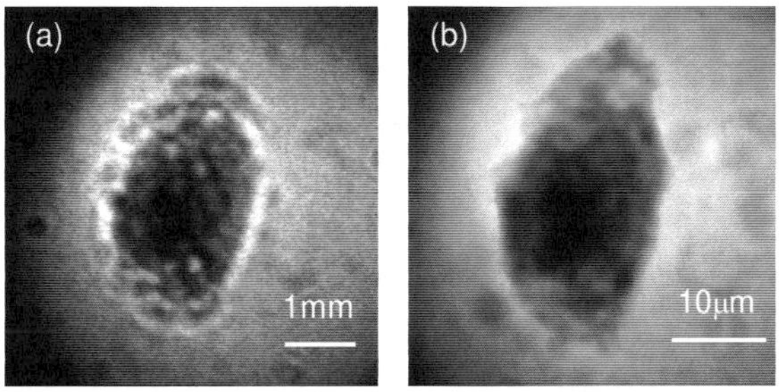

Fig.2 HeLa Cell (a)Projection image (b)Corrected image at the object position.

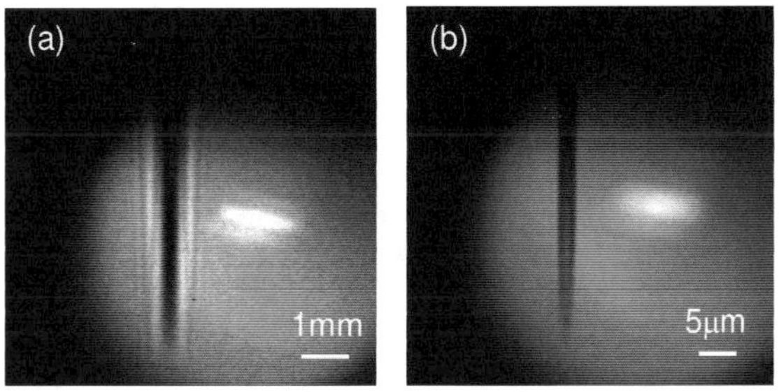

Fig.3 Hollow glass capillary (a)Projection image (b)Corrected image at the object position.

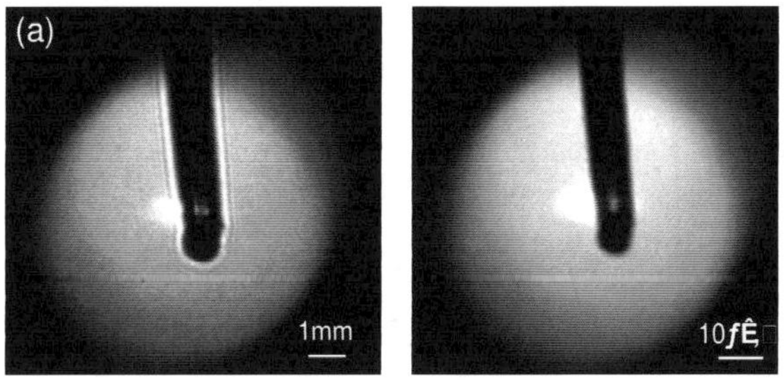

Fig.4 Glass capillary and a latex particles (a)Projection image
(b)Corrected image at the object position.

4. Summary

The blur-correction method with the iteration procedure between Fresnel to inverse Fresnel transformation was developed for the soft X-ray projection microscopy and applied to various specimens. For many of the images the blur was removed adequately and the details of the image became clear. The size of the corrected image per pixel approximated to the physical limit defined by the CCD's pixel size.

References

[1] Mark A Le Gros, Gerry McDermott and Carolyn A Larabell, "X-ray tomography of whole cells", Current Opinion in Structural Biology, Vol.15, pp.193-600, 2005

[2] T. E. Gureyev, A.W. Stevenson, Ya. I. Nesterets, S.W. Wilkins, "Image deblurring by means of defocus", Optics Communications, Vol.240, pp.81-88, 2004

[3] T. Shiina et al, "CT Reconstraction by Diffraction Correction in Soft X-ray Projection Microscopy", Proc. 8th Int. Conf. X-ray Microscopy IPAP Conf. Series Vol.7, pp.363-365, 2005

Development of incident x-ray flux monitor for coherent x-ray diffraction microscopy

Yukio Takahashi[1], Hideto Kubo[2], Hayato Furukawa[2], Kazuto Yamauchi[2,3], Eiichiro Matsubara[4], Tetsuya Ishikawa[5], and Yoshinori Nishino[5]

[1]Frontier Research Base for Global Young Researchers, Frontier Research Center, Graduate School of Engineering, Osaka University, 2-1 Yamada-oka, Osaka 565-0871, Japan
[2]Department of Precision Science and Technology, Graduate School of Engineering, Osaka University, 2-1 Yamada-oka, Suita, Osaka 565-0871, Japan
[3]Research Center for Ultra-precision Science and Technology, Graduate School of Engineering, Osaka University, 2-1 Yamada-oka, Suita, Osaka 565-0871, Japan
[4]Department of Materials Science and Engineering, Kyoto University, Yoshida, Sakyo, Kyoto 606-8501, Japan
[5]RIKEN SPring-8 Center, Kouto, Sayo-cho, Sayo-gun, Hyogo 679-5148, Japan

E-mail: takahashi@wakate.frc.eng.osaka-u.ac.jp

Abstract. An incident x-ray flux monitor for coherent x-ray diffraction microscopy was developed. The intensities of x-rays passing through the sample were measured using an x-ray photodiode, with the simultaneous measurement of the x-ray diffraction intensities of the sample. As a result of the normalization of the x-ray diffraction intensities by the incident x-ray flux determined from the monitor, the fluctuation of the speckle intensities was successfully suppressed.

1. Introduction

Coherent x-ray diffraction microscopy[1] (CXDM) is a promising technique for observing a micrometer-sized sample with nanometer-scale resolution. In CXDM, it is important, for quantitative image reconstruction, to measure the incident x-ray flux. For example, to evaluate the electron-density distribution of a sample by CXDM, the absolute value of the structure factor must be derived using the incident x-ray flux[2]. Two or more items of two-dimensional (2D) diffraction data, e.g., diffraction patterns measured at different incident x-ray angles for a three-dimensional reconstruction[3] or different x-ray energies for elemental identification[4, 5], should be normalized by the incident x-ray flux. In typical CXDM measurements in the transmission scheme, an isolated sample is mounted on a Si_3N_4 membrane. It is necessary not only to measure the x-rays scattered from the sample including the Si_3N_4 membrane (i.e., sample data) but also to measure the x-rays scattered only from the Si_3N_4 membrane (i.e., background data). It takes a few hours to measure a 2D diffraction pattern with a good signal-to-noise ratio. Since the intensities of incident x-rays fluctuate during the period of a few hours, it is necessary to independently determine the incident x-ray flux for both the sample data and the background data to obtain an exact background subtraction. In order to determine the exact incident x-ray flux, the incident intensities must be simultaneously measured with the x-ray diffraction of the

Journal of Physics: Conference Series **186** (2009) 012060 doi:10.1088/1742-6596/186/1/012060

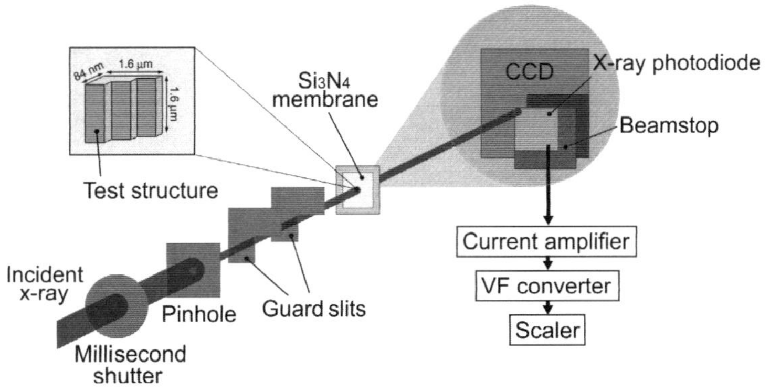

Figure 1. Schematic drawing of the coherent x-ray diffraction microscopy apparatus installed with the incident x-ray flux monitor.

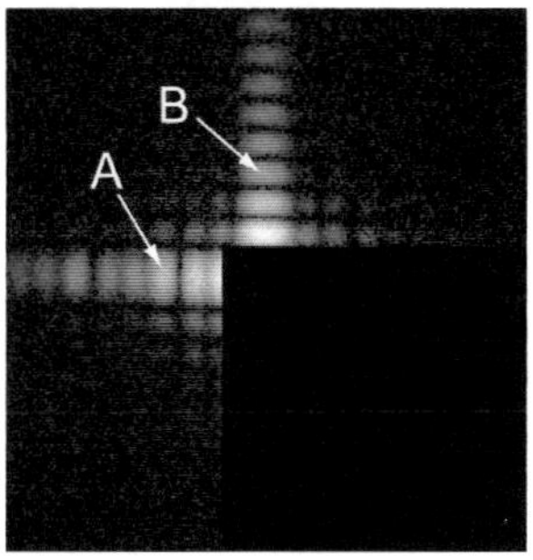

Figure 2. Coherent x-ray diffraction pattern of the test structure in 301×301 pixels.

sample. In this study, an incident x-ray flux monitor for CXDM measurements was developed and its performance was evaluated.

2. Experimental

A test structure was fabricated as a sample on a Si_3N_4 membrane by the focused ion beam deposition of tungsten. Figure 1 shows the schematic drawing of both the sample and the CXDM apparatus installed with the incident x-ray flux monitor. Incident x-rays of 5 keV were irradiated onto the sample through a 20 μm pinhole slit. Two L-shaped guard slits, which were set between the pinhole and the sample, suppressed parasitic scattering x-rays from the pinhole. Forward x-ray diffraction intensities were collected by a charge-couple device (CCD) detector with a pixel size of 20×20 μm^2 placed 2.181 m downstream of the sample. In order to control the x-ray exposure time, a millisecond shutter was set upstream of the pinhole. A direct beamstop was positioned before the CCD detector. An x-ray photodiode with 30-μm-thick Si (International Radiation Detector, INC., AXUVPSV) was mounted on the beamstop. The dead region of the phototiode is less than 50 μm from the beamstop edge. The photodiode detected the x-ray photons passing through the sample, while the CCD measured the x-ray diffraction intensities of the sample. Data signals from the photodiode were processed by a current amplifier, VF converter, and scaler. The performance of the x-ray flux monitor was evaluated. Fifteen independent items of sample data and background data were collected. The background was measured just after each sample measurement. The sample and background data, respectively, were obtained by accumulating 50 and 25 scattering data with 4 s exposure time. The measurement times including the read-out time of the CCD detector were 300 s and 150 s for the sample and the background, respectively. It took $\sim$7000 s to collect all the data.

3. Results and discussion

Figure 2 shows one of the 15 diffraction patterns of the sample. The lower right of the diffraction pattern is the ummeasurable region owing to the presence of the beamstop. The incident x-ray flux determined for each diffraction pattern is plotted in Fig. 3. The flux was normalized by the average value of the 15 flux data. The flux decreases as time passes. It is thought that the decrease is due to not only the fluctuation of the intensities of the optical source but also the drift of the position of the pinhole, the guard slits, and the sample. The repetition number dependence of the integral intensities of speckles indicated by arrows in Fig. 2 is shown in Fig. 3. The intensities were normalized by the average value of 15 intensity data. The magnitude of

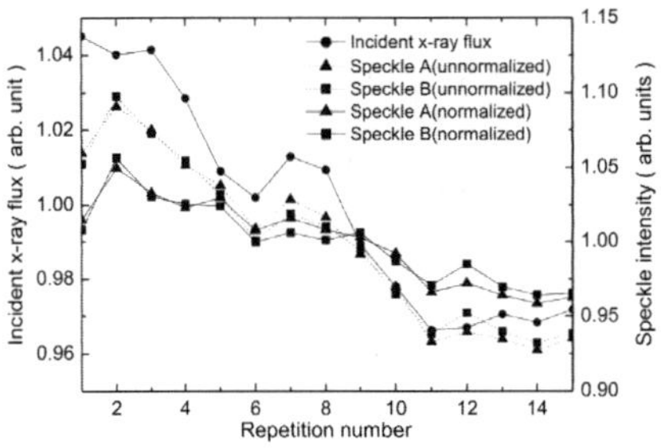

Figure 3. Repetition number dependence of the incident x-ray flux and intensities of speckles shown by arrows in Fig. 2. Each plot was normalized by the average value of 15 data. Speckle intensities shown by solid lines were normalized by the incident x-ray flux.

the noise of speckle intensities was estimated to be ~0.3% from $(N_r^2 + N_D^2 + N_{ph}^2)^{1/2}$, where N_r is the read-out noise, N_D the dark charge noise, and N_{ph} the photon shot noise. The diffraction intensities decrease as the incident x-ray flux decreases. Next, all the sample and background data were normalized based on each incident x-ray flux. The diffraction data from the sample were derived by subtracting the normalized background data from the normalized sample data. The normalized integral intensity of each speckle is plotted in Fig. 3. Comparison with the plot of unnormalized intensities indicates that the fluctuation of speckle intensities is suppressed.

4. Conclusion

In this study, an incident x-ray flux monitor for CXDM was developed. The Si photodiode was mounted on the beamstop. The intensities of x-rays passing through the sample were measured by the photodiode simultaneously with the measurement of the x-ray diffraction intensities. The diffraction data were normalized by the incident x-ray flux. The fluctuation of the normalized diffraction intensities was successfully suppressed in comparison with that of the unnormalized intensities. By using this CXDM system, the quantitative analysis of the electron density distribution of alloys[6] or the evaluation of the elemental distribution of unknown samples will be realized.

Acknowledgments

This research has been carried out at the Frontier Research Base for Global Young Researchers, Osaka University, on the Program of Promotion of Environmental Improvement to Enhance Young Researcher's Independence, the Special Coordination Funds for Promoting Science and Technology of Ministry of Education, Culture, Sports, Science and Technology (MEXT). This work was supported partly by funds from a Grant-in-Aid for Young Scientists (Grant No. 18760499), Scientific Research (B) (Grant No. 18360304, Grant No. 19310084), the "Promotion of X-ray Free Electron Laser Research" of MEXT.

References

[1] Miao J, Charalambous P, Kirz J, Sayer D 1999 *Nature* **400** 342
[2] Miao J, Amonette J M, Nishino Y, Ishikawa T, Hodgson K O 2003 *Phys. Rev.* B **68** 12201
[3] Miao J, Ishikawa T, Johnson B, Anderson E H, Lai B, Hodgson K O 2002 *Phys. Rev. Lett.* B **89** 88303
[4] Song C, Bergstrom R, Ramunno-Johnson D, Jiang H, Paterson D, deJonge M D, McNulty I, Lee J, Wang K L, Miao J 2008 *Phys. Rev. Lett.* **100** 25504
[5] Takahashi Y, Kubo H, Furukawa H. Yamauchi K, Matsubara E, Ishikawa T, Nishino Y 2008 *Phys. Rev. B* **78** 092105
[6] Takahashi Y, Nishino Y, Ishikawa T, Matsubara E 2007 *Appl. Phys. Lett.* **90** 184105

Development of a total reflection double-slit for evaluation of spatial coherence in hard X-ray region

T. Tsuji, T. Koyama, H. Takano, Y. Tsusaka, Y. Kagoshima

Graduate School of Material Science, University of Hyogo, 3-2-1 Kouto, Kamigori, Ako, Hyogo 678-1297, Japan

E-mail:rk07m003@stkt.u-hyogo.ac.jp

Abstract. A new total reflection double-slit for hard X-ray region was developed. It consists of Au narrow stripes deposited on a SiO_2 substrate and can function equivalently as a conventional Young's interferometer by operating with a small grazing incident angle. The Young's interferometer using the total reflection double-slit was constructed at Hyogo-ID BL of SPring-8, and the spatial coherence at 10 keV X-rays was evaluated by analyzing the interference fringes.

1. Introduction

Experimental evaluation of degree of spatial coherence has become important since the advent of the 3rd generation synchrotron radiation facilities where insertion devises mainly operate with low emittance storage rings. High-spatial-coherence hard X-ray beam can be obtained from the small source size with the long propagation length. In Hyogo-ID BL (BL24XU) of SPring-8, we develop various researches, for example, focusing beam optics with Fresnel zone plates (FZP) and a hard X-ray nano-interferometer [1]. In the focusing beam optics, a FZP requires coherent illumination to achieve its diffraction limited focus. The hard X-ray nano-interferometer of wavefront division type also requires highly coherent illumination, and thus coherence property of incident X-ray is important for these experiments. Temporal coherence is defined by energy resolution of monochromater. On the other hand, an interferometer in hard X-ray region is necessary for evaluating spatial coherence.

Many methods for evaluation of spatial coherence in X-ray region have been reported. For example, methods by analyzing a Fresnel diffraction pattern produced by a fiber or a slit [2], by shearing interferometer [3], by two-beam interferometer with prism optics [4], and by Young's interferometer [5,6] has been reported. Degree of spatial coherence is easily measured by using two-beam interferometer like Young's interferometer because fringe visibility is equivalent to it. However, in hard X-ray region a double-slit with fully narrow and thick apertures is required to perform the Young's interferometer because of its high transmittance and small diffraction spread. Generally, that double-slit is difficult to be fabricated, and further a distance between double-slit and observation screen is limited by beam-line geometry. In order to overcome these difficulties for measurement of degree of spatial coherence, we have proposed and developed a new total reflection double-slit (TRDS).

2. Total reflection double-slit (TRDS)

The TRDS consists of two Au narrow stripes deposited on a flat SiO_2 substrate. Reflection only from Au stripes can be extracted by operating the TRDS with a grazing incident angle between critical angles of Au and SiO_2 as shown in figure 1. The grazing incident angle is set at 7.5 mrad where the reflectivity ratio between Au and SiO_2 is maximized. The reflectivity of Au and SiO_2 is 0.608 and 0.002, respectively, and thus the reflection from SiO_2 is enough small to be ignored. Also, aperture sizes configured by Au stripes width are fully reduced by the small grazing incident angle and the reduction ratio is approximately 1/130. Au stripes were deposited to V-figured shape in order that interval between the apertures of double-slit can be changed by selecting illumination area as indicated in figure 2. Parameter of the fabricated TRDS is also shown in figure 2. The effective aperture size of a slit can be obtained to 1.5 μm, and double-slit interval can be changed to 240 μm.

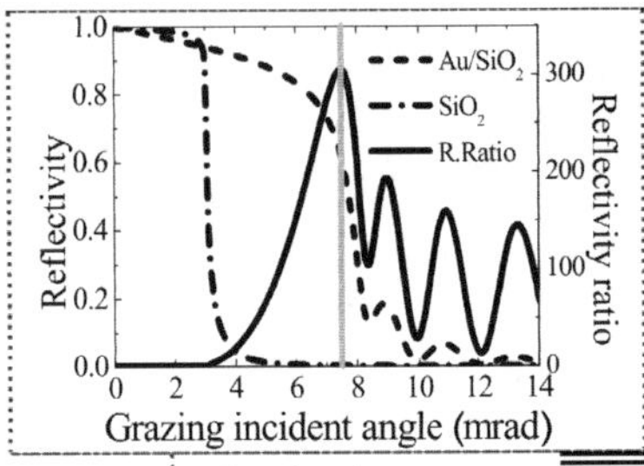

Figure 1. Reflectivity of Au on SiO_2 and SiO_2, and reflectivity ratio between them. The reflectivity ratio is maximized at grazing incident angle of 7.5 mrad.

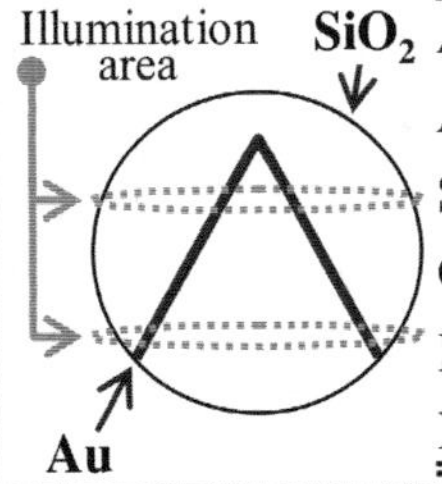

Au stripe width	200 μm
Au stripe thickness	20 nm
Substrate	SiO_2 (optical flat)
Grazing incident angle	7.5 mrad
Effective aperture size	1.5 μm
Range of double-slit interval	~240 μm

Figure 2. Schematic drawing of a TRSD. The double-slit interval is variable by selecting illumination area. Parameters of the fabricated TRDS are shown.

3. Experiment

A hard X-ray Young's interferometer using the TRDS was constructed at Hyogo-ID BL (BL24XU) of SPring-8 as shown in figure 3. The X-ray beam was monochromatized using silicon double crystal monochromator (DCM) at 10 keV with the monochromaticity $\Delta\lambda/\lambda$ of 1.5×10^{-4}. The temporally coherent length l_{coh} defined by $l_{coh} = \lambda^2/\Delta\lambda$, is estimated to 0.83 μm. The maximum optical path difference in the TRDS is fully shorter than that. In order to observe interference fringes, an X-ray zooming tube, that can resolve 0.5 μm structures at 20 keV X-ray, (Hamamatsu Photonics K. K., C5333) was employed. Resolution property of the detector is important to measure accurate fringe visibility. The finer the fringe period is, the lower the resultant fringe visibility is due to the resolution property. Modulation transfer function (MTF) is convenient to correct the effect. The MTF can be obtained from the point spread function (PSF). The PSF can be measured by a fully fine point input generated by X-ray focusing with FZP, and the MTF of the detector is obtained by Fourier transformation of the measured PSF.

An envelope of obtained interference fringes is not uniform, and it is difficult to directly measure the fringe visibility. We analyze the interference fringes using Fourier transformation, and fringe visibility is obtained as an area ratio between two peaks, one by fringe's frequency and the other by the 0th peak, of power spectrum of a fringe image. These obtained visibilities are corrected with the MTF of the detector, and degree of spatial coherence without the effect of the detector is obtained as a function of the double-slit intervals determined by the fringe frequency.

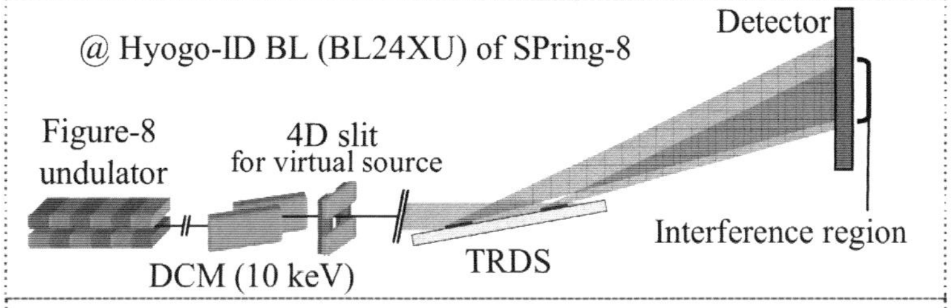

Figure 3. Schematic drawing of the experimental set-up of Young's interferometer with a TRDS.

4. Results

Spatial coherence defined by a virtual source configured by a beam-line 4D slit was evaluated to confirm the performance of the TRDS. When the intensity distribution of the source is assumed as a rectangular function, distribution of the degree of spatial coherence μ at the TRDS is defined as

$$\mu(d) = \left|\frac{\sin(\pi dr/\lambda R)}{\pi dr/\lambda R}\right|$$

by van Cittert-Zernike's theorem, where d is double-slit interval, r is source size, λ is wavelength, and R is source to double-slit distance. Result of measurement, where λ is 0.124 nm and R is 4.1 m, is shown in figure 4. Solid line shows a calculated μ with r of 6 μm, which is obtained by Fraunhofer diffraction image of the virtual source. The experimental result shows good correspondence to theory, and it was confirmed that Young's interferometer using the TRDS worked well.

Spatial coherence at experimental hutch 66.7 m apart from the undulator source was also evaluated. The experimental result in vertical direction is shown in figure 5. The solid line is the best fit of linearly-combined two Gaussian curves. The spatially coherent length assumed as the full width at half maximum (FWHM) is 55 μm. The source size is also calculated to be 46.6 μm (FWHM), whereas the theoretical source size at 10 keV is 16 μm (FWHM). The difference may be due to effect of wavefront deformations caused by beam-line components.

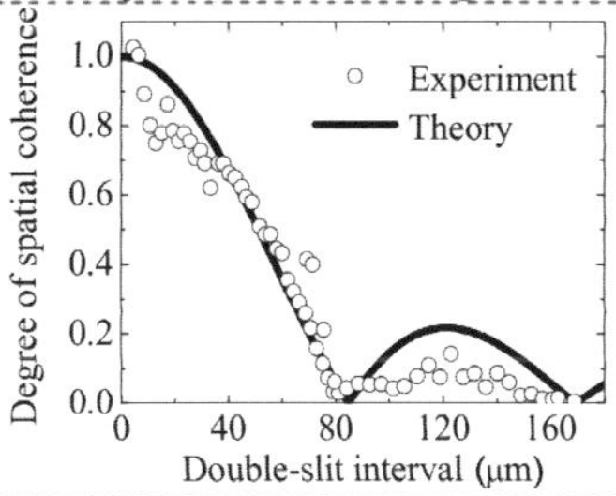

Figure 4. Spatial coherence of virtual source.

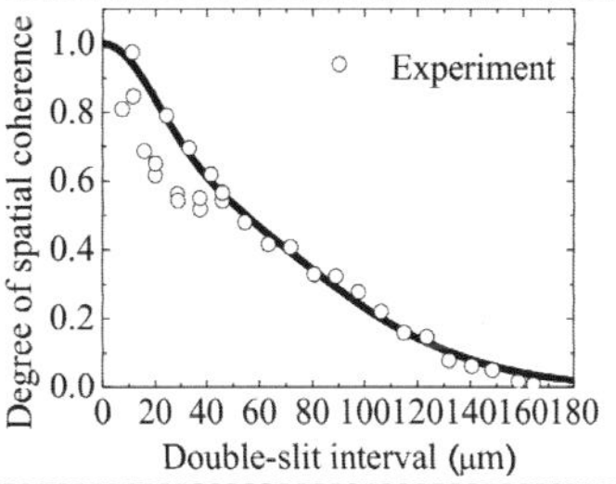

Figure 5. Spatial coherence of undulator source.

References

[1] Koyama T, Tsuji T, Yoshida K, Takano H, Tsusaka Y and Kagoshima Y 2006 *Jpn. J. Appl. Phys.* **43** L1159
[2] Kohn V, Snigireva I and Snigirev A 2000 *Phys. Rev. Lett.* **85** 2746
[3] Pfeiffer F, Bunk O, Schulze-Briese C, Diaz A, Weitkamp T, David C and van der Veen J F 2005 *Phys. Rev. Lett.* **94** 164801
[4] Suzuki Y 2004 *Rev. Sci. Instrum.* **75** 1026
[5] Aoki S and Kikuta S 1974 *Jpn. J. Appl. Phys.* **13** 1385
[6] Leitenberger W and Pietsch U 2007 *J. Synchrotron. Rad.* **14** 196

Germanium and silicon kinoform focusing lenses for hard x-rays

L. Alianelli[1], K.J.S. Sawhney[1], M.K. Tiwari[1], I.P. Dolbnya[1], R. Stevens[2], D.W.K. Jenkins[2], I.M. Loader[2], M.C. Wilson[2] and A. Malik[2]

[1] Diamond Light Source Ltd Didcot - OX11 0DE UK

[2] STFC Central Microstructure Facility Didcot - OX11 0QX UK

Email: lucia.alianelli@diamond.ac.uk

Abstract. Progress in synchrotron radiation instrumentation and successful fabrication of novel x-ray optics enables probing the matter at micro- and nano-scales from virtually any sciences – from biomedical, environmental to material sciences. In-line, compact focusing optics, based on refraction, are designed and fabricated for use in hard and very hard x-ray beamlines at Diamond. The first test results at a Diamond beamline, using micro-focusing silicon and germanium lenses, are discussed in this paper.

1. Introduction

The availability of bright third generation synchrotron sources has driven the development in the past decade of optical devices capable of concentrating the x-ray beam into spots that are less than 100 nm in diameter. Several upgrade programs exist for storage ring renewal capable of lower emittance [1 - 3], with the aim to focus the hard x-ray beam to few nanometres. Exploitation of micro- and nano-focused x-ray beams on a daily basis is seen as a priority by the international synchrotron community. At Diamond Light Source and STFC we work to innovate refractive focusing devices [4, 5] for use with hard and very hard x-rays. We are designing, fabricating and testing lenses that will eventually focus the high intensity monochromatic synchrotron beams to nano-sizes. By employing single crystal silicon and germanium, we are able to reduce scattering of the x-rays that is responsible for the limited resolution of other types of compound refractive lenses. Refractive focusing or collimating optics can be easily kept and used in vacuum hence they do not require any window between optics and sample. Moreover they do not produce diffraction effects, and their use is preferred to the use of slits even when the gain is not very high.

Planar technologies offer high accuracy for the fabrication of planar focusing lenses. Curved surfaces can be made with radii of about 1 μm, therefore focal length in the meter range are achieved with a single refractive surface. A kinoform profile is the most suited for applications requiring high flux and lateral resolution at a fixed energy [6]. The elliptical or parabolic internal wall produces focusing, while the kinoform profile reduces absorption but permits phase conservation between different parts of the electromagnetic wave. The low absorbing kinoform lens is the best candidates for nanofocusing because of the high lens effective aperture. The function of a kinoform lens has been described in the past by several authors [6 - 10]. Typical minimum wall thickness of few micrometers is needed to make an efficient lens. Therefore state-of-the-art hard masks and etching tools are necessary in order to replicate the lens profile with high aspect ratio. For nano-focusing applications, array of refractive surfaces with changing eccentricity have to be carefully simulated and designed in order not to introduce aberrations [11, 12].

2. X-ray lenses fabrication and test

Pre-patterned silicon and germanium lenses were etched by using a modified Bosch process. The maximum depth achieved so far is just above 100 micrometres. SU8 was used as a mask for the Si lens shown in Figure 1. A deposition process to deliver aluminium nitride and silicon oxide hard masks with very high selectivity and purity is being developed for the future lenses. The aim is to be able to etch 500 μm high structures in silicon and germanium. We have made Si lenses for x-ray energy E = 8 to 20 keV, and Ge lenses for E = 20 to 150 keV.

The bending magnet Test Beamline B16 is equipped with an optics table carrying three stages for the optics, the diagnostics, and the detector. The beam from the bending magnet source was imaged by the lens, after diffraction from a double crystal Si 111 monochromator. We used E = 8, 12 keV (Si lenses) and 20 keV (Ge lens) with F = 1000 mm, and beam demagnification in the vertical plane only. The source size FWHM at B16 is 50 μm and, for this experiment, the monochromatic beam before the lenses was unfocused. The experiment was located at 47 m from the source. A profile of the vertical focus from the germanium lens is shown in Figure 2: the derivative of the fluorescence signal, from a gold wire, can be fitted by a Lorentzian with FWHM = 1.06 μm.

In the same experiment we used crossed wafers for 2-dimensional focusing [13] and tested lenses with F = 200 mm and F = 600 mm. The gold wire scans of the beams focused by these lenses give a FWHM of 0.5 μm or less, depending on the method used to smooth the data. We do not include these data here as they suffer from vibrations of the experimental set-up.

The measured one-dimensional gain from the fluorescence data is 32, at least one order of magnitude less than expected. The measurement of the gain *via* the fluorescence scan is not very precise: using measured transmission data T = 50% [7] and an effective aperture of A = 200 microns (Figure 3), we obtain instead G = 100.

3. Future work

Kinoform lenses are efficient focusing optics for hard x-rays. Germanium lenses are presented here for the first time and tests at energy larger than 20 keV (up to 80 keV) are planned at Diamond in the coming months. Improved fabrication methods using modern plasma etchers will be used for the next lenses, which will deliver smoother walls, therefore better gain. There is much scope for improvement of kinoform lens testing as well, for instance using smaller and more coherent sources. Therefore we are confident that higher gain will be achieved during this project.

Acknowledgements

The fabrication of silicon and germanium lenses is funded by STFC Center for Instrumentation and by the STFC grant ST/F001665/1. We acknowledge Diamond Light Source Ltd. for the beamtime at B16.

References

[1] http://www.esrf.eu/AboutUs/Upgrade
[2] http://petra3.desy.de/
[3] http://www.bnl.gov/nsls2/
[4] A. Snigirev, V. Kohn, I. Snigireva, B. Lengeler Nature 384 (1996) 4951.
[5] C.G. Schroer, O. Kurapova, J. Patommel, P. Boye, J. Feldkamp, B. Lengeler, M. Burghammer, C. Riekel, L. Vincze, A. Van der Hart and M. Kuchler Appl. Phys. Lett. 87 (2005) 124103.
[6] K. Evans-Lutterodt, J. Ablett, A. Stein, C. Kao, D. Tennant, F. Klemens, A. Taylor, C. Jacobsen, P. Gammel, H. Huggins, G. Bogart, S. Ustin, and L. Ocola Opt. Expr. 11, 8 (919-926) 2003.
[7] L. Alianelli, K.J.S. Sawhney, D.W.K. Jenkins, I.M. Loader, R. Stevens, A. Snigirev and I. Snigireva SPIE Proceedings 6705 (670507) 2007.

[8] V. Aristov, M. Grigoriev, S. Kuznetov, L. Shabelnikov, V. Yunkin, T. Weitkamp, C. Rau, I. Snigireva, A. Snigirev, M. Hoffmann and E. Voges Appl. Phys. Lett. 77, 24 (2000)

[9] B. Nohammer, J. Hoszowska, A.K. Freund and C. David J. Synchrotron Rad 10 (2003) 168-171.

[10] W. Jark, F. Perennes and M. Matteucci J. Synchrotron Rad. 13 (2006) 239-252.

[11] C.G. Schroer and B. Lengeler Phys. Rev. Lett. 94 (2005) 054802.

[12] K. Evans-Lutterodt, A. Stein, J.M. Ablett, N. Bozovic, A. Taylor and D.M. Tennant Phys. Rev. Lett. 99 (2007) 134801.

[13] L. Alianelli et al, submitted to J. Synchrotron Rad. (2008)

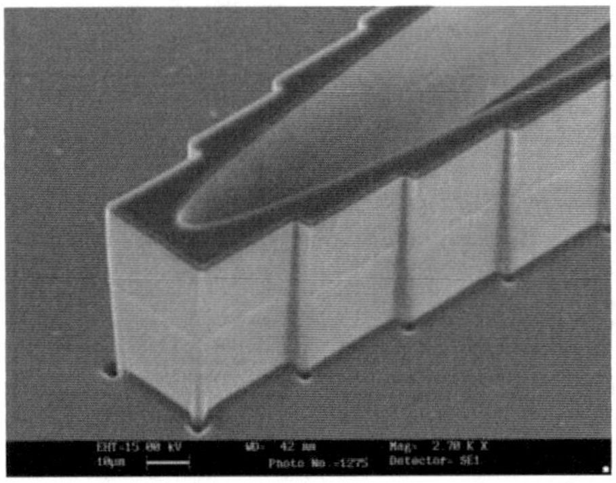

Figure 1. Detail of a Si kinoform lens. The step length L is equal to the ratio of the wavelength to the refractive index decrement. L varies between 20 and 100 μm, depending on energy and material. The step width decreases with increasing the aperture, starting from a few microns at the center. Small widths are difficult to fabricate, therefore some extra thickness is added as can be seen here. The total aperture of lenses reaches 800 μm.

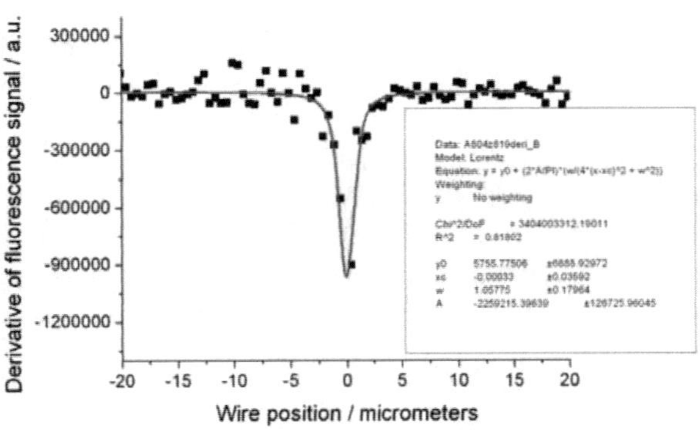

Figure 2. Derivative of the fluorescence signal from a gold wire scan, in the focal plane of a germanium lens, with F = 1000 mm and E = 20 keV. The red line is the fit result, with a Lorentzian curve with FWHM of 1.06 micrometer. The experimental data were not smoothed.

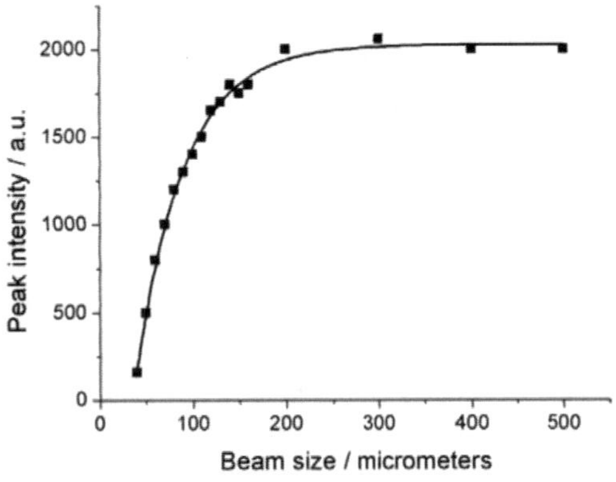

Figure 3. Peak intensity at focused beam versus size of incident beam. These data are recorded with a spatial resolution detector with low resolution, and effective pixel size of 6.5 μm.

Journal of Physics: Conference Series **186** (2009) 012063 doi:10.1088/1742-6596/186/1/012063

Nanofocusing Refractive X-Ray Lenses: Fabrication and Modeling

Pit Boye[1], Jan M. Feldkamp[1], Jens Patommel[1], Andreas Schwab[1], Sandra Stephan[1], Robert Hoppe[1], Christian G. Schroer[1], Manfred Burghammer[2], Christian Riekel[2], Andre van der Hart[3], Matthias Küchler[4]

[1] Institute of Structural Physics, TU Dresden, D-01062 Dresden, Germany
[2] ESRF, B. P. 220, F-38043 Grenoble, France
[3] ISG, Forschungszentrum Jülich, D-52425 Jülich, Germany
[4] Fraunhofer IZM, Reichenhainer Str. 88, D-09126 Chemnitz, Germany

E-mail: boye@xray-lens.de

Abstract. Nanofocusing refractive x-ray lenses (NFLs) form the basis of a hard x-ray scanning microscope. They are characterized by their short focal length (~10 mm at 15 keV to 25 keV) and large numerical aperture, allowing for the generation of hard x-ray nanobeams even at short distances from a synchrotron radiation source. These optics, made out of silicon by electron beam lithography and subsequent deep reactive ion etching, have been shown to focus hard x-rays down to 50 nm. We have modeled these optics, allowing us to characterize slight aberrations and the wave-field properties in the focus by analyzing the beam profile in the far field.

1. Introduction

The short wavelength as well as the large penetration depth in matter make hard x rays highly interesting for microscopy. X-ray analytical techniques, such as x-ray diffraction, fluorescence, and absorption spectroscopy, give information about the crystalline structure, chemical compositions, and the chemical states respectively. The possibility to use these techniques in a scanning x-ray tomograph also allows one to obtain information from inside a specimen without destructive sample preparation and contamination. The great demand for smaller spot sizes to obtain higher spatial resolution led to the development of nanofocusing refractive x-ray lenses (NFLs). With these lenses made out of silicon spot sizes down to 50 nm are accomplished [1, 2]. Further improvement of the fabrication process and the combination with coherent diffraction techniques allows to achieve resolutions of a few nanometers [3].

2. Design and fabrication of nanofocusing lenses

Today a multitude of X-ray optics exist to generate nanobeams in the range down to 50 nm or even below [4, 5]. Among these optics are the nanofocusing lenses (NFLs) which have been developed based on parabolic refractive x-ray lenses (CRLs) made out of aluminum [6] or beryllium [7]. By construction, the focal lengths of the letter lenses is bounded from below by about 100 mm. To achieve the ultimate small beam size with refractive x-ray lenses, just confined by a small diffraction limit, the focal length has to be scaled down significantly compared to

Figure 1. A total of 45 blocks structured into a 4" silicon wafer. Each block holds a different number of lenses (35, 50, 71, 100 and 142 single lenses).

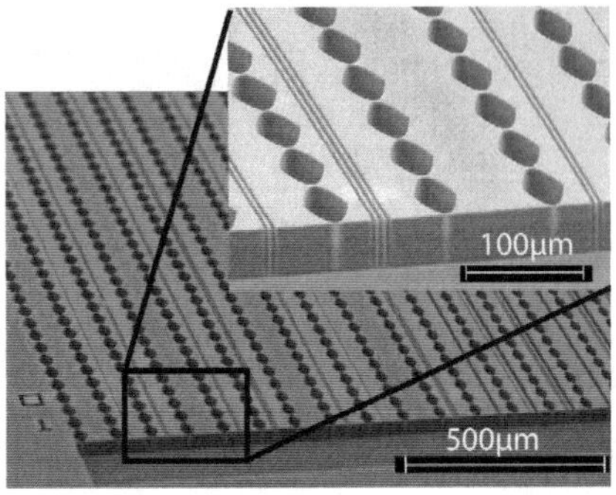

Figure 2. Scanning electron micrograph of a block of NFLs and magnified view on a part of a lens block with a lens depth of 62 μm

that of CRLs. This is achieved with NFLs by reducing the radius of curvature of the the lens surface to the range of a few micrometers. For this reason the nanofocusing lenses are made by electron beam lithography and deep reactive ion etching (see Figures 1 & 2). A stack of parabolic cylinders etched into the lens material focuses just in one dimension. To create a point focus on the sample, a crossed geometry of two planar lenses is required (see Figure 3). A prototype of an x-ray microscope employing this scheme was developed, built and tested by our group in collaboration with beamline ID13 at the ESRF in Grenoble, France (see Figure 4). These lenses are made by 4-inch silicon wafer technology. One wafer holds more than 60000 single structures, with about 45 blocks of lenses (see Figure 1). The different blocks have variable numbers N of single lenses in a row. Each block holds 21 lenses with 3 different radii of curvature R, each corrected in 7 steps for shape displacement by the fabrication process. The radius of curvature lies between 1 μm and 5 μm. Depending on the x-ray energy (from 15 keV to 30 keV), the lens parameters (R and N) can be selected to obtain a focal length between 10 mm and 30 mm. An individual bar code is structured into the wafer in between the lenses to ease the process of alignment. Most of the steps of the fabrication process are well known from semiconductor technology. However, since the performance of this optic is very sensitive to deviations of the lens shape and roughness of the surfaces, a careful selection of process parameters is necessary. We start with a standard silicon wafer with 100 mm diameter and a thickness of about 545 μm. This wafer is wet oxidized to coat it with a silicon dioxide layer of 300 nm thickness. Afterwards the wafer is coated with a 30 nm thick chromium layer. Subsequently, the wafer is spin-coated with a positive e-beam resist and baked on a hot plate. The resist layer has a thickness of 300 nm. The so-prepared wafer is structured by electron beam lithography at 50 kV acceleration voltage and a spot size of about 40 nm. The illuminated structures are developed in a fast developer. The underlying chromium layer is structured in a wet-etch solution. The structured chromium serves as a mask for the silicon dioxide, which is etched in a dry reactive ion etching (standard) process. This process was optimized for silicon dioxide to obtain a highly anisotropic etching process. In the final step, the lens structure is transferred into the silicon bulk material (see Figure 2) by deep reactive ion etching (Bosch process) at a multiplex ICP standard rate etch tool. Process parameters and further details are described in [2].

3. Analysis and modeling of the lens shape

There are many critical points during the manufacturing process of NFLs. The focusing properties of these optics highly depend on the steepness of the side walls and the accuracy

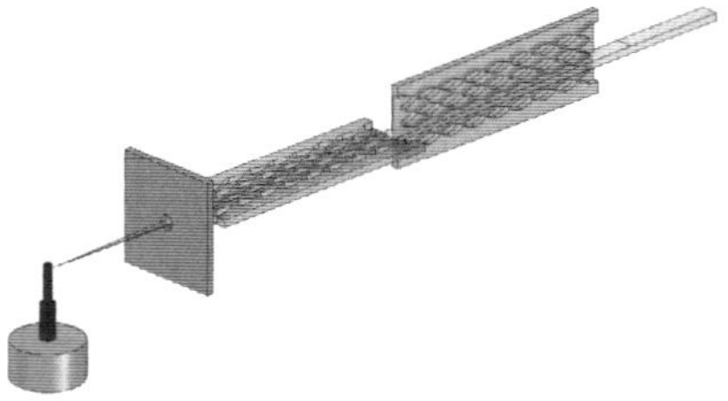

Figure 3. Schematic sketch: A point focus requires two lenses in crossed geometry.

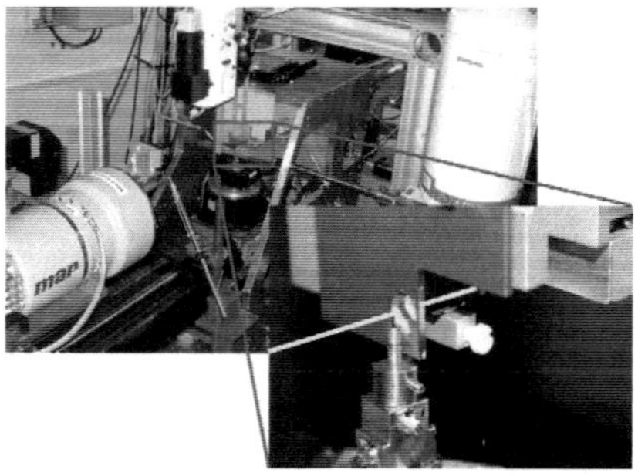

Figure 4. Experimental setup installed at the ID13@ESRF (Grenoble, France).

of the lens shape. If we look at the last manufacturing step - the deep silicon etching process shifts the edges of the lens surface by about 450 nm compared to the mask. This effect is to be compensated by correcting the mask accordingly. In previous NFL production cycles these corrections were underestimated. Therefore, even the largest correction (of 300 nm), could not fully compensate the shape error and the theoretically smallest spot size could not be achieved. Based on this knowledge we modeled our refractive optic inside a scanning microscope. We simulated different under and over-etched lens shapes (also the ideal one) and their influence on the focus size. Especially their effects on the beam profile in the far field were studied. With this tool we are able to characterize our optic (and its aberrations) and the wave field properties in the focus by analyzing the beam profile in the far field. Detailed analysis of the beam profiles in the far field images, recorded with a high resolution CCD-camera, shows that we indeed have a maximal correction that does not fully compensate the over-etching effect. By comparing the modeled and the measured data a shape error of the lens form of a shift by 150 nm was determined. These results are in arrangement with the inherent shift introducing by the Bosch process of 450 nm and our correction of 300 nm in the given case. Recording far field images and taking line scans seems to be a well-suited tool to characterize our optics and measure the aberration of the lens shape. We are going to optimize our manufacturing process accordingly and are expecting significantly improved lens shape and, based on that x-ray spot sizes down to 30 nm [8].

References

[1] Schroer C G, Kurapova O, Patommel J, Boye P, Feldkamp J, Lengeler B, Burghammer M, Riekel C, Vincze L, van der Hart A and Küchler M 2005 *Appl. Phys. Lett.* **87** 124103

[2] Kurapova O, Lengeler B, Schroer C G, Küchler M, Gessner T and van der Hart A 2007 *J. Vac. Sci. Technol. B* **25** 1626–1629

[3] Schroer C G, Boye P, Feldkamp J, Patommel J, Schropp A, Schwab A, Stephan S, Burghammer M, Schöder S and Riekel C 2008 *Phys. Rev. Lett.* **101** 090801

[4] Kang H C, Maser J, Stephenson G B, Liu C, Conley R, Macrander A T and Vogt S 2006 *Phys. Rev. Lett.* **96** 127401

[5] Mimura H, Matsuyama S, Yumoto H, Yamamura K, Sano Y, Shibahara M, Endo K, Mori Y, Nishino Y, Tamasaku K, Yabashi M, Ishikawa T and Yamauchi K 2005 *Japan. J. Appl. Phys.* **44** L539–L542

[6] Lengeler B, Schroer C G, Richwin M, Tümmler J, Drakopoulos M, Snigirev A and Snigireva I 1999 *Appl. Phys. Lett.* **74** 3924–3926

[7] Lengeler B, Schroer C G, Benner B, Gerhardus A, Günzler T F, Kuhlmann M, Meyer J and Zimprich C 2002 *J. Synchrotron Rad.* **9** 119–124

[8] Schroer C G and Lengeler B 2005 *Phys. Rev. Lett.* **94** 054802

Journal of Physics: Conference Series **186** (2009) 012064 doi:10.1088/1742-6596/186/1/012064

Ellipsoidal capillary as condenser for the BESSY full-field x-ray microscope

P Guttmann[1], X Zeng[2], M Feser[2], S Heim[1], W Yun[2] and G Schneider[1]

[1] BESSY m.b.H., Albert-Einstein-Straße 15, 12489 Berlin, Germany

[2] Xradia Inc., 5052 Commercial Circle, Concord, CA 94520 USA

E-Mail: guttmann@bessy.de

Abstract. The BESSY x-ray microscopy group has developed a new full-field x-ray microscope which employs an advanced x-ray optical concept. Traditionally, zone plate based condensers are used in x-ray microscopes providing an energy resolution of only $E/\Delta E \leq 500$. In addition, this conventional monochromator concept requires a pinhole close to the sample restricting the available space for tomography applications. In our new BESSY microscope, a standard monochromator beam line provides a high energy resolution of up to 10,000 which permits NEXAFS studies. An elliptically shaped mono-capillary is used to form the hollow cone illumination necessary for sample illumination and to match the aperture of the objective. Calculations regarding the performance and accuracies needed are presented and characterizations of capillaries especially made for the BESSY soft x-ray microscope are shown. For the first time, we demonstrate that glass capillaries are well suited as condensers in the soft x-ray energy domain. Their focusing efficiency was measured to be 80% which is about an order of magnitude higher than the diffraction efficiency of zone plate based condensers.

1. Introduction

The new optical setup for the BESSY full-field transmission x-ray microscope at the undulator U41 [1] is shown in Figure 1. The main advantage over other full-field soft x-ray microscopes is the absence of a monochromator pinhole close to the sample. Therefore, high spatial resolution and high tilt tomography of cryogenic or heated samples on flat sample holders are possible. As opposed to the zone plate condenser monochromator used so far, monochromatization and object illumination are independent from each other. Without changing or moving the condenser optic a wide energy range is available and the monochromaticity can be tuned. With this new optical setup, spectromicroscopy studies are routinely feasible in the full-field microscope.

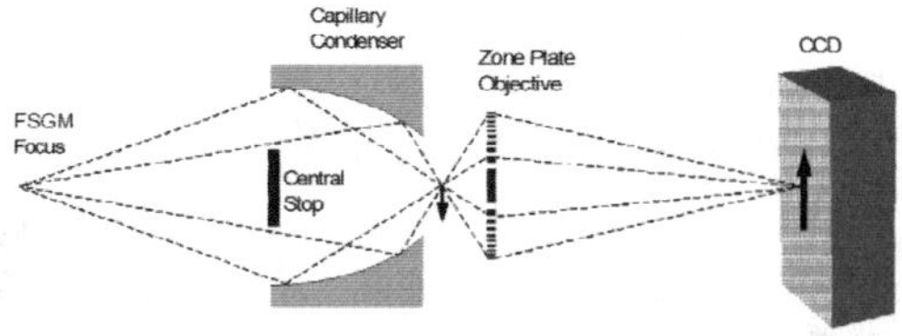

Fig. 1: The new optical setup for the BESSY full-field transmission x-ray microscope at the undulator U41 using a standard monochromator (FSGM) and an ellipsoidal glass capillary as condenser.

2. Ray tracing calculations

The parameters of the ellipsoidal capillary are determined by the beam divergence of the monochromator and by the desired aperture matching to the high resolution objective of the microscope. A homogeneous illumination of the capillary is required for an isotropic distribution of the radiation in the hollow cone illumination of the specimen. Therefore, ray tracing calculations for the monochromator beam line at the specific undulator source were carried out. Figure 2 shows the illumination of the capillary in different distances. Further results from these calculations show that the slope error of the capillary should be below 100 μrad and the alignment accuracy in the tilt angle of the capillary with respect to the incoming beam should be better than 100 μrad.

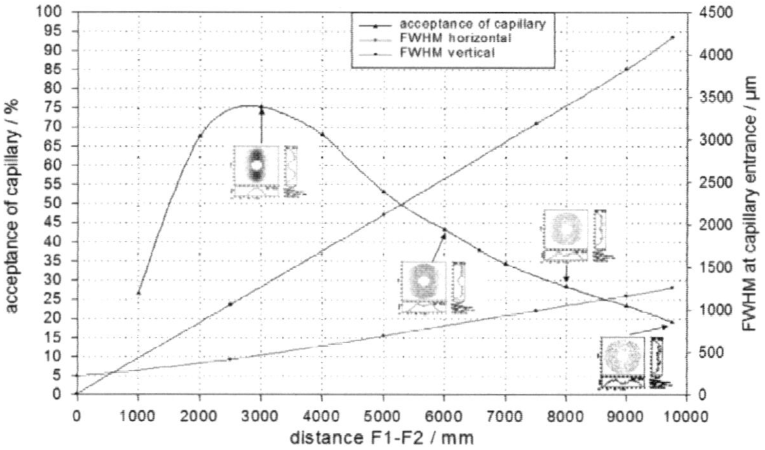

Fig. 2: Determination of the optimum distance F_1-F_2 of the focus of the capillary to the focus of the monochromator.

3. Results

The single-bounce ellipsoidal glass capillary was fabricated and evaluated by optical measurements during and after fabrication by Xradia [2]. The resulting slope error of 80 μrad is well below the limit calculated by ray tracing and is shown in Figure 3.

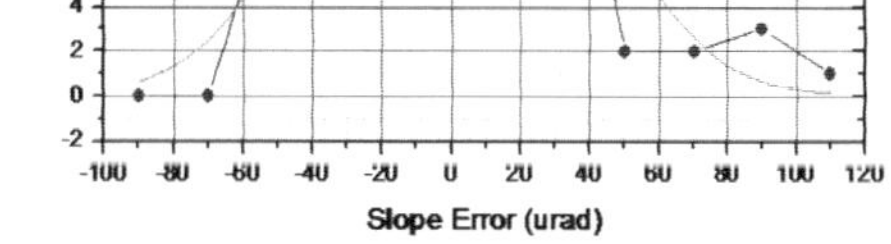

Fig. 3: Slope error of the fabricated single-bounce ellipsoidal capillary measured by optical methods by Xradia.

The x-ray performance of the capillary was tested with the BESSY x-ray microscope at the U41-FSGM beam line at a photon energy of 510 eV. Firstly, the capillary was adjusted to get the smallest possible focus. Figure 4a shows an x-ray image of the focus which is in size 620 nm x 990 nm (FWHM-values). Secondly, a Siemens star test pattern was used to demonstrate the imaging capabilities of the x-ray microscope using a capillary as condenser. To obtain a large homogeneously illuminated object field, the condenser is helically scanned. Figure 4b shows the x-ray image of this test pattern using a micro zone plate objective with 25 nm outermost zone width.

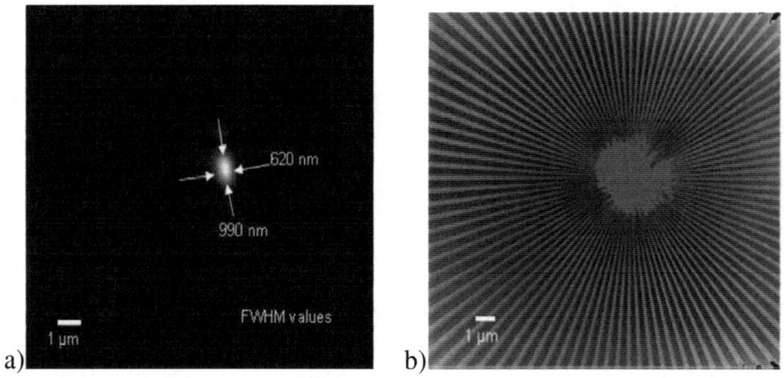

Fig. 4: a) Focus of the capillary imaged by a micro zone plate at $\Delta E/E \approx 9000$, 510 eV, 100 ms exposure time. b) Test pattern imaged by a micro zone plate (both fabricated at BESSY by S. Rehbein, S. Werner) at $\Delta E/E \approx 9000$, 510 eV, 2 s exposure time with scanned condenser.

To analyze the performance of the new condenser, quantitative measurements were performed to measure the efficiency of the capillary. The focusing efficiency of the capillary was measured to be 80% for a photon energy of 510 eV. Note, this value exceeds the efficiency of zone plate condensers by nearly an order of magnitude.

4. Conclusions and outlook

For the first time, it is demonstrated that single-bounce ellipsoidal glass capillaries are well suited as condensers in the soft x-ray domain for full-field transmission x-ray microscopes. We report the highest focusing efficiency of 80% measured so far for soft x-rays. In the near future, it is planned to improve the slope errors of the capillaries and to increase the numerical aperture of the capillaries.

Acknowledgments

This work has been funded by the German Federal Ministry of Education and Research (BMBF) under contract number 05KS4BY1/7 and by the Human Frontier Science Program (HFSP) Research Grant Ref. RGP0053/2005-C.

References

[1] G. Schneider, P. Guttmann, S. Heim, S. Rehbein, D. Eichert, B. Niemann: X-Ray Microscopy at BESSY: From Nano-Tomography to Fs-Imaging, SYNCHROTRON RADIATION INSTRUMENTATION: Ninth International Conference on Synchrotron Radiation Instrumentation (Eds.: J.-Y. Choi, S. Rah), AIP Conference Proceedings **879** (2007), 1291-1294

[2] Zeng X, Duewer F, Feser M, Huang C, Lyon A, Tkachuk A, Yun W., Ellipsoidal and parabolic glass capillaries as condensers for x-ray microscopes, Appl Opt. 2008 May 1;47(13):2376-81

Journal of Physics: Conference Series **186** (2009) 012065 doi:10.1088/1742-6596/186/1/012065

Soft x-ray zone plate fabrication at KTH, Stockholm

A Holmberg, M Lindblom, J Reinspach, M Bertilsson and H M Hertz

Biomedical and X-Ray Physics, Department of Applied Physics, Royal Institute of Technology/Albanova, SE-10691 Stockholm, Sweden

E-mail: anders.holmberg@biox.kth.se

Abstract. We present the status of our zone plate and test object fabrication processes along with the latest fabricated components. With our nickel process, zone plates with outermost zone width of 20 nm and zone height of 90 nm have been fabricated. A gold electroplating process has recently been introduced for the fabrication of test objects. The first result for gold gratings with 70 nm period and 135 nm height is shown.

1. Introduction

The zone plate optics fabrication at the Royal Institute of Technology (KTH) is focused on components for our in-house table-top laser-plasma-based water-window microscope. We have previously fabricated zone plate objectives [1] and condensers [2] that have successfully been implemented in this soft x-ray microscope [3]. The limited flux of the laser-plasma source increases the importance of high-efficiency optics. The emphasis in our zone-plate fabrication program has therefore been equally strong on efficiency as on resolution. Recent process development have included e-beam cross linked SU-8 resist for high-aspect-ratio electroplating mold fabrication [4], in-situ electroplating rate monitoring and end-point detection for accurate plating control [5], and pulse-plating techniques for uniform mold filling [6].

In the present paper we present the latest result with our standard, nickel electroplating-based fabrication process, as well as results with a newly introduced gold electroplating process. The gold process is intended for the structuring of gold test objects for the soft x-ray microscope.

2. Fabrication

2.1. Nickel components

In our standard process we electroplate the x-ray optical material into a mold that is prepared from a tri-layer resist mask. The fabrication process for nickel structures is shown in figure 1. In the first step, a stack of materials are prepared on a 50 nm thick Si_3N_4-membrane. On the membrane, a plating base consisting of 5 nm chromium and 10 nm germanium is vapor deposited. Onto the plating base a tri-layer resist is prepared. It contains a spincoated 100-200 nm thick plating mold material, ARC XL-20 (Brewer Science), a vapor deposited 5 nm titanium hard mask, and a spincoated 25 nm thick ZEP 7000 (Nippon Zeon) e-beam resist.

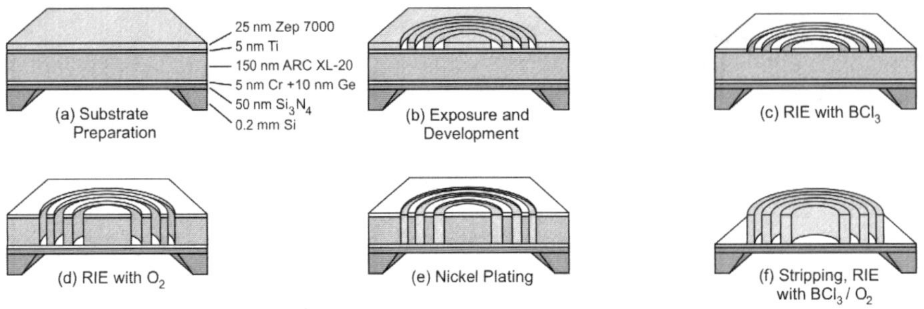

Figure 1. Fabrication process.

The structuring starts with an electron beam lithography exposure at 25 keV (Raith 150) and with a dose of typically 100-200 μC/cm^2. The resist is then developed in hexylacetate for 30 s followed by a 15 s rinse in iso-propanol. The pattern is transferred into the underlying titanium layer by reactive ion etching (RIE) with BCl_3 for 2 min and 40 s, at 15 mTorr, 10 sccm, and 80 W (0.18 W/cm^2). The sample is afterwards rinsed for 15 s in water and 10 s in iso-propanol. This rinse step is necessary to remove chlorine-containing residues that form on the sample during the BCl_3 etching. After that, the mold material is structured down to the electroplating base by O_2 RIE with the titanium as a hard mask. The O_2 etching is done for typically 6 min at 5 mTorr, 10 sccm, and 50 W (0.11 W/cm^2). Both of the RIE etch steps are done in Oxford Instruments Plasmalab 80+ systems with the cathode at room temperature and without any thermal contact to the membrane. The mold is then filled with nickel by electroplating. This is done in a nickel-sulfamate type bath (Enthone Inc., Lectro-nic 10-03s), which is operated at ph 3.25 and 52 °C. The current is adjusted to achieve a plating rate between 25-35 nm/min in the structures. As final step the mold is removed by BCl_3 and O_2 etching with the same parameters as for the mold preparation. This results in the final nickel zone plate on the Si_3N_4-membrane with plating base.

Figure 2 below shows zone plates with outermost zone width dr_N = 25 nm and dr_N = 20 nm. The dr_N = 25 nm zone plate has a diameter of 54 μm and a focal length of 546 μm at λ = 2.48 nm. Zone plates with 25 nm outermost zones are routinely made with nickel heights up to 130 nm in the outer part of the zone plate. With an in-house house diffraction efficiency measurement arrangement [7] we typically measure a 1st order groove diffraction efficiency of ~13% (@ λ = 2.88 nm) for such dr_N = 25 nm zone plate. The dr_N = 20 nm zone plate has a diameter of 55 μm and a focal length of 440 μm at λ = 2.48 nm. This zone plate has so far not been characterized in terms of efficiency or imaging performance.

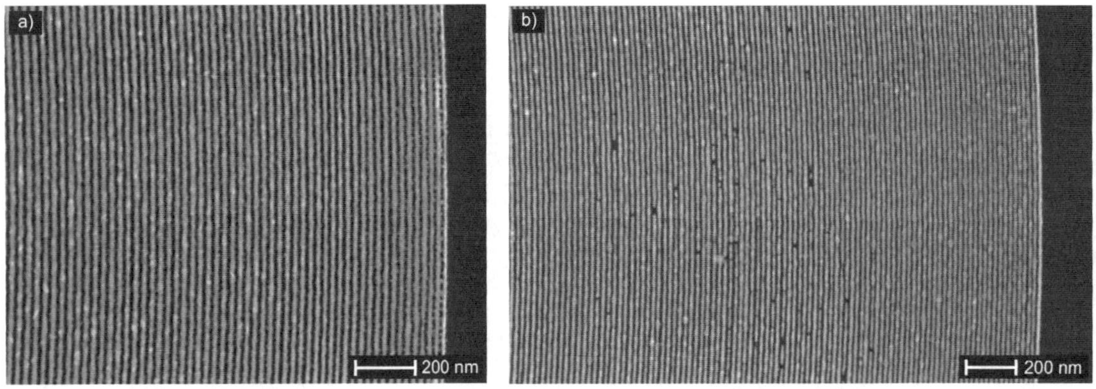

Figure 2. Scanning electron microscope images of a) a dr_N = 25 nm and b) a dr_N = 20 nm zone plate

2.2. Gold components

In addition to the nickel electroplating process, we have also started with gold electroplating. The material stack for the process is identical with the nickel process except from the plating base, where the germanium layer is replaced by a 10 nm thick vapor deposited gold layer. The gold electroplating bath (Enthone Inc., Neutronex 309) is operated at ph 9.50 and 50 °C. The samples are plated with a typical plating rate between 20-30 nm/min.

Figure 3 below shows the results from the first successful experiments. The images show 80 nm and 70 nm period gratings with a height of 135 nm.

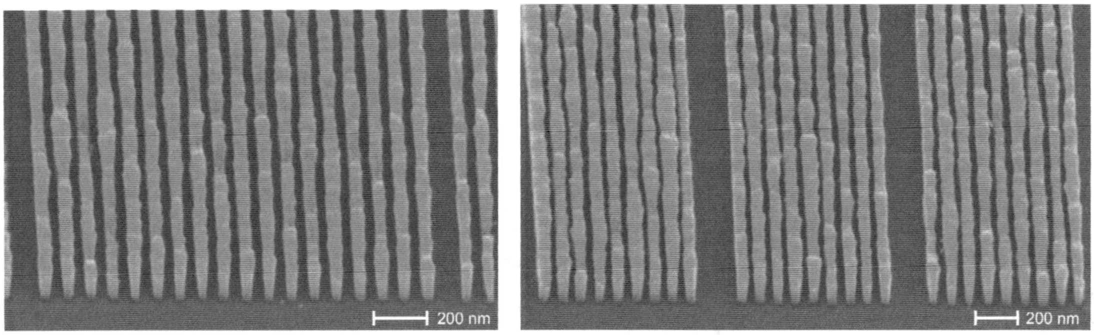

Figure 3. Scanning electron microscope images of gold gratings for use as test objects. The images are taken at a 52 degree angle. The left image shows 80 nm period gratings and the right image shows 70 nm period gratings.

3. Summary

In this paper our recent fabrication development was reviewed. We have presented new nickel zone plates with 20 nm outermost zone width. These will be tested in our table-top microscope in the near future. We have also shown our first results with a gold electroplating fabrication process. Gratings with periods down to 70 nm have been fabricated.

References

[1] Holmberg A, Rehbein S and Hertz H M 2004 *Microel. Engin.* **73-74** 639
[2] Rehbein S, Holmberg A, Johansson G, Jansson P A C and Hertz H M 2004 *J. Vac. Sci. Technol.* B **22** 1118
[3] Takman P A C et al 2007 *J. Microsc.* **226** 175
[4] Lindblom M, Hertz H M and Holmberg A 2007 A *Microel. Engin.* **84** 1136
[5] Holmberg A, Lindblom M and Hertz H M 2006 *J. Vac. Sci. Technol.* B **24** 2592
[6] Lindblom M, Hertz H M, and Holmberg A 2006 *J. Vac. Sci. Technol.* B **24** 2848
[7] Bertilson M C et al 2007 *Rev. Sci. Instrum.* **78** 026103

Performance test of Mo/Si and $MoSi_2$/Si multilayer Laue lenses

T Koyama[1], T Tsuji[1], H Takano[1], Y Kagoshima[1], S Ichimaru[2], T Ohchi[2] and H Takenaka[2]

[1]Graduate School of Material Science, University of Hyogo, 3-2-1 Kouto, Kamigori, Ako, Hyogo 678-1297, Japan
[2]NTT-AT Nanofabrication Corporation, 3-1 Wakamiya, Morinosato, Atsugi, Kanagawa, 243-0124, Japan

E-mail: koyama@sci.u-hyogo.ac.jp

Abstract. A multilayer Laue lens (MLL) as a hard X-ray focusing device has been designed and fabricated. Mo/Si and $MoSi_2$/Si were chosen to form the multilayers owing to their superior properties of high diffraction efficiency and relatively sharp interface between layers. DC magnetron sputtering system was used for deposition of the multilayers. The optical properties of the MLL were measured at BL24XU of SPring-8 with 20-keV X-rays. To confirm the dynamical diffraction effect, far-field diffraction images were taken at various incidence angles and depths. The resultant intensity distributions showed similar structure to those expected by calculations. Further, an almost diffraction-limited beam size of around 30 nm was obtained.

1. Introduction

Development of X-ray optical devices for X-ray microscopy and X-ray microanalysis at a nanometer scale is advancing rapidly in recent years. Focused spot sizes of a few tens of nanometers have been obtained using Kirkpatrick-Baez mirrors [1], compound refractive lenses [2], Fresnel zone plates [3-5], and multilayer Laue lens (MLL) [6-8]. MLL was pioneered by the APS group [6-8], and this technique appears promising with regard to the achievement of a 1-nm focus with a very high diffraction efficiency, even in the hard X-ray region [9]. Their MLL is composed of WSi_2 and Si layers [7]. We recently designed and fabricated an MLL by DC magnetron sputtering, choosing Mo/Si and $MoSi_2$/Si as the multilayer materials owing to their superior properties of high diffraction efficiency at an X-ray energy of 20 keV. In particular, $MoSi_2$/Si multilayer has smaller interdiffusion between the layers and higher heat resistance than Mo/Si multilayer [10]. The optical properties of the MLL due to the dynamical diffraction effect have been investigated theoretically using the coupled wave theory [11,12] and experimentally using undulator radiation from SPring-8. Further, the performance of an MLL as a hard X-ray focusing device was also studied. In this paper, we report the results of these experiments.

A DC magnetron sputtering deposition system was used to fabricate the Mo/Si and $MoSi_2$/Si multilayer with each boundary r_n according to the Fresnel zone configuration which is given by $r_n^2 = nf\lambda + n^2\lambda^2/4$, where λ is the wavelength and f is the focal length. The outermost zones were deposited first onto the flat Si substrate to decrease the layer disturbance, and the deposition was stopped before it reached the thickest central zone. The minimum layer thickness was designed to be 10 nm, while the maximum thickness was designed to be 100 nm. The total multilayer thickness was 4.6 μm, and the number of layers was 248. The focal length was designed to be 1.6 mm for an X-ray energy of 20 keV. The multilayer structure was sectioned and polished to an appropriate depth.

2. Performance test

The optical properties were measured at BL24XU [13] of SPring-8. A Si (111) monochromator was used to select an X-ray energy of 20 keV. The incident beam illuminated an area much larger than the zone structure region using a crossed slit to avoid the edge effect of the slit. A far-field diffraction image was used to measure the relative local diffraction efficiency as a function of the zone position r_n [7,14]. By appropriately mounting obstructions such as slits and blades, only the desired first-order diffraction beam was extracted. The tilt angle and the depth of the MLL were varied in order to study the dynamical diffraction effect, and the corresponding far-field images were captured by the CCD detector. The distance between the MLL and the CCD detector was approximately 30 cm. The depth was determined by measuring the X-ray transmission of the Si substrate of the MLL. Figures 1(a) and 2(a) show the experimentally measured intensity distributions, while figures 1(b) and 2(b) show the corresponding calculated distributions for Mo/Si MLL and $MoSi_2$/Si MLL, respectively. The horizontal axes indicate the tilt angle, which was varied from approximately −1 mrad to 3 mrad. One vertical line was obtained from one diffraction image with a horizontal beam width of 20 μm, and one distribution was composed of approximately 30 diffraction images. As the measured distributions agreed well with the calculated ones, the effects of volume diffraction were confirmed. In figure 1(a), intensity dips, which do not appear in calculation, appear at r_n = 3.6 μm. It can be considered that there is a local imperfection of the MLL structure. This result provides us with evidence whether or not the multilayer structure was properly fabricated and behaved as an MLL.

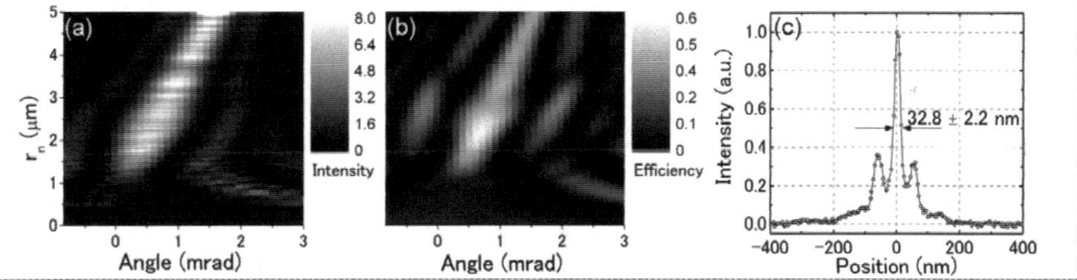

Figure 1. (a), (b) The first order diffraction patterns of the Mo/Si MLL with a depth of 34 μm. (a) Measured intensity distribution. (b) Calculated diffraction efficiency distribution. (c) The measured intensity profile of the best focus point with a depth of 20 μm.

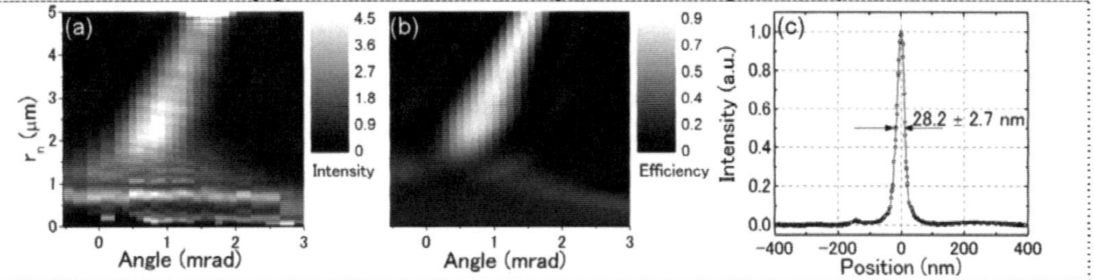

Figure 2. (a), (b) The first order diffraction patterns of the $MoSi_2$/Si MLL with a depth of 38 μm. (a) Measured intensity distribution. (b) Calculated diffraction efficiency distribution. (c) The measured intensity profile of the best focus point with a depth of 15 μm [14].

Next, the focused beam size was measured by the knife-edge method. A platinum wire with a diameter of 200 μm was used as the knife-edge. The intensity of the X-rays scattered by the knife-edge was measured using a PIN diode detector. The direct beam was blocked by a blade placed in front of the detector. The intensity of the scattered X-rays measured along the knife-edge position is proportional to the line spread function of the beam profile [4]. A piezoelectric stage that has a 2-nm resolution in the closed-loop mode was used to measure the beam size precisely. Although the knife-edge was fixed on the piezoelectric stage, the distance between the MLL and the knife-edge changed due to environmental disturbances. In order to avoid such effects, an electrostatic capacitance sensor was used to determine the relative distance between the MLL and the knife-edge. The relative position was measured at the same time that the intensity of the scattered X-rays was measured. The measured intensity distributions at the best focal point are shown in figures 1(c) and 2(c) for Mo/Si MLL and $MoSi_2$/Si MLL, respectively. The obtained beam sizes were 32.8 ± 2.2 nm and 28.2 ± 2.7 nm in the full width at half maximum. Ten measurements were performed in order to confirm the reproducibility. These values were close to the diffraction-limited beam size of 22 nm, which was estimated from the effective numerical aperture of the MLL using Rayleigh's criterion. Measured photon flux was approximately 5×10^7 photons. The results obtained show that the beam size of the Mo/Si MLL is slightly larger than that of the $MoSi_2$/Si MLL, and that the intensity profile of the Mo/Si MLL has strong side peaks. It may be considered that the local imperfection revealed by figure 1(a) effects the intensity profile.

3. Conclusion

We investigated the optical properties of the Mo/Si and $MoSi_2$/Si MLL fabricated by DC magnetron sputtering. The effects of volume diffraction were confirmed by experimentally and theoretically. These results suggest that the $MoSi_2$/Si MLL developed in this study has considerable potential for use as a nanofocusing optical device.

Acknowledgments

This development was supported by SENTAN, JST.

References

[1] Mimura H, Yumoto H, Matsuyama S, Sano Y, Yamamura K, Mori Y, Yabashi M, Nishino Y, Tamasaku K, Ishikawa T and Yamauchi K 2007 *Appl. Phys. Lett.* **90** 051903

[2] Schroer C G, Kurapova O, Patommel J, Boye P, Feldkamp J, Lengeler B, Burghammer M, Riekel C, Vincze L, van der Hart A and Küchler M 2005 *Appl. Phys. Lett.* **87** 124103

[3] Chao W, Harteneck B D, Liddle J A, Anderson E H and Attwood D T 2005 *Nature* **435** 1210

[4] Suzuki Y, Takeuchi A, Takano H and Takenaka H 2005 *Jpn. J. Appl. Phys.* **44** 1994

[5] Yin G C, Song Y F, Tang M T, Chen F R, Liang K S, Duewer F W, Feser M, Yun W and Shieh H P D 2006 *Appl. Phys. Lett.* **89** 221122

[6] Maser J, Stephenson G B, Vogt S, Yun W, Macrander A, Kang H C, Liu C and Conley R 2004 *Proc. SPIE* **5539** 185

[7] Kang H C, Maser J, Stephenson G B, Liu C, Conley R, Macrander A T and Vogt S 2006 *Phys. Rev. Lett.* **96** 127401

[8] Kang H C, Yan H, Winarski R P, Holt M V, Maser J, Liu C, Conley R, Vogt S, Macrander A. T and Stephenson G B 2008 *Appl. Phys. Lett.* **92** 221114

[9] Yan H, Maser J, Macrander A, Shen Q, Vogt S, Stephenson G B and Kang H C 2007 *Phys. Rev. B* **76** 115438

[10] Takenaka H, Kawamura T, Ishii Y and Asagiri S 1995 *J. Appl. Phys.* **78** 5227

[11] Maser J and Schmahl G 1992 *Opt. Commun.* **89** 355

[12] Schneider G 1997 *Appl. Phys. Lett.* **71** 2242

[13] Tsusaka Y, Yokoyama K, Takeda S, Takai K, Kagoshima Y and Matsui J 2001 *Nucl. Instrum. & Methods Phys. Res., Sect. A* **467-468** 670

[14] Koyama T, Ichimaru S, Tsuji T, Takano H, Kagoshima Y, Ohchi T and Takenaka H 2008 *Appl. Phys. Express* **1** 117003

Smart X-Ray Optics

A.G. Michette[1], S.J. Pfauntsch[1], S. Sahraei[1], M. Shand[1], G.R. Morrison[1], D. Hart[1], B. Vojnovic[1], T. Stevenson[2], W. Parkes[2], C. Dunare[2], R. Willingale[3], C. Feldman[3], T. Button[4], D. Zhang[4], D. Rodriguez-Sanmartin[4], H. Wang[5]

[1]King's College London, Department of Physics, Strand, London WC2R 2LS, UK
[2]Scottish Microelectronics Centre, University of Edinburgh, The Kings Buildings, West Mains Road, Edinburgh EH9 3JF, UK
[3] The University of Leicester, Department of Physics and Astronomy, University Road, Leicester LE1 7RH, UK
[4] The University of Birmingham, Department of Metallurgy and Materials, Edgbaston, Birmingham , B15 2TT, UK
[5]University College London, Department of Physics and Astronomy, Gower Street, London WC1E 6BT, UK

E-mail: alan.michette@kcl.ac.uk

Abstract. This paper describes reflective adaptive/active optics for applications including studies of biological radiation damage. The optics work on the polycapillary principle, but use arrays of channels in thin silicon. For optimum performance the x-rays should reflect once off a channel wall in each of two successive arrays. This reduces aberrations since then the Abbe sine condition is approximately satisfied. Adaptivity is achieved by flexing the arrays via piezo actuation, providing further aberration reduction and controllable focal length.

1. Introduction

The UK Smart X-ray Optics programme is developing advanced adaptive x-ray optics; large scale for astronomy and small scale for terrestrial applications, specifically radiation damage studies in biological systems. This paper describes optics for the latter, a Micro-Structured Optical Array (MOA), similar to but more flexible than a polycapillary optic [1]. MOAs use grazing incidence reflection to focus x-rays along aligned channels (figure 1); x-rays from a source S are reflected twice at grazing incidence, first by an unbent array, then by an array curved to a radius R by piezoelectric actuation. An image of the source is formed a distance f from the second array. Changing either array's radius of curvature changes the focal length; by varying the bending across the array, aberrations can be reduced. Arrays can be circular, for 2-D focusing, or linear, providing a line focus from a point source. Work so far has concentrated on the latter for radiobiology at 4.5 keV [2] with 10:1 aspect ratio channels in 2×2 mm arrays of 100–200 μm thick silicon and with $R = 4.7$ cm, $f = 5$ cm.

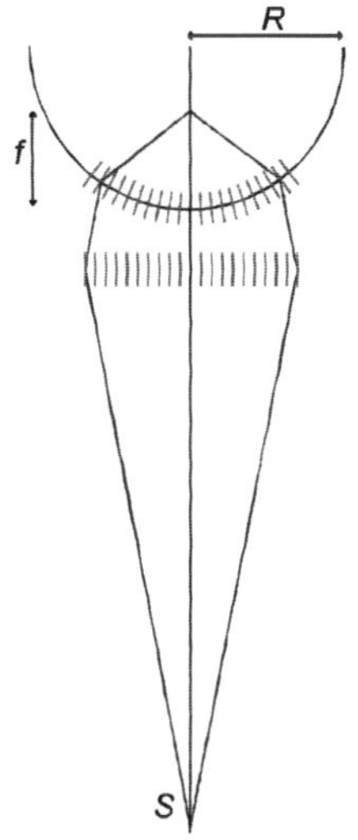

Figure 1. MOA configuration.

2. Modelling

Modelling of the MOAs has been carried out using finite element analysis (FEA), to assess the amount of bending achievable, and ray tracing, to determine the quality of the focus.

2.1. Finite Element Analysis

Array bending has been modelled by FEA using COMSOL Multiphysics™. The actuator arrangement assumed in the modelling is shown in figure 2a. Because of the symmetry the FEA could be carried out on a quarter of the chip as in figure 2b, to minimise computing overheads. The FEA indicates high stress points, as shown in figure 2c, where fracture would occur if the bending radius was too small. Note that only a small number of channels was modelled, due to computing overheads.

The modelling indicates that, for the parameters and actuator arrangement used, it is not possible to obtain the amount of bending needed. This result has been verified by Talysurf measurements on 200 μm thick silicon. The required bending can only be achieved by using very thin silicon and very thin piezo material; neither of these is technically easy but ways of overcoming the problems are currently being addressed.

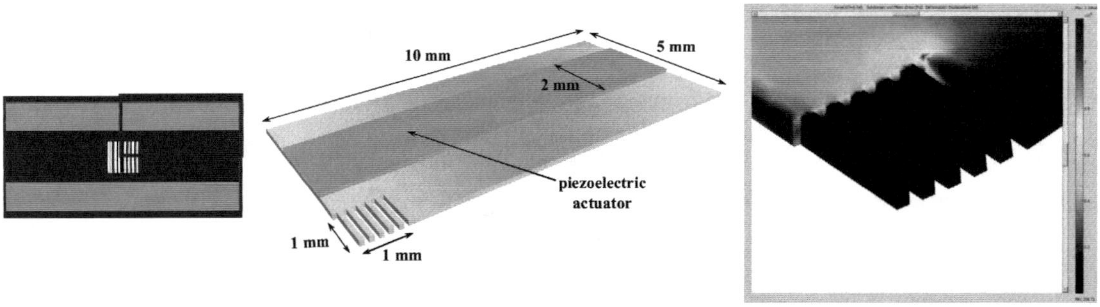

Figure 2. (a, left) The arrangement of the piezo actuators (grey) and the area modelled (rectangle). (b, centre) the geometry of the FEA model. (c, right) FEA result showing the high stress points (light).

2.2. Ray Tracing

A software package, Q, developed at the University of Leicester specifically for grazing incidence optics [1,3], is used for ray tracing the MOAs. Summarising the results briefly, with the design parameters of section 1, a focal line width of 0.3 μm is predicted for a 4.5 keV point source, with 2 μm for a 5 μm diameter source (the source size planned for radiobiology is 1 μm). Figure 3 shows the reduction in focal width using a two-reflection MOA compared to a single reflection. More significantly, for radiobiology, which requires a high throughput of samples, such a MOA should provide over two orders of magnitude focused flux than a zone plate working at the same source demagnification.

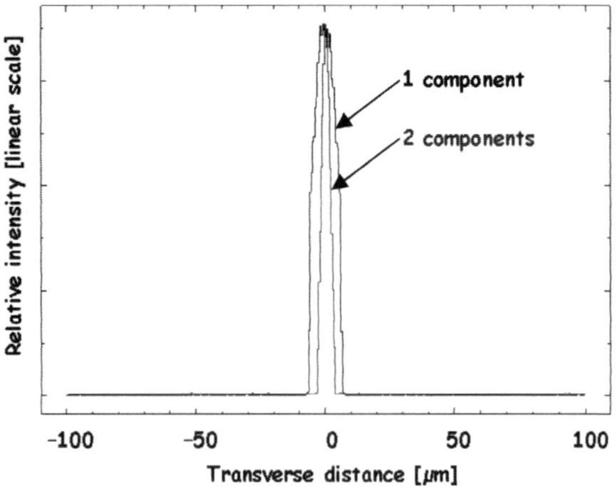

Figure 3. Comparison of the focal widths provided by single and double reflection MOAs for radiobiology.

3. Manufacturing Issues

The results of the modelling discussed above rely on the reflective channel walls being smooth; to provide the 100× advantage over a zone plate, the wall rms roughnesses must be less than about 2 nm. Initial structures were made via the Bosch process of deep silicon etching [4], which uses successive etch and passivate cycles to reduce lateral etching. However, even with post processing (oxidation and removal) it was not possible to reduce the rms roughness below ≈3 nm over small areas at the channel tops, increasing to ≈13 nm at the bottoms, where the channels were also narrower. For larger areas the roughnesses are significantly higher than these values.

Latterly the silicon etching has been done using a directional wet etch process on the <110> crystal surface [5], giving a wall roughness of 1.2 nm, independent of depth in the channel (figure 4). In addition, the channel walls show no significant curvature.

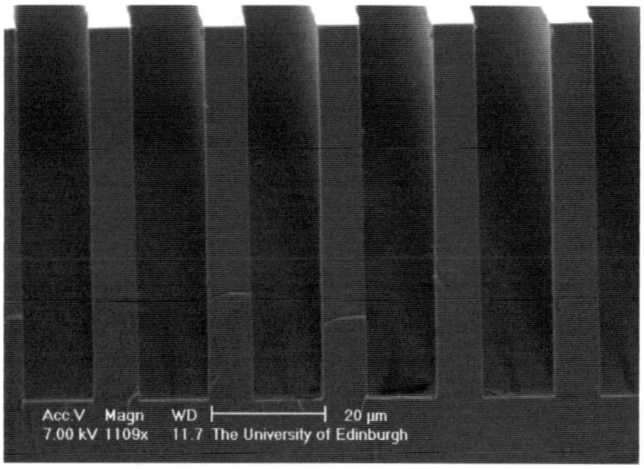

Figure 4. Straight silicon channels made by wet etching. The silicon was not etched fully to allow the structure to be cleaved for examination by electron microscopy. Note that the mark/space ratio is not quite 1:1; this will not affect the optical performance.

4. The Future

As discussed previously, a first application of MOAs will be in studies of radiation induced cancers, providing orders of magnitude more focused flux than a zone plate using a microfocus x-ray source. This improvement will result from the larger geometric aperture and the ability to focus polychromatic radiation. It will allow studies of tissue samples and effects leading to cell mutation (rather than death, which is less important in terms of cancer); neither of these have yet been possible.

The dependence of the wet etching process on crystal orientation means that currently only linear arrays can be manufactured. However, the line focus that such a MOA could provide would be useable in radiobiological studies, since cell samples could be arranged in a linear fashion, and linear irradiation of tissue samples could also be undertaken. Irrespective of that, several technical problems still need to be addressed, including (i) obtaining sufficient bending in one or both arrays, e.g., by using a different actuator arrangement, and (ii) design and manufacture of arrays for 2-D focusing to enable, e.g. irradiation of single cells and for other applications including imaging and lithography.

Acknowledgements
The Smart X-Ray Optics (SXO) consortium – University College London (including the Mullard Space Science Laboratory), King's College London (incorporating part of the former Gray Cancer Institute), the Scottish Microelectronics Centre at the University of Edinburgh, the University of Birmingham, the University of Leicester and STFC Daresbury Laboratory – is funded by the Basic Technology programme of Research Councils UK, grant number D04880X. Silson Ltd and Diamond Light Source Ltd are associate members of SXO. The work is also supported by the ESF COST Action MP0601 "Short Wavelength Laboratory Sources".

References

[1] Michette AG, Button T, Dunare C, Feldman C, Folkard M, Hart D, McFaul C, Morrison GR, Parkes W, Pfauntsch SJ, Powell AK, Rodriguez-Sanmartin D, Sahraei S, Stevenson T, Vojnovic B, Willingale R and Zhang 2007, *Proc. SPIE* **6705** 670502
[2] Folkard M, Prise KM, Michette AG and Vojnovic B 2007 *Nucl. Instrum. Methods Phys. Res.* A **580** 446–450
[3] http://www.star.le.ac.uk/~rw/q_v6/index.html. Last accessed 24 September 2008
[4] Laermer F and Schilp A 1996 US Patent 5498312
[5] Biswas K and Kal S 2006 *Microelectronics Journal* **37** 519–525

Journal of Physics: Conference Series **186** (2009) 012068 doi:10.1088/1742-6596/186/1/012068

High precision surface metrology using a phase retrieval method

Mohammed H. Modi[1,2], Pascal Mercere[1], Mourad Idir[1]

[1]Synchrotron SOLEIL, L'Orme des Merisiers, 91192 Gif Sur Yvette CEDEX France.
[2]X-ray Optics Section, Raja Ramanna Centre for Advanced Technology, Indore 452013 India.

modi.mohammed@yahoo.fr

Abstract. At wavelength metrology offers a unique opportunity to characterize ultra high quality optics of diffraction limited performance. This technique is based on a coherent interaction of incident wave fields reflected from different surface positions. The mirror surface acts as a phase object which introduces a phase error in the outgoing wave fields. Resultant aberrated wavefront modulates the intensity near the focal plane. The modulated intensity profile can be used to retrieve the mirror shape profile using a numerical phase retrieval method which is developed in the present work. This at-wavelength metrology gives a precise shape information of a mirror having a figure error < 1nm which is extremely difficult to measure with conventional metrology tools.

1. Introduction

In the last decade, a significant improvement has been made in the design of electron accelerators in order to reduce the beam emittance and to increase the photon brilliance. This has increased the coherent properties of the beam and has opened up new branches of microscopy and spectroscopy at nano meter length scales. X-ray nano probe is going to be an important tool for future research, hence numerous work have been carried out to develop a nano focusing optics [1,2] of diffraction limited performance.

To generate a diffraction limited nano meter size beam is a big challenge for an optics designer. It needs a very high quality surface with figure errors less than few nm. This essentially requires ability to selectively suppress the most harmful components of a shape error frequency from a given optics. Another challenge is metrology of aspherical shapes having a figure error < few nm over the full mirror length.

We have developed a numerical approach to derive shape errors from a single intensity distribution of x-rays diffracted by a mirror surface at a focal plane. This method is called "at-wavelength metrology" [3] which is based on coherent interaction of incident wave fields reflected from different surface positions. The mirror surface acts as a phase object which introduces a phase error in the incident coherent wave fields. The resultant wavefields at focal distance generated by coherent superposition of all secondary waves are calculated by Fresnel-Kirchhoff diffraction integral [4]. Practical realization of this technique is rather simple provided the availability of coherent x-ray beam. Nevertheless, it needs a sophisticated computational skill to retrieve the phase information from a single intensity data. This technique is free from the effects of environmental variants like temperature, ground vibration etc. It can potentially measure all general class of surfaces with comparable accuracy as achieved with interferometry based techniques.

.

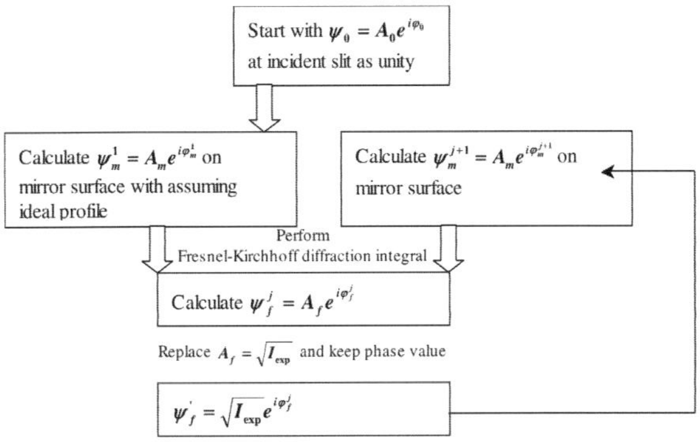

Figure 1. Flowchart of an iterative phase transforms method used in the present work.

Table 1: Parameters of elliptical mirror

Mirror Length	64mm
Focal Length	60mm
Glancing angle	3mrad
Wavelength	0.08nm

2. Phase retrieval method

In the at-wavelength metrology process, a mirror surface is required to be illuminated with a wave front of known shape. The incoming coherent wave field $\psi_0 = A_0 e^{i\varphi_0}$ incident at a distance r_0 on the mirror surface is then reflects from different surface positions with a certain aberrations introduced by shape errors. The aberrated wavefields modulate the intensity profile near the focal plane. This wavefields at a distance r_1 generated by coherent superposition of all secondary waves is calculated by Fresnel-Kirchhoff diffraction integral [3] as

$$\psi_1(r_1) = -\frac{i}{2\lambda}\int_{-\infty}^{+\infty} \psi_0(r_0) \times \frac{e^{ikr_{01}}}{r_{01}} (1 + \cos\gamma) \cdot dx$$

where $r_{01} = r_1 - r_0$, γ is oblique angle of wavefield with respect to aperture normal, k is wave vector and λ is wavelength. The wavefield error φ is connected with surface height error $\boldsymbol{h}$ in grazing incidence geometry by

$$\varphi = 2hk \cdot \sin\theta$$

The above integral formula well explains the interference fringes arising from wavefields reflected from different surface heights and gives a powerful method for figure error analysis in term of "at wavelength metrology"

Interpretation of intensity data measured near the focal plane is not obvious but needs a sophisticated algorithm of phase retrieval. Such algorithms have been developed earlier [5,6] but they are not in the form to be used for general optical metrology. In the present work a numerical algorithm has been developed to retrieve the phase from a single intensity measurement at a focal point. The method is based on the iterative phase transform scheme where the calculated intensity at focal point is replaced in each iteration by experimental values and keeps the calculated phase for next iteration. The loop works until the calculated intensity matches well with the experimentally measured values. Flow chart of the approach has been explained in the Figure 1.

3. Results and discussions

We have used an actual shape error data of a 64 mm length elliptical mirror measured with a long trace profilometer (LTP) to calculate the intensity profile according the parameters given in Table 1.

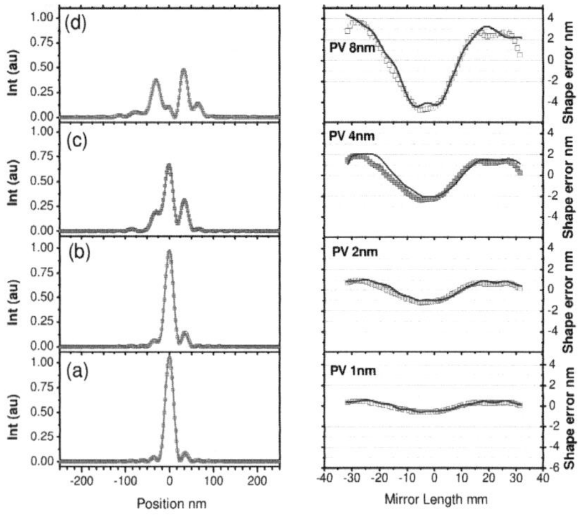

Figure 2: Calculated intensity profiles (*left*) at the focal plane are shown by open circles and corresponding shape error profiles are given in the adjacent curves (*right*) which are shown by open squares. Continuous line in the intensity plot (*left*) is obtained with phase retrieval method and corresponding retrieved shape error profiles are plotted in the adjacent graph.

To test the effectiveness of the developed phase retrieval algorithm this intensity data is used to retrieve the phase information. The intensity profile obtained at 60 mm focal distance assuming zero shape error gives a focus size of 22.86 nm FWHM.

This mirror had an actual shape error of 8 nm peak to valley (PV). The PV height error is rescaled from 1 nm to 8 nm in order to generate different intensity profiles at the focal spot. All these intensity profiles are plotted in Figure 2 (left) and corresponding figure error profiles are also given in the adjacent curves (right).

For the present configuration (see Table I), the Rayleigh criteria of $\lambda/4$ requires the figure error below 3.33 nm PV to achieve the diffraction limited focal spot. It is evident that the intensity distribution at the focal plane remains symmetric for figure error less than 3.33 nm PV. For the figure errors of 4nm and above the intensity profiles suggest a degradation of focusing performance.

To test our phase retrieval scheme the intensity profiles data corresponding to 1nm, 2nm, 4nm and 8nm PV are used to retrieve the corresponding shape error profiles. The algorithm given in above flow chart is implemented in the Labview platform, which operates in Windows environment. In this program the iterative phase transform loop executes till the Chi square value attains a defined figure of merit criterion. For all these four intensity profiles the corresponding shape errors are retrieved in 30-50 iterations. In Figure 2, retrieved intensity and shape curves are shown by continuous lines. The retrieved shape errors are in well agreement with the profiles measured with LTP. This agreement is found to be excellent. The error between true and retrieved profile is found to be less than 0.25nm or better. The results are very encouraging and efforts are underway to improve it further.

4. Conclusions

At wavelength metrology is a promising tool for characterizing optics of diffraction limited performance. The method needs just an intensity data to recover the shape error profile. The technique involves a simple instrumentational approach provided that a coherent beam on any synchrotron source is available. A sophisticated computational skill is essential for efficient phase retrieval. The technique is easy to apply on variety of surfaces and shapes.

5. References

[1] O. Hignette, P. Cloetens, G. Rostaing, P. Bernard, and C. Morawe, 2005 *Rev. Sci. Instrum.* **76** 063709.
[2] A. Snigirev, V. Kohn, I. Snigireva, and B. Lengeler, 1996 *Nature (London)* **384** 49.
[3] H. Yumoto, H. Mimura, S. Matsuyama, et al., 2006 *Rev. Sci. Instrum.* **77** 063712.
[4] M. Born and E. Wolf, Principles of Optics, 7th ed. Cambridge University Press, Cambridge 2001.
[5] J.R. Fienup, 1982 *Appl. Opt.* **21** 2758.
[6] J.R. Fienup, 1993 *Appl. Opt.* **32** 1737.

Journal of Physics: Conference Series **186** (2009) 012069 doi:10.1088/1742-6596/186/1/012069

Highly efficient 2-D nano focusing by an optical system of planar refractive lenses

V. Nazmov[a,b], E. Reznikova[a,b], A. Last[a,b], V. Saile[a,b]

[a)] Institute for Microstructure Technology, Forschungszentrum Karlsruhe GmbH, 76021 Karlsruhe, Germany
[b)] Institute for Microstructure Technology, Karlsruhe University, 76131 Karlsruhe, Germany

Corresponding author's e-mail address: nazmov@imt.fzk.de

Abstract. A lens system of two parabolic refractive X-ray lenses giving an intensity gain over 10^8 for a focal spot size below 10 nm for photon energies above 10 keV is suggested.

1. Introduction

A large number of experiments in X-ray spectroscopy and X-ray imaging with microprobes require, because of their structural complexity, dispersion and separation into voxels below 1 μm [1, 2]. This can be solved using an X-ray beam focused into a nano spot. This requires a decrease in the focal distance of the refractive lens, resulting in a reduced lens aperture, and consequently a reduced photon flux. We develop an X-ray lens system containing two refractive lenses, and providing nano focus spot and high flux to a microprobe, simultaneously.

2. The lens system arrangement

It is known from the theory of refractive optics that the effective aperture *A* for linear focusing lens increases as [3]:

$$A = 0.707\sqrt{\delta \cdot \lambda \cdot f / \beta} \tag{1}$$

where λ – wavelength and β, δ are absorption index and refractive index decrement in the index of refraction $n = 1 - \delta + i\beta$ of the lens material. This requires increasing a focal distance *f*, but at the same time, the focus spot σ will be increased as:

$$\sigma = 0.664\sqrt{\beta \cdot \lambda \cdot f / \delta} \tag{2}$$

One possible solution is to install two refractive lenses, namely the first of them downstream the X-ray beam as a condenser lens collecting the incoming radiation onto the entrance of the second lens called objective lens, as shown in Fig. 1.

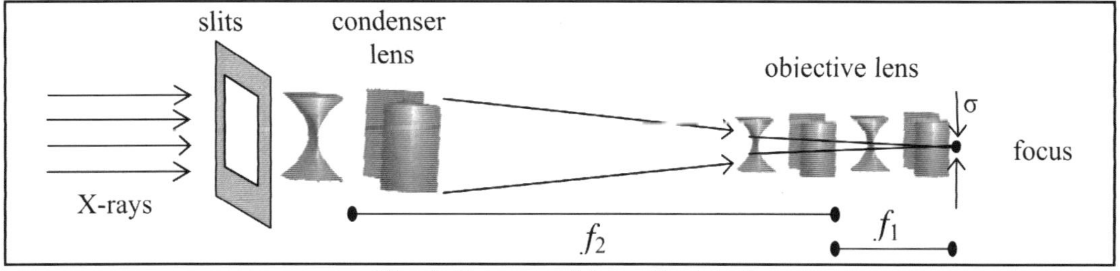

Fig. 1. Schema of the optical design.

The objective lens is positioned at the focal plane of the condenser lens and covers the length of focus waist. This waist length can achieve, for a highly coherent X-ray beam, several millimetres. The resulting intensity gain for a system of two crossed lenses in paraxial approximation is:

$$G = 0.56\frac{f_2 \cdot \delta_1 \cdot \delta_2}{f_1 \cdot \beta_1 \cdot \beta_2} \cdot \exp(-\mu_1 \cdot d_1 - \mu_2 \cdot d_2) \tag{3}$$

where μ – linear absorption coefficient, d – total length of bridges between parabolas, β_1, δ_1, μ_1 and β_2, δ_2, μ_1 belong to the condenser and objective lens material, respectively. Typically, the distance f_2 is – in accordance to the beam line setup (hutch) – between 1 meter and 4 meters. The other values are defined below.

2.1. Lens design

The condenser lens and the objective lens have been developed and manufactured at the Institute for Microstructure Technology (Karlsruhe) by means of LIGA processes including deep X-ray lithography at ANKA synchrotron source [4].

2.1.1. *Condenser lens.* To achieve higher intensity at the focus spot, an X-ray condenser lens has been developed of SU-8 polymer, showing high radiation-damage stability, Fig. 2a. Tilted in two directions i.e. +45 and – 45 degrees, the combination of two linear lenses produces a point like focus spot, Fig. 2b. As 16 lenses are fabricated on one wafer, 16 different photon energy-focal distance combinations can be realized on one wafer.

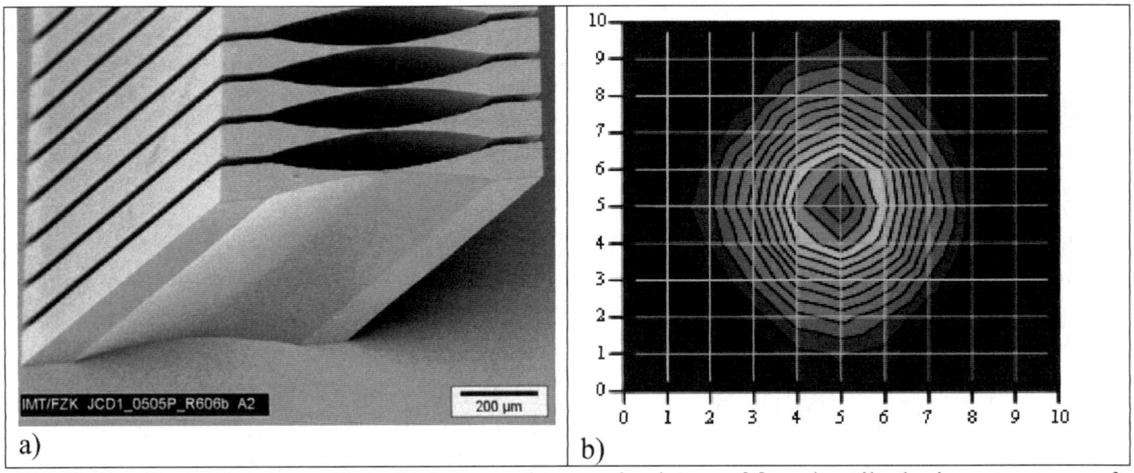

Fig. 2. Condenser lens inspection: a) electron microscopic picture of focusing tilted micro structures of one half of the condenser; b) measured focus spot on the detector screen for a photon energy of 20 keV and focal distance of 870 mm, the image field is 10 × 10 pixels, where 1 pixel = 660 × 660 nm.

2.1.2. *Objective lens.* Lenses with a very small radius of curvature and of heavy material as nickel, providing the shortest possible focal distance have been considered, Fig. 3. A short lens of SU-8 in comparison to a nickel lens becomes significantly more transparent for the outermost rays, for the energy range marked below. For this reason a prolongation of a SU-8 lens due to increasing the radius of curvature to several micrometers offers the possibility to enlarge the lens aperture by using a Fresnel-like truncated profile, Fig. 4.

The compact arrangement of focusing microstructures on a wafer provided by deep X-ray lithography gives several hundred micrometers long nickel lenses and several millimeters long SU-8 lenses. In accordance with formula (3) the resulting intensity gain lies between 10^7 and $2{\cdot}10^8$ for f_2=1 m and f_1 in the millimeter range.

Now, manufactured condenser and objective lenses have been equipped with very precise translation tool for testing.

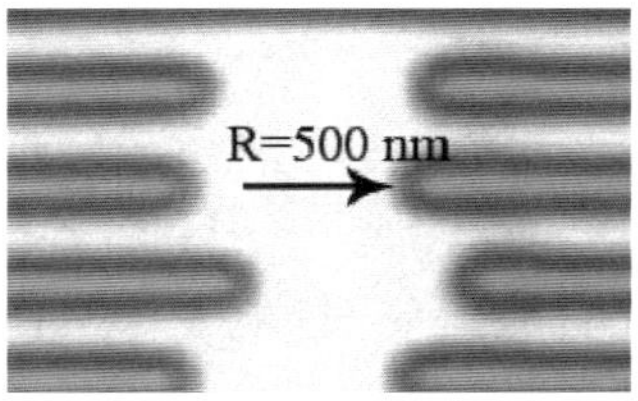

Fig. 3. Light microscopic image of the minimum radius of curvature achieved in SU-8 micro structures: this radius could be realized as radius of curvature of the last focusing elements of linear objective lenses.

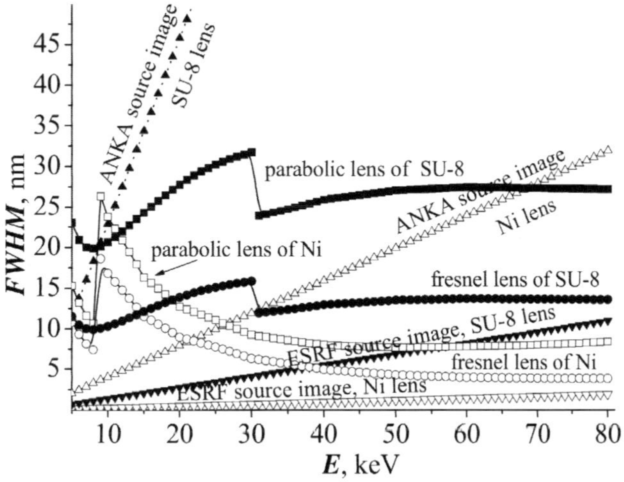

Fig. 4. Calculated diffraction limit and dimension of the demagnified source projection for ESRF (60 μm source size, 60 m long beam line) and ANKA synchrotron sources (200 μm source size, 11 m long beam line), and for an objective lens made of SU-8 respectively nickel as a function of the photon energy.

3. Summary

A new optical system of two refractive planar X-ray lenses has been described. A SU-8 polymer lens precisely structured by means of LIGA allows achieving an extremely high aperture of a condenser lens. For the focusing lens a crossed planar parabolic lens of heavy materials as nickel provides a lower diffraction limit of the spot size and high stability under the increased photon flux. The simulated gain is over one hundred million in the focus spot of several nanometers in size.

References

[1] O'Day P, Carroll S A, Bajt S 2003 Comparison of Synchrotron X-ray Fluorescence Mapping and Micro-XANES to Bulk X-Ray Absorption Spectra in Metal-Contaminated Sediments *Lawrence Livermore National Laboratory Report UCRL-ID-151516* http//www.doc.gov./bridge
[2] Hoppe P 2008 *Astronomical Society of the Pacific Conference Series* **309** 265
[3] Kohn V 2003 *Journal of Experimental and Theoretical Physics* **124** 1-13
[4] Nazmov V, Reznikova E, Somogyi A, Mohr J, Saile V 2004 *Proc. SPIE* **5539** 235-243

Journal of Physics: Conference Series **186** (2009) 012070 doi:10.1088/1742-6596/186/1/012070

Transmission hard X-ray microscope with increased view field using planar refractive objectives and condensers made of SU-8 polymer

Elena Reznikova[1], Timm Weitkamp[2], Vladimir Nazmov[1], Markus Simon[1], Arndt Last[1], Volker Saile[1]

[1] Institute of Microstructure Technology (IMT), Forschungszentrum und Universität Karlsruhe, Postfach 3640, D-76021 Karlsruhe, Germany

[2] European Synchrotron Radiation Facility (ESRF), 38043 Grenoble, France

Corresponding author's e-mail: reznikova@imt.fzk.de

Abstract: Planar X-ray refractive lenses in crossed geometry providing 2D focusing are fabricated from SU-8 polymer using tilted deep X-ray lithography. The profiles of the objective lens elements are parabolic. The lens elements for the condensers were designed with a power of the surface function of 1.5, 1.7 and 2 to vary the field of views for a transmission X-ray microscope (TXM). With these lenses a TXM was set up at the ESRF beamline BM-5 for a photon energy of 15 keV. Different test nanostructures were imaged with this TXM using in-line phase-contrast, with X-ray magnification factors of 29, demonstrated spatial detail resolution of 100 nm (half-period of lines and spaces) and exposure times around 10 s. Further improvement of a TXM based on the SU-8 lenses using refractive condensers with large apertures is considered with the aim to reduce exposure times.

1. Introduction

Since 1974 the development of TXMs based on zone plates resulted in a great progress [1]. Recently the first TXM based on X-ray refractive lenses was installed [2]. We realized this possibility due to the development of tilted deep X-ray lithography fabrication of SU-8 polymer lenses. Planar crossed SU-8 lenses are promising as they allow TXM constructions in air surrounding within a total lengths of about 2 m for large X-ray energies up to E=40 keV which could provide the smallest radiation damage for living biological cells in water solutions. With our current work we improved the resolution of a TXM based on SU-8 lenses and increased the TXM field of view.

2. Setup of TXM with SU-8 refractive lenses

The TXM (Fig. 1) was built-up at the micro optics test bench of 2 meter length at the ESRF BM5 beamline using planar X-ray refractive crossed objective and condenser lenses. To align the TXM optical elements all stages allowed 3-dimensional translations and, for both lenses, tilting around two axes was provided. In order to obtain the largest X-ray magnification factor using the optics bench and the sharp TXM images independent of a position of CCD-based detector the objects were placed in the plane of the virtual focus of the objective lens and the real focus of the condenser. The distance between the synchrotron radiation source and the condenser was 55 m. Planar sets of objective and condenser lenses (Fig. 2, 3) were made at the IMT (Forschungszentrum Karlsruhe, Germany) using lithography beamlines at the ANKA storage ring [2].

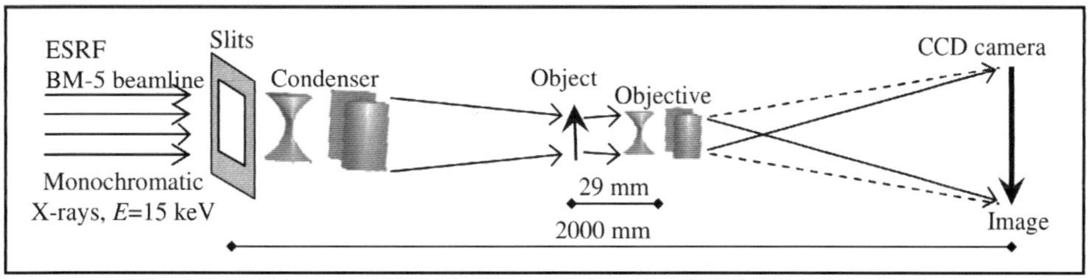

Figure 1. Scheme of TXM assembled at the ESRF BM-5 beamline.

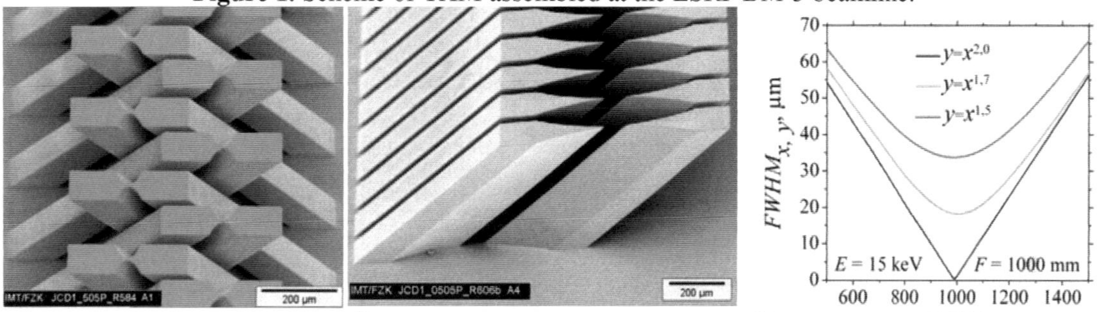

A B C

Figure 2. SEM views of elements of parabolic objective lens (A) and condenser lens (B) with surface function of $y= x^{1.5}$ and calculated $FWHM_{x,\,y}$ of focused beam along the optical axis z for continuous profiles with different function powers (C).

A crossed objective lens (Fig. 2) consists of N= 97 pairs of linear parabolic concave-concave elements focusing in horizontal (H) and vertical (V) directions. The VHVH type of the element arrangement on a Si substrate provides identical virtual and real focal distances. The lens overall length, aperture square and radius of curvature are L= 23 mm, A= 49×49 µm^2 and $R\approx$ 6 µm. For E= 15 keV, the lens focal distance F= 29 mm calculated by formulas (1), where λ is the wavelength, δ and β are the real and imaginary parts of the decrement of the refraction index n from: $n= 1-\delta-i\beta$, was equal to the measured value. The calculated diffraction limit is $R_{dif}\approx$ 30 nm. The object is illuminated by the focal spot of the condenser of VVHH type. To increase the TXM view field, which was several microns in case of the parabolic condenser, the condensers were designed with refractive surfaces fitting by lower power functions (Fig. 3). In case of $y=x^{1.5}$ condenser, its focal spot, a full width at half maximum of intensity ($FWHM_{x,\,y}$), corresponded to the objective lens acceptance.

(1) $F=\dfrac{R}{2\cdot N\cdot\delta(\lambda)}+\dfrac{L}{6}$; $R_{dif}=0.66\cdot\sqrt{\lambda\cdot F\dfrac{\delta(\lambda)}{\beta(\lambda)}}$; For SU-8 polymer: $\delta(\lambda)=1.7564\cdot10^{-6}\cdot\lambda^2$ and

$$\beta(\lambda) = \lambda\cdot10^{-10}\cdot[\mu C(\lambda)\cdot7.03+\mu O(\lambda)\cdot1.74+\mu H(\lambda)\cdot0.67+\mu Sb(\lambda)\cdot0.13+\mu F(\lambda)\cdot0.12+\mu S(\lambda)\cdot0.03]$$

where $[\lambda]$= Å, $[\mu E(\lambda)]$= cm^2/g, $\mu E(\lambda)$ are the mass absorption coefficients tabulated for the chemical element specified in the periodic table.

3. X-ray phase contrast images

Taking into account the X-ray magnification factor of 29 and the 0.7 µm pixel size of the CCD-based detector, the TXM image scale was 24 nm per pixel in case of the test structures and the Xradia scale nanostructures (Fig. 3A and 4B). A fit of the experimental data results in 85 nm half-period of lines and spaces (Fig. 4C). For the Xradia star structure, the TXM scale was 48 nm per binary pixel and the TXM resolution allowed identification of 100 nm wide line & spaces (Fig 4A). The TXM images are presented without flat-field corrections. The Modulation Transfer Functions (MTF) of the images (Fig. 4D, E) and the resolution values differed within the TXM field of view because of a non-uniform illumination of the objects by the condenser lens used in the experiment. Several non-symmetrical defect microstructures existed within the lens. Further improvements of the objective lens quality are required as well. The resolution of the TXM images was in the range from ~ 60 nm to 100 nm.

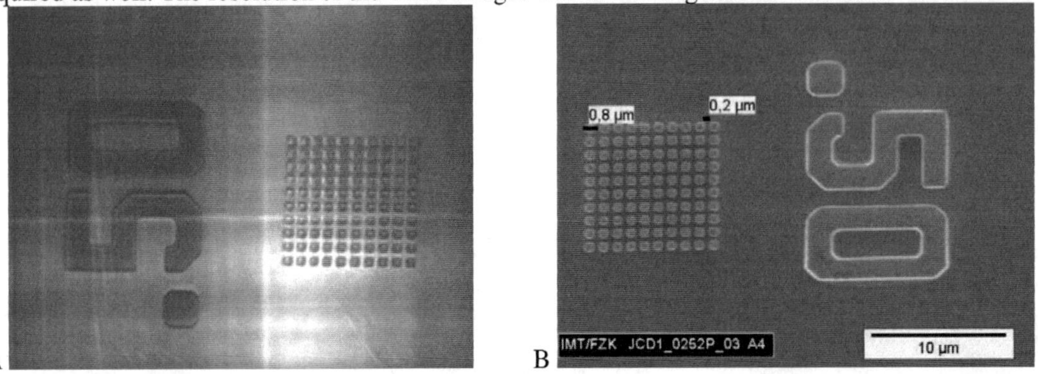

Figure 3. The TXM image of test structures recorded for ~ 10 sec in a real time (A) with the TXM field of view of ~ 100×100 μm^2 and the view of these structures by scanning electron microscope (B).

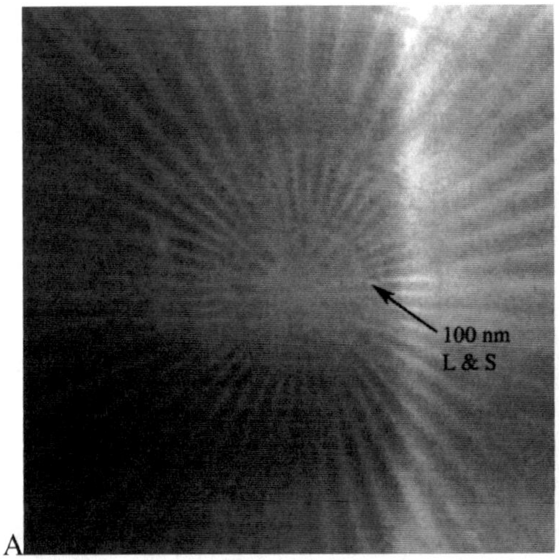

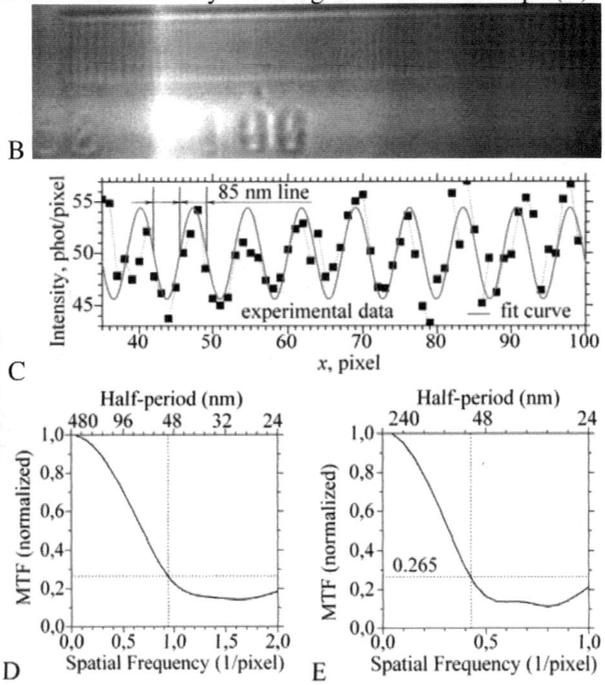

Figure 4. Fragments of TXM images of Xradia star (A) and periodical scale of nanostructures (B) with intensity profile (C) and MTF for these fragments (D, E), correspondingly.

4. Conclusion

The resolution of the TXM installed at ESRF and based on SU-8 polymer refractive lenses has been detected in the range of 100 nm line & space for the photon energy of 15 keV. An increase of the TXM field of view to an area up to ~ 100×100 μm^2 was demonstrated. For further improvements, we plan to produce condensers with large apertures due to considerable truncation of the absorbing volume of the lenses to increase the intensity of an object illumination and TXM magnification factor.

References

[1] W. Chao, B.H. Harteneck, J.A. Liddle, E.H. Anderson, D.T. Attwood. Soft X-ray Microscopy at Spatial Resolution Better than 15 nm. // Nature 435 (2005) 1210-1213.

[2] E. Reznikova, T. Weitkamp, V. Nazmov, A. Last, M. Simon, V. Saile. Investigation of phase contrast hard X-ray microscopy using planar sets of refractive crossed linear parabolic lenses made from SU-8 polymer. // Phisica Status Solidi (a) 204 (2007) 2811-2816.

Fresnel zone plates made by holography in the extreme ultraviolet region

Sankha S. Sarkar[1], Pratap K. Sahoo[1], Harun H. Solak[1, 2,*]
Christian David[1], J. Friso van der Veen[1, 3]
[1] Laboratory for Micro- and Nanotechnology, Paul Scherrer Institut, CH 5232 Villigen-PSI, Switzerland
[2] EULITHA AG, CH 5232 Villigen-PSI, Switzerland
[3] ETH Zürich, 8093 Zürich, Switzerland
*E-mail: harun.solak@psi.ch

Abstract: We present the fabrication of Fresnel zone plates (FZP) by holography that yields FZPs completely free from imperfections such as zone placement errors and finite pixel size effects found in serial writing techniques such as e-beam lithography (EBL). Our holographic scheme is based on the interference of a spherical wave and a plane wave. A partially transparent membrane containing a pinhole is illuminated by a fully coherent plane wave (λ=13.4 nm). The spherical wave, obtained by diffraction from the pinhole, interferes with the plane reference wave that is transmitted by the membrane. The interference pattern, which is a hologram of the pinhole, is recorded on a photoresist film and used to fabricate a FZP that has resolution similar to the size of the pinhole.

1. Introduction
Focussing of x-rays to the nanometer scale has drawn considerable interest in the Fresnel zone plates (FZP) as diffractive optical elements. Currently high-resolution FZPs are almost exclusively fabricated by electron beam lithography (EBL) **[1,2]**. This fabrication method is challenging due to difficulties in the placement of the pattern, the proximity effect and problems related to finite pixel-size and reproducibility. A holographic technique of ZP fabrication in the EUV regime was demonstrated by Solak et al. **[3]**, who obtained frequency-multiplied daughter-FZPs using a master made by e-beam lithography. In this paper, we demonstrate a new approach based on the use of a simple pinhole, where the potential imperfections that may arise from the e-beam writing step are completely avoided. EUV lithography enables the fabrication of sub-50 nm scale patterns fabrication and has potential for achieving higher resolutions.

2. Holographic approach
The principle of the holographic approach, as illustrated in Figure 1, is to let the diverging spherical wave obtained by diffraction from a circular aperture (pinhole) interfere with a plane wave of comparable intensity. The pinhole is fabricated on a semi-transparent membrane which also allows the plane wave to pass through, with an attenuated magnitude. The interference between these two waves gives rise to a hologram of the pinhole which is essentially a zone plate pattern. The pattern is then recorded at a certain distance *z* from the pinhole in a highly sensitive photoresist and subsequently the pattern is transferred to a suitable thin film in order to obtain a working zone plate.

3. Experimental highlights
The pinhole is fabricated **[4]** by EBL (LION LV1, Leica, Jena) on a chromium (23 nm) coated 100 nm thick silicon nitride (Si_3N_4) membrane using polymethyl methacrylate (PMMA) resist on top. Subsequently the pattern is etched (reactive ion etching, RIE) into the Cr and then etched into the membrane. Additional 160 nm of gold is deposited on the membrane which leaves the pinhole open. The diameter of the pinhole (figure 2, inset) after Au deposition is measured to be 300 nm.

The membrane containing the pinhole is used as a mask for the EUV holographic exposure which is performed at the X-ray Interference Lithography (XIL) **[5]** beamline at the Swiss Light Source (SLS). The mask is placed in the path of a spatially coherent EUV plane wave of 13.4 nm central-wavelength (transverse coherence length ~ 2.1mm) with 2.5% bandwidth. The pinhole diffracts the EUV radiation which then interferes with the attenuated plane reference wave transmitted through the rest of the membrane. At a distance of 600 μm from the pinhole mask, the resulting pattern is recorded on another Cr (23 nm) coated silicon nitride membrane (300 nm thick) in a sensitive chemically amplified EUV resist MET 2D **[6]** (Rohm & Haas). The plane wave is attenuated to 0.012% of the incident intensity by the silicon nitride membrane and the Cr and Au layers. Hence the total transmitted intensity is extremely low and the choice of highly sensitive resist is crucial for keeping the exposure time down to practicable levels. In these experiments, the exposure time was on the order of 10 minutes. The exposure system should remain stable regarding drift and vibrations during this period.

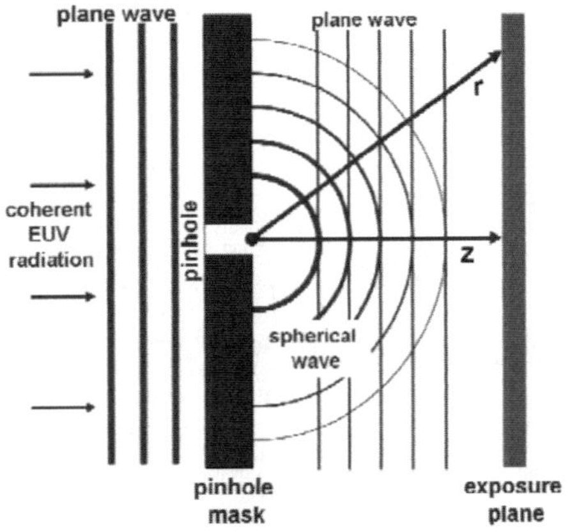

Figure 1. Schematic of the holographic technique. A spherical wave created by pinhole diffraction interferes with the attenuated plane wave partially transmitted through the pinhole mask to produce a FZP pattern at the exposure plane at a distance *z*.

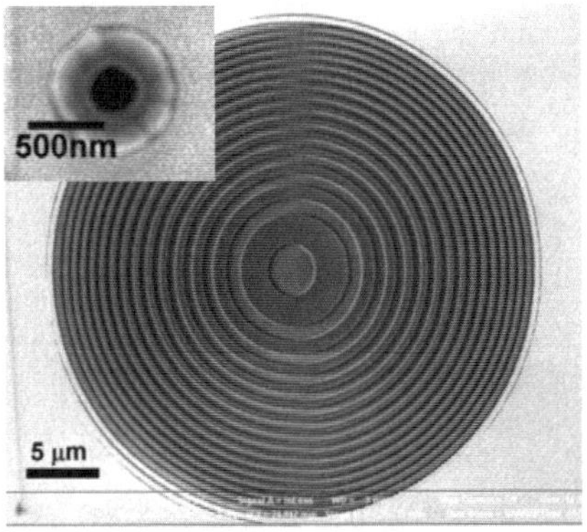

Figure 2. Zone plate fabricated with the holographic technique. The pinhole used for the holographic fabrication is shown in the inset.

The holographic pattern is transferred into the Cr and then etched (RIE) into the Si_3N_4 membrane. A scanning electron microscopy (SEM) image of the structure is shown in figure 2. The diameter of the FZP is 32 μm and the number of zones is 30 (the outermost smallest feature being 260 nm). The micrograph shows an apparent discontinuity (or phase shift) in the zones along a vertical line that goes across the zones, especially in the top half of the picture. We have determined this to be an artefact due to a dynamical charging effect on the insulating Si_3N_4 membrane, as the electron beam of the SEM is scanned across the structure.

4. Performance test:

The focal spot size of the zone plate is measured with a knife edge scan at the same beamline at the same wavelength (13.4 nm) as the holographic exposure which was used to create the FZP. The measured knife edge scan profile is presented in figure 3 and compared with theoretical integrated intensity profile at the focal plane of a lens of same numerical aperture as of the zone plate. The focal spot size is measured to be 300 nm (distance between 90% to 10% of the intensity profile **[7]**) which is comparable to both the outermost zone width (260 nm) and the diameter of the pinhole (300 nm). The good agreement between the theoretically calculated and experimentally measured curves is a testament to the aberration-free nature of the holographically made FZP.

5. Deviation from an ideal zone plate:

The diffraction from a non-ideal pinhole (due to finite thickness and diameter) is studied using a scheme similar to one discussed by Goldberg *et al* **[8]**. The exit field diffracted from a circular cylindrical hole (of 300 nm diameter) embedded in the mask material is simulated by a Finite Difference Time Domain (FDTD) algorithm using the EMExplorer® software package **[9]** and then propagated to the exposure plane by numerically evaluating the Huygens-Fresnel diffraction integral **[10]**. The deviation of the diffracted phase front from an ideal spherical wave (shown in figure 4) shows that the phase differs by $\pi/4$ at a numerical aperture (NA) of 0.06. This corresponds to the NA for the 1st Airy minimum of the intensity diffracted from an ideal pinhole of 300 nm diameter. This limit determines the radial region in which a zone plate pattern could be recorded **[11]**. When the effective zone plate diameter is limited by the 1st minimum of the Airy pattern, the minimum outermost zone width that can be obtained by this technique is given by the pinhole diameter divided by 2.4.

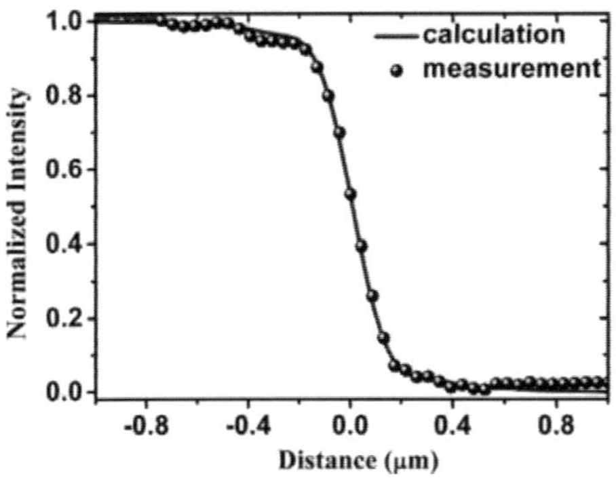

Figure 3. Knife-edge scan profile across the zone plate focal spot. The solid circles represent the measured profile while the solid line represents the calculated theoretical behaviour.

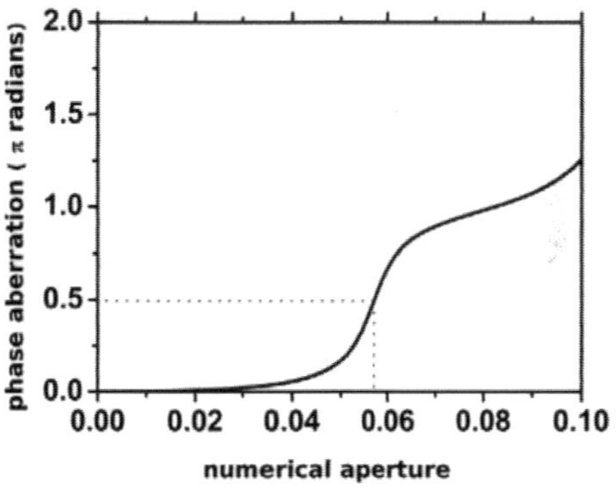

Figure 4. The phase deviation from an ideal spherical wave for a 300 nm diameter pinhole. The straight lines indicate the maximum allowable deviation ($\pi/4$) at NA ~ 0.06

In conclusion, FZPs have been fabricated holographically, by recording the interference pattern formed by a spherical wave and plane wave. The measured performance agrees well with theoretical calculations. Fabrication using EUV light offers the potential for sub-50 nm resolution owing to the short wavelength and the absence of a photoelectron-induced proximity effect in this photon energy range.

The authors would like to thank Swiss National Science Foundation (project: 200021-109678) for financial support. The holographic exposures and the measurements have been performed at the Swiss Light Source, Paul Scherrer Institut, Villigen, Switzerland.

References

[1] W. Chao, B. D. Harteneck, J. A. Liddle, E. H. Anderson, D. T. Attwood 2005 *Nature (London)*, **435**, 1210
[2] K. Jefimovs, J. Vila-Comamala, T. Pilvi, J. Raabe, M. Ritala, C. David 2007 *Phys. Rev. Lett.* **99**, 264801
[3] H. H. Solak, C. David, J. Gobrecht 2004 *Appl. Phys. Lett.* **85**, 2700
[4] S. S. Sarkar, P. K. Sahoo, H H Solak, C. David and J. F. van der Veen 2008 *J. Vac. Sci. Tech. B* (accepted)
[5] H. H. Solak 2006 *J. Phys. D: Appl. Phys.* **39**, R171-R188
[6] J. W. Thackeray et. al. 2007 *J. Photopolym. Sci. Technol.* **20** (3), 411-418
[7] D. Atwood 2000 *Soft X-rays and Extreme Ultraviolet radiation: Principles and Applications*, Cambridge University Press, United Kingdom, p.359
[8] K. A. Goldberg, E. Tejnil, and J. Bokor 1996 *OSA Trends in Optics and Photonics Vol. 4, Extreme Ultraviolet Lithography*, G. D. Kubiak and D. R. Kania (eds.) Washington, D.C.: *Optical Society of America*, 13137.
[9] URL: *http://www.emexplorer.net*
[10] J. W. Goodman 2005 *Introduction to Fourier Optics* (Roberts & Co., Colorado)
[11]M. Young 1972 *J. Opt. Soc. Am.*, **62** (8), 972 (1972)

High energy X-ray nanofocusing by silicon planar lenses

A. Snigirev[1], I. Snigireva[1], M. Grigoriev[2], V. Yunkin[2], M. Di Michiel[1], G. Vaughan[1], V. Kohn[3], S. Kuznetsov[2]
[1] ESRF, 38043 Grenoble, France
[2] IMT RAS, 142432 Chernogolovka, Moscow region, Russia
[3] Kurchatov Institute, 123182 Moscow, Russia

E-mail: snigirev@esrf.fr

Abstract. Optimizing the lens design and improving the technological process, we manufactured X-ray planar compound refractive lenses with vertical sidewalls up to 70 microns deep. The lens surface roughness in the order of 20 nm was attained. The minimal thickness of the material between two individual lenses of 2 μm was realized. The optical tests of the new planar lenses were performed at the ESRF BM05 and ID15 beamlines. The technological breakthrough allows reaching the nanometer focusing. The resolution below 200 nm was measured in the energy region of 15-80 keV. The best resolution of 150 nm was demonstrated at 50 keV X-rays.

1. Introduction

In the recent years there is a great demand for nanometer X-ray beams at energies higher then 20 keV. A number of new applications such as surface and interface scattering, high pressure magnetic scattering, and depth strain analysis using powder micro diffraction were developed. It turned out that X-ray planar refractive lenses made by microfabrication technology are very promising candidates for nanoprobe applications [1-6]. Si microfabrication technologies are well developed and fabrication of lenses with very small parabola radius is straightforward. Microelectronics technology allows to make lens-chips consisting of different integrated lens systems (ILS) with dozens of parallel compound refractive lenses [5-6]. The ILS was proposed in order to simplify the use of refractive lenses in the energy tunable experiments. The design feature of the ILS is that it consists of arrays of refractive lenses aligned in parallel. All lenses in ILS have fixed focal distance that was achieved by varying the numbers of individual lenses in each compound refractive lens. Each lens array is optimized for certain energy but number of arrays taken together allows to cover considerable energy range. To choose the desirable working energy, one can switch from one lens array to another by parallel displacement of ILS in vertical direction.

2. Lens design and manufacturing

The lenses were manufactured by microfabrication process consisting of two main steps, i.e. pattern generation in the hard mask on the silicon surface and pattern transfer into the silicon. The pattern generation was performed by photolithography. A SiO_2 layer (1 μm thick) was used as a highly selective hard mask for pattern transfer into the silicon substrate. The etch process was based on the time multiplexed etching technique, the Bosch process, where the sidewall passivation layer is

Journal of Physics: Conference Series **186** (2009) 012072 doi:10.1088/1742-6596/186/1/012072

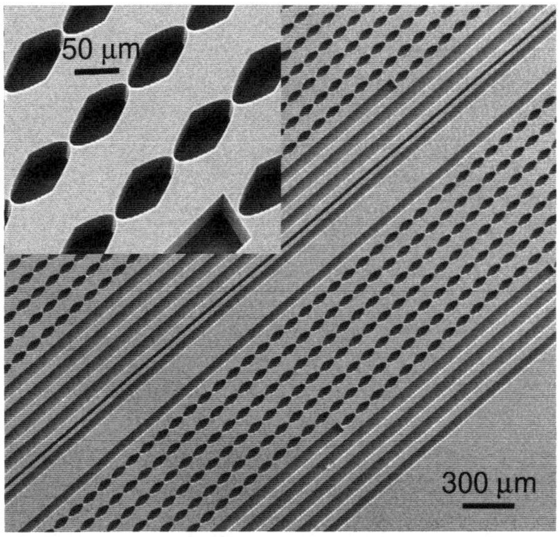

Figure 1. SEM image of ILS and individual lenses (insert)

deposited in a separate step from silicon etching. C_4F_8 and SF_6 were used as the passivation and etch gases, respectively. The mask undercut was regulated by polymer deposition and for some lenses corrections for undercut were done. The scalloping amplitude and pitch were controlled by duration of etching and passivation steps. To reduce the scalloping amplitude and mask edge originated roughness some chips after deep etching were thermally oxidized and then grown SiO_2 was removed in etching solution.

The current ILS design comprises 10 lens arrays optimized for energy range from 10 to 55 keV with the step of 5 keV. Each of these lenses has focal distance of 10cm. The length of the individual lens was 100 µm and their aperture was 50 µm. The web thickness of individual lenses on the optical axis is 2µm. The lens surface roughness was attained in the order of 20 nm. The structures are 50 µm deep. Parameters of the nanolenses are summarized in Table 1. In parallel with ILS, the test structures like bi-lenses are implemented on the Si chip. They can be used for the diagnostics of the beam and verification the focusing properties of nanolenses. SEM image of nanolenses is shown in Fig. 1.

Table 1. Lens parameters

	Energy, keV	Focal distance, mm	Focal depth, mm	Number of lenses	Total lens length, µm	Effect. aperture, µm
1	10	106.6	5.15	12	620	23.38
2	15	103.5	2.33	28	1436	27.58
3	20	99.3	1.3	52	2660	30.63
4	25	101.0	0.89	80	4088	33.71
5	30	100.5	0.65	116	5924	35.69
6	35	99.2	0.52	160	8168	3678
7	40	99.7	0.43	208	10616	37.73
8	45	99.5	0.38	264	13472	38.06
9	50	100.1	0.34	324	16532	38.15
10	55	100.1	0.32	392	20000	37.84

3. Experiment and discussion

Focusing properties of nanolenses were studied at the ESRF BM5 (Micro optics test bench end station) and ID15 beamlines. The complete energy tunability test and the influence of fabrication drawbacks or inaccuracies on focusing properties were performed at the BM5 beamline using X-ray energies from 10 to 30 keV. The focal properties of nanofocusing refractive lenses at energies higher then 50 keV were studied at ID 15 beamline.

At BM05 the lateral size of the microbeam was measured by scanning an Au wire of 200 µm in diameter across the beam vertically, recording the transmission radiation by a PIN-diode detector. All lenses on the chip have shown the focal spot (FWHM) around 200-300 nm. All compound refractive lenses in the ILS focus X-rays at designed distance with accuracy better then 2% when changing the X-ray energy. In order to study the influence of fabrication inaccuracies on focusing properties of nanolenses, we have tested lenses on three Si chips varying the scalloping amplitude and pitch. The measured optical properties of lenses on these chips are presented in Table 2. The best resolution (FWHM) and gain in flux density were measured for the lenses with 20 nm roughness.

Table 2. Summary of the experimental results at BM05

Chip	Energy, keV	Focal distance, mm (theoretical)	Focal distance, mm (measured)	Focal spot FWHM, nm	Transmission, %	Gain
1[a]	15.0	25	26.75	240	20	22.7
	17.5	34	35.50	280	27.1	21.7
2[b]	15.0	25	25.00	180	16.7	25.8
	17.5	34	33.75	220	24.8	35
3[c]	15.0	25	25.00	170	18.6	25.2
	17.5	34	34.00	140	24.6	30

[a] Scalloping amplitude 30 nm and scalloping pitch 400nm; [b] Scalloping amplitude 25 nm and scalloping pitch 250nm; [a] Scalloping amplitude 20 nm and scalloping pitch 400nm

At high energies the lateral size of the microbeam was measured by scanning a 35μm thick Au knife-edge across the beam vertically, recording the intensity by PIN diode detector. Results of the measurements are summarized in Table 3. We would like to stress that 150 nm is the best ever measured focal spot at 50 keV.

Table 3. Summary of the experimental results at ID15

Energy, keV	Focal distance, mm	Number of lenses	Focal spot FWHM, nm (theory)	Focal spot FWHM, nm (experimental)
50	106	324	70	150
60	124	392	80	250
70	172	392	110	250
80	225	392	150	200

4. Conclusion

We have designed, manufactured and experimentally tested the planar refractive lenses for hard X-ray nanofocusing. Planar refractive lenses were designed as Integrated Lens System which combines the advantages of compound refractive optics with the precise alignment possibilities of planar fabrication technology. The lens optical properties as size of the focal spot, gain were performed at ESRF BM05 and ID15 beamlines. The resolution below 200 nm was measured in the energy region of 15-80 keV. The resolution of 150 nm was demonstrated at 50 keV energy, to our knowledge this is the best measured resolution within high energy X-rays. As a next step we would like to realize two-dimensional focusing with a specially developed nanopositioning system. In the future we would like to optimize the lens design in the way to have better focal distance matching for vertically and horizontally focusing lenses.

5. References

[1] Snigirev A, Kohn V, Snigireva I, Lengeler B 1996 *Nature* **384** 49
[2] Aristov V, Grigoriev M, Kuznetsov S, Shabelnikov L, Yunkin V, Weitkamp T, Rau C, Snigireva I, Snigirev A, Hoffmann M, Voges E 2000 *Appl. Phys. Lett.* **77** 4058
[3] Schroer CG, Kurapova O, Patommel J, Boye P, Feldkamp J, Lengeler B, Burghammer M, Riekel C, Vincze L, van der Hart A, Kuchler M 2005 *Appl. Phys. Lett.* **87** 124103
[4] Snigireva I, Snigirev A, Rau C, Weitkamp T, Aristov V, Grigoriev M, Kuznetsov S, Shabelnikov L, Yunkin V, Hoffmann M, Voges E 2001 *Nucl.Instr. & Meth.s* **A 467-468** 982
[5] Snigireva I, Snigirev A, Yunkin V, Drakopoulos M, Grigoriev M, Kuznetsov S, Chukalina M, Hoffmann M, Nuesse D, Voges E 2004 *AIP conference proceedings*, **705**, 708
[6] Snigirev A, Snigireva I, Grigoriev M, Yunkin V, Di Michiel M, Kuznetsov S, Vaughan G 2007 *Proc. of SPIE* **6705**, 670506-1

High energy X-ray Transfocator based on Al parabolic refractive lenses for focusing and collimation

A. Snigirev, I. Snigireva, G. Vaughan, J. Wright, M. Rossat, A. Bytchkov, C. Curfs
ESRF, 38043 Grenoble, France

E-mail: snigirev@esrf.fr

Abstract. We describe an X-ray transfocator - a lens assembly whose focal length can be continuously adjusted by the mechanical movement of one or more groups of individual lenses. The transfocator consists of 7 cylindrical cartridges containing 2, 4, 8, 16, 32, 64 and 128 Al parabolic lenses respectively. Pneumatic actuators are used to bring cartridges in and out of the beam independently providing collimation or focusing at a desired distance downstream. The transfocator is installed in the ID11 extension beamline at 90 m from the source. The tests were performed at the energy range between 25 and 60 keV X-ray energy.

1. Introduction

Refractive lenses are becoming standard elements in synchrotron beamline instrumentation enabling the focusing of high energy radiation from micro- to nano- meter dimensions [1-3]. These devices are simple to align and allow a good working distance between the optics and the sample. Unfortunately, since the index of refraction depends on the X-ray energy, refractive lenses are chromatic and that substantially limits their applicability. Therefore there is a strong demand for the development of energy tunable refractive optical systems when X-ray energy can be tuned in a reasonable range without changing the beam position on the sample.

Recently, in order to simplify the use of refractive lenses in the energy tunable experiments refractive lenses as Integrated Lens Systems were proposed (ILS) [4-5]. The design feature of the ILS is that it consists of arrays of refractive lenses aligned in parallel. All lens arrays in ILS have fixed focal distance that was achieved by varying the numbers of individual lenses in each compound refractive lens. Each lens array is optimized for certain energy but number of arrays taken together allows to cover considerable energy range. To choose the desirable working energy, one can switch from one lens array to another by parallel displacement of ILS in vertical direction.

In this paper for energy tunable applications we propose an X-ray transfocator. This device comprises a lens assembly whose focal distance can be continuously adjusted by the mechanical movement of one or more groups of individual lenses.

2. Transfocator design

The transfocator consists of 7 cylindrical cartridges containing 2, 4, 8, 16, 32, 64 and 128 Al parabolic lenses respectively. Pneumatic actuators are used to bring cartridges in and out of the beam independently changing the number of lenses from 2 to 254 with 2 lenses step. Pneumatic actuators (Festo ref. DSNU-10) have a stroke of 20 mm driven by bi-stable solenoid valve (WIT). Lens cartridge is made of aluminum and has 12mm (10 μm tolerance) and 40 mm (12 μm tolerance) internal and external diameter respectively.

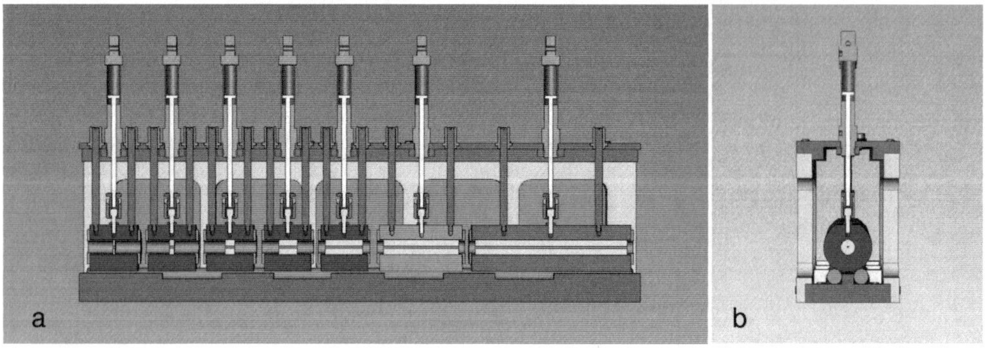

Figure 1. The X-ray transfocator assembly (a) and cross section sketch (b).

The individual lenses are made form polycrystalline aluminum by pressing technique. The aperture of the lens is 1mm; the web thickness is around 50 microns. The radius of parabola apex of the individual lens is 200 microns. The aluminum paraboloids are held and centered in the brass frame of 12 mm in diameter. The thickness of the lens frame is 1 mm.

2-lens-step design corresponds to the change of the focal distance in the order of 60 mm and 15 mm at 20 keV and 50 keV respectively. If focal distance is fixed, the 2 lens step corresponds to the energy change of 300 eV and 100 eV at 20 keV and 50 keV respectively.

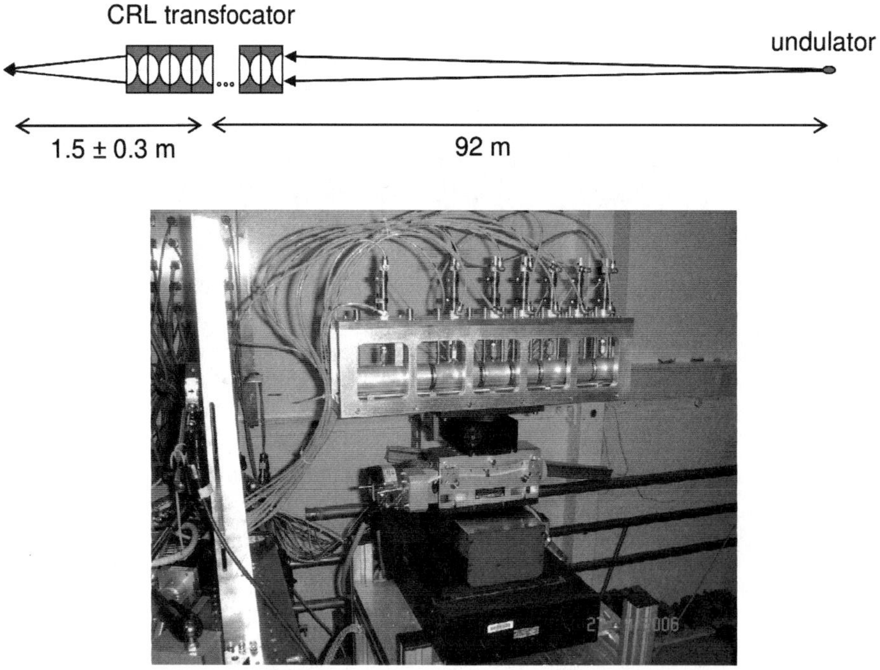

Figure 2. Sketch of the experimental set-up (top) and photo of the X-rat transfocator at the beamline.

3. Experimental test

Experimental test of transfocator was performed at ID 11 ESRF beamline at the energy range between 25 and 60 keV. The transfocator was placed at 92 m from the source. It was mounted on the two translation (Z,Y), one rotation and one tilt stages for lens alignment in the incoming beam. Figure 3a shows the theoretical verses the measured effective aperture when inserting the different cartridges into the beam, measured by the output flux. The near perfect match indicates that the different cartridges are aligned within the necessary tolerance. Figure 3b depicts the dependence of effective aperture versus energy when inserting the lens cartridges keeping the focal distance about 1.5 m.

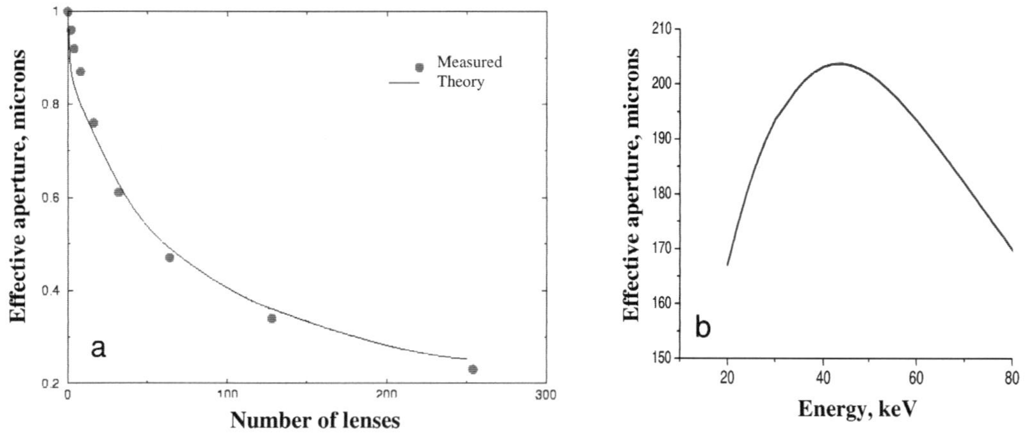

Figure 3. Measured and theoretical performance of transfocator: (a) effective aperture versus number of lenses at 50 keV (calculated and measured) and (b) effective aperture (mm) versus energy at 1.5 m focal length (calculations).

4. Conclusion

The transfocator functioned perfectly within specifications and produced an easily tunable ~micron spot with long term stability. Following this successful development the transfocators with changeable number of lenses are being installed at different beamlines at the ESRF. An in vacuum version is currently under design. By placing such a device near the front end, we will obtain large gains in flux and/or energy resolution, by collimating or condensing the primary beam.

5. References

[1] Snigirev A, Kohn V, Snigireva I, Lengeler B 1996 *Nature* **384** 49

[2] Lengeler B, Schroer C, Tummler J, Benner B, Richwin M, Snigirev A, Snigireva I, Drakopoulos M 1999 *J. Synchrotron Rad..* **6** 1153

[3] Lengeler B, Schroer C, Richwin M, Tummler J, Drakopoulos M, Snigirev A, Snigireva I 1999 *Appl. Phys. Lett.* **74** 3924

[4] Snigireva I, Snigirev A, Yunkin V, Drakopoulos M, Grigoriev M, Kuznetsov S, Chukalina M, Hoffmann M, Nuesse D, Voges E 2004 *AIP conference proceedings*, **705**, 708

[5] Snigirev A, Snigireva I, Grigoriev M, Yunkin V, Di Michiel M, Kuznetsov S, Vaughan G 2007 *Proc. of SPIE* **6705**, 670506-1

Mo/Si and $MoSi_2$/Si nanostructures for multilayer Laue lens

H Takenaka[1], S Ichimaru[1], T Ohchi[1], T Koyama[2], T Tsuji[2], H Takano[2] and Y Kagoshima[2]

[1]NTT-AT Nanofabrication Corporation, 3-1 Wakamiya, Morinosato, Atsugi, Kanagawa, 243-0124, Japan
[2]Graduate School of Material Science, University of Hyogo, 3-2-1 Kouto, Kamigori, Ako, Hyogo 678-1297, Japan

E-mail: hisataka.takenaka@nanofab.ntt-at.co.jp

Abstract. To develop a multilayer Laue lens (MLL), we fabricated depth-graded Mo/Si and $MoSi_2$/Si multilayers with each boundary according to the Fresnel zone configuration. The multilayers were deposited by magnetron sputtering. From the result of SEM image analysis of the multilayer cross sections, $MoSi_2$/Si multilayer had smaller layer-thickness errors than Mo/Si multilayer. In addition, from the result of the focusing test by using 20-keV X-rays, the measured beam size of $MoSi_2$/Si MLL had a small blurring from the diffraction limited beam size. These results suggest that $MoSi_2$/Si multilayer is better suited than Mo/Si multilayer for use as an MLL in hard x-ray nanofocusing.

1. Introduction

Nanometer-sized x-ray beams are desired for x-ray microanalysis and x-ray microscopy at a nanometer scale. For this purpose, development of x-ray optical devices is advancing rapidly in recent years. Focused spot sizes of a few tens of nanometers have been obtained using Kirkpatrick-Baez mirrors [1,2], compound refractive lenses [3], Fresnel zone plates [4-6], and multilayer Laue lens (MLL) [7-9]. MLL was pioneered by the APS group. The MLL consists of multilayer structures and uses the Bragg diffraction effect to improve spatial resolution and diffraction efficiency; thus, this technique appears promising with regard to the achievement of a 1-nm focus with very high diffraction efficiency, even in the hard x-ray region [10]. Their MLL is composed of WSi_2 and Si layers [8]. We recently designed and fabricated an MLL by magnetron sputtering, using Mo/Si and $MoSi_2$/Si as the multilayer materials by their superior properties of high diffraction efficiency. In particular, $MoSi_2$/Si is a likely candidate for good multilayer system because of its excellent thermal properties and sharp interfaces between the Mo-silicide layers and the Si layers [11]. In this paper, we report results of SEM image analysis of these multilayer nanostructures and those of the focusing test for hard x-ray nanofocusing.

2. Fabrication

A magnetron sputtering system was used to fabricate the Mo/Si and $MoSi_2$/Si depth-graded multilayers [12] with each boundary r_n according to the Fresnel zone configuration which is given by $r_n^2 = nf\lambda + n^2\lambda^2/4$, where λ is the wavelength and f is the focal length. The outermost zones were deposited first onto the flat Si substrate to get smooth interfaces around the thinner layer thickness, that is, the layers increased in layer-thickness from the bottom to the surface of these multilayers. In the Mo/Si ($MoSi_2$/Si) multilayer, the minimum layer thickness was designed to be 15 nm (10 nm), while the maximum thickness was designed to be 83 nm (100 nm). The total multilayer thickness was 4.2 μm (4.6 μm), and the number of layers was 165 (248). The focal length was designed to be 2.4 mm (1.6 mm) for an x-ray energy of 20 keV. The multilayers were subsequently sectioned and polished to appropriate depth.

3. Characterization by SEM observations

Figure 1 and figure 2 show the cross sectional SEM images of Mo/Si and $MoSi_2$/Si MLL structures and the measured values of layer thickness versus layer numbers. Each layer thickness was measured by SEM image analysis to compare with the designed layer thickness. The difference from the designed layer thickness was 9.2% RMS in Mo/Si multilayer and 4.0% RMS in $MoSi_2$/Si multilayer. $MoSi_2$/Si multilayer has a smaller layer-thickness error than Mo/Si multilayer in spite of the thinner design thickness. The large layer-thickness disturbance in the Mo/Si multilayer is mainly due to the large interdiffusion and aggregation between the Mo-layers and Si-layers. The shaper interface in $MoSi_2$/Si multilayer is mainly owing to the non-reactive nature of $MoSi_2$ with Si.

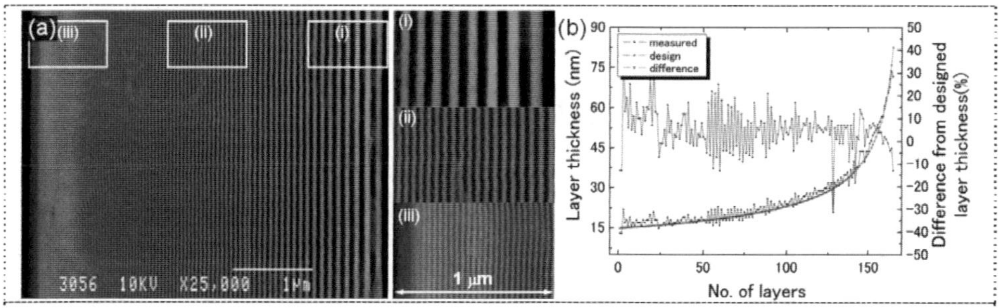

Figure 1. (a) Cross sectional SEM image of a Mo/Si MLL structure. (b) Measured layer thicknesses as a function of layer number obtained from the SEM image. Right vertical axis indicates a difference from a designed layer thickness. RMS value of the difference was 9.2%.

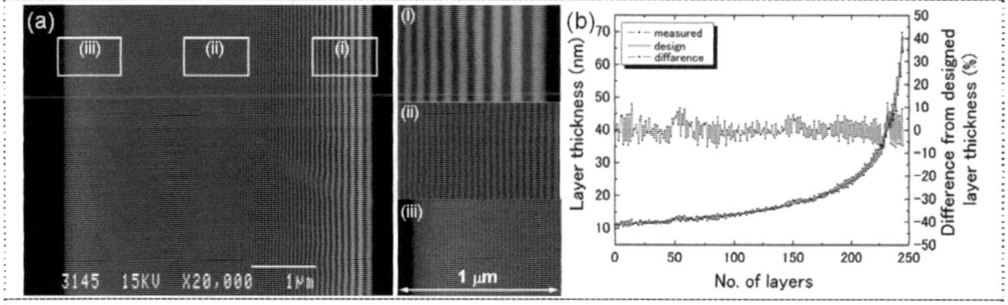

Figure 2. (a) Cross sectional SEM image of a $MoSi_2$/Si MLL structure. (b) Measured layer thicknesses as a function of layer number obtained from the SEM image. Right vertical axis indicates a difference from a designed layer thickness. RMS value of the difference was 4.0%.

Focusing test was performed at Hyogo-ID beamline (BL24XU) [13] of synchrotron radiation facility SPring-8 with 20-keV x-rays. The obtained beam sizes were 50.2 ± 3.9 nm and 28.2 ± 2.7 nm in the full width at half maximum, while the diffraction limited sizes are 36 nm and 22 nm for the Mo/Si and $MoSi_2$/Si MLL, respectively. Ten measurements were performed in order to confirm the reproducibility. The measured beam size of $MoSi_2$/Si MLL had a small blurring from the diffraction limited beam size in comparison with Mo/Si MLL.

4. Conclusion

To develop an MLL, we fabricated Mo/Si and $MoSi_2$/Si depth-graded multilayers with each boundary according to the Fresnel zone configuration. The multilayers were deposited by a magnetron sputtering. From the result of SEM image analysis of the multilayer cross sections, $MoSi_2$/Si multilayer had smaller layer-thickness errors than Mo/Si multilayer. In addition, from the result of the focusing test by using synchrotron radiation, the measured beam size of $MoSi_2$/Si MLL had a small blurring from the diffraction limited beam size. These results suggest that $MoSi_2$/Si multilayer is better suited than Mo/Si multilayer for use as an MLL in hard x-ray nanofocusing.

Acknowledgments

This development was supported by SENTAN, JST.

References

[1] Mimura H, Yumoto H, Matsuyama S, Sano Y, Yamamura K, Mori Y, Yabashi M, Nishino Y, Tamasaku K, Ishikawa T and Yamauchi K 2007 *Appl. Phys. Lett.* **90** 051903
[2] Morawe Ch, Hignette O, Cloetens P, Ludwig W, Borel Ch, Bernard P, and Rommeveaux A, 2006 *Proc. SPIE* **6317** 63170F-1
[3] Schroer C G, Kurapova O, Patommel J, Boye P, Feldkamp J, Lengeler B, Burghammer M, Riekel C, Vincze L, van der Hart A and Küchler M 2005 *Appl. Phys. Lett.* **87** 124103
[4] Chao W, Harteneck B D, Liddle J A, Anderson E H and Attwood D T 2005 *Nature* **435** 1210
[5] Suzuki Y, Takeuchi A, Takano H and Takenaka H 2005 *Jpn. J. Appl. Phys.* **44** 1994
[6] Yin G C, Song Y F, Tang M T, Chen F R, Liang K S, Duewer F W, Feser M, Yun W and Shieh H P D 2006 *Appl. Phys. Lett.* **89** 221122
[7] Maser J, Stephenson G B, Vogt S, Yun W, Macrander A, Kang H C, Liu C and Conley R 2004 *Proc. SPIE* **5539** 185
[8] Kang H C, Maser J, Stephenson G B, Liu C, Conley R, Macrander A T and Vogt S 2006 *Phys. Rev. Lett.* **96** 127401
[9] Kang H C, Yan H, Winarski R P, Holt M V, Maser J, Liu C, Conley R, Vogt S, Macrander A. T and Stephenson G B 2008 *Appl. Phys. Lett.* **92** 221114
[10] Yan H, Maser J, Macrander A, Shen Q, Vogt S, Stephenson G B and Kang H C 2007 *Phys. Rev. B* **76** 115438
[11] Takenaka H, Kawamura T, Ishii Y and Asagiri S 1995 *J. Appl. Phys.* **78** 5227
[12] Takenaka H and Ishii Y 1990 *Proc. SPIE* **1345** 180
[13] Tsusaka Y, Yokoyama K, Takeda S, Takai K, Kagoshima Y and Matsui J 2001 *Nucl. Instrum. & Methods Phys. Res., Sect. A* **467-468** 670

Journal of Physics: Conference Series **186** (2009) 012075 doi:10.1088/1742 6596/186/1/012075

Quasi-kinoform Type Multilayer Zone Plate with High Diffraction Efficiency for High-energy X-rays

S Tamura [1,4], M Yasumoto [2,4], N Kamijo [3,4], K Uesugi [4], A Takeuchi [4], Y Terada [4] and Y Suzuki [4]

[1] Photonics Research Institute, AIST Kansai, Osaka, Japan
[2] Research Institute of Instrumentation Frontier, AIST Tsukuba, Tsukuba, Japan
[3] Kansai Medical University, Osaka, Japan
[4] JASRI/SPring-8, Hyogo, Japan

e-mail: sg-tamura@aist.go.jp

Abstract. Fresnel zone plate (FZP) with high diffraction efficiency leads to high performance X-ray microscopy with the reduction of the radiation damage to biological specimens. In order to attain high diffraction efficiency in high energy X-ray region, we have developed multilevel-type (6-step) multilayer FZPs with the diameter of 70 micron. The efficiencies of two FZPs were evaluated at the BL20XU beamline of SPring-8. For one FZP, the peak efficiency for the 1st-order diffraction of 51 % has been obtained at 70 keV. The efficiencies higher than 40 % have been achieved in the wide energy range of 70-90 keV. That for the 2nd-order diffraction of 46 % has been obtained at 37.5 keV.

1. Introduction

This paper describes a multilayer Fresnel zone plate (FZP). The advantage of the multilayer FZP will be (1) ideally the kinoform-type FZP can be fabricated, (2) it operates in the high energy X-ray region (even more than 100 keV) because of its large "aspect ratio" and also (3) the optical system can easily be constructed (the tilting angle of the FZP is easily adjustable to the optical axis). Previously, we have developed multilevel-type (quasi-kinoform type: 4-step or 6-step) multilayer FZPs and have achieved the efficiency around 50 % at [1, 2]. Such a FZP is structurally equivalent to the multilevel FZP fabricated by a lithographic technique [3]. This paper presents the experimental results on the focusing test of the multilevel-type (6-step) multilayer FZPs in the wide-range of high-energy X-ray region.

2. Experimental details

The schematics of various FZPs (cross-section views) are shown in Fig.1: Fig.1-(a) is the conventional FZP, Fig.1-(b) is the multilevel FZP [2] and Fig.1-(c) is the 6-step multilevel-type multilayer FZP. The multilevel layered-structure (Fig.1-(c)) can be made by DC sputtering method with two sputtering-targets (Cu, Al). The FZP with 6-step structure composed of concentric multilayers of alternating high-Z(i.e., large atomic number) , low-Z (i.e., small atomic number) and four types of composite materials (high-Z, low-Z) layers [1,2].

A total of 72 layers (12 pairs) was deposited. The inner zone width (thickness of the first layer: Al) was 0.14 micrometer and the outermost zone width (Cu) was 0.10 micrometer. The designed focal length was 200 mm at 12.4 keV. Two FZPs have been fabricated: FZP-1 with the thickness of 35 micron, FZP-2 with the thickness of 66 micron. The thickness was controlled by mechanical polishing process.

Fig.1 Schematics of various types of FZPs: (a) conventional FZP, (b) multilevel FZP with 4-step structure, and (c) multilevel-type multilayer FZP with 6-step structure.

The focusing test of the FZPs was performed at the BL20XU undulator beamline of SPring-8: the diffraction efficiency of the 1st- and 2nd-order focus and the beam size were measured. The diffraction efficiency was estimated by comparing the incident beam intensity measured by ion chambers upon the OSA and the total intensity of the focused beam through the OSA [1].

3. Results and discussion

The microscopic images of the multilevel-type (6-step) multilayer FZP were shown in Fig.2: Fig.2-(a) was an image by an optical microscopy and Fig.2-(b), (c) were ones by a SEM. Six kinds of zones can be observed.

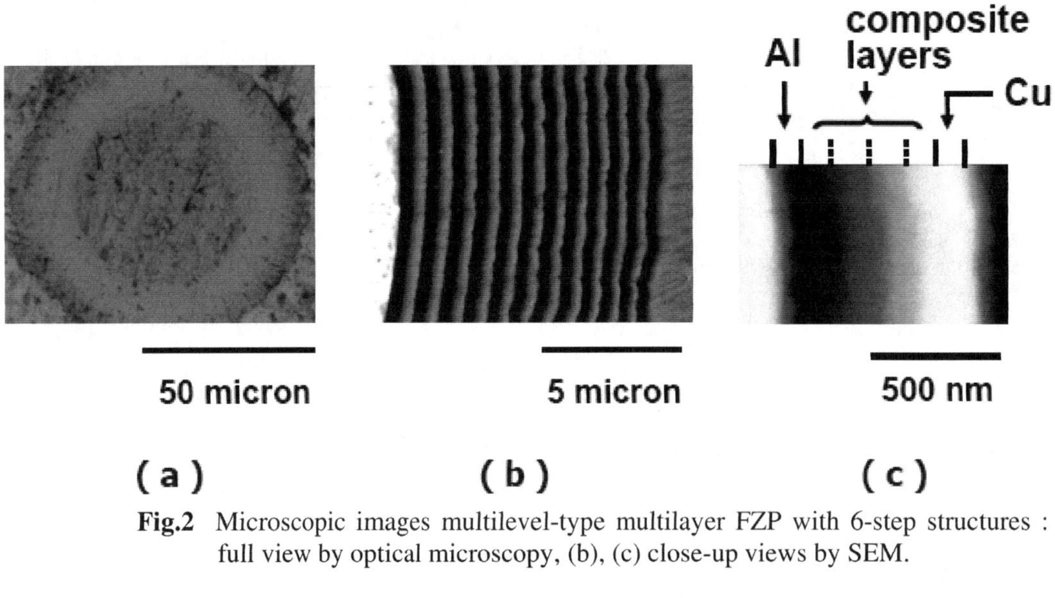

Fig.2 Microscopic images multilevel-type multilayer FZP with 6-step structures : (a) full view by optical microscopy, (b), (c) close-up views by SEM.

Measured diffraction efficiencies for 1st-order diffraction and 2nd-order diffraction of the zoned area have been shown in Fig. 3. For the FZP-1, the peak efficiencies have been obtained at 41.3 keV (52 %) and 21 keV (24 %), respectively. The diffraction efficiencies higher than 45 % have been achieved in the wide energy range of 33-49.5 keV. For the FZP-2, the peak efficiencies have been obtained at 70 keV (51 %) and 37.5 keV (44 %), respectively. The diffraction efficiencies higher than 40 % have been achieved in the wide energy range of 70-90 keV. These results indicate that the composite layers deposited by the co-sputtering have functioned well to achieve high diffraction efficiency. The measured focused beam sizes by the knife-edge scanning have been around 1 micron.

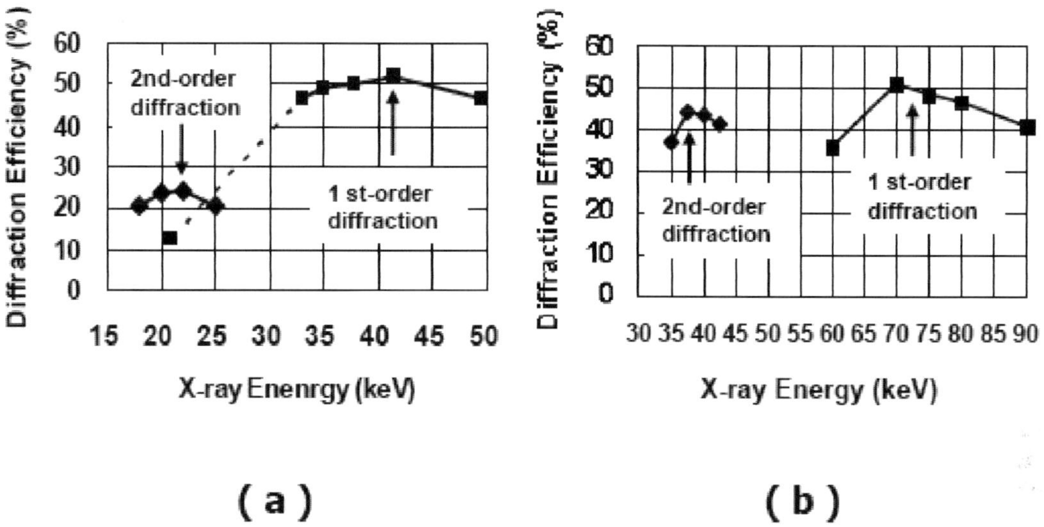

Fig.3 Measured diffraction efficiencies for 1st-order diffraction and 2nd-order diffraction: (a) FZP with 35 micron thickness, (b) FZP with 66 micron thickness.

4. Conclusions

Our experimental results on the fabrication and the focusing test of a multilevel-type (quasi-Kinoform type) multilayer FZP have been shown. The measured diffraction efficiencies of the zoned area of the FZPs have been 50 % at high-energy X-ray region. Efficiencies higher than 40 % have been achieved in the wide energy range. It can be said that the sputtered-sliced technique is effective to fabricate FZP with high diffraction efficiency for use in high-energy X-ray region. At present, we are trying to develop a kinoform-type multilayer FZP.

References

[1] Kamijo N, Suzuki Y, Tamura S, Takeuchi A and Yasumoto M 2006 *IPAP Conference Series 7 (Proc.8th International Conference on X-ray Microscopy)* Institute of Pure and Applied Physics Japan pp 97-99.

[2] Tamura S, Yasumoto M, Kamijo N, Takauchi A, Uesugi K and Suzuki Y 2008 *Vacuum* **82** (in press).

[3] Fabrizio E Di, Romannato F, Gentili M, Cabrini S, Kaulich B, Susini J and Barrett B 1999 *Nature* **401** 895.

Design study of two-aspherical-mirror anastigmat with reduced sensitivities to misalignments: correction of higher-order aberrations

M Toyoda and M Yanagihara
Laboratory of Soft-X-Ray Optics, Institute of Multidisciplinary Research for Advanced Materials, Tohoku University, 2-1-1 Katahira, Aoba-ku, Sendai 980-8577, Japan

E-mail: toyoda@tagen.tohoku.ac.jp

Abstract. A practical design method for correcting fifth and higher order aberration of two-aspherical-mirror optics was developed and applied to an imaging objective for soft-X-ray microscope. A sixth and eighth order deformation was applied to the both mirrors, and the deformation coefficients were optimized numerically to minimize a blur on the image plane. With this method, fifth or higher-order aberrations, which have remarkable effect on a fast soft-X-ray optics of a numerical aperture over 0.1, can be corrected without modifying third-order-aberration coefficients. A third-order aspherical design with a large misalignment tolerance was optimized to have a 50x magnification and a numerical aperture of 0.25, which resulted in a spatial resolution higher than 20 nm with tolerating a mirror decenter of 1 μm.

1. Introduction

The Schwarzschild optics has been used for imaging experiments of microscopy in soft-X-ray region [1], however, the spatial resolution was still limited to a few micron range. The resolution limitation of the Schwarzschild optics can be attributed to its high alignment accuracy required. As Horikawa has pointed out, for imaging by a soft X-ray of 3.98 nm in wavelength, an allowable alignment error of the Schwarzschild mirrors falls within 300 nm for diffraction limit imaging[2]. Therefore, this high sensitivity to misalignments should be the dominant difficulty for implementing the mirror optics under various disturbances such as temperature drifts etc. It is known that this high sensitivity is inherent property of the two-spherical-mirror optics. As Erdos showed, there are only three aplanatic solutions in the two-spherical mirror system, and the Schwarzschild design is a unique solution of a real imaging[3]. This fact prevents an optical designer to seek a new practical solution with reduced sensitivity to misalignments in the spherical system.

The most promising way to overcome this difficulty is to seek for low alignment sensitivity configurations allowing larger misalignments by extending the spherical mirrors to aspherical ones. Since we have additional degree of freedom in the design parameters, wider possibilities with reduced sensitivity can be explored. Recently, we proposed new practical equations formulating third order aberrations in the case of two-aspherical-mirror anastigmat[4], and discovered the four new designs with large misalignment tolerance. The two-concave-mirror system, that is one of the new designs detailed in [4], has achieved reduction in sensitivity by a factor of twelve for the decenter of the mirrors, as compared to the Schwarzschild optics.

In the designing, performance of the optics was analyzed on the basis of the third-order aberration theory. In the case of the optics with relatively large numerical aperture, typically over 0.1, however, the higher-order effect cannot be ignored. In this paper we propose a practical method for correcting fifth and higher order aberration of the two-aspherical-mirror optics with large misalignment tolerance. In the design example, a magnification of -1/50 and a numerical aperture of 0.25 were taken as a typical design parameter.

2. Design method for higher-order aberration correction

In the following analysis, we employ a standard notation used in the optical designing of [4]. The optics we treat here can be represented by an object plane (i=0), the first mirror (i=1), the second mirror (i=2), and an image plane (i=3). The optical axis is taken as x in Cartesian coordinate with an object located in y-z plane. The surface figure of the i-th (i=1 or 2) aspherical mirror can be represented by the following formula, employing a fourth-order deformation coefficient b_i to modify the radius of curvature r_i, as

$$x_i = \frac{y_i^2 + z_i^2}{2r_i} + \frac{1}{8}\left(\frac{1}{r_i^3} + b_i\right)\left(y_i^2 + z_i^2\right)^2 . \qquad (1)$$

In a two-aspherical-mirror anastigmat, configuration parameters and aberration coefficients to the third order can be described by one independent design parameter γ_1, a marginal ray height ratio of the first mirror[4]. Shown in table 1 are the configuration parameters of the two-concave-mirror anastigmat with reduced sensitivities, which was found at γ_1= -0.650.

Table 1. Configuration parameters of the aspherical optics derived by the method in [4].

Radius of curvature	r_1 (mm)	-148.600
	r_2 (mm)	26.677
Mirror separation	d_0 (mm)	1031.748
	d_1 (mm)	-41.649
	d_2 (mm)	9.901
Mirror deformation	b_1 (mm^{-3})	6.985×10^{-6}
	b_2 (mm^{-3})	-2.366×10^{-5}

A lateral aberration for on-axis object as a function of the numerical aperture, i.e., ray fans is shown in figure 1. We see that, in the case of third-order design of Table 1, which is shown by the solid curve, higher-order spherical aberration cannot be ignored when the numerical aperture |NA| exceeds 0.1.

For correction of higher-order aberration, we modify the figure equation by applying sixth and eighth deformation coefficients, c_6 and c_8 as

$$x_i = \frac{y_i^2 + z_i^2}{2r_i} + \frac{1}{8}\left(\frac{1}{r_i^3} + b_i\right)\left(y_i^2 + z_i^2\right)^2 + c_6\left(y_i^2 + z_i^2\right)^3 + c_8\left(y_i^2 + z_i^2\right)^4 . \qquad (2)$$

The higher-order aberration can be corrected by optimizing the deformation coefficients, c_6 and c_8, on both mirrors. It should be noticed that, the third-order aberration coefficients of the system should not be affected by these higher-order terms, since these new terms will generate only fifth or higher-order

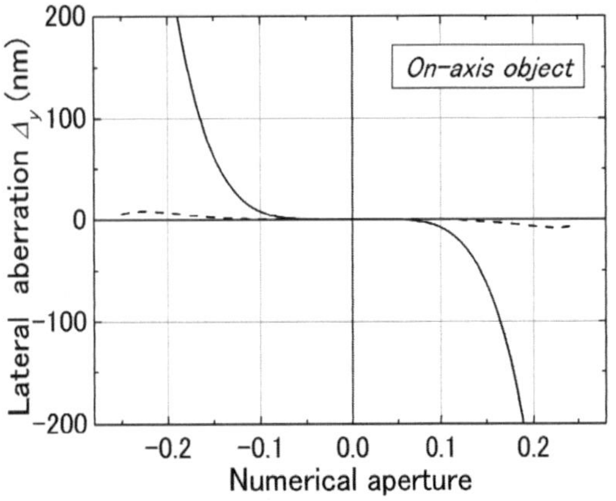

Figure 1. Ray fans plot for on-axis object. The solid curve represents the aberration of the third-order system. As the dashed curve shows, the aberration can be corrected by applying the higher-order deformation terms on (2).

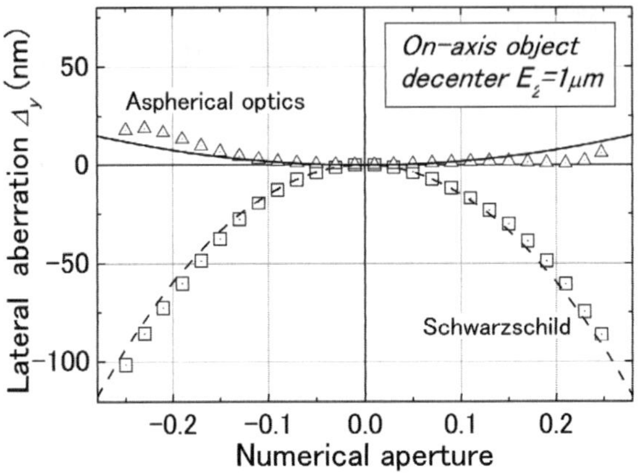

Figure 2. Ray fans plot with a second mirror decenter of E_2=1μm. The symbols represent the aberration obtained by numerical ray tracing. A solid and dashed curve shows corresponding result given by a third-order aberration theory in [4].

lateral aberration on the image plane. Thus, we can correct higher-order aberration of the aspherical anastigmat shown in table 1 by keeping a reduced sensitivity to the mirror misalignment, since the sensitivity depends mainly on third-order coefficients for the spherical aberration I_i and coma II_i on the both mirrors[4].

The deformation coefficients, c_6 and c_8, on both mirrors were optimized numerically to minimize blurs on image plane. The optimization of the mirror deformation was calculated with a conventional ray tracing method by using an optical designing software, CodeV[5]. Typical results of the optimization are shown in figure 1 with the dashed curve. The higher-order effect which is remarkable on the third-order design, shown with solid curve, is clearly corrected below a lateral aberration of 10 nm after the optimization. Sensitivity to the misalignment was also confirmed by a ray tracing method. Shown in figure 2 is the lateral aberration of the higher-order-corrected optics with the second mirror decenter of E_2=1μm. The aspherical-mirror system shown with triangle symbols keeps the aberration below 20 nm, while the Schwarzschild system shown with square symbols is easily influenced by the mirror decenter. The aberrations calculated with exact computer simulations, shown by the symbols, also well coincide with the analytical description of a misaligned optics in [4], which is shown by solid and dashed curves. The results verify the usefulness of our design method for correction of higher-order aberration without modifying the sensitivity to a mirror misalignment.

Practical method for correcting higher-order aberration presented in this paper has been proved to be useful for designing of two-aspherical-mirror anastigmat, particularly, of a soft-X-ray optics where a large misalignment tolerance is essential for stable operation.

References

[1] Toyoda M, Shitani Y, Yanagihara M, Ejima T, Yamamoto M and Watanabe M 2000 *Jpn. J. Appl. Phys.* **39** 1926
[2] Horikawa Y, Mochimaru S, Iketaki Y and Nagai K 1992 *Proc. SPIE* **1720** 217
[3] Erdos P 1959 *J. Opt. Soc. Am.* **49** 877
[4] Toyoda M and Yamamoto M 2006 *Opt. Rev.* **13** 157
[5] http://www.opticalres.com/cv/cvprodds_f.html

Development of a period-by-period EUV multilayer milling system for the final nm figure error correction by 0.1 nm per period

T Tsuru, A Tosaka[1], Y Sakai and M Yamamoto
Division of Soft X-ray Microscopy, CAMS, IMRAM, Tohoku University, 2-1-1 Katahira, Aoba-ku, Sendai 980-8577, Japan
E-mail: tsuru@tagen.tohoku.ac.jp

Abstract. A period-by-period ion milling system has been developed for final reflection wavefront error correction of an imaging EUV multilayer mirror by a stepwise 0.1 nm-per-period correction of the residual nm figure errors. For effective, gentle and uniform milling of the surface areas selected by a template, the system is designed with a rotating substrate holder exposed to a 150 mm-wide ion beam with a dose homogenizer mask plate. For demonstration of the wavefront correction principle, local milling of a dielectric multilayer mirror for visible light was carried out by the system. The wavefront as measured by a phase shifting interferometer showed that the reflection phase was "advanced" by the milling, which formed a geometrical depression at the multilayer mirror surface. This confirmed the physical optics principle of our method and proved the procedure of the method being promising and practical for the accurate reflection phase correction of an EUV multilayer mirror.

1. Introduction

In the extreme ultraviolet (EUV) wavelength region, multilayer mirrors are successfully utilized for various normal incidence imaging optics. For diffraction-limit EUV imaging expected by the multilayer mirror optics, the figure error should be smaller than $\lambda/28$, which is in a sub-nm level for the short EUV wavelength λ of a few tens nm. Although ion beam figuring (IBF) techniques are recently developed for finishing the substrates [1-3], the figure error of a 0.1nm level is still difficult to correct. As a possible solution, we proposed a phase correction method by period-by-period milling of the multilayer mirror surface [4] based on a physical optics principle. Theoretical calculation shows that 1-period removal should compensate a reflection phase error of 6°, which corresponds to the substrate figure error of 0.1 nm.

In this manuscript, a new ion milling system developed for the physical optics phase correction of EUV multilayer mirrors is described briefly. Then, we present an experimental proof of the phase correction principle at a visible wavelength of 633 nm with a dielectric multilayer mirror sample.

2. The period-by-period ion milling system for physical optics phase error correction

For accurate reflection phase error correction based on a physical optics principle, we have developed a period-by-period milling system with an electron cyclotron resonance (ECR) type ion gun enabling

[1] present address: Basic and Applied Science, Division of Science, International College of Arts and Science, Yokohama City University, 22-2 Seto, Kanazawa-ku, Yokohama, 236-0027, Japan

effective, gentle and uniform milling of EUV multilayer mirrors [5]. Photograph of the ion milling system and a cut-off view of the milling process chamber are shown in Fig. 1.

An Ar ion beam of a 150 mm in diameter accelerated at a low 500 V for minimal surface roughening is irradiated through a template having openings for the area of milling. The templates are prepared with the opening patterns for every 0.1 nm contour mapping of the figure errors as measured by our EUV interferometer with a laser produced plasma light source [6]. A multilayer mirror up to ϕ100 mm, together with the template, is mounted on a rotating holder facing to the ion beam.

A constant milling rate over the whole area of a mirror rotating at 200 rpm is realized by a homogenizer mask plate designed for uniform radial distribution of the ion dose within the exposure time. The original profile of the Ar ion current density without the homogenizer mask plate has a quadratic distribution within ϕ100 mm with an ion current density of 0.1 mA/cm^2 at the center (r=0 mm) and of 40% less 0.06 mA/cm^2 at the peripheral (r=50 mm). With the homogenizer mask plate designed, ion dose has been successfully controlled for uniform radial distribution. Uniform milling rates of 2.5 nm/min for Mo and 4.3 nm/min for Si can be realized with the homogenizer mask plate. The rate of 1-period removal within 2 min is practical enough for repeating the process of 0.1 nm/period correction by 10 times to cover 1 nm figure error in total.

Since contamination of multilayer mirrors by sputtered material of the acceleration grid, the correction mask and the template would be difficult to avoid, these elements are made of Mo or Si for transparency at the EUV wavelength of 13.5 nm. For the milling period detection, sensing systems of visible emission and/or reflection spectrum, total light intensity and the drain current are equipped to distinguish milling materials during ion beam irradiation. Further details of the system will be described in a separate publication.

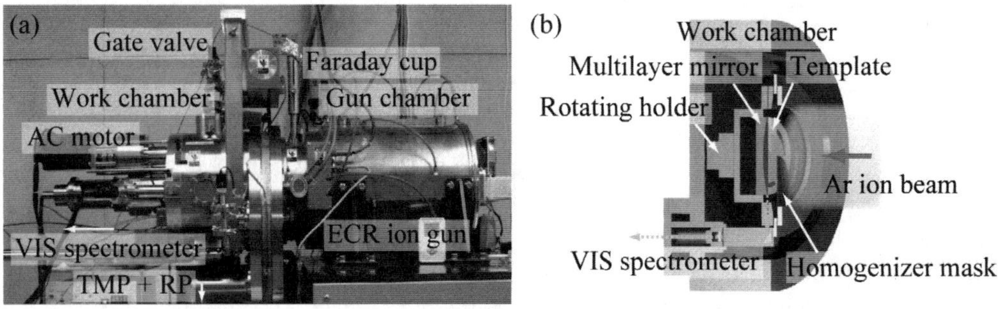

Figure 1. (a) The multilayer ion milling system. (b) Cut off view of the work chamber.

3. Experimental proof of the correction principle at a visible wavelength

For the proof of the reflection phase correction principle specific to a multilayer mirror, local areal milling of a 10 periods Ta_2O_5/SiO_2 multilayer mirror for the wavelength of 633 nm was carried out by our ion milling system.

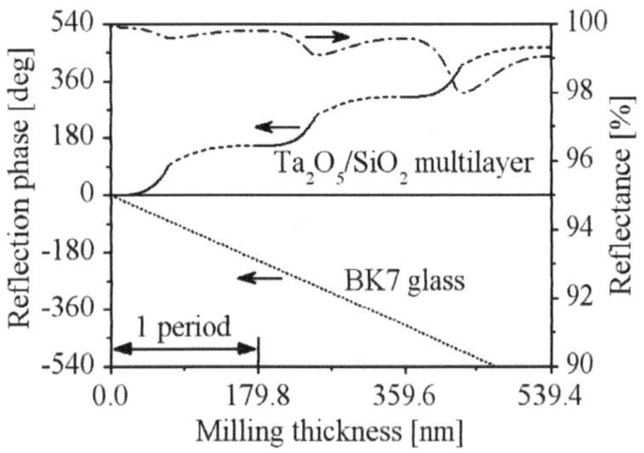

Figure 2. Theoretical reflection phase by surface milling of 10 period Ta_2O_5 (solid curve) /SiO_2 (broken curve) multilayer mirror for visible light and of a bare BK7 glass substrate (dotted line). Reflectance variation of the multilayer mirror is also shown. For calculation, the refractive indices n and the layer thickness d of 2.196 and 72.0 nm for Ta_2O_5 layers, 1.468 and 107.8 nm for SiO_2 layers and 1.515 for a BK7 glass substrate were assumed, respectively.

Fig. 2 shows theoretical expectation of reflection phase and reflectance calculated as a function of milling thickness from the surface. Normal incidence and a wavelength of 633 nm are assumed in the calculation. Calculation method of multilayer milling is described in Ref. [4] in detail. Since the refractive indices of layer materials are larger than 1 in the visible region, the reflection phase at the milling area, specifically of a multilayer mirror, shifts to "positive" direction, whereas the phase of a bare substrate shifts naturally to "negative" direction by the delay as shown in Fig. 2. The "positive" phase shift is caused by the removal of the layer material as in the case of a Schmidt phase correction plate. The surface layers thus behave as a phase correction layer since the multilayer reflection by the constructive interference is a volume effect of the layers underneath.

After the ion milling at a template opening of ϕ5 mm for 30 min by the Ar ion accelerated at 500V, the reflection phases of the dielectric multilayer mirror and BK7 glass plate were measured by a phase-shifting interferometer (Zygo mark II) at the wavelength of 633 nm. The measured optical pass differences (OPD) of the glass plate and the multilayer are shown in Figs. 3 (a) and (b), respectively.

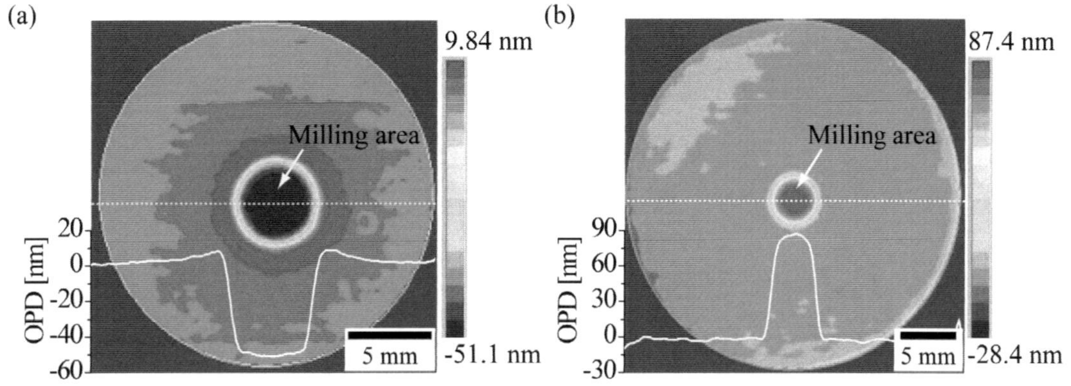

Figure 3. Measured optical pass differences (OPD) maps of (a) BK7 glass and (b) 10 period Ta_2O_5/SiO_2 multilayer mirror by the phase-shifting interferometer (Zygo mark II) after local millings for 30 min in the area of ϕ5 mm by the ion milling system we developed. Line profiles of OPD at dotted lines are also shown.

In referring to the reflection phase at the untreated top-surface of the multilayer, the phase at the milling area "proceeds" as seen in Fig. 3 (b). The multilayer surface locally removed by the milling shows wavefront protrusion as predicted in Fig. 2. This is in good contrast to the substrate case Fig. 3 (a) showing depression by the milling. These results confirmed that our physical optics method of phase correction, specifically for the multilayer mirror, is promising and practical for the reflection phase correction with 0.1 nm accuracy at the EUV. The correction factor of (1-n) being equal to 1/70 should result such a high accuracy by the Mo/Si multilayer milling for the final reflection phase correction at the EUV.

4. References

[1] Schindler A, Hänsel T, Nickel A, Frost F, Thomas H J, Neumann H, Seidenkranz G, Schwabe R, Franz T, Gürtler S, Görsch S, Bogatz A and Rauschenbach B 2005 *Proc. Int. Conf. on Leading Edge Manufacturing in 21st century (Nagoya, Japan, 19-22 October, 2005)* vol 2005 (Tokyo, The Japan Society of Mechanical Engineers) pp 589-594

[2] Drueding T W, Fawcett S C, Wilson S R and Bifano T G 1995 *Opt. Eng.* **34** 3565

[3] Savvides N 2006 *J. Appl. Phys.* **99** 094912

[4] Yamamoto M 2001 *Nucl. Instrum. Methods A* **467-468** 1282

[5] Tosaka A, Sakai Y, Tsuru T and Yamamoto M 2008 *J. Vac. Soc. Jpn.* **51** 115 in Japanese.

[6] Yamamoto M, Hatano T and Furudate M 2001 *Opt. Precis. Eng.* **9** 405

Journal of Physics: Conference Series **186** (2009) 012078 doi:10.1088/1742-6596/186/1/012078

Advanced X-Ray Diffractive Optics

J Vila-Comamala[1], K Jefimovs[1,2], T Pilvi[3], M Ritala[3], S S Sarkar[1], H H Solak[1], V A Guzenko[1], M Stampanoni[1,4], F Marone[1], J Raabe[1], G Tzvetkov[1], R H Fink[5], D Grolimund[1], C N Borca[1], B Kaulich[6] and C David[1]

[1]Paul Scherrer Institut, CH-5232 Villigen PSI, Switzerland
[2]EMPA, CH-8600 Dübendorf, Switzerland
[3]Department of Chemistry, FIN-00014 University of Helsinki, Finland
[4]Institute for Biomedical Engineering, CH-8092 University and ETH Zürich, Switzerland
[5]Friedrich-Alexander Universität, D-91058 Erlangen, Germany
[6]ELETTRA Sincrotrone, I-34012 Basovizza, Italy

E-mail: joan.vila@psi.ch

Abstract. X-ray microscopy greatly benefits from the advances in x-ray optics. At the Paul Scherrer Institut, developments in x-ray diffractive optics include the manufacture and optimization of Fresnel zone plates (FZPs) and diffractive optical elements for both soft and hard x-ray regimes. In particular, we demonstrate here a novel method for the production of ultra-high resolution FZPs. This technique is based on the deposition of a zone plate material (iridium) onto the sidewalls of a prepatterned template structure (silicon) by atomic layer deposition. This approach overcomes the limitations due to electron-beam writing of dense patterns in FZP fabrication and provides a clear route to push the resolution into sub-10 nm regime. A FZP fabricated by this method was used to resolve test structures with 12 nm lines and spaces at the scanning transmission x-ray microscope of the PolLux beamline of the Swiss Light Source at 1.2 keV photon energy.

1. Introduction

New developments in x-ray optics are improving the performance of x-ray microscopes, which in turn benefit a lot the investigation of both inorganic and biological materials. At the Paul Scherrer Institut, research activities in x-ray diffractive optics include the manufacture of Fresnel zone plates (FZPs) and diffractive optical elements for both soft and hard x-rays by several methods: Au-electroplated FZPs useful in the multi-keV range have been produced [1, 2] and a new type of diffractive optical elements to be used as condenser in full-field transmission x-ray microscope has been prepared [3, 4]. For soft x-rays, silicon-etched FZPs that can withstand high radiation heat loads have been manufactured [5] and a new technique to produce the FZP patterns using extreme ultraviolet holographic lithography is under development [6]. In particular, we want to introduce here a novel technique for the fabrication of FZPs with extremely narrow outermost zone widths. In a FZP-based microscope, there exists an intimate relationship between the spatial resolution and the outermost zone width of the FZP, the two of them being essentially comparable. Due to manufacturing limitations, the spatial resolution of x-ray microscopes is currently stagnating at 25-20 nm, and higher resolution has only been reported in very few exceptional cases [7, 8]. The method we present can overcome one of the main limitations during the manufacture of FZPs and it provides a clear route to push the spatial resolution of x-ray microscopes to the sub-10 nm regime.

2. Zone-doubling technique for high resolution FZPs

Up to date, x-ray FZPs are mostly prepared by electron-beam lithography tools. Even though these systems are capable of writing patterns with spot sizes and position accuracies in the sub-10 nm range, the final size of the patterned structure is mainly depending on the range of the secondary electrons created in the sensitive resist layer. This effect is particularly detrimental when writing dense patters of lines such as gratings or FZPs, and it explains the stagnation at 25-20 nm in terms of spatial resolution of FZP-based x-ray microscopy. Here we report on a simple nanofabrication method that overcomes the difficulty of high density patterning. Since this approach only requires a single lithography step and no need for alignment, this method is very reproducible. Moreover, it allows the fabrication of structures with extremely high aspect ratios. It is based on a deposition of thin layer of a high refractive index material onto the sidewalls of template made of a low-index material. This leads to a doubling of the effective line density of the deposited material comparing to that of the template, which in turn improves the spatial resolution of the FZP by a factor of two.

The fabrication steps are schematically shown in Figure 1. The process starts with the electron-beam exposure of the FZP pattern on a high resolution negative tone resist. Then, the pattern is transferred to an intermediate chromium layer which is used as mask during the reactive ion etching process to create the template structure on a previously prepared silicon membrane. The final step in this fabrication technique requires a highly conformal deposition of the high refractive index material with accurate control of the thickness. Because of its unique self-limiting growth mechanism, atomic layer deposition [9] is the best choice for this purpose and was employed in this work. A cross section of the 15 nm wide outermost zones of a zone-doubled FZP is shown in Figure 2a. The aspect ratio of these structures is around 12.

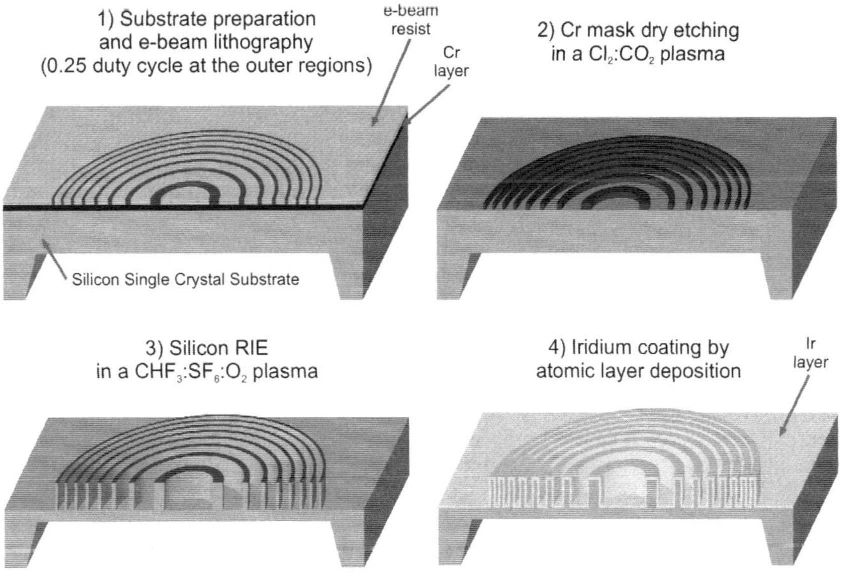

Figure 1. Zone-doubled FZP manufacturing steps: 1) electron-beam lithography, 2) pattern transfer to chromium layer by reactive ion etching, 3) pattern transfer to silicon by reactive ion etching and 4) iridium coating by atomic layer deposition.

In addition, theoretical calculations [8] show that the proposed focusing zone-doubled structures can provide diffraction efficiencies slightly below but comparable to those of ordinary FZPs (~10—20 %). For instance, 9.1% diffraction efficiency is expected for a zone-doubled FZP made of silicon and iridium, considering a diameter of 100 μm and an outermost zone width of 20 nm at 1.0 keV. Further calculations also show that the zone-doubled structures are more efficient at higher energies, for which the absorption of the silicon template is highly reduced.

3. High resolution scanning transmission x-ray microscopy

To demonstrate the feasibility of the proposed fabrication technique a zone-doubled FZP with an outermost zone width of 15 nm and a diameter of 100 μm was produced and tested at the scanning transmission x-ray microscope (STXM) of the PolLux beamline of the Swiss Light Source at 1.2 keV photon energy. During the experiments, a GaAs/AlGaAs heterostructure containing several line widths ranging from 40 nm down to 9 nm was used as a test sample. Figure 2b shows the STXM image of the sample where the smallest visible features consist of 3 lines of 12 nm width.

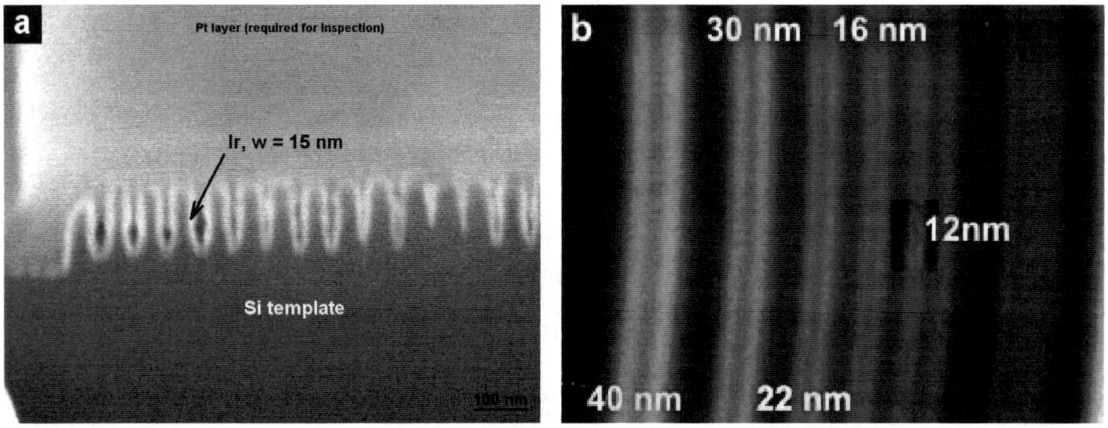

Figure 2. a) Cross section of zone-doubled FZP with an outermost zone width of 15 nm. b) STXM image of GaAs/AlGaAs heterostructure with several line widths. Smallest visible features consist of 3 lines of 12 nm width.

In conclusion, we have developed a novel technique that frees us from the present limitation of electron-beam lithography in the production of FZPs. It improves the spatial resolution by a factor of 2 and provides a clear route to push the resolution of FZP-based x-ray microscopy into the sub-10 nm regime.

References

[1] Jefimovs K, Bunk O, Pfeiffer F, Grolimund D and van der Veen J F 2007 *Microelectron. Eng.* **84** 1467
[2] Borca C N, Grolimund D, Willimann M, Meyer B, Jefimovs K, Vila-Comamala J and David C 2008 *These proceedings*
[3] Jefimovs K, Vila-Comamala J, Stampanoni M, Kaulich B and David C 2008 *J. Synchrotron Rad.* **15** 106
[4] Stampanoni M, Marone F, Mikuljan G, Jefimovs K, Trtik P, Vila-Comamala J, David C and Abela R 2008 *These proceedings*
[5] Vila-Comamala J, Jefimovs K, Raabe J, Kaulich B and David C 2008 *Microelectron. Eng.* **85** 1241
[6] Sarkar S S, Sahoo P K, H H Solak C D and van der Veen J F 2008 *J. Vac. Sci. Tech. B (accepted)*
[7] Chao W, Harteneck B D, Liddle J A, Anderson E H and Attwood D T 2005 *Nature* **435** 1210
[8] Jefimovs K, Vila-Comamala J, Pilvi T, Raabe J, Ritala M and David C 2007 *Phys. Rev. Lett.* **99** 264801
[9] Leskelä M and Ritala M 2003 *Angew. Chem., Int. Ed.* **42** 5548

Journal of Physics: Conference Series **186** (2009) 012079 doi:10.1088/1742-6596/186/1/012079

Towards stacked zone plates

S Werner[1, 2], S Rehbein[1], P Guttman[1], S Heim[1] and G Schneider[1]

[1] BESSY m.b.H., Albert-Einstein-Straße 15, 12489 Berlin, Germany

[2] Electronic mail: werner@bessy.de

Abstract. Fresnel zone plates are the key optical elements for soft and hard x-ray microscopy. For short exposure times and minimum radiation load of the specimen the diffraction efficiency of the zone plate objectives has to be maximized. As the efficiency strongly depends on the height of the diffracting zone structures the achievable aspect ratio of the nanostructures determines these limits. To reach aspect ratios $\geq$ 20:1 for high efficient optics we propose to superimpose zone plates on top of each other. With this multiplication approach the final aspect ratio is only limited by the number of stacked zone plate layers. For the stack process several nanostructuring process steps have to be developed and/or improved. Our results show for the first time two layers of zone plates stacked on top of each other.

1. Introduction

Most modern x-ray microscopes use Fresnel zone plates as x-ray objectives. By using the first-order of diffraction for imaging, the efficiency of high resolution x-ray optics is limited by the aspect ratio of the nanostructures which can be fabricated with existing nanostructuring processes. So far the highest reported aspect ratio for zone plates with 20 nm outermost zone width is about 9:1 [1] which is close to be optimal for water window wavelengths, but too low for efficient hard x-ray optics. To overcome this limitation we propose to superimpose several zone plate layers on top of each other. With this so-called stack process only the number of stacked zone plates limits the achievable aspect ratio.

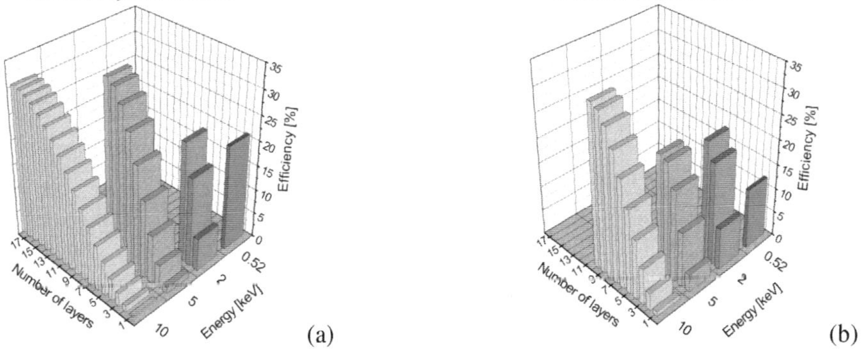

Fig. 1: Calculated first order diffraction efficiency for (a) nickel and (b) gold layers with 200 nm thickness (electrodynamic theory, 20 nm zones parallel to optical axis).

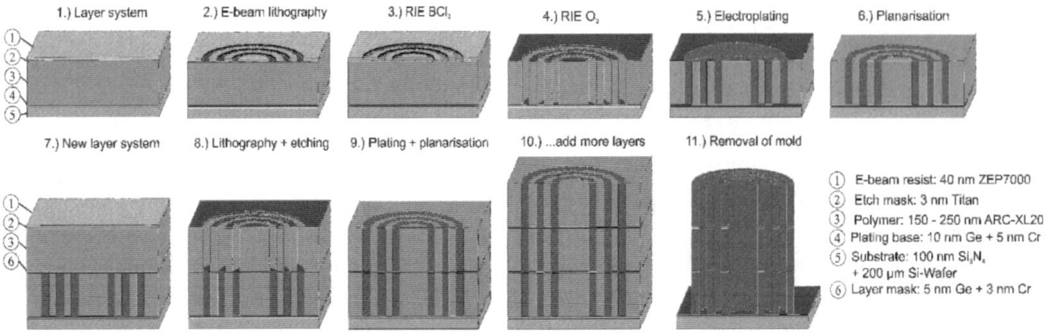

Fig. 2: Proposed stack process for zone plates with high aspect ratios and high diffraction efficiency.

Figure 1 shows a calculation of the first-order diffraction efficiency for nickel and gold zone plates as a function of the photon energy and the number of stacked zone plate layers. For the calculation, we assumed a thickness of each single layer of 200 nm and an outermost zone width of 20 nm [2]. It is shown that the efficiency significantly increases at higher photon energies with each layer added to the zone plate stack. The comparison of both materials shows similar achievable diffraction efficiencies, but different numbers of layers are required to obtain the maximum diffraction efficiency. At 5 keV and 10 keV photon energy, nickel requires significantly more layers than gold for the maximum efficiencies. Therefore, gold is the preferred material for developing stacked zone plates for hard x-rays.

2. Stack process

The proposed stack process for high aspect ratio zone plates is shown in Figure 2. Each single layer for the final stacked zone plate is fabricated with a Tri-Level-Process which is based on electron beam lithography, reactive ion etching and electroplating [3, 4, 5].

One important step for the stacking is to fill up the plating mold completely. Therefore, we developed a process step based on chemical mechanical polishing to remove overlaying nickel after overplating the mold [2]. For the planarisation the plating mold has to be hardened. This is done by cross-linking the polymer with 5 keV electrons. The polymer was sufficiently hardened for the CMP process after cross-linking with an electron dose exceeding 900 mC/cm². With the CMP process nickel structures with widths down to 25 nm were successfully polished.

After planarisation a new layer system is added on top of the polished zone plate followed by a second e-beam lithography step. The overlay exposure is done by using alignment marks located on the substrate. With this method the second zone plate pattern can be well aligned to the nickel zone plate beneath. Afterwards, the pattern is developed and transferred into the second polymer layer. A second plating and polishing step leads to the final in thickness doubled nickel zone plate. If the aspect ratio has to be increased further more zone plate layers can be added onto the zone plate stack.

3. Recent Results

In first experiments two layers of nickel zone plates and linear gratings were stacked on top of each other. The height of the single layers was about 160 nm corresponding to a total height of 320 nm for the stacked structures. After electroplating of the second layer, a trench was cut into the nanostructures by FIB to visualize the cross section (thanks to Y. Ritz and D. Chumakov, AMD, Dresden). Figure 3 shows a SEM micrograph of two layers of stacked linear gratings with 30 and 25 nm line widths. Both gratings show an overlay mismatch of about 20 nm. The corresponding total aspect ratios are 10.5 and 12.5 for the 30 and 25 nm lines. The stacked zone plate with an outermost zone width of 30 nm is shown in Figure 4. A mismatch of about 30 nm between the first and the second zone plate layer is

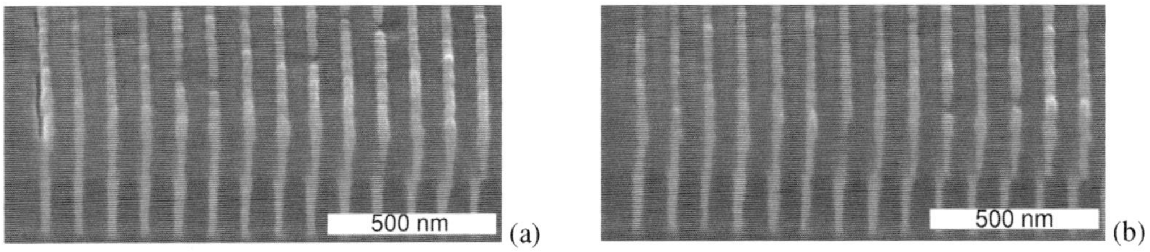

Fig. 3: SEM micrograph of two layers of stacked linear nickel gratings with L:S = 1:3 and (a) 30 nm and (b) 25 nm line width.

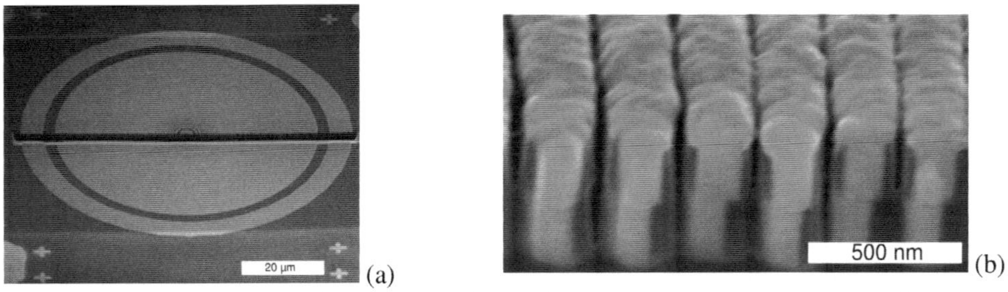

Fig. 4: SEM micrograph of a stacked nickel zone plate with 30 nm outermost zone width. (a) Overview with cross marks and FIB produced trench. (b) Stacked zones are slightly overplated.

observed. The second layer has not been polished yet and is still slightly overplated.

4. Summary and Outlook

With the stack process, the achievable aspect ratios of high resolution zone plates are only depending on the number of stacked zone plates. Our first stacking experiments show a mismatch of about 30 nm between the second and the first zone plate layer. The required overlay accuracy is $\leq 2dr_n/7$ to avoid a degradation of the zone plate efficiency and resolution [6]. This means our achieved overlay accuracy has to be improved in order to fabricate efficient high resolution zone plates. Recently, an overlay error of about 2 nm was reported [7] showing that a sufficient accuracy is possible with advanced e-beam lithography. The polished zone plates are fabricated on a thick substrate due to the mechanical stability for the CMP process. For using these zone plates as x-ray optics the substrate has to be thinned by wet chemical etching. A process to protect the structures during the backside etching is under development.

Acknowledgments

This work was supported by the EU within the 6th framework program under contract No. RII3-CT-2004-506008 and the BMBF/VDI within the "Labor-Röntgenmikroskop" project under contract No. 13N8915.

References

[1] M. Peuker, Appl. Phys. Lett. 78 (2208), 2001
[2] S. Rehbein, P. Guttmann, S. Werner and G. Schneider, J. Vac. Sci. Technol. B 25 (1789), 2007
[3] M. Peuker, Appl. Phys. Lett. 78, 2208 (2001)
[4] G. Schneider, T. Schliebe, and H. Aschoff, J. Vac. Sci. Technol. B 13,2809 (1995)
[5] A. Holmberg, S. Rehbein, and H. M. Hertz, Microelectron. Eng. 73–74, 639 (2004)
[6] G. Schneider, Appl. Phys. Lett. 73, 599 (1998)
[7] W.Chao, B. D. Harteneck, J. A. Liddle, E. H. Andersson and D. T. Attwood, Nature (London) 435, 1210 (2005)

Stitching interferometric measurement system for hard x-ray nanofocusing mirrors

Hirokatsu Yumoto[1,2], Hidekazu Mimura[1], Soichiro Handa[1], Takashi Kimura[1], Satoshi Matsuyama[1], Yasuhisa Sano[1], Makina Yabashi[2], Yoshinori Nishino[3], Kenji Tamasaku[3], Tetsuya Ishikawa[2,3], Kazuto Yamauchi[1]

[1]Graduate School of Engineering, Osaka University, 2-1 Yamada-oka, Suita, Osaka 565-0871, Japan

[2]SPring-8/Japan Synchrotron Radiation Research Institute (JASRI), 1-1-1, Kouto, Sayo-cho Sayo-gun, Hyogo 679-5198, Japan

[3]SPring-8/RIKEN, 1-1-1 Kouto, Sayo-cho, Sayo-gun, Hyogo 679-5148, Japan

E-mail: yumoto@spring8.or.jp

Abstract. We have developed an advanced type of relative angle determinable stitching interferometry (RADSI) for hard x-ray nanofocusing mirrors. The surface metrology system consists of two types of optical interferometers. One is a microscope Michelson interferometer, which measures the local figure profile of the mirror surface, and the other is a large-area Fizeau interferometer, which determines the stitching angles between neighboring shots using a flat mirror. The surface figure profile of each of the one-shots and the stitching angles are measured at the same time. Precise measurement of local surface area and the stitching angles of 1×10^{-8} rad order were achieved using the developed system. To demonstrate the accuracy of the developed measurement system, we measured an elliptically figured mirror for hard x-ray nanofocusing. The mirror's surface shape has a curvature radius of a few meters. As a result, measurement reproducibility and reliability of less than PV 3 nm was achieved by using a 25 nm focusing mirror.

1. Introduction

Third-generation synchrotron radiation facilities press on with leading-edge research in many fields such as nano-science, medicine, biology, materials, and chemical science. Optical components for focusing x-ray beams are indispensable elements for such application fields.

We have been developing the hard-x-ray nano-focusing mirror optics[1,2] and have applied them to scanning x-ray fluorescence microscope[3] with the high spatial resolution of 30 nm. These mirrors were manufactured by using precision machining method of computer-controlled elastic emission machining (EEM)[4] and plasma chemical vaporization machining (PCVM)[5]. As metrology for mirror surface figure, we developed stitching interferometry technique of microstitching interferometry (MSI)[6] and relative angle determinable stitching interferometry (RADSI)[7].

In this article, we introduce an advanced type of RADSI. Performance test for estimating reproducibility and reliability was demonstrated for evaluating the developed system.

2. Microscope interferometer type RADSI

Conventional stitching method to minimize the superposition error between neighboring surface data does not have enough accuracy to measure x-ray focusing mirror because it causes the angle-error integration. To solve this problem, the measurement system was developed; in which stitching angle is measured at the same time surface figure is measured. In the developed system, local area of test mirror surface is measured by microscope interferometer and posture of test mirror (stitching angle) is determined by measuring the standard flat mirror posture by large-area interferometer. To measure local figure profile of x-ray mirror with a high degree of accuracy and to determine stitching angle to an accuracy of 1×10^{-8} rad, the system consists of optical interferometer heads and high precision stages showed in Fig. 1 was constructed. This instrument has a large-area Fizeau type interferometer (ZYGO: GPI XP HR) for determining the posture of test surface by measuring the standard flat mirror and a Michelson type microscope interferometer (ZYGO: NewView 300) for measuring the surface figure profile of test mirror. This instrument is installed on a vibration isolation table in a soundproof chamber and these are established in a temperature-controlled room within ± 0.1 degrees centigrade. PC placed outside of the soundproof chamber performs the operation of the instrument and automatic measurement is possible.

Figure 1. Overview of developed measurement instrument for stitching interferometry

3. Performance assessment of microscope interferometer type RADSI

We examined a measurement accuracy of developed system. Fig. 2 shows optical arrangement of the evaluated elliptical figured mirror. This mirror is designed for 1-km-long beamline (BL29XUL) of SPring-8. Mirror length is 50 mm, incident angle at the mirror center is 7.2 mrad, and working distance is 25 mm. Designed ideal focal spot size is 25 nm (FWHM) at an x-ray energy of 7 keV under total reflection condition and less than 10 nm (FWHM) at an x-ray energy of 20 keV with multilayer coating. Fig. 3 shows surface figure profile of designed mirror. This mirror has 5–26 m of radius of curvature. The mirror was fabricated by using a computer-controlled precision figuring system of EEM and the developed measurement system. The mirror surface figure error is removed as much as possible within the range of reproducibility. Fig. 4 shows the evaluation result of measurement reproducibility and reliability. Reproducibility is estimated by the differential profile from the average data of continuously measured three surface figure profiles. Reliability is estimated by the differential profile between two measured surface profiles in normal and inverted mirror positions. As a result of evaluation, measurement reproducibility and reliability of less than PV 3 nm were achieved using the developed measurement system.

Journal of Physics: Conference Series **186** (2009) 012080 doi:10.1088/1742-6596/186/1/012080

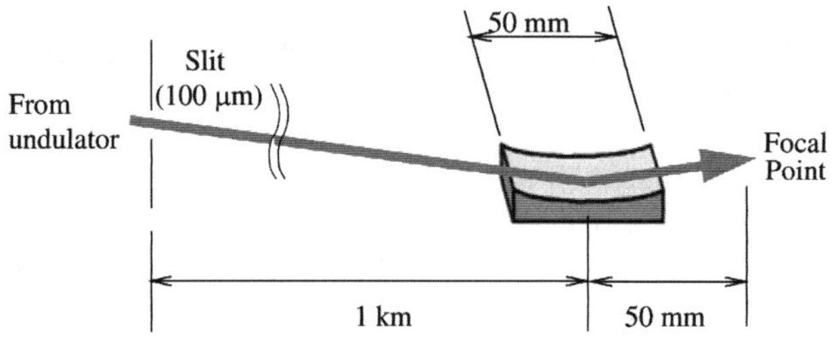

Figure 2. Optical arrangement of evaluated elliptical figured mirror

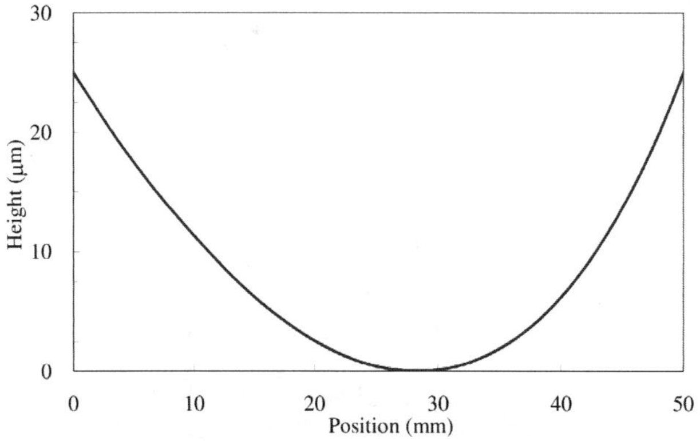

Figure 3. Surface figure profile of designed focusing mirror

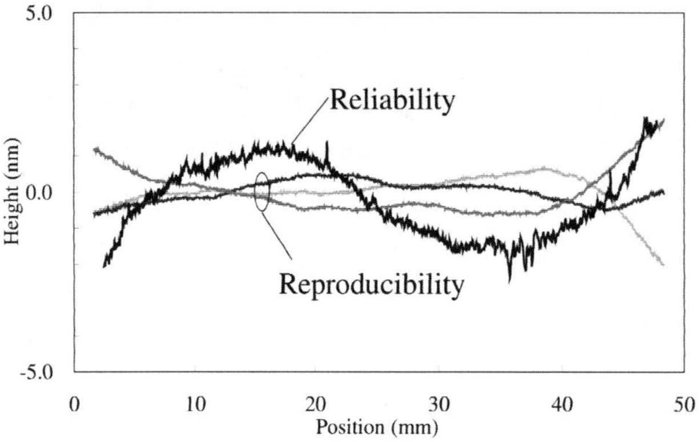

Figure 4. Evaluation result of measurement reproducibility and reliability

References

[1] H. Yumoto *et al.* 2005 *Rev. Sci. Instrum.* **76**, 063708
[2] H. Mimura *et al.* 2007 *Appl. Phys. Lett.* 90 051903
[3] S. Matsuyama *et al.* 2006 *Rev. Sci. Instrum.* **77**, 103102
[4] K. Yamauchi, H. Mimura, K. Inagaki, and Y. Mori 2002 *Rev. Sci. Instrum.* **73**(11) 4028
[5] K. Yamamura *et al.* 2003 *Rev. Sci. Instrum.* **74**(10) 4549
[6] K. Yamauchi *et al.* 2003 *Rev. Sci. Instrum.* **74** (5), 2894
[7] H. Mimura *et al.* 2005 *Rev. Sci. Instrum.* **76** 045102

Full-field transmission x-ray microscopy for bio-imaging

J C Andrews[1,7], S Brennan[1], Y Liu[2], P Pianetta[1], E A C Almeida[3], M C H van der Meulen[4], Z Wu[2], Z Mester[5], L Ouerdane[5], J Gelb[6], M Feser[6], J Rudati[6], A Tkachuk[6] and W Yun[6]

[1] Stanford Synchrotron Radiation Laboratory, Menlo Park CA 94025, USA
[2] Institute of High Energy Physics, Beijing, China
[3] NASA Ames Research Center, Moffett Field, CA 94035, USA
[4] Cornell University, Ithaca NY 14853, USA
[5] Institute for National Measurement Standards, Ottawa ON K1A0R6, Canada
[6] Xradia Inc., Concord CA 94520, USA

Abstract. A full-field hard-x-ray microscope at SSRL has successfully imaged samples of biological and environmental origin at 40 nm resolution. Phase contrast imaging of trabeculae from a female mouse tibia, loaded *in vivo* to study the effects of weight-bearing on bone structure, revealed a complex network of osteocytes and canaliculi. Imaging of cordgrass roots exposed to mercury revealed nanoparticles with strong absorption contrast. 3D tomography of yeast cells grown in selenium rich media showed internal structure.

1. Introduction

The full-field transmission x-ray microscope at SSRL [1] has been used to obtain images at 40 nm resolution of mouse bone, plant roots, and yeast cells, as described below.

2. Phase contrast imaging of mouse cancellous bone

The mammalian skeleton consists of bone, roughly a composite of collagen and mineral seeded with cells. Osteocytes, bone forming cells trapped within the tissue, are connected through a vast three-dimensional network of cellular processes known as canaliculi. As a result, osteocytes are presumed to act as a sensor network within the tissue that regulates remodeling in response to a variety of biophysical stimuli. Viewing the structure/morphology of the osteocytic network including the processes is desirable both for developing a mechanistic understanding of the regulation of healthy bone and to understand the adaptive changes that occur due to anabolic and catabolic stimuli. In collaboration with Eduardo Almeida of NASA Ames Research Center and Marjolein van der Meulen of Cornell University, mouse bone tissue was examined following experiments to either increase or decrease mechanical stimuli to the skeleton in vivo.

The TXM mosaic view of the mouse tibia trabecula (figure 1a) was obtained by raster scanning successive full-field images taken in phase contrast (14μm square fields at 8 keV). Higher resolution images were acquired of cells within the tissue, revealing elliptical osteocytic lacunae with emanating canaliculi (figure 1b,c) whose diameter approaches the resolution of the TXM. The use of phase contrast was critical: the lacunae were barely evident at 8 keV with absorption imaging. This level of detail, within the intact tissue structure, will be studied further to determine the effect of loading conditions on cell

[7] Corresponding author. E-mail: jandrews@slac.stanford.edu

morphology, orientation and interaction via processes to understand how mechanical stimuli regulate bone mass.

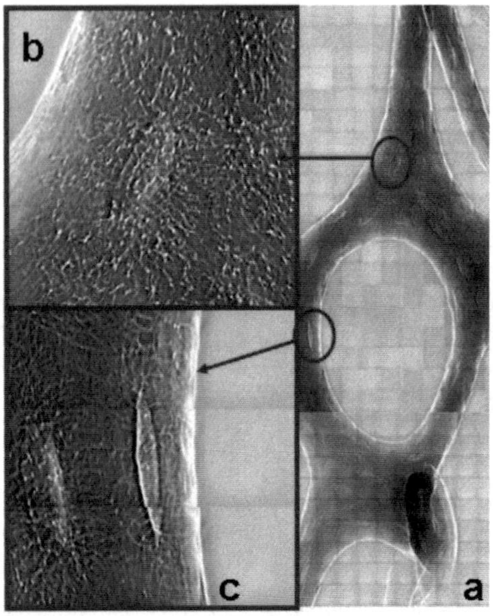

Figure 1: (a) Mosaic phase contrast image at 8 keV of mouse trabecula subjected to in vivo loading of the tibia. Detail images of individual osteocyte lacunae (b) and (c) show network of cellular processes (canaliculi) that facilitate communication and transport between cells.

3. Absorption contrast and tomography of mercury nanoparticles in plant roots

Spartina foliosa is a wetland cordgrass prevalent in San Francisco Bay, where past mercury mining has resulted in mercury contamination. Microorganisms in the rhizosphere of *S. foliosa* and other wetland plants are known to have a significant role in mercury methylation [2]. Methylated mercury can be biomagnified up the food web, resulting in levels in sport fish up to one million times greater than in surrounding waters and resulting in advisories to limit fish intake. Understanding the uptake and methylation of mercury in the plant rhizosphere can yield insight into ways to manage mercury contamination.

Figure 2 is a mosaic in absorption contrast of a root hair from *S. foliosa* that was exposed to 1 ppm Hg (as $HgCl_2$) hydroponically for one week. The dark particles have been confirmed to be Hg with microprobe analysis. Figures 2a-2c are 2D stills from 3D tomography on a group of nanoparticles, which show that the particles are on the root surface, because the side closer to the root (the right side of (a) and the left side of (c)) do not contain particles, only the front (b) away from the root surface. The lower-density material surrounding the particles may be biofilm. Figure 2d is a 2D still from tomography of one of the particles. A slice from the tomographic reconstruction of this particle (figure 2e) reveals that it is hollow, consistent with microorganisms that have a monolayer of Hg on the surface.

4. Tomography of yeast cells grown in Se-rich media

Selenized yeast is one of the primary sources of organoselenium in the nutrition supplement market, and better understanding of the distribution of Se in the yeast is of great interest. In collaboration with Zoltan Mester and Laurent Ouerdane of the Institute for National Measurement Standards in Canada, we obtained images of yeast cells that had been grown in Se-rich media. A wild-type strain of baker's yeast, *Saccharomyces cerevisiae*, was grown on synthetic defined media optimized for *S. cerevisiae*, prepared to control sulfur during yeast growth. After growth in media containing 0.1 mM sodium selenate, yeast cells were fixed using glutaraldehyde buffer and spotted on a Si_3N_4 membrane for x-ray microscopy. TXM results indicate that the cells have excellent contrast (figure 3a), to a degree previously seen only in the water window (low-energy x-ray region). 3D tomography (stills in figure 3b and c) reveals a view of the inner cell in addition to the outer shell.

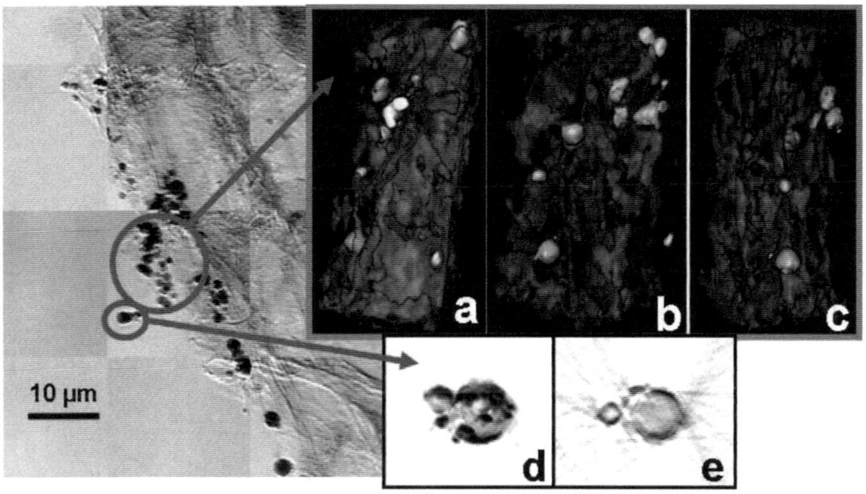

Figure 2: Absorption contrast mosaic image at 9 keV of cordgrass root exposed to mercury, and 2D stills (a-d) of tomographic reconstructions of Hg- containing nanoparticles (2-sec images taken every 1 degree from -70 to 70 degrees at 8 keV). In a-c highest absorption is rendered in white for best clarity, in d-e highest absorption is black. Slice from reconstruction of particle in (d) is shown in (e). In-house iterative reconstruction viewed with Amira.

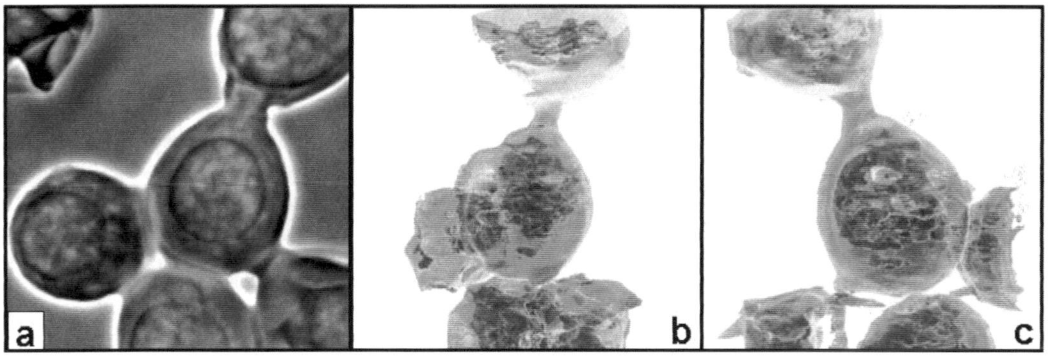

Figure 3: Zernike phase contrast TXM at 8 keV of yeast cells grown in Se-rich media. (a) Average of 20 2-sec scans. In (b) and (c) single images at two different viewing angles from a 3D tomographic reconstruction (8-sec images taken every 5 degrees from -75 to 75 degrees at 8 keV) show transparent view of inner cell. In-house iterative reconstruction viewed with Amira.

5. Conclusion

In summary, the TXM at beam line 6-2 at SSRL has successfully imaged samples with biological and environmental applications. With this method we have been able to discern features not previously apparent with other methods.

Acknowledgments

This work has been supported by NIH/NIBIB grant number R01-EB004321. SSRL is supported by the Department of Energy, Office of Basic Energy Sciences.

References

[1] Andrews J C *et al.*, this proceedings
[2] Mauro J B, Guimaraes J R D, Hintelmann H, Watras C J, Haack E A and Coelho-Souza S A 2002 *Anal. Bioanal. Chem.* **374** 983

Integration of X-ray micro tomography and fluorescence for applications on natural building stones

J Dewanckele[1, 3], V Cnudde[1], M Boone[2], D Van Loo[2], Y De Witte[2], K Pieters[1], J Vlassenbroeck[2], M Dierick[2], B Masschaele[2], L Van Hoorebeke[2] and P Jacobs[1]

[1] Department of Geology and Soil Science, Ghent University, Krijgslaan 281/S8, B-9000 Gent, Belgium
[2] Department of Subatomic and Radiation Physics, Ghent University, Proeftuinstraat 86, B-9000 Gent, Belgium

E-mail: Jan.Dewanckele@UGent.be

Abstract. X-ray computed tomography (CT) is an excellent, non-destructive analysis tool for characterising many different materials. In geosciences, 3D visualisation is becoming of prime importance in characterising internal structures of various rock types. It enables new approaches in petrophysical research of rock components, including pore and mineral distribution. Although CT provides a lot of information, this technique is limited concerning information on chemical element distribution. X-ray fluorescence (XRF) on the other hand is an excellent technique to obtain the missing information on chemical properties. At the recently established "Centre for X-ray Tomography" of Ghent University (UGCT) a micro- and nanoCT scanner has been constructed. It is expected that by combination of high-resolution CT and XRF it will be possible to characterise the spatial mineral and element distribution. The combination of both techniques has been applied on natural building stones, in order to get a better insight into some geological parameters (porosity, pore structure, mineral distribution, colour, grain orientation, etc.). Afterwards, the integration of the Morpho+ software tool provides us a 3D quantification of the resulting data.

1. Introduction

All monuments built with natural building stones are sooner or later affected by weathering processes. Most of the stones are formed under relative high temperature and pressure conditions in the earth's crust. Placement of those stones in buildings results into a thermodynamically unstable situation. Exposure of the stone to chemical, biological and physical weathering disintegrates the material and can lead to problems concerning the stability of the building. The use of alternative stone or replacing stone might solve problems of stone decay, but is not always a desired solution. Geological parameters that differ from those of the original stone may result in dissimilar weathering processes and thus different patina. Understanding the properties of natural building stone on micro-scale and the knowledge of its mineralogy are crucial for the assessment of weathering degree. To obtain important information on natural building stones, non-destructive tools as X-ray computed tomography (CT) and X-ray fluorescence (XRF) can be applied [1]. Without any form of sample preparation it is possible to

[3] Corresponding author: Tel: +32/(0)9 264 46 84; www.ugct.ugent.be

obtain a full 3D model of the samples' internal structure within a few minutes. The mineral distribution of the Balegem sandstone and the Massangis Roche Claire limestone has been visualized and quantified. Iron bearing minerals are potential spot producers, due to the oxidation of the iron; so its distribution is an important aspect in the evaluation of natural building stones.

2. Methods and instrumentation

To perform CT, digital radiographs of the sample are made from different orientations by rotating the sample along the scan axis from 0 to 360 degrees. The sample, in this case a natural building stone, is positioned between the X-ray source and the detector. In only one radiograph, the information of the sample is summed along the rays of the X-ray beam. This can already reveal some information, but to obtain a full 3D model, 800 radiographies were taken. Afterwards the projected data has been reconstructed, which produces horizontal cross-sections of a sample [2]. X-ray CT is based on the X-ray transmission information and from Beer´s law it follows that:

$$I_x = I_0 \cdot e^{-\mu x} \qquad (1)$$

where I_x is the measured intensity of the X-rays, I_0 is the emitted intensity, μ is the linear attenuation coefficient and x is the thickness of the object [3]. The attenuation of X-rays is strongly increasing with higher mass numbers and a few centimetres of natural building stone can already attenuate the entire X-ray beam intensity. The resolution we can obtain depends on the resolution of the detector, the spot size of the X-ray source and the magnification. The latter is expressed as the ratio of the source detector distance (SDD) to the source object distance (SOD) and depends on the largest diameter of the sample. In case of natural building stone, with a diameter of 8 mm, we obtain a resolution of 15 μm. Despite constant improvement of CT research and its valuable contribution to detect physical properties (like porosity changes), there is still missing information on chemical properties. To obtain those missing information add-ons like XRF can be applied.

The experiments were carried out at the Centre for X-ray CT at the Ghent University (Belgium) [4]. The X-ray tube of this high-resolution CT scanner is an open-type device with dual head (Feinfocus®, FXE-160.51). The head can produce a voltage between 10-160 kV, but for natural building stones a voltage of 130 kV with a power of 8 W was chosen, with an Al-filter to block the low energetic X-rays. As X-ray detector, a Varian 2520V Paxscan was used. The CT data is processed with the in-house developed reconstruction software Octopus. 3D morphological analysis was performed using the software Morpho+ which is able to segment the volume using advanced tresholding and to label, classify and separate tools. For 3D volume rendering, VGStudioMax from Volume Graphics was used. XRF analyses are made with the X-PIPS™ Detector (Canberra). The flexible set up of the scanner area at the UGCT allows using the same tube of the CT scanner as source for the XRF measurements.

3. Experimental data and results

With the aid of Morpho+ it is possible to obtain 3D data from the reconstructed 2D slices. The iron bearing minerals (glauconite) inside the Balegem sandstone and the Massangis Roche Claire limestone were segmented and labelled. Since the iron bearing minerals have similar grey values, they can be segmented as one object. Because the scans were performed with a resolution of 15 μm, fine dispersed grains cannot be visualized. Our 3D analysis shows that the limestone consists of 0.007% iron bearing minerals and the Balegem sandstone of 0.13%. The visualization of the mineral distribution is shown in figure 1. The XRF analyses were done with an integration time of 500 s and a tube voltage of 60 kV. Only surface detection of iron was possible, because the secondary X-rays cannot penetrate through the stone. The graphs in figure 2 clearly reveal the high iron concentration originating from the glauconite in the Balegem sandstone and the calcium rich content in the Massangis Roche Claire limestone. Due to their different mineral composition, the two natural building stones will have different weathering patterns.

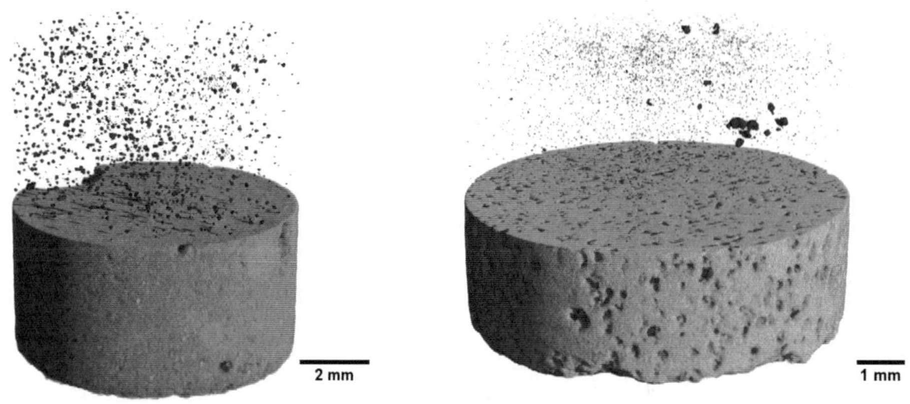

Figure 1. Visualization of the iron bearing minerals in the Balegem sandstone (left) and the Massangis Roche Claire limestone (right). Resolution of the scan is 15 μm.

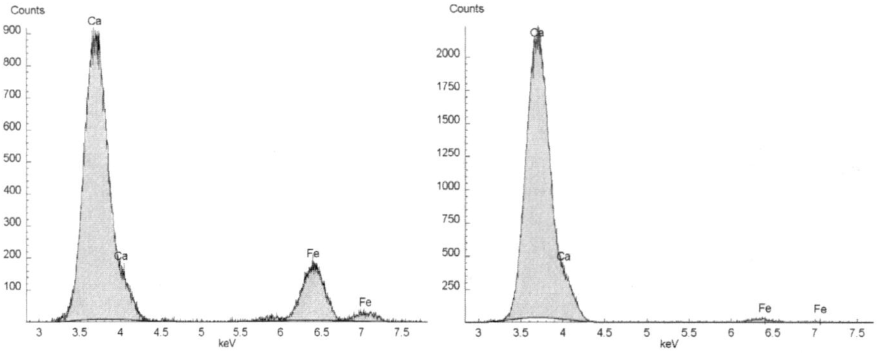

Figure 2. Surface sample XRF spectra of the Balegem (left) and the Massangis Roche Claire (right).

4. Conclusions

An X-ray CT system is a powerful tool to visualize the internal structure of natural building stones, in a non-destructive way. To obtain missing information concerning chemical properties, XRF can be applied. CT provides the possibility to compare some geological parameters of the replacing stone with the original material, to assist the restoration in welfare to our historical buildings. Materials with high glauconite content will have weathering patterns different from other stones. Due to the conversion of glauconite into goethite, the Balegem sandstone acquires a rust-colour pattern, while the Massangis Roche Claire remains white. Topics like porosity inside stone samples can be visualized and quantified with the assistance of Morpho+. Mineral distribution of replacing materials can be compared with the original material in a non-destructive way and new building materials can be characterized and their weathering qualities assessed.

References

[1] Cnudde V and Jacobs P 2004 *Env. Geol.* **46** 477-85

[2] Vlassenbroeck J, Dierick M, Masschaele B, Cnudde V, Van Hoorebeke L, Jacobs P 2007 *Nucl. Inst. Meth. in Phy. Res.* A **580** 442-45

[3] Ketcham R, Carlson W 2001 Comput. Geosci. **27** 381-400

[4] Masschaele B, Cnudde V, Dierick M, Jacobs P, Van Hoorebeke L, Vlassenbroeck J 2007 *Nucl. Inst. Meth. in Phy. Res.* A. **580** 266-69

X-ray fluorescence microscopy of olfactory receptor neurons

T. Dučić, E. Breunig[1], D. Schild[1], J. Herbst, E. Nováková, J. Susini[2], R. Tucoulu[2], T. Salditt

Institute for X-ray Physics, Georg-August-University, Friedrich-Hund-Pl. 1, 37077 Göttingen, Germany; [1] Department of Molecular Neurophysiology, Georg-August University Göttingen; [2] European Synchrotron Radiation Facility ESRF, 6 rue Jules Horowitz, 38043 Grenoble, France

E-mail: tducic@gwdg.de

Abstract. We report a x-ray fluorescence microscopy study of cells and tissues from the olfactory system of *Xenopus laevis*. In this experiment we focus on sample preparation and experimental issues, and present first results of fluorescence maps of the elemental distribution of Cl, K, Ca, P, S and Na both in individual isolated neural cells and in cross-sections of the same tissue.

1. Introduction

The sense of smell is concerned with the parallel and simultaneous detection of a multitude of molecular structures. Olfactory signal processing includes transport of odorants through the mucus to the transduction compartments, either stereocilia or microvilli, issuing from the dendrite of an olfactory sensory cell into the mucus [1]. As both stereocilia and microvilli have diameters between 100 and 200 nm, conventional optical microscopy is not sufficient to elucidate their fine structure. The olfactorial signal pathways presents a tremendous analytical challenge, regarding chemical and spatial resolution. The aim of the present investigations was to see to which extent x-ray fluorescence microscopy and x-ray spectromicroscopy is suitable at the presently achievable resolution and sensitivity to address this problem. In order to understand complex biomedical issues involving inorganic elements, the chemical species (chemical specification) of the elements must be considered: oxidation state, coordination, and/or complex or molecular structure [2, 3].

2. Sample preparation x-ray microscopy experiments

X-ray fluorescence and x-ray spectromicroscopy are in principle well suited methods to achieve these high requirements to study the biological structures of interest. However, a major challenge, is given by the need to prevent highly mobile molecular species such as ions, from rearrangement during sample fixation and preparations. Preservation of biological samples from X-ray damage during the measurements is a second important issue, excluding high resolution experiments in aqueous solution without fixation. While cryo microscopy is the best solution towards the goals of structural isomorphism and reduced radiation damage, it is also technically the most challenging one. Here we used freeze drying method after rapid cryogenic plunging as well as embedding samples in styrene-methacrylate mixture and cutting 1 μm thin slices for analysis, see Fig 1. As experimental model we used cells and tissues from the olfactory system of *Xenopus laevis*, in which both kinds of transduction compartments are co-exist.

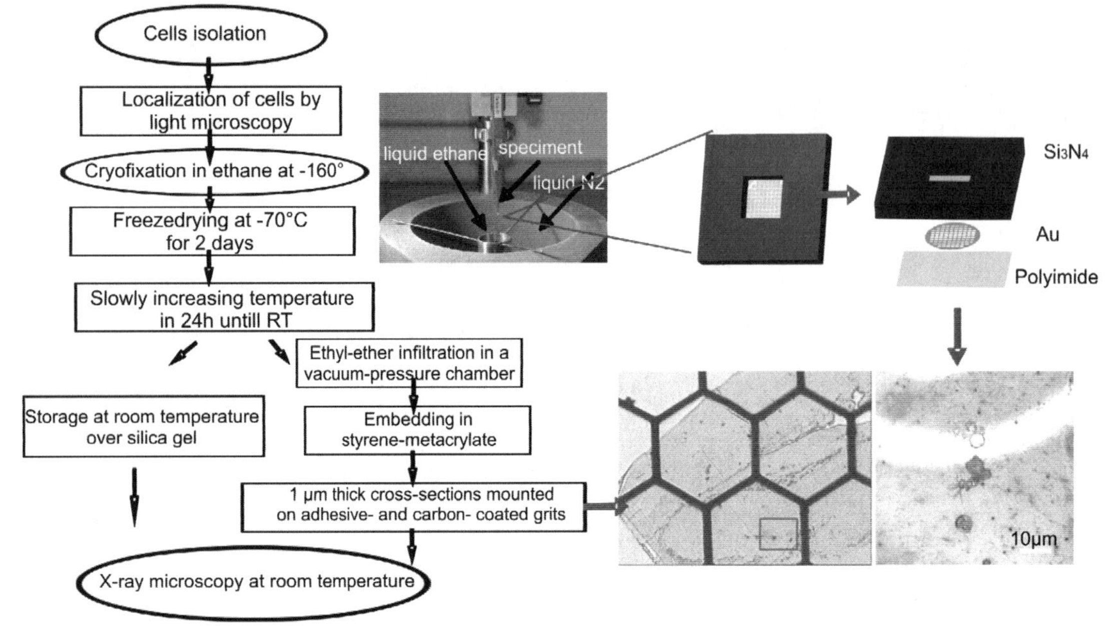

Figure 1: Two analytical procedures for cell preparation and X-ray fluorescence measurement at room temperature. After freeze-drying protocol, cell and tissue samples were prepared in two different ways: direct measurement without any fixation or with the classical electron microscopy embedding preparation using a styrene-methacrylate mixture.

Cryo-preserving and freeze drying method: The sample of the olfactory receptor neurons (ORNs) were freshly isolated from the nasal epithelium of larvae *Xenopus laevis* (stage 54-56)using mechanical trituration in divalent free buffer solution and short incubation in papain [4, 5]. Isolated cells were carefully transferred into a Ringer solution containing 0.05 M ammonium acetate buffer, pH 7.8 (1:1 w:w). The cells were suspended in 5 μl solution were placed on a 200 nm thick Si_3N_4 membrane, which had been coated previously with 0.1 % laminin and 0.01 % poly-L-lysine. All cells used in this study were clearly identified, prior to the measurements as ORNs on the basis of their characteristic morphology, having a single thick dendrite with a knob-like swelling from which emanated 3-10 cilia. Cells naturally attached to the substrate after approximately 3 min. The samples were rapidly frozen in liquid ethane cooled in liquid nitrogen, and subsequently quickly transferred in liquid nitrogen until freeze-drying at -70°C for two days. Samples are then very slowly warmed up to room temperature over 24 h and stored over silica gel in a desiccator.

Embedding the neural tissue for X-ray microscopy: Samples isolated from nasal epithelium of larvae *Xenopus laevis* were rapidly frozen in a mixture of propane: isopentane (2:1) cooled with liquid nitrogen to -196°C in an aluminum mesh and freeze-dried at -70°C for three days and stored at room temperature over silica gel in a desiccator. Freeze-dried samples were infiltrated with ether in a vacuum-pressure chamber and embedded in styrene-methacrylate using a technique specifically developed for analysis of diffusible elements. 1 μm thick sections were cut dry by glass knives, mounted on adhesive 100-mesh hexagonal Cu grids (G-100 hex, Canemco and Marivac Inc.), coated with carbon, and stored over silica gel.

X-ray fluorescence: The elemental distribution and elemental speciation of isolated neural receptor cells with intact ciliae was carried out by scanning x-ray microscopy, using the combined capabilities of the scanning transmission microscope (STXM) at ESRF ID21 beamline, as well as the nanoprobe imaging system at ID22NI beamline, in two seperate beamtimes. The specification of these beamlines are given elsewhere [6, 7]. The following ions are of special importance in the neural receptor cells: Cl, K, Ca, P, S, Na as well as the trace elements Zn, Fe and Cu. The best fluorescence results, without x-ray damage, have been obtained at a photon energy of 4.1keV energy at ID21. Typical fluorescence maps of individual cells (analyzed with PyMca software) and cross section of the embedded olfactory epithelium are shown in

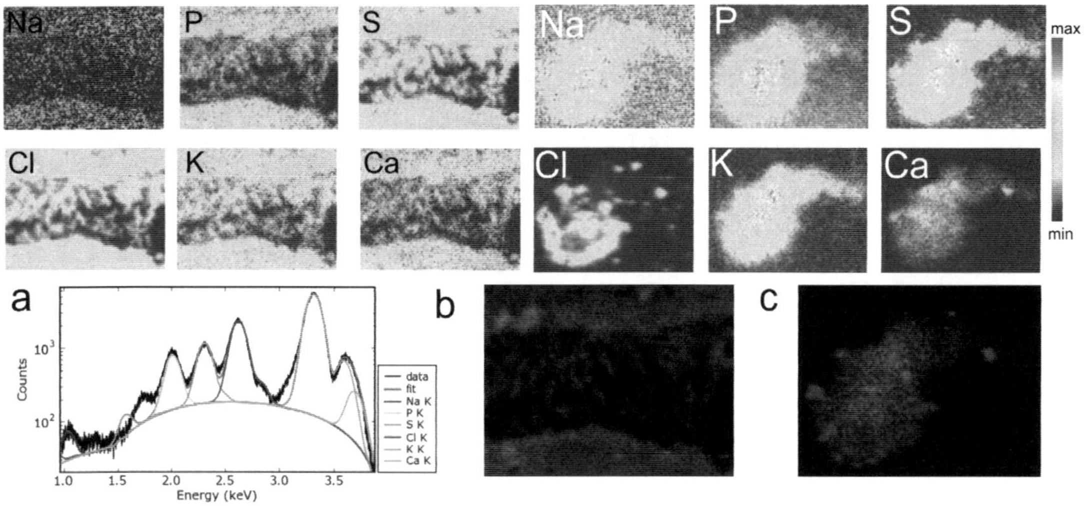

Figure 2: Na, P, S, Cl, K and Ca distribution and overlapped map of K (red) and Cl (blue) of the embedded tissue of neural receptor cells on the left part (marked with black letters); the zoom in the small part (26 x 19 μm), image 130 x 96 pixels and a dwell time 400 ms per pixel at E=4.1 keV. On the right side (marked with white letters) is shown the element distribution a individual isolated neural receptor cell; image 25 x 15 μm 100 x 70 pixels with a dwell time 900 ms per pixel at E=4.1 keV. Typical sum spectra (a), the distribution and overlapped map of K (red) and Cl (blue) in the embedded tissue (b) and individual isolated neural receptor cell(c).

Fig. 2. The distribution and concentration of chloride in this system is relevant in view of signal transduction [4]. The main result of this study was the localization of Cl, mainly in the soma of the receptror cell, but also in patches along the dendrite. We hypothesise that these patches of high chloride concentration may correspond to acidic intracellular compartments. This can be deduced in particular from the single cell maps (Fig. 2). While Cl was expected to be located primarily to the cilia, it appears an intracellular gradient and possibly a transport mechanism exist in this kind of receptor cell. This is consistent with recent immunocytochemical results indicating Na^+-K^+-$2Cl^-$ co-transporters to be located to the somata and dendrites [8]. Chloride was also observed to accumulate in globular structures close to the plasma membrane. The formation of spurious NaCl salt as an artefact of the preparation can be excluded from the simultaneously measured Na map. In our analysis Na, K and Ca were distributed to variable extents in the different parts of the cell, their distributions showing some overlap (Fig. 2). These differences show that the distribution is not dominated by a single functional architecture such as homogeneously expressed and membrane associated transporters or co-transporters. The comparison of the two presented methods of sample preparation shows that individual ORNs are better suited for elemental mappings than tissue slices. On the other hand, thin tissue slices can potentially help to address questions such as the odorant transport within the mucus. Even if the preparation of dissociated ORN is more invasive, it certainly gives reliable results as to the element distributions.

Acknowledgements: The work was supported by DFG Research Center for Molecular Physiology of the Brain (CMPB), Germany and ESRF, France. Thanks to K. Giewekemeyer for great cooperation.

References

[1] Gennerich A and Schild D 2002 *Biophys. J.* **83** 510–522
[2] Fahrni C 2007 *Current Opinion in Chemical Biology* **11** 121–127
[3] Bacquart T, Deves G, Carmona A, Tucoulou R, Bohic S and Ortega R 2007 *Anal. Chem* **79** 7353–59
[4] Manzini I, Peters F and Schild D 2002 *J. Neurosci. Meth.* **121** 159–167
[5] Manzini I and Schild D 2003 *J. Physiol.* **551** 115–123
[6] Susini J, Somogyi A, Barrett R, Salomé M, Bohic S, Fayard B, Eichert D, Dhez O, Bleuet P, Martinez-Criado G and Tucoulou R 2004 *Proc. of International Symposium on Portable Synchrotron Light Sources and Advanced Applications (AIP Conference Proceedings, Japan, 2004)* **716** 18
[7] Sole V, Papillon E, Cotte M, Walter P and Susini J 2007 *Spectrochimica acta B-atomic spectroscopy* **62** 63–68
[8] Reisert J, Lai J, Yau K W and Bradley J 2005 *Neuron* **45** 553–561

Journal of Physics: Conference Series **186** (2009) 012084 doi:10.1088/1742-6596/186/1/012084

High resolution micro-XRF maps of iron oxides inside sensory dendrites of putative avian magnetoreceptors

G. Falkenberg[1]*, Ge. Fleissner[2], Gue. Fleissner[2], K. Schuchardt[2], M. Kühbacher[3], E. Chalmin[4], K. Janssens[5]

[1] HASYLAB at DESY Hamburg, Germany
[2] Institute of Cell Biology and Neurosciences, Goethe-University Frankfurt a. M., Germany
[3] Department of Molecular Trace Element Research in the Life Sciences, Helmholtz Centre Berlin for Materials and Energy, Berlin, Germany
[4] ID21 ESRF, Grenoble, France
[5] Department of Chemistry, University Antwerp, Belgium

Abstract. Iron mineral containing sensory dendrites in the inner lining of the upper beak of homing pigeons [1] and various bird species [2] are the first candidate structures for an avian magnetic field receptor. A new concept of magnetoreception [3, 4] is based on detailed ultrastructural optical and electron microscopy analyses in combination with synchrotron radiation microscopic X-ray fluorescence analysis (micro-XRF) and microscopic X-ray absorption near edge structures (micro-XANES). Several behavioral experiments [5, 6] and first mathematical simulations [6] affirm our avian magnetoreceptor model. The iron minerals inside the dendrites are housed in three different subcellular compartments (bullets, platelets, vesicles), which could be clearly resolved and identified by electron microscopy on ultrathin sections [1, 3]. Micro-XRF and micro-XANES data obtained at HASYLAB beamline L added information about the elemental distribution and Fe speciation [3], but are averaged over the complete dendrite due to limited spatial resolution. Here we present recently performed micro-XRF maps with sub-micrometer resolution (ESRF ID21), which reveal for the first time subcellular structural information from almost bulk-like dendrite sample material. Due to the thickness of 30 μm the microarchitecture of the dendrites can be considered as undisturbed and artefacts introduced by sectioning might be widely reduced.

1. Introduction

Though magnetoreception could be established as a "sixth" sense in animals from different phyla mainly by behavioral tests, only little is known on the receptor- and neurophysiological basis of magnetoreception [for review 6]. This might be attributed to a principle lack of both: studies which apply receptorphysiological principles to a putative magnetoreceptor candidate and also critical physical check-ups concerning its function under natural conditions. Therefore, hypotheses rather than facts govern the field and basically two mechanisms are proposed as a basis for magnetic field guided orientation and possibly interact for a safe tool to find a distant goal [6]. A photopigment-based magnetoreceptor within the retina might provide a compass, an iron-mineral-based receptor within the beak could serve as a magnetometer. Experiments for testing the retinal magnetoreceptor are mainly behavioral tests or physical simulations of magnetic field effects on the assumed key molecule, cryptochrome. A detailed cellular localization and the nature of magnetic field induced receptor mechanisms are still unknown. In this paper, we will deal only with the second mechanism, the iron mineral based magnetoreception. In upper beaks of different bird species, homing pigeons, migratory and non-migratory birds, we could describe highly specialized sensory dendrites [1, 2, 3], which match all histological and physical prerequisites as the smallest units of a receptor system sensing the gradual changes of the Earth magnetic field [4]. But we still must specify the stimulus-response function[1] for a sound characterization of the physiological features and the biological meaning of this system. A major difficulty originates from the broad range of different dimensions (cm down to nm) of the particles and components involved and the unsolved question concerning e.g. their relative position: (1) Within the beak we find six areas (each about 300×15×15 μm^3 wide) with multiple iron containing dendrites.

* To whom all correspondence should be addressed: gerald.falkenberg@desy.de

[1] Detailed knowledge of the putative threshold and saturation values as well as the dynamics of the proposed magnetoreceptor are indispensable for all further studies as inadequate stimuli might destroy or at least temporarily constrain the system and conceal its biological function.

The 3-dimensional arrangement of the dendrites within these areas indicates that they may form the structural basis for a 3-axial magnetometer. But how precise is their alignment? Is their a bias? Are the dendrites within the bilateral areas parallel to each other or tilted for a better spatial resolution? (2) So far, we have found all dendrites of identical shape and size (20×5 μm^2) and equipped with the same iron containing compartments: 10-15 bullets (diameter 1 μm), a vesicle (diameter 5 μm) and two parallel sets of chains of crystals (each 1×1×0.1 μm^3). But as we can observe all these particles in ultrathin sections only, we do not know their in situ alignment, which would be necessary to better evaluate the magnetic forces between the iron containing components. (3) The bullets consist of nanomagnets (about 4 to 7 nm). So far we do not have a precise idea of their assembly. Magnetic field effects may vary when they are free to move as single crystals compared to an organization in small interdependent chains.

All histological procedures and physicochemical analyses, which have been applied in our project before, provided either highly resolved details in ultrathin sections or rather coarse overviews in thick sections. By means of ID21 at ESRF we now have started pilot experiments with XRF-mappings combining both: in rather thick sections (30 μm)[2] - thus avoiding structural distortion by histological procedures - we have analyzed large dendritic areas of the skin with a high resolution, which would allow an identification of the microarchitecture. The lateral resolution is maintained because the Fe containing structures are much smaller than the section thickness. Further experiments might help to identify the specific iron oxides inside the dendrites and clearly discriminate them from the iron generally occurring within the surrounding tissue and blood capillaries.

2. Results and Discussion

In histological sections of the inner lining of the upper beak of a homing pigeon distinct areas, sensory dendrites, with high concentrations of iron could be found by means of Prussian Blue staining. This qualitative method has served as a preselection of samples, which were mapped by μ-XRF-methods for a quantitative analysis of the dendritic iron minerals [3]. XRF-mappings of 10 μm thick sections at beamline L at HASYLAB (DESY Hamburg) of an axonal bundle can depict different minerals, which are indicative for nervous tissue (e.g. Ca) colocalized with iron (Fe) concentrated at several sites (Fig. 1). But iron is also present all over the tissue strand (weak tint). Here we face two difficulties, when we try to specify the iron within distinct compartments: Due to the focal size (about 5 μm) of the XRF-beam it is not (yet) possible to discriminate between the iron of certain particles (bullets or platelets with 1 μm diameter) within the sensory dendrite, which we assume as essential for magnetoreceptive function, and even individual dendrites are not resolved within the dendritic field. Iron regularly occurring in any other cellular compartment of the surrounding nervous and connective tissue cells as part of their iron metabolism, e.g. hemoglobin, cytochrome, ferritin will also add to the measurement.

At ID21 (ESRF) we were able to test thicker sections (30 μm) with higher resolution (Fig. 2). The μ-XRF-mappings now clearly show networks of iron containing structures, which resemble in their size and shape (5 to 20 μm) the known sensory dendrites. For the first time this thick section also shows their microarchitecture with a main alignment of the dendrites in parallel to the rim of the nervous strand. Few dendrites, however, deviate from this direction in an angle of about ±45 ° (Fig. 2C). In our previous papers we have provided evidence that each dendrite may sense only one direction of the magnetic field [3, 4]. Several dendrites with a different alignment in the skin are necessary to enable the system to monitor different directions of the magnetic field. Previously, we could show six dendritic areas with a perpendicular orientation of the dendrites [3, 4]. But we did not know how strict or uniform this 3D-microarchitecture might be in situ, as all data so far stem from serial sections of the skin – with the disadvantage of an ambiguous distortion or displacement. Sensory systems, in general, avoid step-like stimulus-response functions, as this would yield an all-or-none signal and prevent a gradual excitation according to smooth changes of the adequate stimulus. Based on the ID21 data, we

[2] In previous papers [1,3] we have published the caveats of the processing of this material in detail, also the parallel investigation of unstained test sections in parallel to Prussian Blue stained control sections in order to find the relevant dendritic areas of the skin. For comparison, parallel sections were mapped at beamline L at HASYLAB and then inspected at ID21 at ESRF.

now predict that the entity of the iron containing dendrites deliver a transition curve with a shallow slope. The system may monitor gradual changes of the 3 axis of the earth magnetic field.

Further experiments at ID21 or similar X-ray microprobe beamlines will hopefully disclose the subcellular micro-architecture in greater detail, i.e. the relative position of the iron-containing compartments, once the analyses can be performed without a disintegration of the structural context, and also identify the specificity of iron minerals within single bullets, platelets or vesicles.

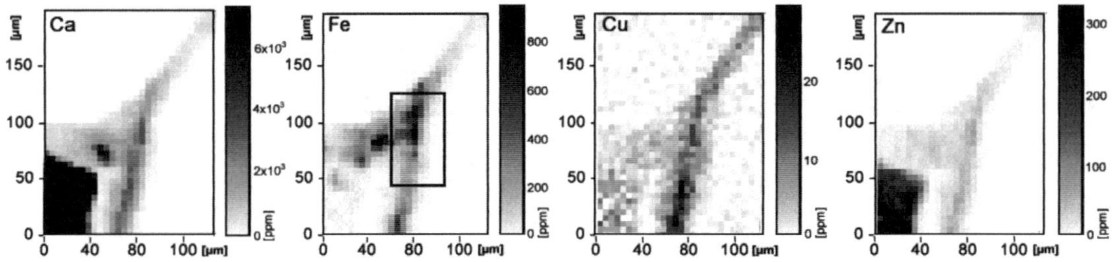

Fig. 1 Micro-XRF element mappings at beamline L at HASYLAB (beam size 5×5 µm², $1x10^{10}$ ph/s, 17 keV). The measurements confirm the distribution of Fe within or next to nervous tissue according to the distribution of Ca, Cu and Zn. [10 µm thick paraffin section, unstained. The Ca intensity color scale has been truncated at 1/20 of the maximum value. The square indicates the site of the ID21 measurements in a parallel section (see Fig. 2)].

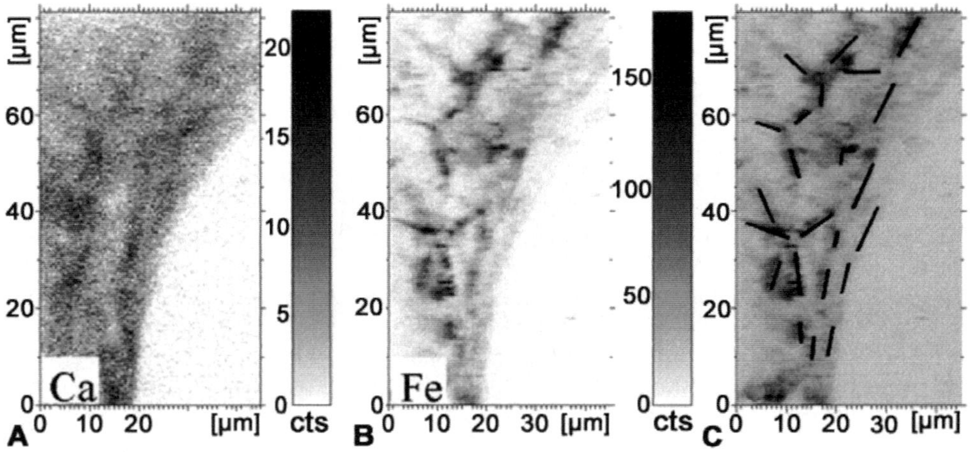

Fig. 2 µ-XRF mappings at ID21 at ESRF (beam size 0.3x1 µm², $1x10^{9}$ ph/s, 7.5 keV). Iron maps with submicrometer lateral resolution reveal structures within the dendritic field, where Ca (Fig. 2A) and Fe (Fig. 2B) are strictly colocalized in a quantity above background level. Fig. 2C: The orientation of these structures which we suggest to assign to individual dendrites (indicated by black bars) scatters around a main direction of the axon bundle. [30 µm thick paraffin section, unstained].

3. Acknowledgement. The project has received grants of the DFG (Fl 177/15-1, 16-1), of HASYLAB (I-05-94, I-05-95), of ESRF (EC-159), of the Freunde und Förderer der Goethe-University Frankfurt and the Stiftung Polytechnische Gesellschaft Frankfurt. The general regulations, policies, and principles given in the "Guide for the Care and Use of Laboratory Animals" were followed.

4. References

[1] Fleissner Ge. et al. 2003 *J. Comp. Neurol.* **458** 350-360
[2] Stahl B et al 2007 *HASYLAB Ann Report 2006 (DESY)* 1338-1339
[3] Fleissner Ge et al 2007 *Naturwissenschaften* **94** 631-642
[4] Fleissner Ge et al 2007 *J. Ornithol.* **148** 643-648
[5] Thalau P et al 2007 *Naturwissenschaften* **84** 813-819
[6] Wiltschko R and Wiltschko W 2006 *Bioessays* **28** 157-168
[7] Solov'yov A and Greiner W 2007 *Biophys. J.* **93** 1493-14509

STXM Search for Carbonate in Samples of Comet 81P/Wild 2

G. J. Flynn[1], S. Wirick[2], L. P. Keller[3] and C. Jacobsen[2]

1) Dept of Physics, SUNY-Plattsburgh, 101 Broad St., Plattsburgh, NY 12901 USA
2) Dept. of Physics, SUNY Stony Brook, Stony Brook, NY 11794 USA
3) NASA Johnson Space Center, Houston, TX 77058 USA

george.flynn@plattsburgh.edu

Abstract. We employed a Scanning Transmission X-ray Microscope (STXM) to search for carbonate in samples of dust from Comet 81P/Wild 2 collected by NASA's Stardust spacecraft. Although carbonate was not expected in Wild 2 because it generally forms by aqueous processing, which is not thought to occur on comets, we identified two sub-micron carbonate grains in Wild 2 samples. The results demonstrate the effectiveness of the STXM as a search and identification tool for rare minerals that exhibit distinctive X-ray Absorption Near-Edge Structure (XANES) spectra.

1. Introduction

Comets are generally believed to have formed in a very cold region of our Solar System, at a temperature below the freezing point of CO_2. When they formed they trapped the original dust grains of our Solar System in the ice, where these grains have been preserved in cold-storage for the last 4.5+ billion years.

On Earth, carbonates are generally deposited from liquid water. Carbonates are common in hydrous meteorites and in hydrous interplanetary dust particles (IDPs), where they are believed to have formed by parent-body aqueous processing. Since simple models of cometary evolution involve no aqueous processing, carbonates were generally presumed not to occur in comets. However, Toppani et al. [1] have performed experiments that indicate carbonate can be formed by non-equilibrium condensation in circumstellar environments where water is present as a vapor, not as a liquid.

2. Carbonate in comets

Toppani's experiments suggest carbonate might have condensed in cold regions of the Solar Nebula, and might be present in comets.

2.1 Prior Results on Carbonate in Comets

The two VEGA spacecraft and the Giotto spacecraft that flew through the coma of Comet 1P/Halley in 1986 determined the elemental composition of Halley dust particles using impact ionization mass spectrometers. The dust particles were separated into groups based on their element abundance patterns. One group consisted of particles with high Mg but relatively low Si. Many of these grains also had high C/S ratios, indicating they were not Mg-sulfides. The low Si and S content caused Fomenkova et al. [2] to identify this group as Mg-carbonate grains. A weak infrared emission feature at 6.8 μm, attributed to carbonate, was also observed in Comet Halley [3]

A projectile from the Deep Impact spacecraft struck Comet 9P/Tempel 1, producing a dust plume that was examined using the infrared spectrograph on the Spitzer Space Telescope. Lisse et al. [4] associated a

Journal of Physics: Conference Series **186** (2009) 012085 doi:10.1088/1742-6596/186/1/012085

strong emission between 6.5 and 7.2 μm with carbonate. After subtraction of strong silicate features, they identified weaker features suggesting that both Mg- and Fe-carbonate were present [4].

2.2. Carbonate Outside Our Solar System

Astronomical infrared observation provides the opportunity to detect minerals that condensed in the gaseous nebula around other star systems. Infrared spectra of the grains in the planetary nebula NGC 6302 and NGC 6537 from the Infrared Space Observatory have been interpreted by Kemper et al. [5] as showing diopside ($CaMgSi_2O_6$), dolomite ($CaMg(CO_3)_2$ -- a carbonate), and water ice. The spectrum indicates NGC 6302 is so hot that its emission peaks in the ultraviolet, so this planetary nebula does not contain water in the liquid phase. In addition, Ceccarelli et al. [6] reported the detection of carbonate in the protostar IRAS4. Both astronomical results remain controversial, in part because a relatively broad absorption feature is fit with a mixture of the absorptions from several reference materials, so no distinctive feature from carbonate alone was detected. Hofmeister et al. [7] provides a more complete discussion of this controversy.

2.3 Stardust Samples of Comet 81P/Wild 2

NASA's Stardust spacecraft flew through the dust coma of Comet 81P/Wild 2 in January 2004, collecting dust by impact into ultra-low density silica aerogel [8]. These samples were delivered to Earth in January 2006, providing the first opportunity for laboratory examination of dust identified as being from a comet. The Wild 2 particles provide the first unambiguous samples of the early dust of our Solar System, preserved at low temperature in the comet since its formation.

3. Wild 2 Dust Characterization

Fragments from Wild 2 dust particles were extracted from the aerogel. Some fragments were embedded in epoxy or elemental S, prepared as ultramicrotome slices, and made available for our carbonate search.

3.1 The Scanning Transmission X-ray Microscope (STXM) as a Carbonate Identification Instrument.

We used a STXM to perform X0ray Absorption Near-Edge Structure (XANES) spectroscopy on Wild 2 samples. Pre-edge absorptions in C-, N-, and O-XANES spectra result from x-ray induced electron transitions from the core to higher electron energy levels. The energy of each pre-edge absorption is diagnostic of the functional group, and thus provides an indication of the interatomic bonding or mineralogy of the sample. For example, C-bearing minerals – graphite, diamond, C_{60}, and carbonate – each have a unique, easily identifiable C-XANES spectrum (as shown in Figure 1). In particular, carbonate has only a single, strong absorption, from the $C\text{-}O_3$ bond, near 290 eV, that distinguishes carbonate from most other carbon-bearing minerals.

Carbonate grains as small as ~50 nm can be located using the STXM to an ultramicrotome slice of the sample at two energies, one on the carbonate absorption and a second at an energy slightly above the absorption feature. Pixels showing significantly greater absorption on the $C\text{-}O_3$ peak can be investigated further by obtaining a full C-XANES spectrum. However, we elected to acquire a full image stack [9] over the energy range from 280 to 310 eV, Carbonate grains are immediately visible as a bright spot in the images near the $C\text{-}O_3$ absorption energy, and the C-XANES spectrum can be extracted from the image sequence. STXM imaging provides a rapid and reliable technique to locate and identify carbonate.

3.2 Carbonate Identification in Wild 2 Fragments

We have examined 14 ultramicrotome slices from Wild 2 fragments, 7 embedded in epoxy and 9 embedded in S. Figure 2 shows an ~200 nm carbonate found in Wild 2 particle FC4,0,3,1,2 using the STXM on beamline X1A of the National Synchrotron Light Source (Brookhaven National Laboratory). It is difficult to assess the total area of Wild 2 material we examined in the 14 sections because Wild 2 material is intermixed with compressed and/or melted aerogel in most sections. However, it is clear that the two sub-micron carbonates we identified constitute only a very small fraction of the total Wild 2 material we examined.

Figure 1: C-XANES spectra of C_{60}, CVD diamond, calcite, graphite, and amorphous carbon standards.

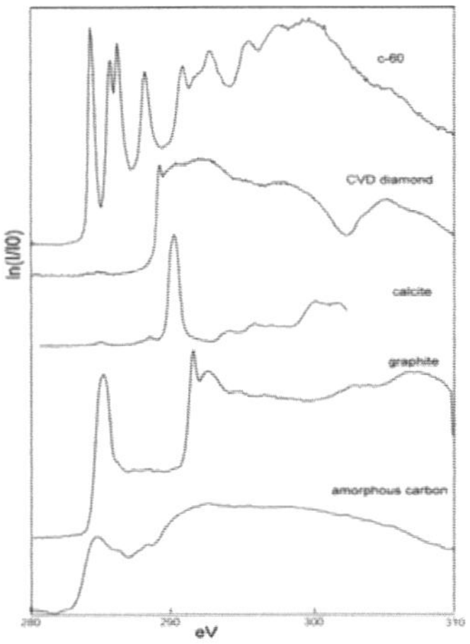

Figure 2: STXM image and C-XANES spectrum of an ~200 nm carbonate (circled) found in Wild 2 particle FC4,0,3,1,2.

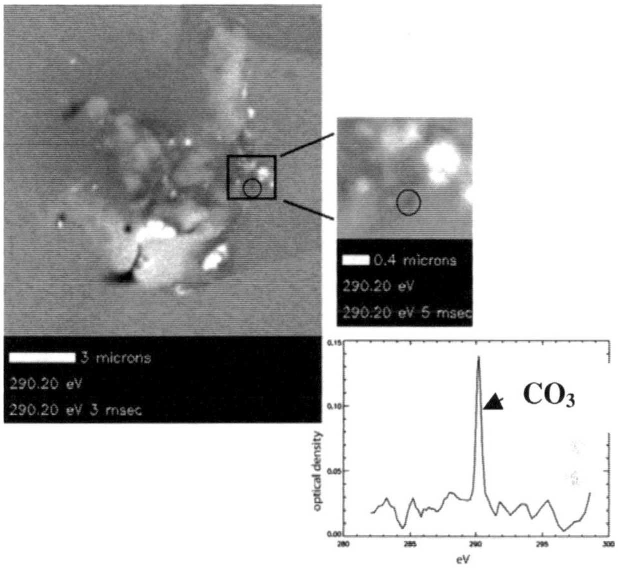

4. Conclusions

We have identified two carbonate grains, each smaller than 300 nm in the Wild 2 sections examined to date. The area of carbonate is very small compared to the total area of Wild 2 material we examined,. Thus carbonate appears to be a rare component of Wild 2 grains. However, particle capture into aerogel at 6 km/sec is size selective. It appears that "components larger than micron-size were often well preserved, whereas smaller or finer-grained components were strongly modified" by capture [7]. If all the carbonate in Wild 2 is sub-micron in size, while silicate minerals and sulfides several microns in size were observed, the mass-fraction of carbonate in the impacting particles may never be determined because most were modified on capture. Our observations set only a lower limit on the carbonate abundance.

Our results demonstrate the effectiveness of employing a STXM to search for carbonate. Because other rare phases, including diamonds, fullerenes, nitrides, and metal-oxides, also have distinct C-, N-, or O-XANES spectra, STXMs can also be employed to search efficiently for these minerals.

Acknowledgements:
This work was supported by a NASA research grant NNX07AM85G. Support for the upgrade of the X1A STXM was provided by a NASA SRLIDAP grant. Use of the National Synchrotron Light Source, Brookhaven National Laboratory, was supported by the U.S. Department of Energy, Office of Science, Office of Basic Energy Sciences, under Contract No. DE-AC02-98CH10886.

References

[1] Toppani, A. et al. (2005) *Nature,* **437**, *1121-1124.*
[2] Fomenkova, M. N.; Kerridge, J. F.; Marti, K.; McFadden, L.-A. (1992) *Science* **258**, 266-269.
[3] Bergman, J. D. et al. (1987) *Astron. Astrophys.* **187**, 617. [
[4] Lisse et al. (2006) *Science*, **313**, 635-640 and Supporting Online Material.
[5] Kemper, F. et al. (2002) *Astronomy & Astrophysics*, **394**, 679-690.
[6] Ceccarelli, C. et al. (2002) Ast*ronomy & Astrophysics*, **395**, L29-L33.
[7] Hoffmeister, et al., (2004), *Geochemica et Cosmochimica Acta*, **68**, 4485-4503
[8] Brownlee, D. et al. (2006) *Science*, **314**, 1711-1716.
[9] Jacobsen, C. et al. (2000) *Journal of Microscopy,* **197**, 173-184.

Fluorescence imaging of *Dictyostelium discoideum* with a hard X-ray nanoprobe

K. Giewekemeyer[1,#], M. Hantke[1], C. Beta[2,3], R. Tucoulou[4] and T. Salditt[1]

[1]Institute for X-ray Physics, Georg-August-University, Friedrich-Hund-Pl. 1, 37077 Göttingen, Germany
[2]Max-Planck-Insitute for Dynamics and Self-Organization, Am Faßberg 11, 37077 Göttingen, Germany
[3]Institute of Physics and Astronomy, University of Potsdam, Karl-Liebknecht-Str. 24/25, 14476 Potsdam/Golm, Germany
[4]European Synchrotron Radiation Facility (ESRF), BP 220, 38043 Grenoble, France

E-mail: #k.giewek@phys.uni-goettingen.de

Abstract. The preparation and a novel sample environment for X-ray based imaging of freeze-dried *Dictyostelium discoideum* cells are presented. As a first application a fluorescence imaging experiment with a nanofocused hard X-ray beam has been performed. The successful preparation was verified in elemental mappings with sub-200nm resolution, which allowed for the isolation of several ionic components specific to the cell body.

1. Introduction

The ability of cells to respond to an external chemical stimulus by directed motion (*chemotaxis*) is an essential requirement for cellular differentiation and complex cooperative behaviour. It has been studied very successfully on the molecular level in the eukaryotic amoeba *Dictyostelium discoideum* [1]. However, an equally-detailed view of the structure and dynamics on a (sub)cellular level, i.e. in the range of 10-500 nm, has not been achieved yet.

Here we report on the preparation and a novel sample environment for X-ray-based imaging of freeze-dried *Dictyostelium* cells. The successful preparation is demonstrated in an X-ray fluorescence imaging experiment at the ESRF (Grenoble) paving the way for a general application to X-ray imaging experiments of biological cells. X-ray fluorescence imaging has been proved a powerfull imaging technique in the recent past due to its specificity to any given metal ion and the possibility to measure ion concentrations of various elements for the same sample without staining simultaneously at high resolution [2].

2. Instruments and Methods

Cells of the *Dictyostelium discoideum* wild-type strain AX2-214 were used in our experiments. The cells were cultivated in shaking suspension of HL5 nutrient medium (7 g/L yeast extract, 14 g/L peptone, 0.5 g/L KH2PO4, 0.5 g/L Na2HPO4). Prior to preparation, cells were transferred to a Petri dish for transport. They were allowed to attach to the bottom surface of the Petri dish for at least 1h. Before transfer to the sample holder the supernatant nutrient solution was removed and cells were resuspended in 1-2 ml phosphate buffer (2 g/L KH2PO4, 0.36 g/L Na2HPO4 · 2H2O, pH 6.0). As sample holders microfabricated polyimide foils of 10-20 μm thickness and ca. $300 \times 800\,\mu$m lateral dimensions (Mitegen, USA, cf. also [3]) were used. After cleaning in an ultrasonic bath of methanol and immediatly prior to use the substrates were rendered hydrophilic by a 5-min-treatment in a plasma cleaner (Harrick Plasma, USA). A drop of cell suspension was placed onto the substrate, which was then injected

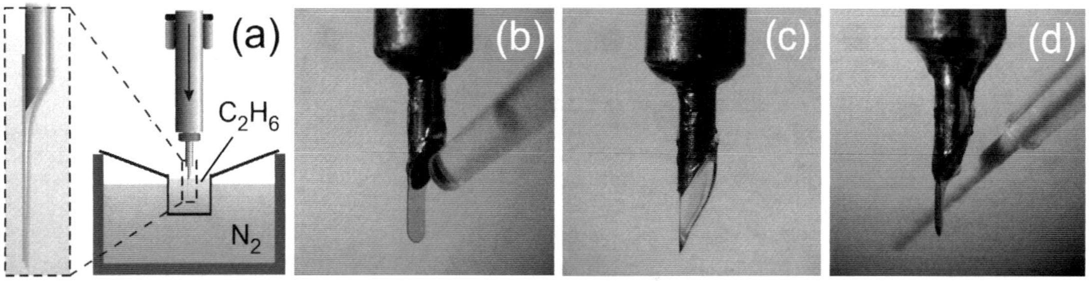

Figure 1: Cell preparation by rapid freezing in liquid ethane (a). A small metal container filled with liquid ethane is kept in a bath of liquid nitrogen. The sample holder consisting of the polyimide foil on a metal pin is mounted onto a magnetic base, which is attached to a metal rod movable in the vertical direction. Before plunging the cell suspension is transferred onto the foil with a micropipet (b), and after a resting time of 1-3 min (c) the supernatant suspension is soaked up with a thin paper rod (d). Immediatly afterwards, the metal rod is released, so that the polyimide tip is injected into the liquid ethane by gravitational force. During this process, the cells attached to the foil are covered with a thin water film of less than 1 μm thickness at the tip (magnified inset in (a)) allowing for rapid cooling of the solution into a glassy state.

into liquid ethane to prevent crystalization [4] (cf. Fig. 1). After transfer into liquid nitrogen the sample holders were placed in a freeze-drier (Christ, Germany) and removed after 40h under vaccum (0.001 mbar) and -76 °C (temperature of the condenser).

The experiment was performed at the ID22NI undulator beamline of the ESRF (Grenoble, France) [5]. Quasi-monochromatic radiation of photon energy 17.5 keV was generated in the so-called pink mode using the intrinsic monochromaticity of the undulator source. The energy bandpass of the incoming radiation was further reduced to about $\Delta\lambda/\lambda \sim 0.02$ by two-dimensional focusing of the beam with a Kirckpatrick-Baez (KB) mirror system and a flat, horizontally deflecting Pd-coated Si-mirror for higher-harmonics rejection. The focus of the KB-mirror system was characterized by translation of an Au stripe of a nano-fabricated test pattern (Xradia, USA) recording both the transmitted intensity by a diode and the Au-L_α fluorescence by a silicon drift detector (Vortex-EX, SII NanoTechnology Inc., Japan). The measured focal spot size was 160 nm (FHWM) in the vertical and 146 nm in the horizontal direction, respectively. The total intensity in the focal spot was on the order of 10^{11}cps. The sample was placed directly into the focal plane of the KB system and translated by two piezo stages in two directions prependicular to the beam. During the scanning process the fluorescence signal was detected with the silicon drift detector (see above). Before and after the experiment optical micrographs of the sample could be obtained with an online microscope at the beamline. Great care was taken on limiting the radiation dose impinging on the sample by scanning only previously non-illuminated regions on the substrate. Optical fluorescent micrographs of the illuminated regions of the sample were obtained after the experiment with a fluorescent microscope (Z1 Observer, Zeiss, Germany) at an excitation wavelength of 590 nm.

3. Results and Discussion

An X-ray fluorescence mapping of a freeze-dried *Dictyostelium* cell and the corresponding cumulative energy spectrum obtained from the total mapping area are shown in Fig. 2a and c. Energy calibration and further processing were performed with the freely available software package PyMCA [6]: As a first step the total spectrum was analyzed by fitting all contributing lines to yield a fit to the measured curve. Using the same fit parameters the spectra of each scan point were fitted individually in the "batch fitting" mode of the software and summarized as elemental mappings of the illuminated area. Several elements specific to the cellular interior with respect to the intercellular area could be identified. For the two most prominent elements, i.e. K-K_α and Zn-K_α, a combined elemental mapping is shown in Fig. 2a. As visible in the graph, the K and Zn content of the cell is well localized to the cellular area as visible in the optical (fluorescence) micrographs (cf. Fig. 2b). This indicates a successfull preparation process without early disruption of the cell membrane.

As visible in the cumulative spectrum shown in Fig. 2 many parasitic fluorescent lines are excited by air-scattered high-energy photons interacting with the metal rod of the sample holder. The unambiguous attribution of the K and Zn fluorescence contributions to cellular material is supported by a comparison of two spectra resulting from areas with and without cellular

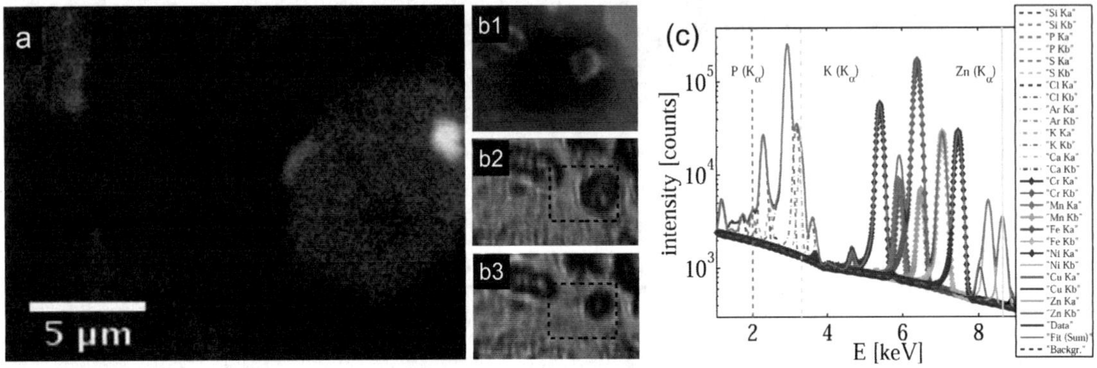

Figure 2: (a) Combined fluorescent yield of the K-K_α (red) and the Zn-K_α (green) lines for a freeze-dried *Dictyostelium* cell (on the right), on a logarithmic scale. (b1) Optical fluorescence micrograph of the illuminated area shown in (a). As the autofluorescence of the subtrate decreases during strong x-ray exposure the illuminated area can be localized easily. Contrast has been enhanced for better visibility. (b2 and b3) Optical micrographs of the imaged region on the sample before and after the experiment taken with the online-microscope at the beamline. A significant shrinking of the cell on the right is observable, however, the sample was slightly out of focus here, making a quantification of the shrinking effect very difficult. (c) Total fluorescence spectrum of the illuminated sample areas shown in (a) and (b) as well as corresponding fit (energy resolution (FWHM) ca. 150 eV). The strongest lines specific of the cell are indicated by vertical lines.

material on the substrate (cf. Fig. 3). As further elements specific to the cellular interior Ca and P are identified.

The problem of radiation damage is regarded as the major effect ultimately limiting the resolution in X-ray imaging experiments [7]. Here the effect of radiation exposure was visible in optical micrographs by an overall shrinking of the cellular material and a reduced autofluorescence signal of the entire substrate. The total flux incident on the sample was 10^{12} photons per μm^2 (area of the cell ca. $48\,\mu m^2$, step size 100 nm, photons per step ca. 10^{10}). Although the observed damage appears to be rather small given these numbers, and although very high resolutions might be possible with lower dose [7], optimized experimental conditions, i.e. cryogenic cooling of the sample, would be desirable here. This can be readily achieved with the presented sample environment.

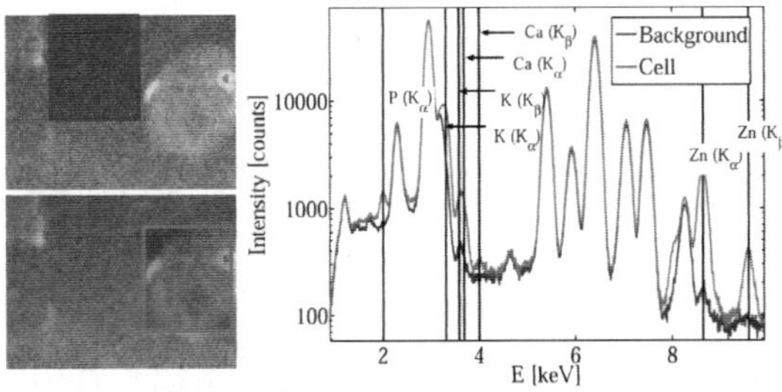

Figure 3: (right) Spectra obtained from the rectangular selections depicted in the cumulative x-ray fluorescent micrographs on the left: The selection of the cellular area (red) and a representative background area (blue) have the same size. Several fluorescence lines have been inserted into the spectral graph to mark the elements that are specific to the sample (Zn, K, P, Ca). The Ca-fluorescence is at the limit of the detection level.

Acknowledgements: The ESRF is ackgnowledged for beamtime. T. Ducic is thanked for valuable discussions and advice. Financial support by the German Science foundation through SFB "Nanoscale Photonic Imaging" is gratefully acknowledged.

References

[1] Van Haastert P J M and Devreotes P N 2004 *Nat. Rev. Mol. Cell Biol.* **5** 626–634
[2] Yang L, McRae R, Henary M M, Patel R, Lai B, Vogt S and Fahrni C J 2005 *Proc. Nat. Ac. Sc.* **102** 11179–11184
[3] Thorne R E, Stum Z, Kmetko J, O'Neill K and Gillilan R 2003 *J. Appl. Cryst.* **36** 1455–1460
[4] Dubochet J, Adrian M, Chang J J, Homo J C, Lepault J, McDowall A W and Schultz P 1988 *Quart. Rev. Biophys.* **21** 129–228
[5] Hignette O, Cloetens P, Rostaing G, Bernard P and Morawe C 2005 *Rev. of Sc. Instr.* **76** 063709
[6] Sole V, Papillon E, Cotte M, Walter P and Susini J 2007 *Spectr. Act. B: At. Spect.* **62** 63–68
[7] Shen Q, Bazarov I and Thibault P 2004 *J. Synchr. Rad.* **11** 432–438

3D quantification of brain microvessels exposed to heavy particle radiation

C Hintermüller[1], J S Coats[2], A Obenaus[2], G Nelson[2], T Krucker[3] and M Stampanoni[1,4]

[1]Swiss Light Source, Paul Scherrer Institut, CH-5232 Villigen, Switzerland
[2]Loma Linda University, Dept of Radiation Medicine, CSP A-1010 Loma Linda, CA, USA
[3]Novartis Institutes for Biomedical Research, Cambridge, MA, USA
[4]Institute for Biomedical Engineering, University and ETH Zürich, 8092 Zürich, Switzerland

E-mail: christoph.hintermueller@psi.ch

Abstract. Space radiation with high energy particles and cosmic rays presents a significant hazard to spaceflight crews. Recent reviews of the health risk to astronauts from ionizing radiation concluded to establish a level of risk which may indicate the possible performance decrements and decreased latency of late dysfunction syndromes (LDS) of the brain. A hierarchical imaging approach developed at ETH Zürich and PSI, which relies on synchrotron based X-ray Tomographic Microscopy (SRXTM), was used to visualize and analyze 3D vascular structures down to the capillary level in their precise anatomical context. Various morphological parameters, such as overall vessel volume, vessel thickness and spacing, are extracted to characterize the vascular structure within a region of interest. For a first quantification of the effect of high energy particles on the vasculature we scanned a set of 6 animals, all of same age. The animals were irradiated with 1 Gy, 2 Gy and 4 Gy of 600MeV ^{56}Fe heavy particles simulating the space radiation environment. We found that with increasing dose the diameter of vessels and the overall vessel volume are decreased whereas the vessel spacing is increased. As these parameters reflect blood flow in three-dimensional space they can be used as indicators for the degree of vascular efficiency which can have an impact on the function and development of lung tissue or tumors.

1. Introduction

Space radiation with high energy particles and cosmic rays presents significant health hazards to spaceflight crews. There is a need to establish a level of risk associated with exposure to space radiation and the potential development of late dysfunction syndromes (LDS) such as Alzheimers Disease (AD) and stroke. This risk assessment is similar to increases found for carcinogenesis after radiation exposure. To evaluate the structural changes in the CNS micro-vasculature after radiation exposures similar to those expected in long term space flight we use an established genetic model of Alzheimer disease.

2. Methods

The temporal progression of the vascular changes was studied using a transgenic (tg) Alzheimer disease mouse model overexpressing the amyloid percussor protein (APP). Vascular corrosion casts (VCC) (fig. 1.b), produced three weeks after 0, 1, 2, 4 Gy ^{56}Fe radiation exposure (fig. 1.a), were imaged (fig. 1.c, 1.e) with synchrotron radiation based X-ray tomographic

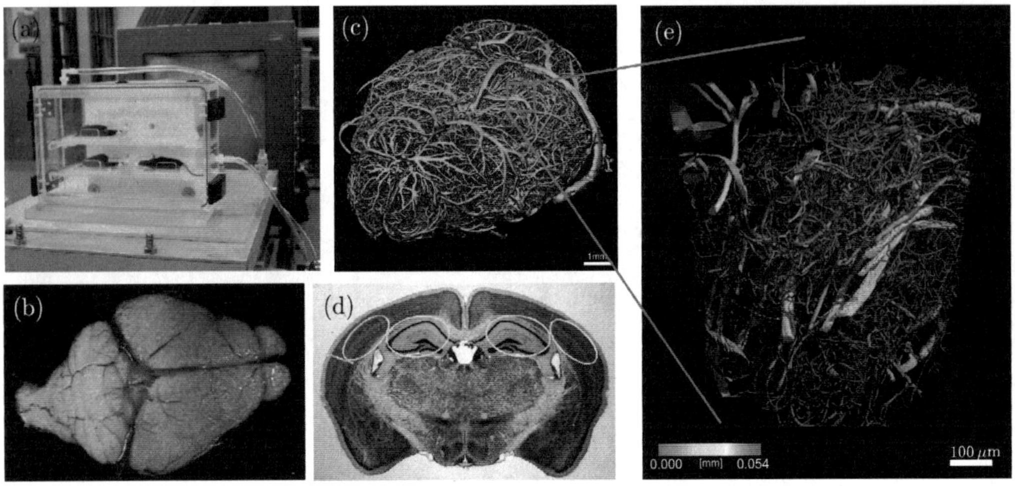

Figure 1. (a) Irradiation of animals at Nasa Space Radiation Laboratory at Brookhaven National Laboratory; (b) Vascular corrosion cast produced 3 weeks after irradiation; (c) 3D rendering of the brain vasculature scanned with a pixel size of 12 μm at the TOMCAT beamline of the Swiss Light Source (SLS) at Paul Scherrer Institute; (d) Regions targeted for morphometric evaluation of the vasculature; (e) 3D rendering of region from left posterior hippocampus with a pixelsize of 0.78 μm scanned at the TOMCAT beamline. Vessel thickness has been color coded.

microscopy (SRXTM). Hippocampal and cortical morphological parameters including vessel thickness, volume, spacing, length and number of vessels per volume were calculated in APP mice.

3. Results

A previous study [1] has shown significant alterations in the micro-vasculature structures of APP tg mice. In our work, quantification of heavy particle (^{56}Fe) irradiation on the micro-vasculature revealed that as dose increased the vessel diameter and the overall vessel volume decreased (Fig. 2, 3), thus increasing the inter-vessel spacing. Quantitative analysis indicated that with increasing irradiation dose more vessels with diameters of $7-30\,\mu$m are lost. In contrast, the number and volume of capillaries was increased in animals irradiated with 2 Gy.

4. Discussion and Outlook

The dose dependent decrement in vessel density indicates an incremental vascular insufficiency. Part of the vascular loss appears to be compensated by processes like angiogenesis and/or vascular re-modeling (Fig. 2, 3). Thus, alterations in the vasculature after exposure to radiation can be important factors of neuro-degenerative diseases. How the radiation induced changes evolve over time is part of the ongoing research.

Acknowledgement

Supported by a NASA NSCOR # NNJ04HC9OG

References

[1] Heinzer S, Krucker T, Stampanoni M, Abela R, Meyer E P, Schuler A, Schneider P and Müller R 2006 *NeuroImage* **32** 626–636

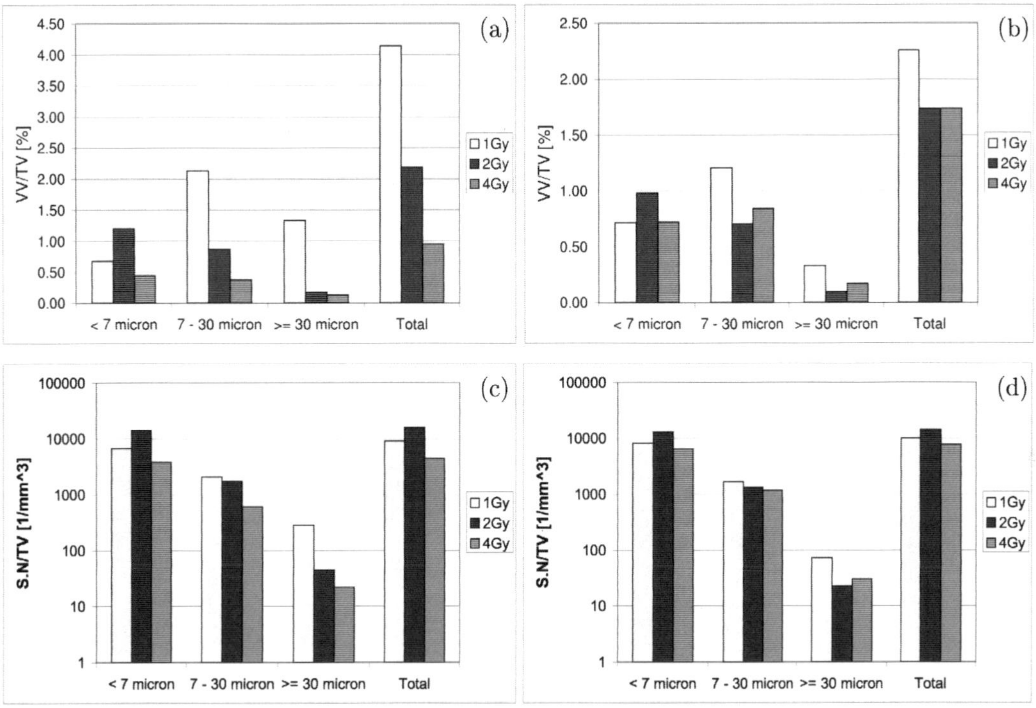

Figure 2. The plots (a) and (b) depict the ratios between the vessel volume (VV) and the tissue volume (TV) for a region of interest (ROI) located in the right anterior hippocampus (RAHPC) (a) and the average of all analyzed ROI (b). The corresponding numbers of vessels (S.N) per TV are plotted in (c) for RAHPC and (d) for all ROI. For the animals irradiated with 2 Gy ^{56}Fe the volume and number of capillaries (Diameter $\leq 7\,\mu$m) is increased. It appears that part of the loss of vessels is compensated by processes like angiogenesis and vascular remodeling.

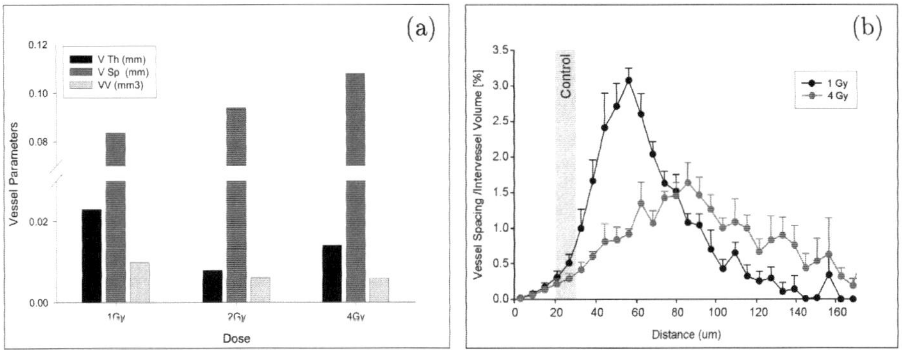

Figure 3. Dose dependent changes in brain vasculature: (a) Vessel spacing (V Sp) increases with dose while vessel volume (VV) and thickness (V Th) decrease; (b) The histograms of the distribution of the vessel spacing reveal a shift to increasing vessel distances per volume with increased doses;

Visualization of tumor vessels using synchrotron radiation-based micro computed tomography

Sabrina Lang[1], Marco Dominietto[2], and Bert Müller[1]

[1]Biomaterials Science Center, University of Basel, Switzerland, [2]Institute for Biomedical Engineering, ETH Zürich, Switzerland

sabrina.lang@unibas.ch

Abstract. The visualization of the vascular network is a prerequisite to understand the formation of cancerous tissue. Synchrotron radiation-based micro computed tomography (SRμCT) provides the necessary micrometer resolution, to image the vessels down to the 3 to 5 μm wide capillaries, the contrast differences between vessels and surrounding tissue, however, are too weak. Thus, in the present study highly X-ray absorbing barium sulfate micro-particles, which cannot penetrate through healthy vessel walls, were injected into the vessels. The acquired tomography data show that the stained vessels of the tumor tissue have different morphology and density with respect to the surrounding healthy tissue. The vessels of the cancerous tissue often exhibit spiral shapes.

The vascular structure of tumors differs from that of healthy tissue. To understand the formation of cancerous tissue, the vascular network of tumors should be uncovered down to the capillary level. Standard synchrotron radiation-based micro computed tomography (SRμCT) in absorption contrast mode provides the necessary micrometer resolution [1] even for centimeter-sized tumors [2]. The visualization of a vessel tree, however, also requires sufficient contrast. Because the tissue consists mainly of water and low absorbing species, SRμCT does not lead to significant X-ray absorption differences between vessels and surrounding tissue. So, the successful application of absorption contrast tomography requires dedicated tissue preparation procedures including embedding [3] and erosion casting [4]. The more common procedure is the use of staining materials such as the incorporation of barium sulfate into the vessels [1,2,5].
The present study is based on the injection of a barium sulfate suspension with a grain size of 0.5 to 1 μm and a concentration of 80 g/l via the left ventricle of the heart of mice under anesthesia using a peristaltic pump. The mice contained C51 or U87 tumors grown during two to three weeks until they clearly emerged to be easily extracted post mortem. The tumors were transferred to Eppendorf tubes filled with 4% formalin for fixation. For the tomography measurements at the beamline TOMCAT (SLS at PSI, Switzerland) using the photon energy of 18 keV (bandwidth 2% to 3%) the container was fixed on the high-precision manipulator that rotated the tumor from 0° to 180° in steps of 0.12° to record 1501 projections. It should be mentioned that continuous irradiation of the specimen caused the formation of bubbles. To master this serious problem, the shutter in front of the specimen was closed during CCD-readout. Hence, the irradiation could be interrupted by 0.1 s between the exposure periods of 0.3 s per projection. The conventional filtered back-projection algorithm served for the reconstruction. Because the specimen was significantly larger than the field of view, local tomography of the inner part was performed, which yielded relative local X-ray absorption coefficients. The 3D

Journal of Physics: Conference Series **186** (2009) 012088 doi:10.1088/1742-6596/186/1/012088

representations of the vessel morphology were generated by means of VG Studio MAX 1.2.1 (Volume Graphics, Heidelberg, Germany).
The homogeneous staining of the capillaries using barium sulfate micro- and nanometer sized particles is demanding, since the carrier medium has to exhibit low viscosity that associates with density differences to the barium sulfate. Consequently, sedimentation phenomena cannot be avoided, which result in an inhomogeneous sulfate particle distribution of the vessel trees [1,5]. The smaller vessels often appear interrupted [1]. In order to improve the situation, the suspension should be suitably selected. The sedimentation velocity v_{sed}, derived from Strokes law [6], should be as small as possible. It depends on the particle's grain radius r_p , the density difference between the carrier medium and the particles ρ - ρ_p, the viscosity of the fluid η and the gravitation constant g:

$$v_{sed} = \frac{2}{9}\frac{(\rho - \rho_p) \cdot r_p^2 \cdot g}{\eta}$$

The most important parameter is the barium sulfate particle size. It has not only to be smaller than the smallest capillary, but well below one micrometer to reach low enough sedimentation velocities. Much smaller nanometer-sized particles, however, show a strong tendency to form clusters, which stop the perfusion through the blood vessel tree. Accordingly, globular barium sulfate particles with a narrow sub-micrometer size distribution provide most homogeneous vessel staining.
Healthy and cancerous tissue can be straightforwardly differentiated, since the fast grown tumor tissue exhibits a much higher density of capillaries than the surrounding healthy vascular structure. In several regions, however, the opposite behavior has been observed [2]. The explanation lies most probably in the necrosis of the inner part of the tumor. It is hypothesized that the path from the arteries to the veins within the necrotic part of the tumor is not intact anymore. Furthermore, there are several indications for damages of the vessel walls. First, already in alive mouse the tumor becomes dark red in the advanced stages indicating extended regions of blood coagulation. Second, the applied pressure for the injection of the barium sulfate suspension generates leakage as experimentally found at several sites of the tumor tissue [2] and demonstrated in Fig. 1a.
Nevertheless, blood vessels in the cancerous tissue with diameters down to 11 μm could be clearly identified. The morphology of the vessel tree and the related shape of individual vessels within the cancerous tissue significantly differ from the healthy parts. The 3D representation in Fig. 1b demonstrates that many vessels show a spiral shape, which belongs to typical signs of cancer tissue [7]. These details cannot be identified using in vivo magnetic resonance imaging, where only vessels with diameters down to about 100 μm come to light.
The quantitative analysis of the barium sulfate stained vessels of the tumor, however, remains questionable, since significant parts are not stained and parameters such as the bifurcation probability versus vessel diameter cannot be meaningfully extracted. Therefore, the value of our study for the validation of computer simulations on the tumor formation [8] is limited. Here, it is highly desirable to improve the spatial resolution of phase contrast techniques that usually offer enough contrast to visualize the vessel tree without any stain even in formalin solution [2].

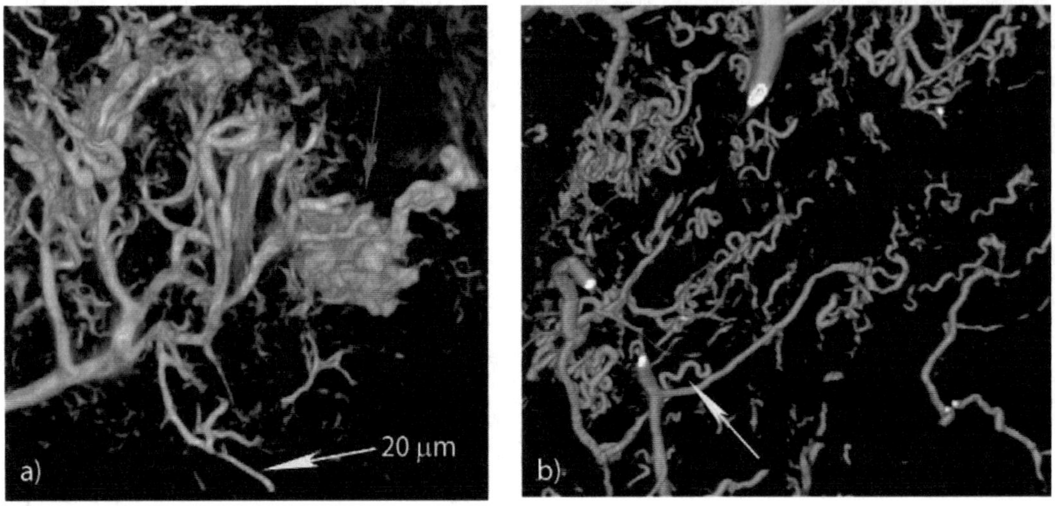

Fig. 1: a) The 3D representation shows the stained vessels. The yellow-colored arrow denotes an about 20 μm-wide vessel. The red-colored arrow indicates a barium sulfate accumulation associated with the contrast agent as the result of a damaged vessel wall. b) The 3D image demonstrates that many capillaries in the tumor exhibit a spiral shape as exemplarily indicated by the yellow-colored arrow.

Acknowledgement
The authors acknowledge the support of F. Marone and M. Stampanoni (SLS at PSI, Switzerland) during data acquisition and reconstruction within the project 20070856. The Swiss National Science Foundation financially supported the study through the National Center of Competence and Research 'Computer-Aided and Image-Guided Medical Interventions' (NCCR Co-Me).

References
[1] Müller, B., Fischer, J., Dietz, U., Thurner, P., Beckmann, F., "Blood vessel staining in the myocardium for 3D visualization down to the smallest capillaries", *Nucl. Instrum. Meth. Phys. Res. B* **246**, 254-261 (2006).

[2] Müller, B., Lang, S., Dominietto, M., Rudin, M., Schulz, G., Deyhle, H., Germann, M., Pfeiffer, F., David, C., Weitkamp, T., "High-resolution tomographic imaging of microvessels", *New developments in X-ray tomography VI, Stock, S., Ed., Proc. SPIE* **7078**, 70780B (2008).

[3] Müller, B., Germann, M., Jeanmonod, D., Morel, A., "Three-dimensional assessment of brain tissue morphology", *New developments in X-ray tomography V, Bonse, U., Ed., Proc. SPIE* **6318**, 631803 (2006).

[4] Krucker, T., Lang, A., Meyer, E.P., "New polyurethane-based material for vascular corrosion casting with improved physical and imaging characteristics", *Microsc. Res. Tech.* **69**, 138-147 (2006).

[5] Plouraboué, F., Cloetens, P., Fonta, C., Steyer, A., Lauwers, F., Marc-Vergnes, J.-P., "X-ray high-resolution vascular network imaging",*J. Microsc.* **215**, 139–148 (2004).

[6] Stokes, G.G., *Proc. Camb. Phil. Soc.* **1856**, 9 , 5.

[7] Bullitt, E., Gerig, G., Pizer, S.M., Lin, W., Aylward, S.R., "Measuring Tortuosity of the Intracerebral Vasculature from MRA Images", *IEEE-TMI* **22,** 1163-1171 (2003).

[8] Lloyd, B.A., Szczerba, D., Szekely, G., "A Coupled Finite Element Model of Tumor Growth and Vascularization", *LNCS* **4792**, 874 (2007).

Journal of Physics: Conference Series **186** (2009) 012089 doi:10.1088/1742-6596/186/1/012089

Mapping molecular orientation in dry and wet *Nephila clavipes* dragline spider silk

Thierry Lefèvre,[1] Michel Pézolet,[1] Glyn Cooper,[2] Daniel Hernández Cruz,[2] Marcia M. West,[2] Martin Obst,[2] Adam P. Hitchcock,[2] Chithra Karunakaran[3] and **Konstantine Kaznatcheev[3]**

1. Département de Chimie, Université Laval, Québec, QC, G1V 0A6 Canada

2. BIMR, McMaster University, Hamilton, ON L8S 4M1 Canada

3. Canadian Light Source, University of Saskatchewan, Saskatoon, SK S7N 0X4 Canada

E-mail: aph@mcmaster.ca

Abstract. The alignment of β-sheets within spider dragline silk fibers is an important factor in their tensile strength and extensibility. We are using linear dichroism of the C 1s → π^*_{amide} transition measured using scanning transmission X-ray microscopy (STXM) to generate quantitative maps of the $\langle P_2 \rangle$ orientation parameters with 30 nm spatial resolution. Here we have extended these measurements from dry samples to samples with partial or full hydration. A device for monitoring and controlling the humidity of a sample in the STXM is described and used to measure the effect of saturated humidity on a section of *N. clavipes* dragline spider silk. The microstructure and distributions of molecular orientation change considerably with hydration in ways consistent with the supercontraction observed in free standing dragline spider silk. The STXM results are compared to infrared and Raman microscopy results.

1. Introduction

Spider silk is among nature's most highly engineered structural materials, in some cases achieving combinations of strength and toughness that are unmatched by high-performance synthetic fibers. Spiders produce several different types of silk with mechanical properties that are optimized for specific functions. Dragline silk is a remarkable high toughness fiber that combines both stiffness and extensibility. These properties are due to its block copolymer structure with alternating alanine-rich hard segments containing β-sheets and glycine-rich soft segments. The alignment of β-sheets within the fiber is an important factor in the tensile strength of spider dragline silk. To establish fundamental structure-property relationships in spider silks, it is necessary to characterize the microstructure by measuring the degree of orientation of the β-sheets throughout the fiber.

Scanning transmission X-ray microscopy (STXM) provides quantitative maps of chemical species and orientation with 30 nm spatial resolution. We have shown that the spatial distribution of the linear dichroism of the C 1s → π^*_{amide} transition in silk fibers can be used to investigate the degree of orientation of the β-sheets in various types of silk fibers [1-3]. Here we extend these measurements

Journal of Physics: Conference Series **186** (2009) 012089 doi:10.1088/1742-6596/186/1/012089

from dry samples, to samples with partial or full hydration. A device for varying and monitoring the humidity of a sample in the STXM microscope is described and first results shown.

2. Experimental

Although wet cell designs have been published [5-7] and used in many soft X-ray spectromicroscopy studies, to our knowledge this is the first device for *in situ* humidity control. Figure 1a is a photo of the first version of the humid cell used to carry out the measurements described in this paper. Fig. 1b shows the second implementation of the cell, which integrates a miniature relative humidity sensor (Honeywell Hycal HIH-4000) directly with the sample, provides much more flexibility for sample mounting, and includes a thermocouple to monitor temperature. Custom electronics were fabricated to display and control the relative humidity in the sample cell. The signal of the humidity sensor is compared to the humidity set point and the difference is used to control a miniature pneutronics proportional valve (Parker VSONC-5S11-VD-F8) which admits humid helium into the sample cell. The electronic circuit is calibrated to a particular humidity sensor and valve. In the first controller, that was used for the results reported herein, the humidity in the cell could be decreased only by allowing humid gas to pass from the cell to the exterior of the STXM chamber, via a return line. The newer version of the control system is much more responsive since it allows simultaneous control of a second valve that admits dry helium into the cell if the humidity set point is below the sensor reading.

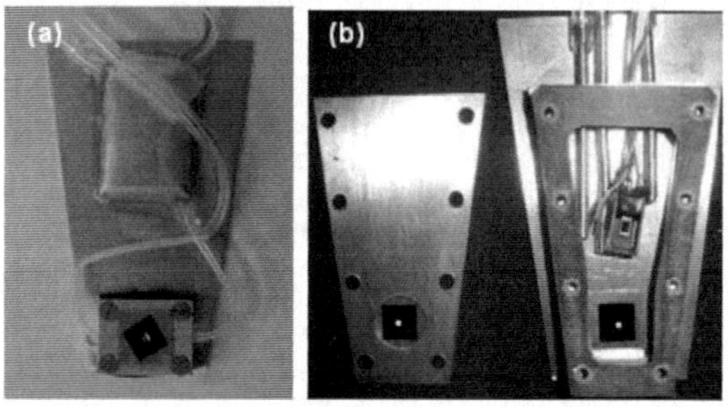

Figure 1. (a) First version of humid cell with the humidity sensor only connected to sample compartment via thin tubing. (b) Mark II humid cell with sensor, dry and humid He ports and thermocouple integrated into the sample region.

The humid cell is built on a standard STXM sample plate which gives reproducible kinematic mounting in the STXM microscopes at ALS, CLS and SLS. So far the cell has only been used at the STXM on CLS beamline 10ID1 but a similar device is available at ALS STXM532. The CLS STXM source is an elliptically polarizing undulator (EPU) in which all 4 quadrants can be displaced, thus allowing tuned X-rays with 99% linear polarization to be produced with arbitrary inclination from -90° to +90° relative the plane of the ring [8]. This facilitates studies of linear dichroic signals since a complete and quantitative study can be performed without having to rotate the sample.

3. Results

Figure 2 displays an optical image of the longitudinal microtomed section of a full hydrated *N. clavipes* dragline silk fiber embedded in epoxy, along with STXM images recorded at 288.5 eV of parts of this fiber in the low humidity ('dry') and saturated humidity ('wet') states.

$<P_2>$ orientation maps for the dry and wet fiber were derived from 3-image sequences measured with the E-vector aligned parallel and perpendicular to the fiber axis as described elsewhere [3,4]. Figure 3 plots the derived $<P_2>$ maps on the same scale. Figure 3 also compares the histograms of the $<P_2>$ values for the dry and wet states. $<P_2>$ is 1.0 for perfect parallel, -0.5 for perfect perpendicular, and 0 for fully random orientation. The negative $<P_2>$ values observed indicate that the carbonyl groups of the silk proteins are preferentially oriented perpendicular to the fiber axis so that the peptide chains of the β-sheets are mostly aligned along the

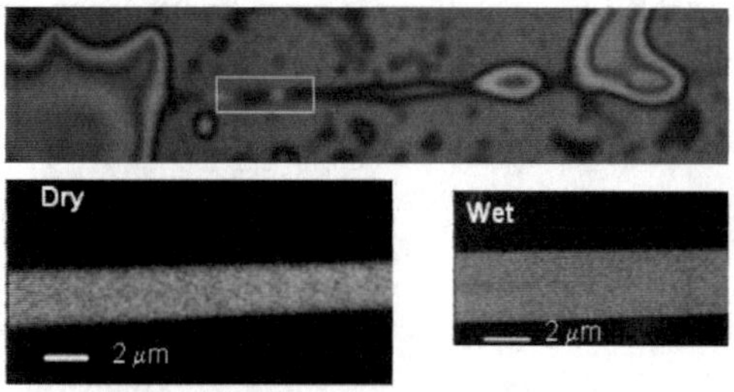

Figure 2 (upper) Optical image and (lower) STXM OD images (at 288.2 eV) of dry and fully water saturated *N. clavipes* dragline silk fiber embedded in epoxy.

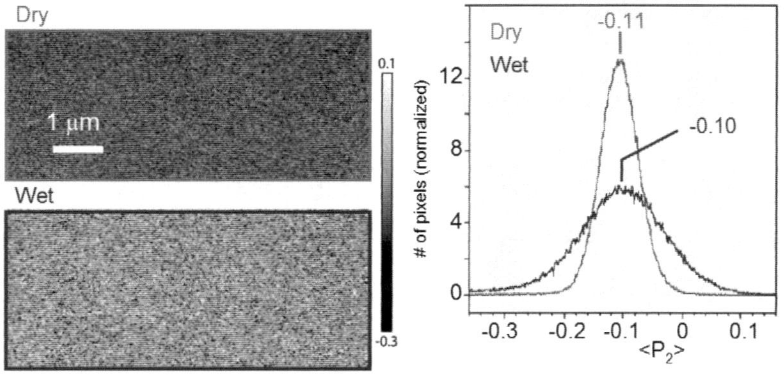

Figure 3. (left) $<P_2>$ orientation maps of dry and wet fiber sections displayed on the same intensity scale. (right) Comparison of histograms of $<P_2>$ values for the dry (blue) and wet (red) fibers.

fiber axis, as known from other studies. The dry fiber presents a very fine homogeneous microstructure of moderately oriented domains. The addition of water to the fiber drastically affects the microstructure of the fiber. The microstructure is still homogeneous throughout the sample, but it has a much coarser texture and a broader range of $<P_2>$ values. This effect is clearly observed by comparing the distribution of $<P_2>$ values for the fiber in the dry and wet states (Figure 3). Both distributions display a Gaussian shape centered at a $<P_2>$ value of about -0.1. However, the distribution is much broader for the fiber in the wet state compared to that observed for the dry state. These results show that water induces a major reorganization of the protein chains, with some carbonyl groups displaying higher and some lower orientation than in the dry fiber.

When unrestrained dragline silk is immersed in water, it undergoes supercontraction, shrinks by approximately 40–50%, and its mechanical properties change markedly. The initial stiffness drops by three orders of magnitude, and the material becomes rubber-like [9]. This effect has been associated with the decrease of inter-chain interactions in the amorphous matrix in the presence of water, leading to a loss of rigidity. Polarized Raman spectromicroscopy studies at Université Laval on unrestrained *N. clavipes* dragline silk supercontracted by about 30% show that supercontraction does not alter significantly the average orientation of the silk proteins but changes slightly their conformation. The present STXM results show that, even for a thin section of dragline silk embedded in a polymer matrix, water induces effects similar to those resulting from supercontraction.

Acknowledgements
Research supported by NSERC, Canadian Foundation for Innovation, and the Canada Research Chair program, STXM data was measured at CLS-SM (BL10ID1) (supported by NSERC, CIHR, NRC and U. Saskatchewan) and at the ALS STXM532 (ALS is supported by U.S. Department of Energy).

References
[1] Hernández Cruz D, Hitchcock A P, West M M, Rousseau M-E and Pézolet M 2006 *Biomacromolecules* **7** 836
[2] Rousseau M-E, Hernández Cruz D, West M M, Hitchcock A P and Pézolet M 2007 *J. Am. Chem. Soc.* **129** 3897
[3] Hernández-Cruz D, Hitchcock A P, Tyliszczak T, Rousseau M-E and Pézolet M 2007 *Rev. Sci. Inst.* **78** 033703
[4] Pézolet M et al. 2008, *2007 Canadian Light Source Annual Report* 118
[5] Williams S et al. 1995 *Rev. Sci. Inst.* **66** 1271
[6] Stollberg H, Guttmann P, Takman P A C and Hertz H M 2007 *J. Microscopy*, **225** 80
[7] Hanhan S et al. 2007, *2006 Canadian Light Source Annual Report* 117
[8] He F, Sigrist M, Summers T, Obst M and Hitchcock A P 2007 *Nucl. Inst. Meth. A* **582** 103
[9] Gosline J M, Guerette P A, Ortlepp C S and Savage K N 1999 *J. Exp. Biol.* **202** 3295

Multi method (STXM, µFT-IR, µXAFS) approach to identify uranium speciation in argillites from the Lodève Basin (France)

P Michel, T Schäfer, M A Denecke, B Brendebach, K Dardenne, F Huber, J Rothe

Forschungszentrum Karlsruhe, Institut für Nukleare Entsorgung (INE), P.O. Box 3640, D-76021 Karlsruhe, Germany

E-mail: pauline.michel@ine.fzk.de

Abstract. Spatially resolved spectroscopic investigations with a nanometer to micrometer-scale resolution (µ-XRF, µ-XAFS, STXM and µ-FTIR) focused on the U speciation in two clay-rich samples originating from the Autunian shales (Permian Lodève Basin; France) a natural uranium deposit associated with organic matter (bitumen). The goal of this study was to determine the U oxidation state in the samples and to ascertain if any correlation between the U distribution and that of other selected elements present in the clay (K, Ca, Ti, Fe, Zr and C) exists. Identification of uranium associated with organic matter and/or mineral phases (e.g. iron oxides, clay minerals) can potentially give insight into the mechanism of uranium immobilization in this sedimentary formation. µ-XRF and µ-XAFS measured in confocal geometry show that the U is present as U(IV) and its distribution is not correlated to areas with a high Fe content, but rather correlates with the distribution of lighter elements. U L3 µ-EXAFS recorded at areas with high U concentration reveals that U in these hot spots likely present as a nanocrystalline uraninite (UO_2) – like phase. Clay fractions of the samples characterized by XRD are mainly composed of chlorite and illite. STXM carbon K-edge, potassium L-edge and iron L-edge investigations conducted on sulphur-embedded microtomes show two types of organic matter present, one positively correlated with the elements K and Fe and another of pure organic composition building fracture infill. The µ-FTIR measurements clearly reveal a direct association of organic matter with clay minerals. From combined results a tentative hypothesis of U immobilization is put forward.

1. Introduction

Micro-focused synchrotron radiation techniques are becoming an increasingly used tool in environmental and particularly nuclear waste disposal research to investigate metal mobility controlling processes in geological systems. There are a number of reasons for this but primarily they are driven by the need to characterize metal speciation in heterogeneous natural systems [1]. Investigations of actinide geological transport in the context of nuclear waste disposal are especially challenging, as an accurate prognosis demands process understanding over a geological and not historical time domain. One strategy for meeting this challenge is through investigations on natural analogs (NA's), by comparing future assessment time periods with past history where these NA's help to put these timeframes into human context. Claystone formations or argillites are currently favored as deep geological host rocks i.e. in France and Switzerland. The very similar sedimentary sequence found in the uranium ore body (Lodève Basin, Massif Central southern France) investigated in this study might be suitable as an analogue for a spent fuel repository in claystone formations. High uranium concentrations formed around 173±6 and 108±5 Ma ago (Alpine and Pyrenean formation) [2] have been observed in the fluvio to deltaic Autunian deposits and more particularly in the Mas d'Alary faulted zone. Two types of fluids can be differentiated; (a) fluids generated during the burial and/or (b) hydrothermal fluids which may have transported metallic ions, notably U(VI). The U mineralization is closely associated with bituminous organic matter and a generic link between the natural deposited bitumen and the migrated organic material found in the Permian petroleum source rocks could be demonstrated [3]. The goal of this study was to determine (a) the oxidation state and speciation of uranium found in enriched regions and (b) to elucidate if kerogen has acted as a reactive reductant [4] or secondary reductants such as iron(II)-bearing minerals formed through anaerobic degradation of the organic matter [5] may have been responsible for uranium immobilization and accumulation.

2. Materials & Methods

The two uranium ore samples investigated in this study have been provided by CREGU (Nancy, France) originating from the contact zone adjacent to the fault zone (sample 15) and the breccia facies within the fault zone (sample 31) of horizon 21 (Mas d'Alary uranium mine; Lodève). For STXM and μFTIR measurements using transmission mode ~100 nm thick sulphur embedded microtomes on TEM grids were prepared (MVA, Inc., Norcross, GA, USA). μ-XRF and μ-XANES measurements are recorded at Beamline L at the Hamburger Synchrotronstrahlungslabor (HASYLAB) or at the INE-Beamline at the Ångströmquelle Karlsruhe, ANKA [6]. A confocal irradiation-detection geometry is used, providing added depth information and allowing probing sample volumes below the surface. Elemental distribution maps are obtained by means of scanning μ-XRF [7]. For details on the measurement set-ups used at HASYLAB or the INE-beamline at ANKA it is referred to [8]. STXM and μFT-IR are conducted at the National Light Source Synchrotron (NSLS), Brookhaven National Laboratories (BNL) New York, beamline X1A and U10B, respectively. For details on the measurements conditions it is referred to [8].

3. Results and Discussion

In summary, application of various spatially resolved spectroscopic methods have been successful in characterizing the uranium speciation in sediment samples from the Lodève Basin and in evaluating if organic matter or Fe^{2+} mineral phases were responsible for uranium immobilization/enrichment in the sediment. μ-XANES and μ-EXAFS results show that uranium in both sediment samples studies is found in the tetravalent state, most probably in the form of nano-crystalline UO_2-like phase. The correlations between elements obtained by means of different techniques allow concluding the following:The uranium distribution in sample 15, originating from the contact zone adjacent to the fault zone is positively correlated to the distribution of lighter weight elements (Ca, K, Ti). We deduce a correlation between clay minerals of illite- type and organic matter in the sample based on the observation that potassium (as indicator element) is found associated with both organic carbon and Fe and the observed spatial coincidence between μ-FTIR clay OH vibrations and various vibrations of organic functional groups (Figure 1).

Sample 31, originating from the breccia facies within the fault zone, also exhibits a positive correlation between uranium and the elements K, Ti and a negative or anti-correlation between the distribution of uranium and those of Ca and iron. In sample volumes of highest potassium content, no Fe Kα signal is observed. Two distinguishable clusters of organic material in sample 31 are identified (Figure 1). The first type is associated with fracture infill and lacks K and Fe signatures, but displays a detectable carbonate peak in its C(1s) XANES. The second cluster type of organic material is associated with discrete mineral aggregates, exhibiting a spatial coincidence between carbon absorption and Fe and K absorption. Because an anti-correlation between uranium and Ca/Fe exists and because carbonates (likely Ca-containing) are associated with organics in the fracture infill we conclude that the fracture infill in this sample is not associated with uranium.

From these observations we attempt to put forward a tentative hypothesis for the mechanism of uranium immobilization. Due to the lack of the correlation between uranium and Fe, we exclude Fe-minerals as dominant reductant during immobilization of U(VI) to U(IV). Combining the knowledge that uranium is found associated with potassium and that clay minerals of illite-type are associated with organic matter we conclude that organic material associated with clay minerals might have played a role as reductant thereby reducing incoming groundwater dissolved U(VI) to less soluble U(IV). This hypothesis remains speculative; numerous unanswered questions remain. For example, what role did the clay play? Did it act as a catalyst [9] or was it merely an anchor for the organic material? In addition, how do we explain the apparent discrepancy between the spatial relationship between potassium and iron distributions observed in the μ-XRF maps with that from STXM results? To help refine our understanding of the redox partner involved in the reduction of U(VI) in this sediments combined μ-XRD/μ-XRF studies are planned.

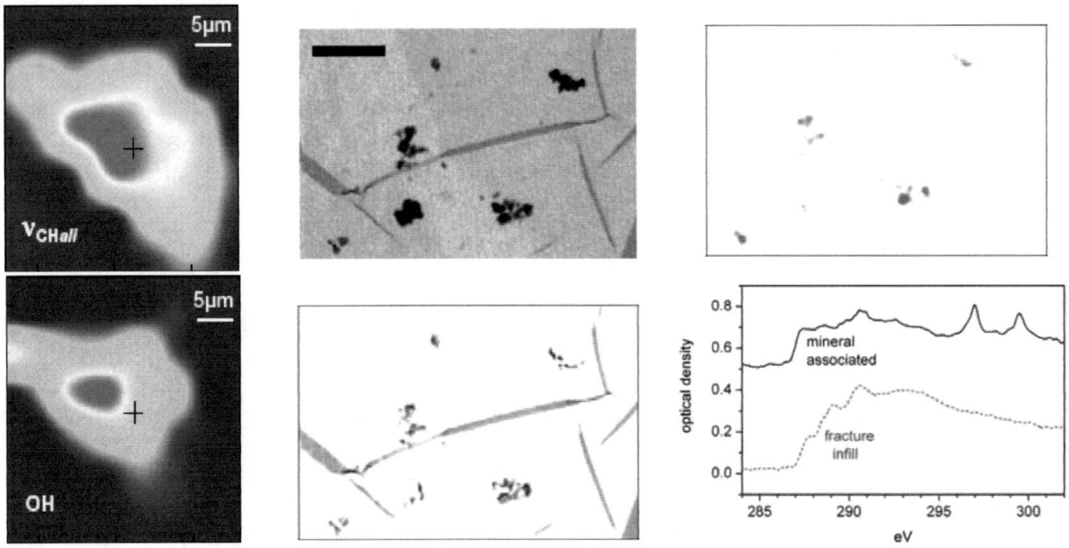

Figure 1. (Left) μFT-IR maps of aliphatics (νCH*ali*; 3000-2800 cm^{-1}) and OH vibrations of clay minerals (3720-3680 cm^{-1}). (Middle) STXM absorption image at 290eV and carbon distribution map (blue); (Right) Fe $L_{2,3}$-edge distribution map and spectra of the two C(1s) XANES clusters found (sample 31). High colour intensity indicates high element content. The scale bar represents 10μm.

4. Acknowledgement

This work was financed through the Integrated project IP 'FUNMIG' and the EURATOM Intra-European fellowship 'COMACK' (FI6W-CT-2006-036493).

References

[1] Liu C, Zachara J M, Yantasee W, Majors P D and McKinley J P 2006 *Water Resour. Res.* **42** W12420
[2] Lancelot J, de Saint André B and de la Boisse H 1984 *Mineral. Deposita* **19** 44-53
[3] Landais P 1996 *Ore Geology Reviews* **11**(1-3) 33-51
[4] Hartog N, Griffioen J, and van Bergen P F **2005** *Chem. Geol.* **216**(1-2) 113-131
[5] Jakobsen R and D Postma 1999 *Geochim. Cosmochim. Acta* **63**(1) 137-151
[6] Denecke M A, Rothe J, Dardenne K, H. Blank H, and Hormes J 2005 *Phys. Scripta* **T115** 1001
[7] Denecke M A, Janssens K, Proost K, Rothe J and Noseck U 2005 *ES&T* **39**(7) 2049-2058
[8] Michel P, Denecke M A, Schäfer T, Brendebach B, Dardenne K, Rothe J, Huber F, Elie M, and Buckau G (in review) In: 'Actinide-XAS-2008'. Vol NEA, sponsored by OECD-NEA.
[9] Giaquinta D M, Soderholm L, Yuchs S E, and Wasserman S R 1997 *Radiochim. Acta* **76** 113

Journal of Physics: Conference Series **186** (2009) 012091 doi:10.1088/1742-6596/186/1/012091

Soft X-Ray Spectromicroscopy of Wood Fibre Composites

A.G. Michette[1], C. Phanopoulos[2], R.J. Newell[1], C. McFaul[1], S.J. Pfauntsch[1], G. Pans[2], S. Wirick[3]

[1]King's College London, Department of Physics, Strand, London WC2R 2LS, UK
[2]Huntsman Polyurethanes, Everslaan 45, 3078 Everberg, Belgium
[3]NSLS, Brookhaven National Laboratory, Upton, NY 11973-5000, USA

E-mail: alan.michette@kcl.ac.uk

Abstract. Understanding the structural properties of wood fibre composites, widely used in building and other applications, is an important area of research in devising robust new materials for a range of environmental conditions, including those encountered in developing countries. Wood fibre composites are prepared by mixing fibres with prescribed amounts of glue, but exactly where the glue distributes in relation to the fibre components is unclear. Characterisation of the bonding conditions has largely been a matter of trial and error, which is time consuming and imprecise. In soft x-ray spectromicroscopy experiments carried out at the NSLS, a range of wood fibre composites, with different glues and glue loadings, was studied in order to attempt to quantify the relative glue content in different components of the composite.

1. Introduction

Wood composite materials, consisting of wood fibres bonded together by resin, are increasingly being used in construction and transportation as well as in applications such as packaging and furniture. Several resins can be used in such applications, with more durable, higher strength options being employed when necessary and when the value of the composite warrants the higher cost. Isocyanate based resins are particularly useful as binders and adhesives. They provide strength, durability and moisture resistance at very low loading rates. To further improve their performance it is necessary to understand the resin distribution within the wood: cell walls, pits, cell lumen, middle lamellae; and the chemical reactions that occur in each region: polyurea, ring substitution reaction products, urethanes. Understanding these issues will aid the development of more efficient, more durable binders, reducing the energy consumption of panel production, the transportation of large volumes of materials, the need for replacement panels and thus will help in the management of forest product feedstocks.

Conventional methods of characterising the distribution and chemistry of the resin, including fluorescence (possibly using dyes), light microscopy and measurement of mechanical properties provide insufficient component contrast, poor chemical signal and/or poor spatial resolution. It has been shown previously that soft x-ray spectromicroscopy is suitable for studying the penetration of glues into wood samples [1]. Here, its use in an investigation of the the isocyanate/wood interphase region in wood fibre composites is described.

2. Measurements

Wood fibre composite samples, with different isocyanate-based resins and resin concentrations, and reference samples of the component materials were prepared as follows. The sampled were (i) cut to ~2×2×10 mm sections, (ii) immersed in embedding medium and left to set, (iii) cured at 80°C for 12 hours, (iv) sectioned to ~0.2 µm and (v) floated onto silicon nitride supports. The reference samples included cellulose, lignin, untreated fibres, embedding medium, the various resins and resin components. The samples were imaged around the C and O absorption edges using the NSLS X1A microscope. Only the C edge images are discussed here.

Figure 1 shows a series of images of a wood fibre composite sample taken at different energies around the carbon absorption edge. The striations in the first image are a result of a beam artefact. High resolution point spectra, at image points identified by eye, were also obtained, and used to determine the relative component concentrations.

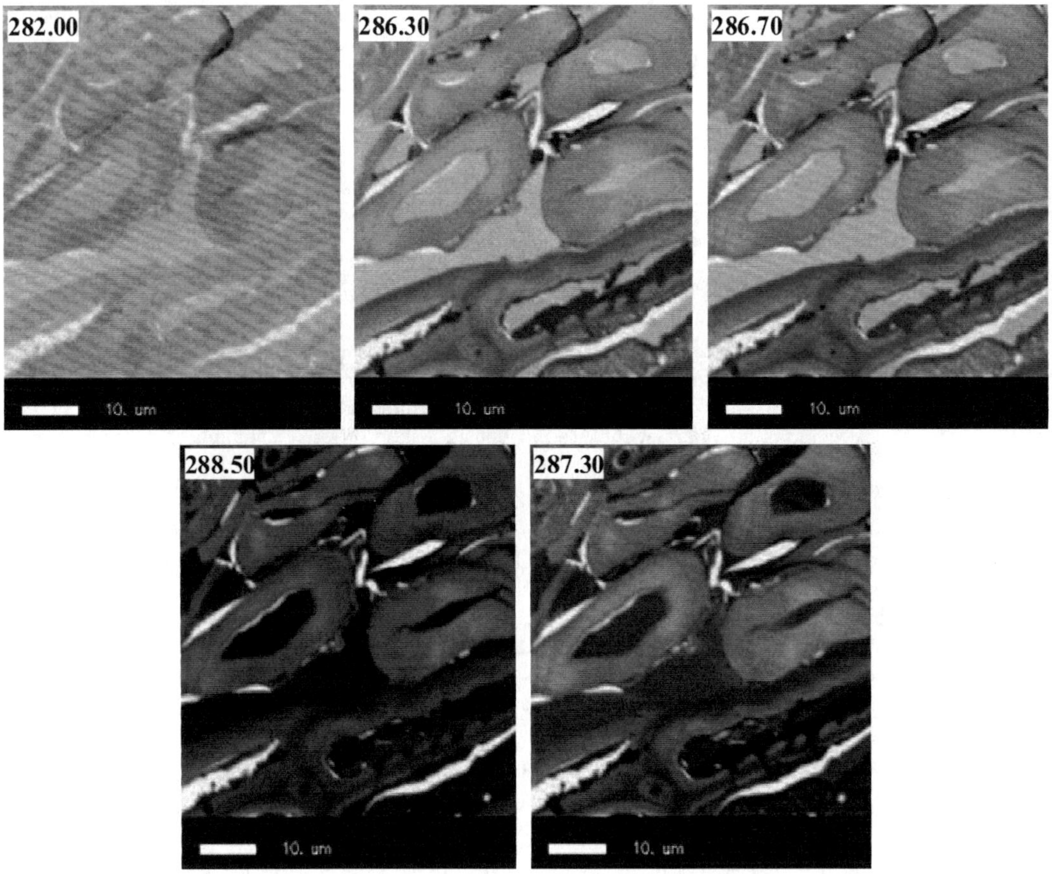

Figure 1. Images of a wood fibre composite sample taken at several energies around the carbon absorption edge (the energies are indicated, in electronvolts, by the numbers at the top left of each image.

3. Results and Conclusions

Figure 2(a) is an example of a point spectrum of a wood fibre, and Figure 2(b) that of a typical cured isocyanate-based resin. The lines are the overall and component fits. The fits were made to Pearson IV functions, which are intermediate between Gaussians and Lorentzians along with an asymmetry factor [2].

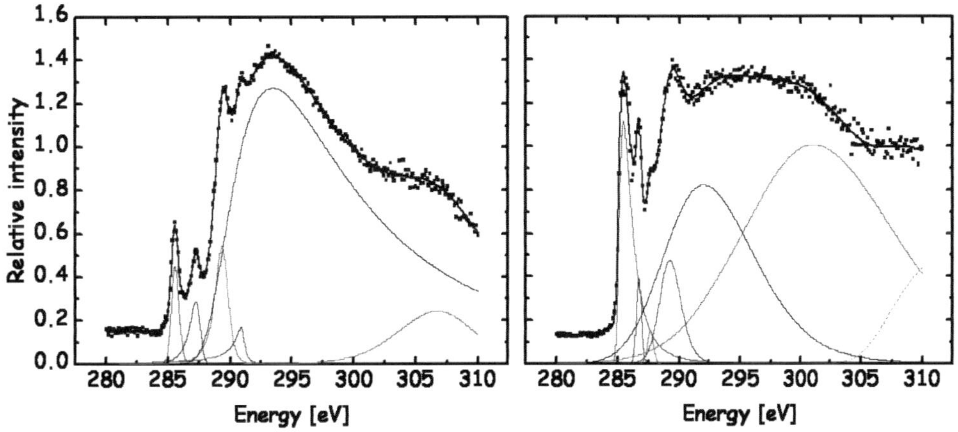

Figure 2. (a, left) Measured (points) and fitted (lines) point spectra of a fibre. The component and overall fits are shown. (b, right) The same for a resin sample.

Fits to the point spectra of the composite and reference samples give the resin concentrations in different areas shown in table 1. The results suggest fast resin penetration into selective regions, distributed by droplet number and size, not smearing or contact. The resin seems to have specific strong affinity for the cell walls, the inner cell wall surface and the interconnecting cell pits. It is not yet clear whether high concentration of well distributed resin is needed for good performance or if only good distribution, at lower concentration, is sufficient. The chemical nature of the resin in different wood regions also needs further study.

Table 1. Fitted resin concentrations in different fibre areas and for different overall loadings.

Overall resin loading	Resin concentration in Main cell wall	Inner cell wall	Cellular pit
2%	4.9±1.5%	4.6±0.5%	6±3%
4%	6±1%	1%	9±3%
6%	4.1±1.5%	4±2%	6±1%

References

[1] Buckley CJ, Phanopoulos C, Khaleque N, Engel A, Holwill MEJ and Michette AG 2002 *Holzforschung* **56** 215–222

[2] Pfauntsch SJ 2001 Developments in soft x-ray laboratory systems for microscopy and cellular probing (PhD thesis: London University)

Journal of Physics: Conference Series **186** (2009) 012092 doi:10.1088/1742-6596/186/1/012092

X-ray microscopy for neural circuit reconstruction

Haruo Mizutani[1], Yoshihiro Takeda[2] , Atsushi Momose[2], Akihisa Takeuchi[3], Toshihisa Takagi[1]

[1]Science Integration Program, University of Tokyo, Chiba, Japan
[2]Department of Advanced Materials Science, University of Tokyo, Chiba, Japan
[3]Japan Synchrotron Radiation Research Institute/SPring-8, Hyogo, Japan

E-mail: mizutani@dpc.scint.u-tokyo.ac.jp

Abstract. Neural circuits in the central nervous system build our various higher brain functions. However, little is known about mechanisms underlying neuronal information processing in the brain. Anatomical graph structures of real neural networks will provide us with fundamental views to elucidate them. We aim at developing a three-dimensional atlas of neural circuits using high resolution hard X-ray microscopy by synchrotron radiation. We stained neurons of a complete brain from a mouse by the Golgi-Cox method. The heavy metals used in our procedure enhanced X-ray absorption and phase contrast. 3D images of fibriform axons and dendrites of various neurons were reconstructed by back projection. X-ray microscopy with a Talbot interferometer demonstrated finer histological structures of pyramidal neurons in the hippocampus and the cerebral cortex. This observation probably serves as a foundation for achieving a mammalian Connectome Project (identifying complete wiring diagrams of the brain) with X-ray nano-tomography in the near future.

1. Introduction

The human brain is composed of complex neural networks that include millions of neurons connected with billions of synapses. Detailed maps of synaptic connectivity will be needed if we are to understand how the brain underlies behavior and how brain malfunctions underlie behavioral disorders. However, there is no evidence of what kinds of neuronal networks are built on the cellular and molecular level. Our purpose is to identify and quantify the network morphology and connections, neuron by neuron, and synapse by synapse namely "The Connectome Project" [1, 2, 3]. The project aims at deciphering all synaptic connections between all neurons and clarifying wiring diagrams of the neuronal networks.

X-ray microscopy performs such high spatial resolution of the synapse, that it has achieved a spatial resolution of 15 nm due to recent developments in optics research [4]. X-ray imaging has higher spatial resolution than light microscopy, since the X-ray wavelength is much shorter than that of visible light. A neuronal connection called a “Synapse” could be observed by using X-ray microscopy and similarly with an electron microscopy [5]. We will be able to identify a great number of “Synapses” in the brain. On the other hand, hard X-ray microscopy can be transmitted through thick materials allowing observation of inner structures without destruction, similar to MRI. It is not necessary to make a number of ultra-thin slices (50 nm) to observe large-scale brain regions. Thus, X-ray computed tomography combined with microscopy can clarify the three dimensional neuronal

Journal of Physics: Conference Series **186** (2009) 012092 doi:10.1088/1742-6596/186/1/012092

fibriform structures (axons and dendrites) and synapses with isometric high resolution. And we will enable the identification of complicated neural circuits and synaptic connections at the same time. We can expect to decipher wiring diagrams in large volumes of the brain.

2. Materials and Methods

2.1. Tissue Preparation

All experiments were performed in accordance with the guidelines of the Physiological Society of Japan. 8-week-old C57BL/6J mice were fixed by perfusion in 4% paraformaldehyde and 1% glutaraldehyde under halothane anesthesia and their brains quickly removed and placed in cacodylate buffer. The tissue was then left in the fixative for overnight at 4°C. Several neurons of a complete brain were stained by the Golgi-Cox method [6]. After rinsing in the buffer, the tissue was then dehydrated in a graded series of ethanol and propylene oxide. The tissue was then embedded in Epon 812 resin and polymerized at 70°C for 2 days. Rods 200 μm on a side were cut with a saw microtome and set on the rotary stage to observe with the X-ray micro-tomography system.

2.2. X-ray microscopy

X-ray projection images were acquired using a transmission hard X-ray microscope at the BL20XU of SPring-8. The microscope was equipped with Fresnel zone plate condenser and objective with 100 nm outer zone width and 320 μm diameter. Data were collected using X-rays with an energy of 12.4 keV (0.1 nm), and images were recorded using an X-ray CCD camera. The magnification of the system was about 19. Since the pixel size of the detector was 4.34 um, converted pixel size was estimated to be approximately 230 nm. An X-ray Talbot interferometer for X-ray phase imaging and tomography was constructed using an amplitude grating of gold pattern, 8 μm in pitch and 30 μm in height developed by X-ray lithography and gold electroplating [7]. The effective area of the grating was 20 mm x 20 mm. Figure 1 shows a setup layout of the optical system.

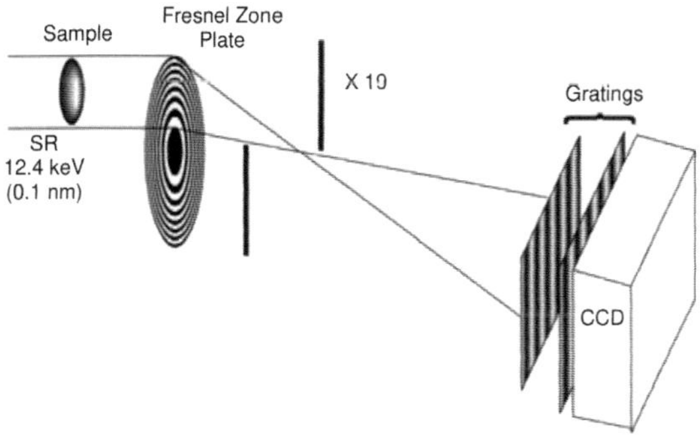

Figure 1. A schematic layout of the X-ray microscope optical system. The system consists of a sample, a fresnel zone plate and gratings for Talbot interferometry. Then the magnified X-ray image is obtained by an X-ray image detector. When the X-ray energy is 12.4 keV (wave length = 0.1 nm), the magnification of the system is about 19. Since the pixel size of the detector was 4.34 μm, the converted pixel size was estimated to be approximately 230 nm.

3. Results

We collected X-ray projection images of several neurons stained with mercury. The spatial resolutions of the images are 500 nm with absorption contrasts (Figure 2-A) and 800nm with the phase contrasts (Figure 2-B). Images produced with phase contrast have much sharper outlines of neurons than those produced by absorption contrast. We reconstructed 3D neurons of the mouse brain by using phase contrast images(Figure 2-C), that showed neuronal somata, axons and dendrites in various areas of the brain; such as the hippocampus CA1 region (Figure 2-A, B) and the cortical layer 2/3 (Figure 2-C). The measuring time was about 3 hours a sample and the data size was about 1.1 GB.

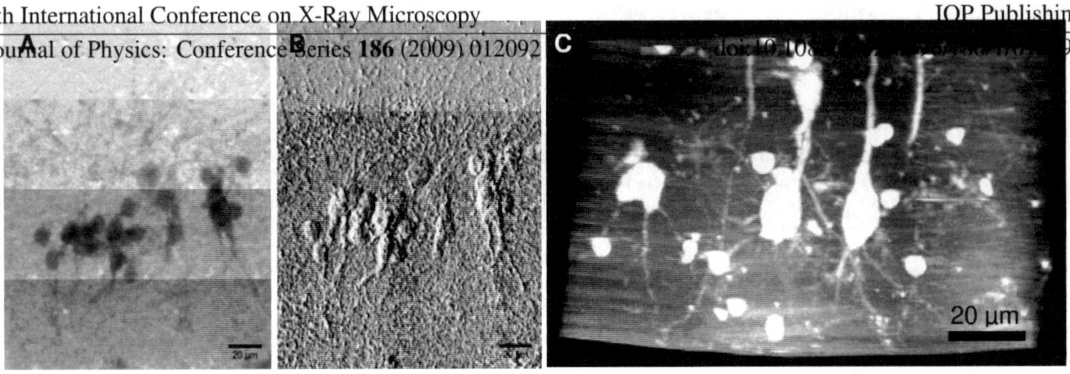

Figure 2. A, B, X-ray microscopy images of absorption contrast and phase contrast of mouse hippocampal pyramidal neurons in the CA1 region respectively. One image was taken by the X-ray micro-tomography system with a Talbot interferometer (right, B) and the other without it (left, A). Spatial resolution is about 800 nm, pixel size is 230 nm. The field of view is about 320 µm max. The scale bar is 20 µm. C, a three-dimensional rendering view of the pyramidal neurons in the cortical layer 2/3 of the mouse brain. This image was reconstructed by CT algorithm and rendered as a 3D view. This was taken by the X-ray micro-tomography system with a Talbot interferometer.

4. Discussion

Recent research shows that the best spatial resolution is better than 15 nm using soft X-ray microscopy. It can, in principle, observe neuronal connections of the mammalian brain with the same proficiency as electron microscopy. However, the X-ray microscopy has still not been applied to nervous tissue slices or blocks to identify the "Synapses" of the brain. We need to test a practical high resolution unit (nearly 10 nm) by utilizing better optics using both soft and hard X-ray microscopy. Focal depth is also a critical parameter to determine the sample thickness with the micro-tomography. Higher energy (hard X-ray) has the appropriate focal depth for thick biomaterials in CT measurement. To develop suitable staining methods for X-ray imaging, *en bloc* electron staining is expected to be useful for the synapse identification at higher resolutions. We also need the uptake of heavy metals into all neurons or the ones with neuronal activity. Field of view depends on the X-ray CCD camera performance. There is a need for a higher performance camera with more than 64 x 10^6 pixels and 10^4 dynamic range.

Extraction of all neurons from 3D images, the analysis of anatomical structures and synaptic connections by computer vision, will require algorithms for 3D image processing; such as feature extraction, region growing and principle component analysis in order to identify and quantify neurons and synapses from large-scale 3D images. We are not currently able to elucidate information processing mechanisms of the mammalian brain or its component parts. One of the critical reasons is due to the fact that anatomical neural wiring diagrams and connection topology are still unclear in synapse-resolution. Functions and characteristics of real neuronal networks are calculated and compared with neural circuits in various brain areas. We will develop a database containing of all wiring diagrams of the mouse brain. Anyone can access the database and acquire the wiring data in areas of interest. Our ultimate purpose is to simulate the mammalian brain activities and behaviors with spatiotemporal information such as neural circuits and firings.

References

[1] Lichtman J W, Livet J and Sanes J R 2008 *Nature Reviews Neuroscience* **9** 417
[2] Smith S J 2007 *Current Opinion in Neurobiology* **17** 601
[3] Lichtman J W, Livet J and Sanes J R 2008 *Nature Reviews Neuroscience* **9** 417
[4] Chao W L, Harteneck B D, Liddle J A, Anderson E H and Attwood D T 2005 *Nature* **435** 1210
[5] Denk W and Horstmann H 2004 *Plos Biology* **2** 1900
[6] Gibb R and Kolb B 1998 *Journal of Neuroscience Methods* **79** 1
[7] Momose A, Yashiro W, Moritake M, Takeda Y, Uesugi K , Takeuchi A , Suzuki Y , Tanaka M and Hattori T 2006 *Proc. SPIE* **6318** 63180T

Observation of iodine in oceanic plankton by scanning x-ray fluorescence micro-tomography

T Ohigashi[1], Y Terada[2], A Takeuchi[2], K Uesugi[2] and K Kubokawa[3]

[1] Research Organization of Science & Engineering, Ritsumeikan University, 1-1-1 Noji Higashi, Kusatsu, Shiga 525-8577, Japan

[2] Research and Utilization Division, JASRI, 1-1-1 Sayo-cho, Sayo, Hyogo 679-5198, Japan

[3] Ocean Research Institute, University of Tokyo, 1-15-1, Minamidai, Nakano, Tokyo 164-8639, Japan

E-mail: ohigashi@fc.ritsumei.ac.jp

Abstract. 3-dimensional distribution of iodine of the female oceanic zooplankton, *Paracalanus parves* s.l., was observed by using scanning x-ray fluorescence micro-tomography. The distribution of iodine along the ventral border of the zooplankton was observed to be concentrated in seven regions between the head and the bases of the swimming legs. Then, 3-dimensional distributions of iron and zinc were also observed.

1. Introduction

There is widely supported hypothesis that life originated in the ocean [1] and the hypothesis was borne out experimentally [2]. The environment of the ocean, such as contained elements, could greatly affect development of the life. As one of the environmental features of the ocean, the ocean water is relatively rich in iodine (58 μg/kg [3]). Iodine is one of the essential trace elements for higher animals to maintain their life. Iodine is mainly accumulated in a thyroid gland to be secreted as thyroid hormone, which is generally known for a relationship with basal metabolism. In the process of evolution, the primitive life in the ocean, such as zooplankton, could accumulate iodine in their environment for preliminary stage of the thyroid gland. Therefore, the distribution of iodine may be a key to explain the evolution. Then, x-ray fluorescence computer tomography (XFCT) is a proper method to observe the distributions of the specific elements of a sample 3-dimensionally without any destructive process [4,5].

In this study, we noticed copepods as the typical oceanic zooplankton because the class of the copepods account for 70% of the oceanic zooplankton and distributes all over the world. The 3-dimensional distribution of iodine and the other elements of the calanoid copepod, *Paracalanus parvus* s.l., were observed by the XFCT to specify the preliminary organs or tissues of the thyroid gland.

2. Experiments

2.1. Sample

The oceanic zooplankton, *Paracalanus parvus* s.l. (female), was used as a sample. This calanoid copepod is widely distributed in epipelagic zone in coastal area throughout the world. In Japan, the copepod is distributed all coastal area [6].

The copepod sample was collected at Sagami bay (Kanagawa, Japan) and was immediately pre-fixed in 2% glutaraldehyde and 2.5% paraformaldehyde. After preservation for 1 week, the copepod was post-fixed in 1% osmium tetroxide solution. The fixed copepod was dehydrated through a graded ethanol series to 100% ethanol and was embedded in the Epon-812 resin.

2.2. Measurement

The sample was measured by using a scanning x-ray fluorescence microscope system at the BL37XU in the SPring-8 (Harima, Japan). The x-ray energy of 10 keV was used as an incident beam to avoid exciting osmium L-line. The x-ray fluorescence of calcium, iron, zinc and iodine (L-line) were detected. The spot size of the microbeam was 3.0 (H) × 1.5 (V) μm^2. The sample was scanned as follows: horizontal direction of 3 μm × 125 steps, the step of rotation angle of 2.4°, the helical pitch of 15 μm/rotation and 41 helical rotations, and the exposure time of 0.2 sec/point. In total, it took 71 hours to measure whole body of the copepod.

2.3. Reconstruction of x-ray fluorescence micro-tomography

Sinograms of the cross sections were extracted by using constant speed helical fullscan with interpolation method from the scanning data of each element [7]. The reconstructions were performed by simultaneous algebraic reconstruction technique (SART) with taking into account the absorption effects of the incident beam and the x-ray fluorescence inside the sample. In this experiment, distribution of linear absorption coefficient (LAC) of the incident beam (10 keV) was obtained by measuring I_0 and I, but those of the LACs of the measured x-ray fluorescence were not obtained. Therefore, the resin was regarded as main absorber of the sample to calculate the distribution of the LACs of the x-ray fluorescence by approximation. Thus, the modified distribution of the LAC of the incident beam, which was multiplied by the ratio of the LAC of the PMMA ($C_5H_8O_2$, ρ: 1.19 g/cm^3) at the energies of the incident beam and the x-ray fluorescence, was used.

The median filter of 3×3 was used to reduce artifacts of the reconstructed FXCT images. 3-dimensional images were made from vertical stacks of the filtered FXCT images by using ImageJ (http://rsb.info.nih.gov/ij/).

3. Results

3-dimensional elemental distributions of iron, zinc and iodine are shown in figures 1. The reconstructed volume is 372×372×585 μm^3 and the voxel size is 3×3×3 μm^3. In these 3-dimensional images, the distribution of iron in figure 1(a) is like speckle and is relatively high concentration in the exoskeleton. The distribution of zinc in figure 1(b) clearly shows muscles. High concentration structure near the center of the image is an ovary (also shown in figure 2(a)). Iodine in figure 1(c) is mainly accumulated at 7 parts in the metasome. These distributions are along the ventral border. The longest one (upper right part of the image) is in the head. Judging from its position, this structure has

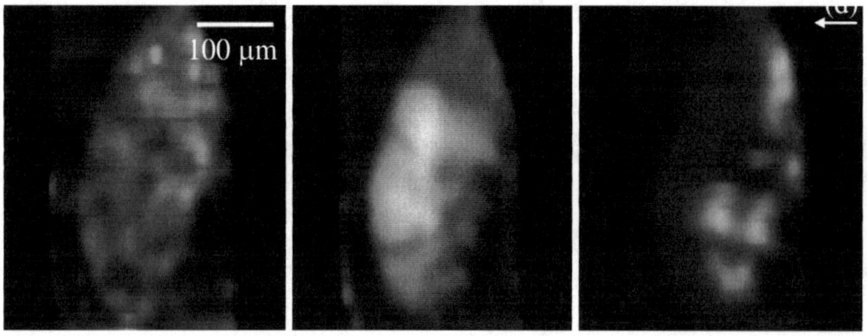

Figures 1. Lateral view of 3-dimensional distributions of (a) iron, (b) zinc and (c) iodine L-line. A head is on the upside and abdomen is toward right side in the image.

possibility to be a nerve cord [8]. The other 6 parts are seen at bases of swimming legs.

Cross section images of the distributions of zinc (shown in gray scale) and iodine (shown in dotted shapes) are shown in figures 2. Longitudinal positions of these cross section images are indicated by arrows in figure 1(c). In figure 2(a), the gut is seen at the centre of the cross section images like a hole. The ovary has high concentration and is observed around the gut. In figures 2(b), 2(c) and 2(d), the dense distributions of iodine coincide with those of zinc. On the other hand, in figure 2(a), the distributions of iodine and zinc do not coincide. These comparisons imply that the tissue which contains dense iodine in the head is different from the other parts.

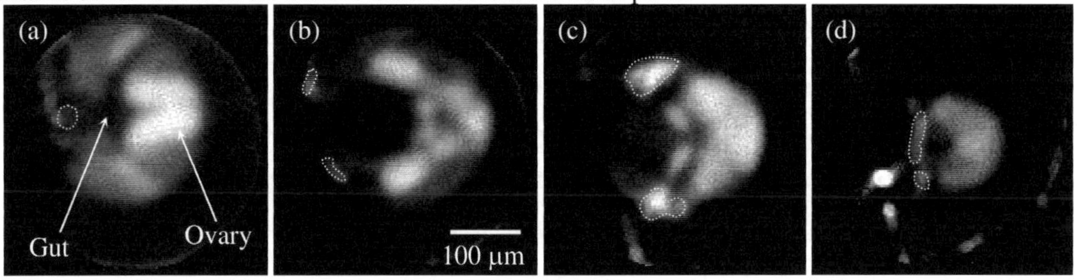

Figures 2. The cross section images of the distributions of zinc (gray scale) and iodine (dotted shapes). Longitudinal positions of the cross section images correspond to arrows in figure 1(c).

4. Conclusion

The 3-dimensional distribution of iodine was observed at the ventral border of the head and at the bases of the swimming legs by using the FXCT. Then, the distributions of the other elements were helpful information to identify the anatomical structures. For further identification of the distribution of iodine and the organs or the structures, more anatomical investigation of the calanoid copepod is needed.

Acknowledgements

The synchrotron radiation experiments were performed at the BL37XU in the SPring-8 with the approval of the Japan Synchrotron Radiation Research Institute (JASRI) (Proposal No. 2007A2066).

The calanoid copepod samples were supplied by professor Syuhei Nishida (Ocean Research Institute, University of Tokyo).

References

[1] Oparin A I 1955 *The Origin of Life* (Moscow: Foreign Languages Pub. House)
[2] Miller S L 1953 *Science* **117** 528-9
[3] Nozaki Y 2001 *Encyclopedia of Ocean Science* vol 2 ed Steele J H, Thorpe S A and Turekian K K (San Diego: Academic Press) pp 840-5
[4] Golosio B, Somogyi A, Simionovici A, Bleuet P and Lemelle L 2004 *App. Phys. Lett.* **84** 2199-201
[5] Watanabe N, Yamamoto K, Takano H, Ohigashi T, Yokosuka H, Aota T and Aoki S 2001 *Nucl. Instr. Meth.* A **467-468** 837-40
[6] Ueda H 1997 *An Illustrated Guide to Marine Plankton in Japan* ed Chihara M and Murano M (Tokyo: Tokai Univ. Press) pp 844–51
[7] Crawford C R and King K F 1990 *Med. Phys.* **17** 967-82
[8] Marshall S M and Orr A P 1972 *The Biology of a Marine Copepod* reprint (Berlin: Springer-Verlag) pp 11-24

Journal of Physics: Conference Series **186** (2009) 012094 doi:10.1088/1742-6596/186/1/012094

STXM and LSLM investigation of Eu(III) induced humic acid colloid aggregation

M Plaschke[1], J Rothe, R Klenze, J Wissler[2], A Naber[2]
Forschungszentrum Karlsruhe, Institut für Nukleare Entsorgung, Postfach 3640, D-76021 Karlsruhe
[2]Universität Karlsruhe, Institut für Angewandte Physik, Wolfgang-Gaede-Strasse 1, D-76131 Karlsruhe

E-mail: plaschke@ine.fzk.de

Abstract. Humic acids (HA) are potentially important in binding traces of actinides or lanthanides, thus affecting their transport in aquatic systems. Eu(III) induced HA colloid aggregation has been investigated by a combination of Scanning Transmission X-ray Microscopy (STXM) and Laser Scanning Luminescence Microscopy (LSLM). Both methods reveal the same aggregate morphology - optically dense zones embedded in a matrix of less dense material observed by STXM correspond to areas with increased Eu(III) luminescence yield in the LSLM micrographs. From these comparative measurements we infer the enrichment of Eu(III) cations in the optically dense zones. These areas also exhibit a C 1s-NEXAFS signature strongly differing from the signal extracted from the less dense areas. Spectral filtering of Eu(III) luminescence lines corresponding to valence transitions affected by organic acid complexation indicates that Eu(III) cations are more strongly bound in the dense zones.

1. Introduction

Humic acids (HA) represent a class of organic residues left over from biodegradation of living organisms. These materials possess a high density of chemically reactive sites and regulate transport and retention of nutrients or contaminants in the environment. HA are generally characterized as colloid sized biopolymers with high structural diversity - depending on their origin and formation history. HA have been subject of intense research in the past, trying to assess possible risks posed by organic acid macromolecules acting as carriers for toxic metal ions (M^{n+}), e.g., actinide cations in the case of nuclear waste disposal [1]. The role of proton-exchanging functional groups such as carboxylic or phenolic groups is investigated in an attempt to quantify HA complexation behaviour. The exact molecular structure of M^{n+}-HA complexes is unknown. Typical HA characteristics like the varying complexation strength (described in terms of stability or complexation constants) or quasi-irreversible reaction kinetics observed for certain HA fractions (sometimes identified as binding by 'strong sites') are not well understood based on bulk characterization techniques.

Carbon K-edge NEXAFS spectroscopy as provided at a Scanning Transmission (soft) X-ray Microscope (STXM) is well suited for the investigation of metal cation HA interaction phenomena. Chemical speciation of HA using C 1s-NEXAFS provides information of structural subunits based on

[1] To whom any correspondence should be addressed.

characteristic spectral features due to C 1s → π*/σ* transition resonances [2]. At the same time, STXM allows to image aggregate morphology of fully hydrated specimen at spatial resolutions down to 50 nm. A distinct complexation effect visible in the C 1s-NEXAFS of, e.g., Eu(III) loaded HA aggregates (where the lanthanide is used to simulate trivalent actinides) has been proposed comparing results obtained for Aldrich HA with those for polyacrylic acid (PAA) [3]. Complementary to the STXM spectromicroscopy of the carbonaceous material, confocal laser scanning luminescence microscopy (LSLM) allows to directly map the metal distribution in Eu(III)-HA aggregates, revealing metal distribution patterns correlating with STXM carbon density maps [4]. Applying optical filtering of the Eu(III) luminescence yield enables us to gain additional spectroscopic information on the complexation mode of the metal cation in the different zones.

2. Materials and methods

Eu(III)-HA samples are prepared according to a procedure described in the literature with a final HA concentration (Aldrich, purified) of 100 mg/L and a Eu(III) concentration of 3×10^{-4} mol/L at pH 5.0 [3]. Eu(III)-HA aqueous suspensions are investigated as thin films between two silicon nitride membranes (STXM) or two cleaned microscope cover slides (LSLM). The sample preparation techniques are described in detail in a previous paper. Soft X-ray spectromicroscopy investigations are performed at the STXM endstation (X-1A Outboard-STXM) at the National Synchrotron Light Source, Brookhaven, NY. Details of the experimental procedures and STXM image stack data processing have been given elsewhere [5]. A LSLM based on an inverse optical microscope (TE2000-U, Nikon GmbH, Germany) is adapted for spatially resolved detection of the weak Eu(III) cation luminescence [4]. Resonant Eu(III) laser excitation is accomplished by a continuous-wave diode laser (DL-SHG 110, Toptica, Germany). The laser wavelength is adjustable over a range of several nanometers at a central wavelength of 394 nm and is fine-tuned to the maximum of the narrow Eu(III) $^7F_0 \rightarrow {}^5L_6$ (λ=393.5 nm) absorption line. The Eu(III) luminescence is recorded by an avalanche photo diode using different emission filters (AHF Analysentechnik, Tübingen, Germany) with the center wavelength set to organic HA luminescence emission (565 nm) and the two dominant Eu(III) emission lines ($^5D_0 \rightarrow {}^7F_1$ at 592 nm and $^5D_0 \rightarrow {}^7F_2$ at 615 nm). All emission filters have a fwhm of 20 nm.

3. Results and discussion

The aggregate morphology of Eu(III)-HA recorded by STXM at 288.5 eV (figure 1a) reveals optically dense material ('dense zones') embedded in a matrix of less dense material ('light zones'). Distinct spectral differences of the C 1s-NEXAFS extracted from the dense and light zones of Eu(III)-HA have been previously discussed [3].

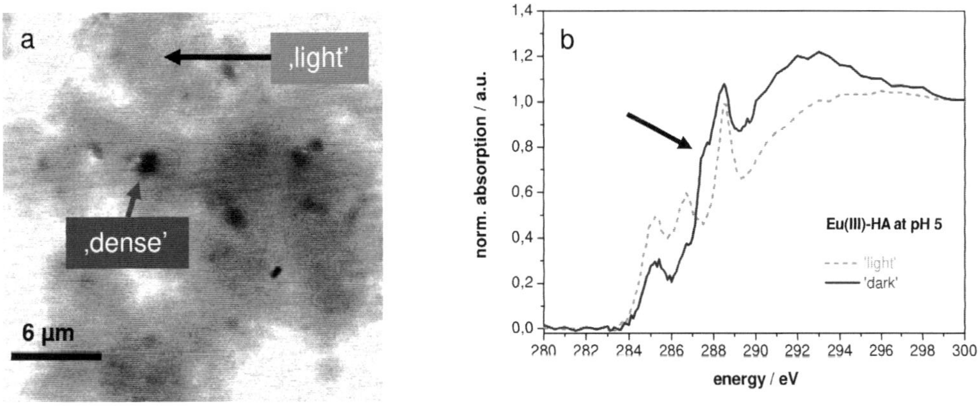

Figure 1. **(a)** STXM image of a Eu(III)-HA aggregate at pH 5 (recorded at 288.5 eV) and **(b)** the corresponding NEXAFS spectra, see text.

The strong shoulder at ~287.5 eV (cf. arrow in figure 1b) adjacent to the carboxylic peak at 288.5 eV in the dense material results from a specific metal ion interaction with the carboxyl functional groups. This interpretation is meanwhile corroborated by quantum chemical calculations [6]. The spatially resolved Eu(III) luminescence intensity obtained for a similar Eu(III)-HA aggregate is shown in figure 2a. Patches with increased luminescence intensity (bright spots in Figure 2a,d) are embedded in a matrix of less fluorescent material. From the similar morphologies obtained by STXM and LSLM we infer the enrichment of Eu(III) cations in the optically dense zones obtained by STXM [4]. In order to discriminate contributions from HA and the complexed Eu(III) cations to the total luminescence yield we apply different emission filters with the center wavelength set to the organic HA luminescence emission (565 nm) and the two dominant Eu(III) metal cation emission lines ($^5D_0 \rightarrow {}^7F_1$ at 592 nm and $^5D_0 \rightarrow {}^7F_2$ at 615 nm) (cf. [7] for details on Eu(III)-HA luminescence spectroscopy). The corresponding LSLM images are depicted in figure 2b-d. The dashed line in figure 2a is crossing a highly luminescent spot (cf. circled area) - corresponding to the dense zones in the STXM micrograph - and a region of less luminescent matrix - corresponding to the lighter zones in the STXM micrograph. The fluorescence intensities along this line are depicted in figure 2e-h for the different filter settings.

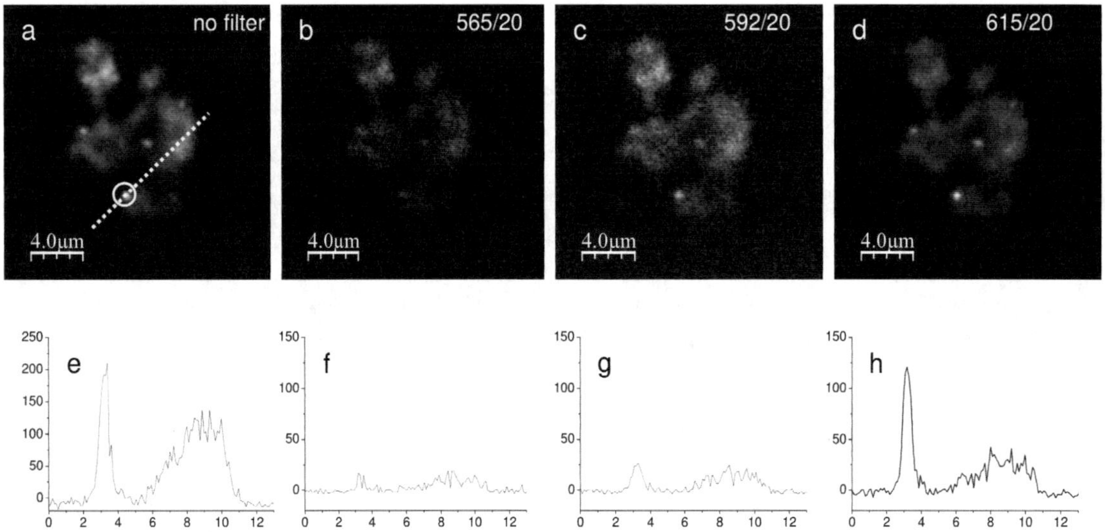

Figure 2. (a-d) LSLM images of a Eu(III)-HA aggregate at pH 5 ((a) recorded without emission filter and (b-d) filters set to 565, 592, and 615 nm) and **(e-h)** the corresponding fluorescence intensities (y-axis, a.u.) as line profiles (x-axis, µm) along the dashed line in (a), see text.

Coarse spectral information for these two sample regions can be derived from the integrated peak intensities obtained with the different filter settings (figure 3). The generally weak luminescence signal at 565 nm (figure 2b) corresponds to organic HA fluorescence [7]. The $^5D_0 \rightarrow {}^7F_2$ transition (hypersensitive band) at 615 nm is known to indicate changes in the Eu(III) coordination geometry. This transition gains intensity upon complex formation while the adjacent $^5D_0 \rightarrow {}^7F_1$ transition is not affected by complexation. We observe that the intensity ratio $^5D_0 \rightarrow {}^7F_2$ (615 nm) / $^5D_0 \rightarrow {}^7F_1$ (592 nm) is strongly increased in the dense, highly luminescent zones compared to the light, less luminescent zones. The following conclusions can be drawn from these observations: i) the aggregate morphology in the unfiltered LSLM image is determined by the Eu(III) distribution in Eu(III)-HA (cf.figures 2a,b); ii) Eu(III) is enriched in the confined bright spots corresponding to the STXM dense regions; iii) from the LSLM intensity ratios of the dominant Eu(III) emission lines different complexation mechanisms in dense and light zones can be inferred. The most plausible explanation for this observation is a higher density and/or strength of cation binding sites in the optically dense HA fraction.

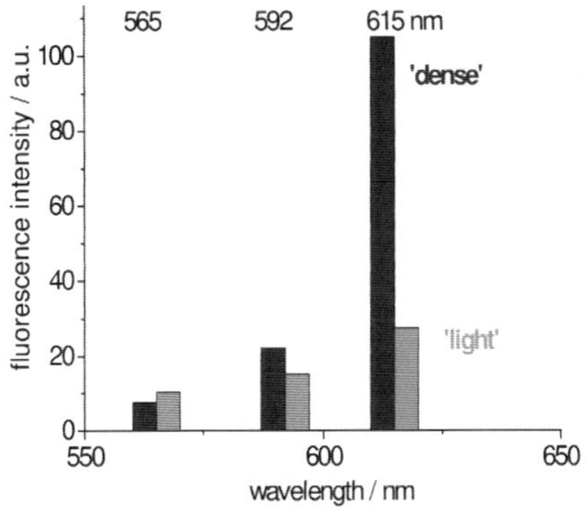

Figure 3. Fluorescence intensities in the area with high Eu(III) luminescence yield (corresponding to STXM dense zones) and lower Eu(III) luminescence yield (corresponding to STXM light zones) shown in Fig. 2 (emission filters set to 565, 592, and 615 nm, see text).

The combination of two spectromicroscopy methods probing the organic ligand (STXM/C 1s-NEXAFS) and the metal center (LSLM/luminescence spectroscopy) gives further evidence to our previous interpretation of metal ion induced aggregation and segregation in Eu(III)-HA. There are at least two different HA fractions. Eu(III) is enriched and/or more strongly bound in a minority fraction containing higher densities of complexing sites. This fraction is likely to play a dominant role in HA colloid mediated transport of lanthanide and actinide cations in the hydrosphere.

Acknowledgments

We would like to thank the Helmholtz Association of National Research Centers, Bonn, Germany for financial support within the Virtual Institute of Functional Properties of Aquatic Interfaces. We are grateful for beam-time allotment by BNL/NSLS. STXM data was taken using the X-1A STXM developed by the group of Janos Kirz and Chris Jacobsen at SUNY Stony Brook, with support from the Office of Biological and Environmental Research, U.S. DoE under contract DE-FG02-89ER60858, and the NSF under grant DBI-9605045.

References

[1] Reiller P E, Evans N D M and Szabó G 2008 *Radiochim. Acta* **96** 345
[2] Hitchcock A P, Urquhart S G and Rightor E G 1992 *J. Phys. Chem.* **96** 8736
[3] Plaschke M, Rothe J, Denecke M A and Fanghänel Th 2004 *J. Electron Spectrosc. Relat. Phenom.* **135** 53
[4] Naber A, Plaschke M, Rothe J, Hofmann H and Fanghänel Th 2006 *J. Electron Spectrosc. Relat. Phenom.* **153** 71
[5] Jacobsen C, Flynn G, Wirick S and Zimba C 2000 *J. Microsc.* **197(2)** 173
[6] Armbruster M K, Schimmelpfennig B, Plaschke M, Rothe J, Denecke M A and R. Klenze 2008 *J. Electron Spectrosc. Relat. Phenom.*, submitted
[7] Monsallier J-M, Scherbaum F J, Buckau G, Kim J I, Kumke M U, Specht C H, Frimmel F H, 2001 *J. Photochem. Photobiol. A* **138** 55

The application of STXM and µFT-IR to characterize sedimentary organic matter in the Opalinus clay

T Schäfer[1], P Michel[1] and F Claret[2]

[1] Forschungszentrum Karlsruhe, Institut für Nukleare Entsorgung (INE), P.O. Box 3640, D-76021 Karlsruhe, Germany

[2] Bureau de Recherche Géologique et Minière, 3 Avenue Claude Guillemin, BP 36009, 45060 Orléans cedex 2, France

E-mail: thorsten.schaefer@ine.fzk.de

Abstract. Clay-rich formations with low organic carbon content are considered as host formations for radioactive waste disposal. Examples are the Callovo-Oxfordian argillite (France) and the Opalinus Clay (Switzerland). The natural organic matter can have an impact on the chemical clay stability and can be the source of strongly radionuclide complexing hydrophilic compounds. The investigated clay sample originates from the Opalinus clay (Benken). 1.1 wt. % of the total organic carbon inventory could be released by alkaline extraction. C-, K-, Ca- edge Scanning Transmission X-Ray microscopy (STXM) and synchrotron µFT-IR microcopy was used for identification of the spatial distribution of clay organic matter and mineral dependent organic matter functionality. The results show the complexity of the organic inventory and could be used to identify possible sources of hydrophilic organic compounds.

1. Introduction

Natural organic matter (NOM) including humic substances (humic and fulvic acids) dissolved in surface/ groundwaters are strong metal complexing agents and can therefore enhance the mobility of contaminants through colloidal transport [1]. The oxygen- and sulfur- containing functional groups of humic substances can vary depending on the origin of humic substances even in different groundwater depths of the same geological site [2]. Geochemical gradients (redox zoning) will also influence the functionality of humic substances altering their redox properties / redox buffering capacity [3]. Carbon K-edge STXM combined with synchrotron-based µFTIR is a very powerful tool to unravel this organic matter heterogeneity under nanometer to micrometer resolution even in very low carbon environments (< 1mg/L DOC) as has been demonstrated in ultra-oligothropic lakes [4]. Using rock microtome sections it is also possible to derive information on sedimentary (immobile) organic matter (SOC) functionality and its distribution in low carbon clay-rich formations as demonstrated for the Callovo- Oxfordian argillite which is considered as possible host rock formation for the deep geo-logical disposal of high level nuclear waste [5]. Residual organic matter in diagenetically overprinted shales or argillites of marine/terrestrial origin still shows some extractable hydrophilic compounds [6].

The objective of this investigation is to compare XANES spectra measured on the C K-edge, K L-edge and Ca L-edge with synchrotron based mid-infrared microscopy in order to determine (a) the spatial distribution and functional group composition of organic matter in the Opalinus clay (OPA) and (b) localize possible sources of hydrophilic organic compounds in the undisturbed sediments.

2. Materials & Methods

The Opalinus Clay (OPA) sample (579.19-579.45m) of the borehole Benken was provided by the Swiss National Cooperative for the Disposal of Radioactive Waste (NAGRA). The total organic carbon (TOC) concentration in the clay fraction of the Opalinus clay is < 0.4 wt. %. The maximum burial temperature in the OPA is between 59° C and 78° C [7]. The untreated rock sample and the clay fraction were embedded in sulfur and ultra-microtomed, for details see [6]. An aliquot of the clay fraction was alkaline extracted [8] and yielded ~1.1 wt. % FA- type material and no quantifiable amount of HA in the OPA. FA sample preparation for STXM analysis was performed by drying a 0.5 to 1 μL droplet of the suspension on a 100 nm thick Si_3N_4 window. STXM measurements were performed at the National Synchrotron Light Source (NSLS), Brookhaven National Laboratories (BNL) in New York, undulator beamline X1A1. Details on the beamline set-up, the measurement conditions as well as the post processing by principle component (PCA), cluster and target spectra analysis can be found in [5, 9]. All carbon XANES spectra are treated according to the de-convolution procedure described in [6, 10] for semi-quantitative analysis of functional group content. Infrared measurements were performed at the U10B beamline (NSLS) using a Spectra-Tech Continuum IR microscope coupled to a Nicolet Magna 860 FTIR. Spectra were collected in transmission mode with 10 x 10 μm aperture, 1024 scans in the wavenumber range 650-4000 cm^{-1} with 4 cm^{-1} resolution. Data post-processing included multivariate statistics analysis (CytoSpec software).

3. Results and Discussion

The combination of the C-, K- and Ca- spectra gives the possibility to differentiate between organic association to carbonates (strong Ca K-edge and carbon K-edge π^* resonance at 290.2 eV), illitic clays (potassium $L_{2,3}$-edge), smectitic clays (Ca-edge and carbon K-edge absorption without 290.2 eV carbonate peak) in the whole rock. Analysis of the OPA microtome section reveals organic matter in regions, where also significant K and Ca absorption indicative for illite/ mixed layer minerals (MLM) intergrown with nanocrystalline carbonate appears (Fig. 1). Both clusters identified are very similar and show a low aromaticity (Σ (benzoquinone-type, $C_{arom=Carom}$, phenol-type = 13-18%) with a very low amount of C_{arom} associated oxygen containing functional groups (phenol-type 0-2%) and a high aliphaticity (24-25%). This observation raises the question if the illite/MLM region could serve as source for the isolated FA like organics in the OPA. From the target spectra map the illite/MLM region can be identified with a high certainty (σ = 0.0195) as source for the isolated FA. Differences between the combination of the first four principal components (PC) out of PCA and the target spectra (isolated FA) are a lower absorption in the phenol-type group and C_{aliph} region and a higher absorption in the carboxyl type group for the isolated FA indicating a higher overall hydrophilicity of FA in line with the extraction behavior.

The IR spectra extracted from the clay-rich region of the OPA microtome reveals kaolinite group mineral (3696, 3669 and 3651 cm^{-1} beside the 3619 cm^{-1} inner hydroxyl group band), a weak OH-stretching feature of trioctrahedral chlorites (3666 cm^{-1}), Si-O or Al-OH vibrations of quartz and/or alumosilicates (1076 cm^{-1}, 1039 cm^{-1}) and carbonates (asymmetrical stretch of CO_3^{-2} around 1426 cm^{-1}). The band at 1163 cm^{-1} can be assigned to the C-C bond in aliphatics and polymeric substances or to a Si-O-C bond, which might indicate a organic linkage to the clay Si-O groups [11]. Characteristic features for organic components include a broad band of phenols, alcohols and carboxylic OH near 3430-3383 cm^{-1}, aromatic CH groups by a pronounced shoulder at 3073 cm^{-1}, aliphatic CH_2 and CH_3 bands (2960, 2922, 2857, 1460 and 1345 cm^{-1}), carboxylic bands of ketones, acids or esters (1753 cm^{-1} and 1260 cm^{-1}), a peptitic amide II band at 1575cm^{-1} and the band at 1631 cm^{-1} for mostly aromatic C=C. In the amorphous organic matter regions a very high absorption for organic OH around 3300 cm^{-1}, C_{arom}-H, aliphatic CH_2 and CH_3 bands, peptitic amide I and amide II bands (1650 and 1540cm^{-1}) and the band for aromatic C=C could be identified. Striking is furthermore the rather high transmittance in the carboxylic band region of ketones, acids or esters at 1753 cm^{-1} indicating a loss of oxygen containing functional groups and increase in hydrophobicity/polymerization of this material.

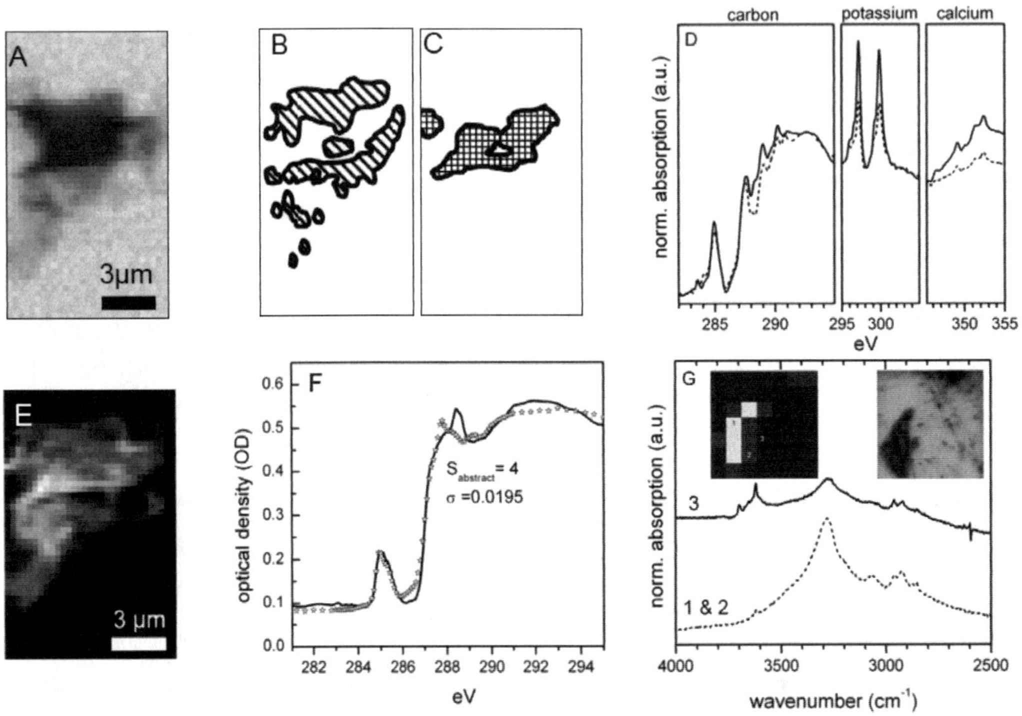

Figure 1. STXM analysis of the OPA microtome. (A) Absorption image at 280 eV, (B, C) The two cluster regions identified, (D) Spectra of the clusters found (cluster B solid line, cluster C dashed line), (E) Target spectra map of the STXM data (light grey colors indicate good correlation), (F) Isolated FA target spectra (solid line) and the fit using the first four significant principle components $S_{abstract}$ out of the PCA analysis (open stars), (G) μFT-IR average spectra of the three clusters found by k-mean clustering (visible microscope image and cluster regions inserted; square size 10x10μm).

4. Acknowledgement

This work was financed through the Integrated project IP 'FUNMIG' and the EURATOM Intra-European fellowship 'COMACK' (FI6W-CT-2006-036493).

References

[1] Kretzschmar R and Schäfer T 2005 *Elements* **1** 205-210
[2] Schäfer T, Buckau G, Artinger R, Kim J I , Geyer S, Wolf M, Bleam W F, Wirick S and Jacobsen C 2005 *Org. Geochem.* **36** 567-582
[3] Einsiedl F, Mayer B and Schäfer T 2008 *Environ. Sci. Technol.* **42** 2439-2444
[4] Schäfer T, Chanudet V, Claret F and Filella M 2007 *Environ. Sci. Technol.* **41** 7864-7869
[5] Schäfer T, Michel P, Claret F, Beetz T, Wirick S and Jacobsen C 2008 *J. Electron. Spectrosc. Relat. Phenom.* **doi:10.1016/j.elspec.2008.05.007**
[6] Khalili F 1990 *Fuel* **69**(2) 151-156
[7] Nagra 2003 Technical report NTB 02-05, Nagra, Wettingen (Switzerland).
[8] Claret F, Schäfer T, Rabung T, Wolf M, Bauer A and Buckau G 2005 *Appl. Geochem.* **20** 1158
[9] Lerotic M, Jacobsen C, Schäfer T and Vogt S 2004 *Ultramicroscopy* **100**(1-2) 35-57
[10] Schäfer T, Hertkorn N, Artinger R, Claret F and Bauer A 2003 *J. Phys. IV* **104** 409-412
[11] Naidja A, Huang P M, Anderson D W, van Kessel C 2002 *Appl. Spectros.* **56**(3) 318-324

X-Ray Imaging of Radioresistant *Deinococcus radiodurans*

K Takemoto[1], I Narumi[2], K Satoh[2], H Namba[3] and H Kihara[1]
[1] Department of Physics, Kansai Medical University, 18-89 Uyamahigashi, Hirakata, Osaka, 573-1136, Japan
[2] Quantum Beam Science Directorate, Japan Atomic Energy Agency, 1233 Watanuki, Takasaki, 370-1292, Japan
[3] Department of Physical Science, Ritsumeikan University, 1-1-1 Noji-Higashi Kusatsu, Shiga, 525-8577, Japan

E-mail: takemoto@makino.kmu.ac.jp

Abstract. *Deinococcus radiodurans* has been known to withstand radiation levels up to 1,000 times than that would kill normal human cells. To cope radiation damage during soft X-ray observation of living cells, *D. radiodurans* incubated with tellurium oxyanions was used as the X-ray microscopy sample. The first observation was successfully performed. In combination of antifreeze solution and subzero temperature, along with carbon window, the cell observation will be more closely to the living condition.

1. Introduction

Soft X-ray microscope is expected to be one of the promising tools for observing living cells and tissues with nm order resolution. However, in order to observe living cells, there are several problems to be solved. The most serious problem is radiation damage. Doses of about 10^4 Gy exhibit immediate morphological changes. Based on Sayer's approach [1, 2], the dose necessary to image the 50 nm protein structure in a 10 μm water layer is 10^5 Gy.

Deinococcus radiodurans has been known to withstand radiation levels up to 1,000 times than that would kill normal human cells. Exponential-phase cultures of *D. radiodurans* survive 5,000 Gy of γ radiation. *D. radiodurans* belongs to a unique family of eubacteria characterized by an exceptional capacity to withstand the usually lethal effects of DNA-damaging agents including ionizing radiation, UV light and desiccation [3]. *D. radiodurans* also exhibits high resistance to tellurium oxyanions [4]. Oxyanions of tellurium, like tellurite (TeO_3^{-2}) and tellurate (TeO_4^{-2}), are highly toxic for most bacteria. However, tellurite-resistant bacteria do exist in nature and they often reduce tellurite to its elemental and less toxic form Te^0 that is accumulated as black deposits inside the cell. Tellurite toxicity results from its ability to act as a strong oxidizing agent to generate reactive oxygen species (ROS) in cell [5]. *D. radiodurans* genome possesses a *ter* operon (*terZ-ABDEF*) which confers tellurite resistance [4]. It has been shown that *Rhodobacter capsulatus* cells incubated with potassium tellurite exhibit increased superoxide dismutase (SOD) activity and increased resistance to tellurite when exposed to the O_2^- generator paraquat [6]. On the other hand, it has been shown that loss of *sod* gene renders *D. radiodurans* sensitive to ionizing radiation [7]. These facts lead us to use *D. radiodurans* cells incubated with tellurium oxyanions for soft X-ray examination in order to make living cells observation successful by reducing radiation damage.

2. Materials and methods

Journal of Physics: Conference Series **186** (2009) 012096 doi:10.1088/1742-6596/186/1/012096

2.1. Sample Preparation

Deinococcus radiodurans strain R_1 (ATCC13939) was grown at 30 °C on TGY agar (0.5 % Bacto tryptone, 0.3 % Bacto yeast extract, 0.1 % glucose and 1.5 % Bacto agar) supplemented with 250 μg/ml of potassium tellurate. One loop of colony was resuspended in 1 ml of MilliQ water, and 1 μl of the cell suspension was placed on a grid with poly-vinyl-formale (PVF) membrane and air-dried.

2.2. X-ray Microscopy

X-ray microscopic observation was performed at beamline BL12 [8, 9]. Optical elements of the X-ray microscope used for imaging and focusing are zone plates. In conjunction with a pinhole (20 μm in diameter), a condenser zone plate (CZP, diameter: 9 mm, outermost zone width: 53.7 nm, number of the zones: 41,890) acts as a dispersive and focusing element. The expected energy resolution ($E/\Delta E$) is about 160. Outermost zone width of ZP defines the achievable spatial resolution. The objective zone plate (OZP) had an outermost zone width of 45 nm. The achieved resolution is c.a. 70 nm judging from the knife-edge estimation (20 % - 80 %). The magnified image was recorded on a peltier-cooled, back-illuminated X-ray CCD camera.

3. Results and Discussion

The most serious problem during the observation of living cells using X-ray microscopy is radiation damage of cellular components including DNA, protein and membrane lipid. In this study, *D. radiodurans* that possesses highly efficient mechanisms to protect oxidative damage was selected as the sample of X-ray microscopy observation. Figure 1 shows a transmission electron microscopic image of *D. radiodurans* in which four cells gather together, forming a tetrad with septa. ROS such as O_2^- are generated during the radiolysis of water, an indirect effect of ionizing radiation. Cells become vulnerable to damage, a condition referred to as oxidative stress, when ROS exceed the capacity of endogenous scavengers to eliminate them. As a defense against ROS, organisms contain two major antioxidant enzymes; SOD and catalase [10]. Therefore, one way to reduce radiation damage of cells is to enhance activity of radical scavengers prior to irradiation. In this study, *D. radiodurans* grown in medium containing tellurate was used for X-ray microscopy observation to enhance intracellular SOD activity. Figure 2 shows a typical X-ray microscopic image of *D. radiodurans* cells. Observation wavelength was 2.3 nm and exposure time was 2 min. Electron microscopy and electron spectroscopy imaging have shown that tellurite resistant bacteria contain intracellular crystals of black metallic tellurium [4]. In Figure 2, each cell with spherical-shaped structure had a dense body that assumed to be derived from accumulated tellurium in cytoplasm.

In order to actually observe a cell in the living state, the observation method needs to be optimized. Cryogenic technique promises reduction of radiation damage of bio-specimens. In the case of *D. radiodurans*, the observation in cryogenic temperature is unnecessary for reducing radiation damage. Using antifreeze solution instead of water, it is possible to decrease the temperature of biological specimens to subzero temperature without freezing. In the above case, the spectral range "4.4 nm < λ < 5 nm" that follows immediately after the K-absorption edges of carbon seems to be more attractive than the water window. In the wavelength range so-called "carbon window", organic materials show strong penetration of X-ray, while carbon is relatively non-absorbing. Combined with metal labeling, metal binding site is clearly visualized like fluorescence of optical microscopy observation as shown in figure 3. In addition, because the antifreeze solution is a major free-radical scavenger, it is possible for specimens to obtain higher radiotolerance. Using these ideas, the cell observation will be more closely to the living condition.

4. Conclusion

Soft X-ray microscopy was applied to the visualization of *D. radiodurans*. The first X-ray microscopic observation of air-dried cells at room temperature was successfully performed. In combination of antifreeze solution and subzero temperature, along with carbon window, the cell observation will be more closely to the living condition.

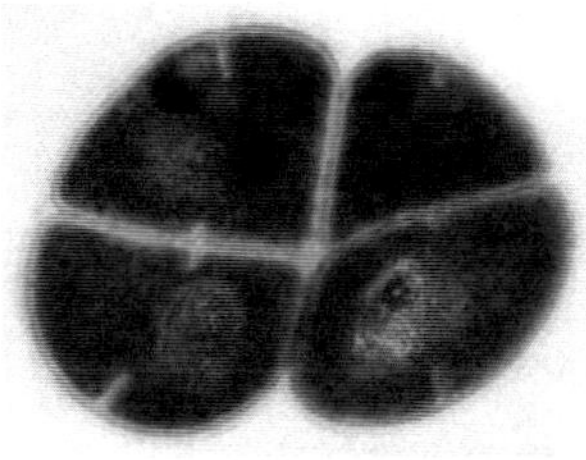

Figure 1. Transmission electron microscopic image of *D. radiodurans*.

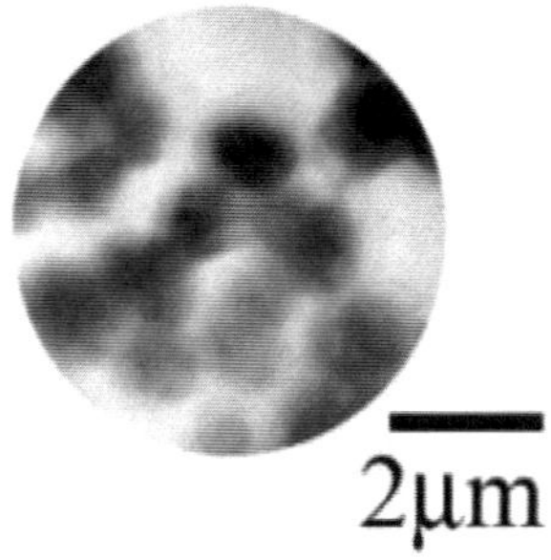

Figure 2. X-ray microscopic image of *D. radiodurans*. Observation wavelength is 2.3 nm and exposure time is 2 min.

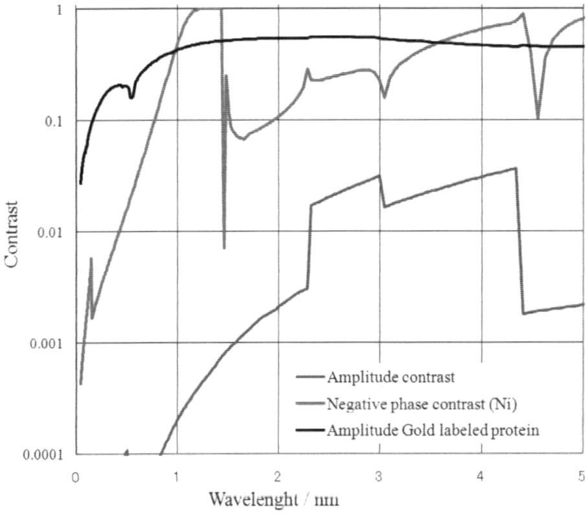

Figure 3. Amplitude contrast and phase contrast of 50 nm protein structure with composition $C_{94}H_{139}N_{24}O_{31}S$ and density 1.35 g/cm^3 in ethylene glycol (EGOH). In labeled protein, 50 nm gold structure is used as a labeling element. Calculation is performed assuming an ideal imaging with a nickel phase plate in back focal plane [11].

Acknowledgments
We are grateful to Ritsumeikan University SR Center staffs for their support. Dr. D. Rudolph *et al.* at Göttingen group and Dr. D. Attwood *et al.* at Lawrence Berkeley National Laboratory are appreciated for the use of a condenser zone plate and an objective zone plate, respectively.

References

[1] Sayre D, Kirz J, Feder R, Kim D M and Spiller E 1977 *Science* **196** 1339-1340
[2] Sayre D, Kirz J, Feder R, Kim D M and Spiller E 1977*Ultramicroscopy* **2** 337-341
[3] Narumi I 2003 *Trend Microbiol.* **11** 422-425
[4] Taylor D E 1999 *Trends Microbiol.* **7** 111-115
[5] Pérez J M, Calderón I L, Arenas F A, Fuentes D E, Pradenas G A, Fuentes E L, Sandoval J M, Castro M E, Elias A O and Vasquez C C 2007 *PLoS ONE* **2** e211
[6] Borsetti F, Tremaroli V, Michelacci F, Borghese R, Winterstein C, Daldal F and Zannoni D 2005 *Res. Microbiol.* **156** 807-813
[7] Markillie L M, Varnum S M, Hradecky P and Wong K K 1999 *J. Bacteriol.* **181** 666-669
[8] Hirai A, Takemoto K, Nishino K, Watanabe N, Anderson E, Attwood D, Kern D, Hettwer M, Rudolph D, Aoki S, Nakayama Y and Kihara H 1998 *J. Synchrotron Rad* **5** 1102-1104
[9] Hirai A, Takemoto K, Nishino K, Niemann B, Hettwer M, Rudolph D, Anderson E, Attwood D, Kern D P, Nakayama Y and Kihara H 1999 *Jpn. J. Appl. Phys.* **38** 274-278
[10] Imlay J A 2003 *Annu. Rev. Microbiol.* **57** 395-418
[11] Rudolph D, Schmahl G, Niemann B 1990 *in Modern Microscopies, Techniques and Applications*, Edited by Michette P D, Plenum Press, London, 59-67

X-ray imaging of Picoplankton in Lake Biwa by Soft X-ray Microscope at Ritsumeikan University SR Center

K Takemoto[1], S Ichise[2], M Ichikawa[1], H Namba[3] and H Kihara[1]
[1] Department of Physics, Kansai Medical University, 18-89 Uyamahigashi, Hirakata, Osaka, 573-1136, Japan
[2] Lake Biwa Environmental Research Institute. 5-34 Yanagasaki, Otsu, Shiga, 520-0022, Japan
[3] Department of Physical Science, Ritsumeikan University, 1-1-1, Noji-Higashi Kusatsu, Shiga, 525-8577, Japan

E-mail: takemoto@makino.kmu.ac.jp

Abstract. Lake Biwa is the largest lake in Japan. Recently, its chemical oxygen demand (COD) index is increasing in spite of a decrease in the values of biochemical oxygen demand (BOD) index. In order to elucidate the mechanism, the first X-ray of a microstructure of picoplanktons inhabiting Lake Biwa was taken. After chemical fixation treatment, the laboratory-cultured *Synechococcus* cells were observed by soft X-ray microscope (BL12) of Ritsumeikan University SR Center. Each cell shows a dark sub-micron core. The low contrast region around the core can be interpreted as agar layers. The cells were estimated as 0.7μm in diameter and agar layers were estimate as 1.2μm in diameter. Namely the agar layer increases the quantification of organic compounds by factor of 1.7.

1. Introduction

Lake Biwa, located in the centre of Shiga prefecture, is the largest lake in Japan. It has a surface of 674 km^2 and is 103m deep (at the deepest point). It was divided into two parts, northern basin (NB) and southern basin (SB) at the narrowest point as boundary. The NB has less population than the SB, and NB is filled with nature. The lake Biwa has 460 inflowing rivers, but there are only two outflows, the Seta River and the Biwako Canal, and provides water for 14 million people in the vicinity and downstream urban regions including Kyoto and Osaka. In the past, Lake Biwa was a typical oligotrophic lake. The quality of the ecosystem has been rapidly reduced. In accordance with the guidelines issued by the OECD, the NB is classified as mesotrophic and the SB has gradually shifted from mesotrophic to eutrophic.

The water quality of Lake Biwa has been regularly surveyed since 1966. In early summer of 1989, a dense bloom of picophytoplankton occurred in the NB. During this bloom, the transparency of the lake decreased to less than half (1.2 m) of that observed during the same period in 1988 [1]. In addition, recently, the chemical oxygen demand (COD) index is increasing in spite of a decrease in the values of biochemical oxygen demand (BOD) index. These results suggest that an organic matter which is hard to decompose underwater has been increasing. Picophytoplankton is considered as an important source of the organic matter. We have started imaging of the picophytoplankton for investigating the microstructures of the picophytoplankton inhabiting Lake Biwa by X-ray microscope.

2. Picophytopkankton

Plankton consists of any drifting organisms that inhabit the pelagic zone of oceans or bodies of fresh water. They are defined by their ecological niche rather than their genetic classification. They are also often described in terms of size. Photosynthetic picoplankton (picophytoplankton) is the fraction of the plankton performing photosynthesis composed by cells between 0.2 and 2 μm. It is especially important in the central oligotrophic regions. Because of its very small size, it is difficult to study by classical methods such as optical microscopy. Since the 1970s, the spread of the epifluorescence microscopy has allowed to detect picophytoplankton such as *Synechococcus* which possess phycoerythrin.

3. Materials and methods

Laboratory-cultured *Synechococcus* cells, photosynthetic picoplankton, were isolated by centrifugation at 6,200 rpm for 3 min at room temperature. The cells were fixed with 1 % glutaraldehyde for 30 min at room temperature. After fixation, *Synechococcus* cells suspension was placed on a copper mesh with polyvinyl formbar (PVF) membrane and air-dried. When approaching a wet sample observation, a wet sample holder was used [2]. It consists of 2 plates and each plate consists of a thin polyimide film (thickness < 300 nm) and a metal support thin metal plate. *Synechococcus* cells suspension was sandwiched between them. X-ray observation was performed at soft X-ray microscope beam line BL12 of Ritsumeikan University SR Center. Observing wavelength is 2.3 nm and exposure time is 5 min.

4. Results and Discussion

Figure1 shows air-dried and wet *Synechococcus* cells. In all micrographs, each cell is clearly distinguishable. Spherical cells and cocoon form cells were observed. The cell of the cocoon form is on cell division. Each cell has a dark sub-micron core. Since a *Synechococcus* cell is covered with agar layer, the low contrast region around the core can be interpreted as agar layer. From figure 1 (b), the cell was estimated as 0.7 μm in diameter and agar content was estimated as 1.2 μm in diameter.

Using the image contrast, fraction of carbon was estimated under the following assumptions: (1) during the drying process, a cell size maintains. Although agar size on the grid plane also maintains, thickness of agar layer is reduced. Therefore the rate of volume contraction of agar is K. (2) Cell consists of protein. ($C_{94}H_{139}N_{24}O_{31}S$, μ_{cell} = 1.81 / μm at 2.3 nm, ρ=1.35g/cm^3) (3) Agar consists of agarose. ($C_{12}H_{12}O_9$, μ_{agal} = 0.95 / μm at 2.3 nm, ρ=1.54g/cm^3) Figure 2 shows X-ray image intensity profile of PP cells. The X-ray transparency ratio is obtained as A: B = 2:7. Using these values, rate of volume contraction of agar layer, K, is estimated as about 80 %. Quantity of organic matter is converted into that of carbon. In case of protein, carbon is 94 per mol and in case of agarose, carbon is 12 per mol. Agar layer makes the amount of carbon of a cell increases 4.5 - 7 times. Namely the agar layer increases the quantity of organic compounds by maximum factor of 7. Quantification of agar content is required for assessing the water quality in Lake Biwa. Improving a sample preparation and observation technique, agar content will be able to be quantified more correctly.

5. Conclusions

In Lake Biwa, COD index is increasing in spite of a decrease in the values of BOD index. Picophytoplankton is considered as an important source of non-biodegradable organic compounds. Picophytoplankton was observed by the soft X-ray microscope.

1. Each *Synechoccoucs* cells were successfully observed with high resolution. In addition, agar content around the cell was also observed.
2. The cell size was estimated as 0.7 μm in diameter and agar layer was estimated as 1.2 μm in diameter. The agar layer increases the quantification of organic compounds by maximum factor of 7.

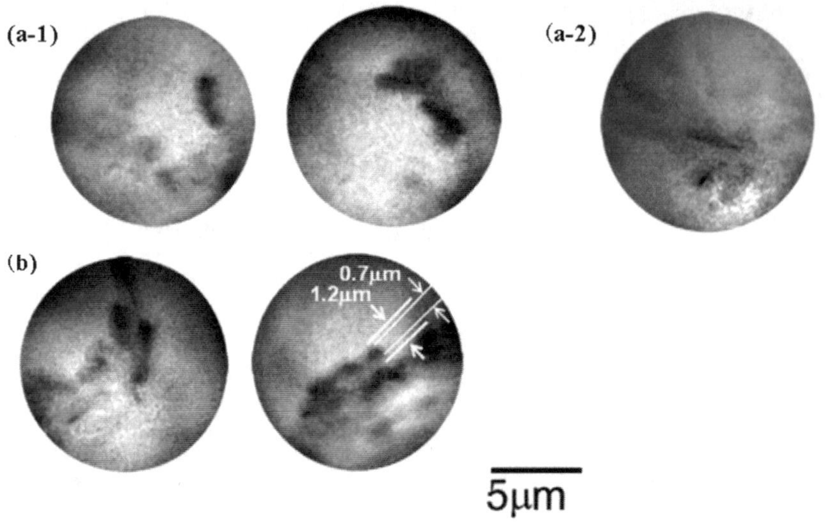

Figure 1. X-ray microscopic images of laboratory-cultured *Synechococcus* PGS and PP cells. (a-1) Dried and (a-2) wet *Synechococcus* PGS cells, (b) dried *Synechococcus* PP cells.

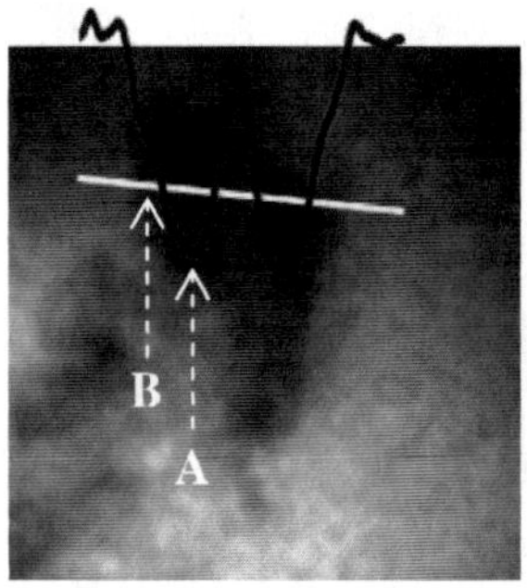

Figure 2. X-ray image intensity profile of PP cells. Digitally enlarged image of figure 1 (b) left side is used for the estimation. A is a cell with agar content. B is agar only.

Acknowledgments
We are grateful to Ritsumeikan University SR Center staffs for their support. Dr. D. Rudolph *et al.* at Göttingen group and Dr. D. Attwood *et al.* at Lawrence Berkeley National Laboratory are appreciated for the use of a condenser zone plate and an objective zone plate, respectively.

References
[1] Eguchi M, Oketa T, Miyamoto N, Maeda H and Kawai A 1996 *J. Plankton Res* **18** 539-550
[2] Takemoto K, Watanabe N, Hirai A, Nakayama Y and Kihara H 1998 in "X-ray Microscopy and Spectromicroscopy (eds. J. Thieme, G. Schmahl, D. Rudolph, & E. Umbach)", Springer-Verlag, Berlin, pp.I-129 - I-134

Imaging of Chromosomes at Nanometer-Scale Resolution Using Soft X-Ray Microscope at Ritsumeikan University SR Center

A Yamamoto[1], K Takemoto[2], I Komura [3], H Namba[3] and H Kihara[2]
[1] Nagahama Institute of Bio-Science and Technology, 1266, Tamura-cho, Nagahama, Shiga, 526-0829, Japan
[2] Department of Physics, Kansai Medical University, 18-89 Uyamahigashi, Hirakata, Osaka, 573-1136, Japan
[3] Department of Physical Science, Ritsumeikan University, 1-1-1, Noji-Higashi Kusatsu, Shiga, 525-8577, Japan

E-mail: takemoto@makino.kmu.ac.jp

Abstract. In order to clarify the process of condensation and distribution of a chromosome with 70 nm resolution, process of cell division was observed by soft X-ray microscopy. During prometaphase of mouse fibroblast cell line NIH3T3, each chromosome is clearly visualized at high contrast. Furthermore, thickness of a chromosome is not uniform but varied from 150 nm to 750 nm. This is considered as a condensing step of chromatin fiber occurring in the prometaphase. During metaphase and anaphase, each chromosome was also observed. The chromosomes were thicker than those in prometaphase. This result implies that during metaphase and anaphase the chromosomes condense and become more compact.

1. Introduction

The chromosomes contain DNA which carries genetic information essential to the cell. The chromosomes, which have replicated during the S phase, become compact transportable forms. They are separated and moved to opposite ends of the cell. Each daughter cell exactly inherits one set of the chromosomes. The mechanisms of chromosome condensation and distribution are the fundamentals of a cell cycle. Knowledges of them are consequently of great importance for an understanding of the continuity of genetic information to the next generation. The fundamental structural unit of chromatin is an assemblage, called the nucleosome [1]. Although chromosome is organized into the chromatin fiber consisting of nucleosome arrays, the packaging arrangement and structure of the higher levels of chromatin fiber organization is not well understood. In this paper, we describe results of soft X-ray microscopic observation of chromosomes. The structure of chromatin varies considerably as the cell progresses through the cell cycle. In each phase, microstructure of condensed chromatin fibers and chromosomes were observed by light and X-ray microscopy.

2. Materials and Methods

2.1. Sample Preparation

Mouse fibroblast cell line, NIH3T3 cells were cultured in DMEM medium in 5 % CO_2 at 37°C. Cells were attached to polyimide membrane and cultured for 24 hours. The attached cells were fixed in 2%

paraformaldehyde in 0.1 M Na-phosphate buffer (pH 7.4) (PB) for 5 min and washed three times in PB, and permeabilized with 0.3 % Triton X-100 for 30 min. In order to observe chromosomes by light microscopy, the cells were stained with acetic orcein. After air-dried, cells in mitotic stage were selected under a light microscope, and the same cells were observed under an X-ray microscope.

2.2. X-ray Microscopy

X-ray microscopic observation was performed at beamline BL12 [2, 3]. The optical element of the X-ray microscope is zone plates used for imaging or focusing. The outermost zone width and aspect ratio define the achievable spatial resolution and diffraction efficiency, respectively. The objective zone plate (OZP) had an outermost zone width of 45 nm. The achieved resolution is c.a. 70 nm judging from the knife-edge estimation (20 % - 80 %). The expected energy resolution ($E/\Delta E$) is about 160. Size of view filed is about $10 \times 10 \mu m^2$. For image acquisition, a sample is scanned in a plane perpendicular to the optical axis. The motion is provided by compact stepping motor actuators (Sigma Koki Co. LTD.) driven linear translation stage. They are located outside the vacuum chamber. A minimum step size is 500 nm and working range is several mm. The magnified image is recorded on a Peltier-cooled, back-illuminated X-ray CCD camera (C4880-21-24WD: Hamamatsu Photonics) with 512×512 pixels and a pixel size is $24 \times 24 \mu m^2$.

3. Results and Discussion

Figure 1 shows a paraformaldehyde-fixed and air-dried 3T3 cell at prometaphase in mitosis. By staining with acetic orcein, each chromosome is imaged under an optical microscope (in figure 1a). Orcein dye ($C_{28}H_{24}N_2O_7$,) doesn't work as an active dye for X-ray microscopy. In soft X-ray region, DNA and protein absorb strongly X-ray, and the chromosomes are also visible as X-ray dense structure. The chromosomes were clearly imaged under an X-ray microscope as shown in figure 1b.

Figure 2 shows an optical and X-ray microscopic images of the same 3T3 cell as figure 1 with higher magnification. Due to the limitation of the conventional light microscope, it is difficult to get clear images at high magnification. On the other hand, in figure 2c, it is easy for us to distingish each chromosome. The thickness of a chromosome is not uniform but varied from 150 nm to 750 nm. Chromatin fiber is the unit of the chromosome. The chromosome with 750 nm in thickness is considered to be almost at the final level of packaging of chromosome. The thinner fiber is considered to be a condensing chromatin fiber in the low level compaction.

Figure 3 and 4 show paraformaldehyde-fixed and air-dried 3T3 cell at metaphase and at anaphase respectively. Each chromosome is also visible as X-ray dense structure (figure 3c and 4c). However the chromosome is not clearly imaged under the X-ray microscope, probably by multiple overlapping of the chromosomes.

The thickness of chromatin is estimated to be 870 nm (figure 3c) in metaphase, and 900 nm (figure 4c) in anaphase respectively. During metaphase, they condense into the fibers about 900 nm in width. It indicates that metaphase and anaphase chromosomes condense and become more compact ones.

4. Summary

In order to reveal the packaging arrangement and structure of the highest levels of chromatin fiber organization, X-ray microscopic observations on the cell division of NIH3T3 cells were performed. In prometaphase, each chromosome is clearly visualized with high contrast. Thickness of the chromosome varied from 150 nm to 750 nm. This is considered as a condensing process of chromatin fiber. In metaphase and anaphase, each chromosome could be also observed under the X-ray microscope. The thickness of the chromosomes was above 900 nm. This result implies that during metaphase and anaphase the chromosomes condense and become more compact.

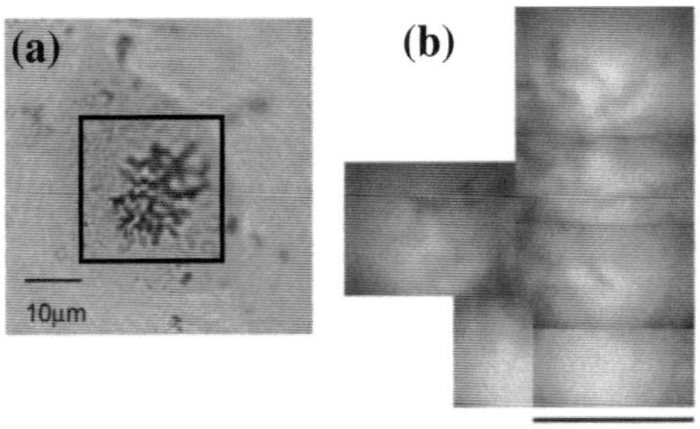

Figure 1. Optical and X-ray microscopic images of a NIH3T3 cell at prometaphase in mitosis. (a) Optical microscopic image, (b) X-ray microscopic image corresponding to square region of (a) at 2.3 nm. Exposure time was 120 s. Scale bar: 10 μm.

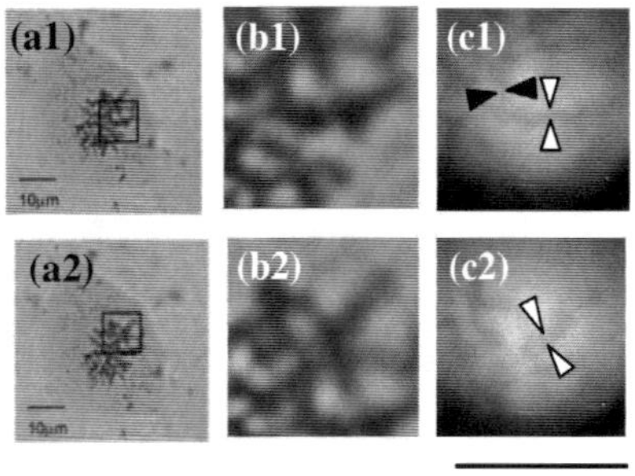

Figure 2. Enlarged images of figure 1. (a) Optical microscopic images. (b) Digitally enlarged images of (a), (c) X-ray microscopic images corresponding to (b) at 2.3nm. Pairs of arrowheads indicate chromosome thickness. (c1) Black arrowheads: 150 nm, white arrowheads: 470 nm, (c2) arrowheads: 750 nm. Exposure time was 120 s. Scale bar: 10 μm.

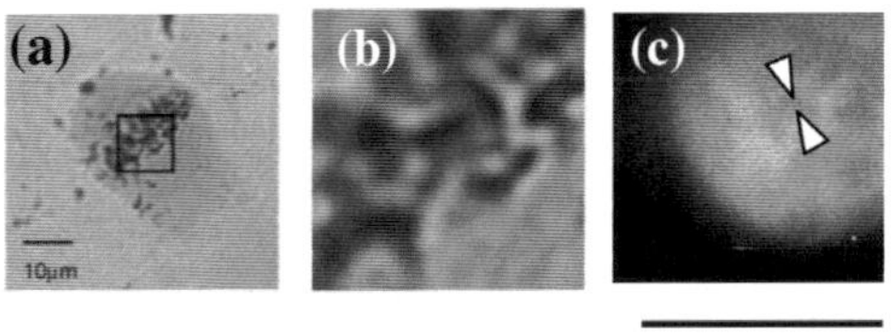

Figure 3. Optical and X-ray microscopic images of a NIH3T3 cell at metaphase in mitosis. (a) Optical microscopic image, (b) Digitally enlarged image of (a), (c) X-ray microscopic image corresponding to (b) at 2.3nm. Pair of arrowheads indicates chromosome of 870 nm in thickness. Exposure time was 120s. Scale bar: 10 μm.

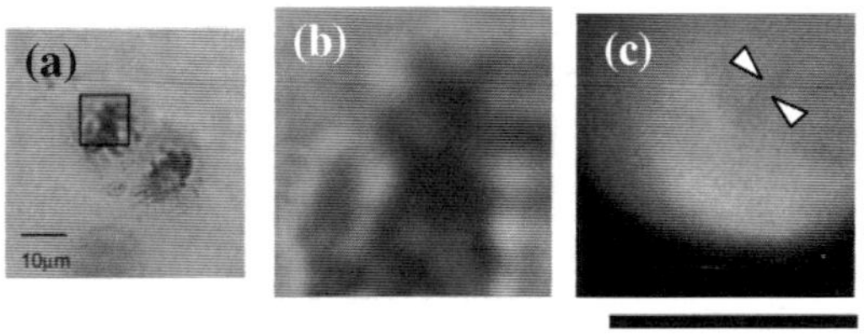

Figure 4. Optical and X-ray microscopic images of a NIH3T3 cell in anaphase. (a) Optical microscopic image, (b) Digitally-enlarged image of (a), (c) X-ray microscopic image corresponding to (b) at 2.3 nm. Pairs of arrowheads indicates chromosome of 900 nm in thickness. Exposure time was 120 s. Scale bar: 5 μm.

Acknowledgments

We are grateful to Ritsumeikan University SR Center staffs for their support. Dr. D. Rudolph at Göttingen group and Dr. D. Attwood at Lawrence Berkeley National Laboratory are appreciated for the use of a condenser zone plate and an objective zone plate, respectively.

References

[1] Kornberg RD 1974 *Science* **184** 868-871
[2] Hirai A, Takemoto K, Nishino K, Watanabe N, Anderson E, Attwood D, Kern D, Hettwer M, Rudolph D, Aoki S, Nakayama Y and Kihara H 1998 *J. Synchrotron Rad.* **5** 1102-1104
[3] Hirai A, Takemoto K, Nishino K, Niemann B, Hettwer M, Rudolph D, Anderson E, Attwood D, Kern D P, Nakayama Y and Kihara H 1999 *Jpn. J. Appl. Phys.* **38** 274-278

Journal of Physics: Conference Series **186** (2009) 012099 doi:10.1088/1742-6596/186/1/012099

Observation of Immuno-Labeled Cells at High Resolution Using Soft X-Ray Microscope at Ritsumeikan University SR Center

A Yamamoto[1], K Takemoto[2], T Fukui[3], Y Yoshimura[3], K Okuno[4], H Namba[3] and H Kihara[2]

[1] Nagahama Institute of Bio-Science and Technology, 1266, Tamura-cho, Nagahama, Shiga, 526-0829, Japan

[2] Department of Physics, Kansai Medical University, 18-89 Uyamahigashi, Hirakata, Osaka, 573-1136, Japan

[3] Department of Physical Science, Ritsumeikan University, 1-1-1, Noji-Higashi, Kusatsu, Shiga, 525-8577, Japan

[4] SR Center, Ritsumeikan University, 1-1-1, Noji-Higashi Kusatsu, Shiga, 525-8577, Japan

E-mail: takemoto@makino.kmu.ac.jp

Abstract. Mouse fibroblast cell line NIH3T3 cells were labeled with the heavy metal (silver and gold) and observed intracellular structure under an X-ray microscope. Microtubules, Golgi apparatus and early endosomes of NIH3T3 cells were stained with immuno-gold nanoparticles, and immuno-staining was intensified by silver or gold enhancement procedure. Using a transmission soft X-ray microscope beamline (BL12) at Ritsumeikan University SR center, we observed immuno-stained NIH3T3 cells with several wavelengths just below and above oxygen edge (λ = 2.32 nm). Using this method, cytoskeleton (microtubules) and organelles (Golgi apparatus and early endosomes) were successfully imaged with high resolution. Thus, immuno-gold silver and gold enhancement technique is useful for specific labeling of intracellular structure under an X-ray microscope.

1. Introduction

Although generally labeling is not required for soft X-ray microscopy, a special labeling technique is quite useful. It can enhance contrast at the specific site, which gives us a clue to observe functionally important structures. In this report, we describe results of soft X-ray contrast enhancement of the specific sites of the cells by heavy-metal labeling. As the first step, we labeled mouse fibroblast cell line NIH3T3 cells with the use of silver enhancement [1] or gold enhancement technique [2] for immuno-gold cytochemistry which has been developed for transmission electron microscopy (TEM). NIH3T3 cells were reacted with antibodies against α tubulin (constituent of microtubule), proteins localizing on Golgi apparatus or endosomes and, then with secondary antibody conjugated with small gold particles (~1 nm). The immuno-staining was intensified by depositing silver or gold on the surface of the gold particles by silver or gold enhancement procedure. Using a transmission soft X-ray microscope at BL12, we examined immuno-stained NIH3T3 cells with several wavelengths just below and above the oxygen edge (λ = 2.32 nm).

2. Materials and Methods

Mouse fibroblast cell line, NIH3T3 cells were cultured in DMEM medium. Cells were attached to polyvinyl formbar (PVF) membrane on Ni grids after poly-L-lysine treatment. The attached cells were fixed in 4% paraformaldehyde in 0.1 M Na-phosphate buffer (pH 7.4) (PB) and washed in PB.

Cells on the membrane were permeabilized in 0.1% Triton-X 100 in PB, and incubated in PB containing 0.005% saponin for blocking. Cells were then treated with mouse monoclonal antibodies (anti- α tubulin, Sigma or anti- GM130, Transduction Lab.) or with rabbit polyclonal antibody (anti-EEA1, Acris) overnight. Then, the cells were washed, and incubated with goat anti-rabbit or anti-mouse IgG that was conjugated to colloidal gold (1.4 nm in diameter) in the blocking solution. Cells were washed with PB, and fixed with 1% glutaraldehyde in PB. After washing, the gold labeling was intensified by using a silver enhancement kit (HQ silver, Nanoprobes) in the dark or by a gold enhancement kit (GoldEnhance, Nanoprobes) in the light. After washing in distilled water, cells were air-dried. X-ray observation was performed at soft X-ray microscope beam line BL12 of Ritsumeikan University SR Center [3].

3. Results and Discussion

Figure1 shows paraformaldehyde-fixed and air-dried 3T3 cells. 3T3 cells on the PVF membrane showed polygonal shape and extended pseudopodia (arrows in figure 1a and b) as fine projection from them. The fine pseudopodia were clearly imaged under an X-ray microscope. The nucleus was also visible as X-ray dense structure (N in figure 1a and b). X-ray lucent vacuole-like structure was observed in the cytoplasm (arrowhead in figure 1a and b). At present, it remains to be solved whether this vacuole-like structure corresponds to some cell structure or is an artifact produced by the fixation and air-dry processes. In unstained cells, we could not distinguish other cellular structures. Next, we observed 3T3 cells immuno-stained with immuno-gold and silver or gold enhancement technique using antibodies against microtubule, Golgi apparatus and endosomes.

Microtubule is tubular structure about 25 nm in diameter, and plays important roles in cell shape regulation, in cell division and in intracellular traffic as one of cytoskeletons of cells. Figure 2 shows tubulin-stained 3T3 cells. Tubulin is the constituent protein of microtubules. Under light microscope, silver enhancement procedure makes stained-tubulin brown (figure 2a, dark color). X-ray microscope imaged many microtubules as fibers radically expanding from the centrosome around the nucleus towards the cell surface (figure 2b). As a result of adjusting the image contrast based on each microtubule, structures of the cytoplasm reconstructed by stitching of small parts were somewhat obscure, but extension of microtubules is clearly visible (figure 2b, white arrows). This result shows that cellular structure (such as microtubule; 25 nm in diameter) which has smaller size than the resolution of the X-ray microscope can be imaged by the imuno-labeling by heavy metal.

Golgi apparatus is an organelle, playing important roles as a center of intracellular traffic and locates near the nucleus. Figure 3 shows silver enhancement of immuno-gold staining of 3T3 cells using antibody against GM130 (Golgi matrix protein 130), which was localized on the Golgi apparatus. Under a light microscope, Golgi apparatus was observed near the nucleus in brown color (figure 3a, dark color). As shown in figure 3b, Golgi apparatus was imaged as X-ray dense structure though the image was somewhat obscure by overlapping with the nucleus. Golgi apparatus was also visualized as X-ray dense structure by gold enhancement procedure of immuno-gold staining of GM130 (data not shown).

The endosome is an organelle, playing important roles in the traffic and the sorting in endocytosis, and consists of early endosome and late endosome. Figure 4 shows gold enhancement for EEA1 (early endosome antigen 1), which was localized on early endosomes. Early endosomes were imaged under X-ray microscope as small dot-like structures (figure 4b). We could only distinguish limited numbers of early endosomes in a cell. This suggests immuno-labeling by this antibody was not sufficiently strong to stain all of the early endosomes under an X-ray microscope. Early endosomes were also observed as small dot-like structures by silver enhancement for EEA1 (data not shown).

4. Summary

It was shown that immuno-gold silver enhancement technique and immno-gold enhancement technique is useful for specific labeling of intracellular structure under an X-ray microscope. Using this method, cytoskeleton (microtubules) and organelles (Golgi apparatus and early endosomes) were successfully observed with high resolution. The improvement of X-ray contrast may be achieved by examining antibodies and staining procedure, and enable us much more intimate analyses of the cell structure under an X-ray microscope.

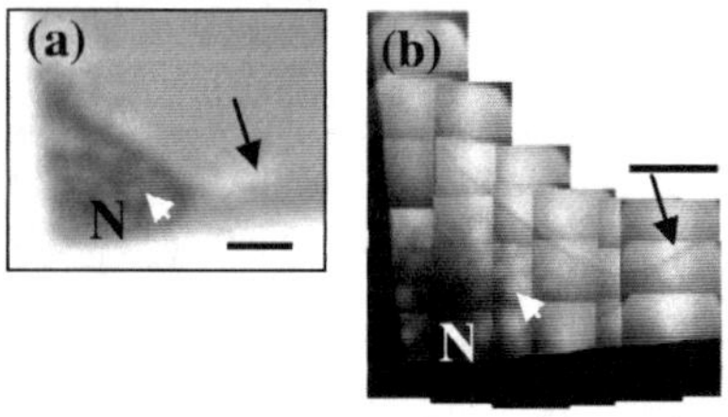

Figure 1. A NIH3T3 cell. (a) Light microscopic image, (b) X-ray microscopic image at 2.4 nm. Exposure time was 120 s. An arrow shows a projection of the cell. An arrowhead indicates vacuole-like structure. N; nucleus, scale bar is 10 μm.

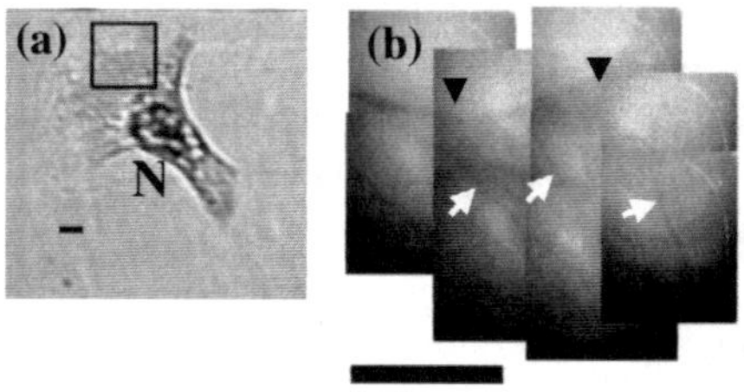

Figure 2. A NIH3T3 cell immuno-labeled by anti-tubulin antibody using immunogold-silver enhancement technique. (a) Light microscopic image, (b) X-ray microscopic image at 2.3 nm. Exposure time was 60 s. The rectangular region of (a) is shown in (b). White arrows in (b) show microtubules. Arrowheads show cell surface. N; nucleus, scale bar is 10 μm.

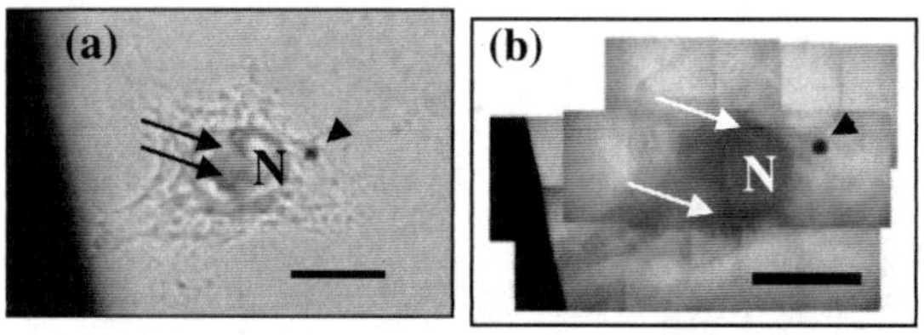

Figure 3. A NIH3T3 cell immuno-labeled by anti-GM130 (Golgi protein) antibody using immunogold-silver enhancement technique. (a) Light microscopic image, (b) X-ray microscopic image at 2.3 nm. Exposure time was 60 s. Arrows show Golgi region. An arrowhead shows a dot produced artifactually. It is not meaningful contrast, and does not correspond to the any cellular structures. N; nucleus, scale bar is 10 μm.

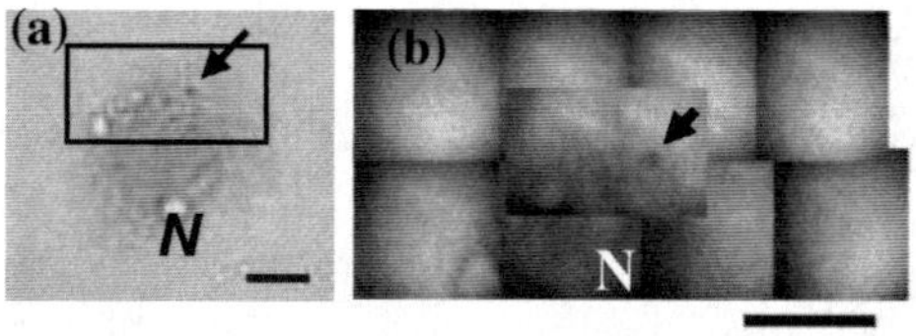

Figure 4. A NIH3T3 cell immuno-labeled by anti-EEA1 (endosome protein) antibody using immunogold-silver enhancement technique. (a) Light microscopic image, (b) X-ray microscopic image at 2.3 nm. Exposure time was 60 s. The rectangular region of (a) is shown in (b). An arrow shows an endosome. N; nucleus, scale bar is 10 μm.

Acknowledgments

We are grateful to Ritsumeikan University SR Center staffs for their support. Dr. D. Rudolph at Göttingen group and Dr. D. Attwood at Lawrence Berkeley National Laboratory are appreciated for the use of a condenser zone plate and an objective zone plate, respectively.

References

[1] Masaki R , Kameyama K and Yamamoto A 2003 *J Biochem (Tokyo)* **134** 415-426
[2] Hainfeld J. F and Powell R D 2000 *J. Histochem. Cytochem.* **48** 471-480
[3] Hirai A, Takemoto K, Nishino K, Watanabe N, Anderson E, Attwood D, Kern D, Hettwer M, Rudolph D, Aoki S, Nakayama Y and Kihara H 1998 *J. Synchrotron Rad* **5** 1102-1104

Imaging Study on Acupuncture points

X H Yan[1,2], X Y Zhang[1,2], C L Liu[3], R S Dang[4], M ANDO[5,6], H SUGIYAMA[6], H S Chen[7] and G H Ding[2,8]

[1] Synchrotron Radiation Research Center, Physics Department, and Surface Physics Laboratory (State Key Laboratory) of Fudan University, Shanghai 200433, China
[2] Shanghai Research Center of Acupuncture and Meridian, Shanghai 201203, China
[3] Physics Department of Yancheng Teachers College, Yancheng 224002, China
[4] Second Military Medical University, Shanghai 200433, China
[5] DDS center, Research Institute for Science and Technology, Tokyo University of Science, Yamasaki 2541, Noda, Chiba 278-8510, Japan
[6] Photon Factory, Institute of Material Structure Science, High Energy Accelerator Research Organization, Oho 1-1, Tsukuba, Ibaraki 305-0801, Japan
[7] Institute of Modern Physics, Fudan University, Shanghai 200433, China
[8] Department of Mechanics and Engineering Science of Fudan University, Shanghai 200433, China

E-mail: xy-zhang@fudan.edu.cn

Abstract. The topographic structures of acupuncture points were investigated by using the synchrotron radiation based Dark Field Image (DFI) method. Four following acupuncture points were studied: Sanyinjiao, Neiguan, Zusanli and Tianshu. We have found that at acupuncture point regions there exists the accumulation of micro-vessels. The images taken in the surrounding tissue out of the acupuncture points do not show such kind of structure. It is the first time to reveal directly the specific structure of acupuncture points by X-ray imaging.

1. Introduction
When people talk about Chinese medicine, the meridians and acupuncture [1-2] must be mentioned. They have unique effects on illness. To acupuncture some special points on human body, which are called acupuncture point (abbr. as acupoint) and distribute along meridians, can cure many kinds of disease or reduce pains. Unfortunately, until now no one can give a scientific explanation about how the acupuncture takes part in the process of treatment. Something is interesting, but strange. To make clear the mechanism of the acupuncture, many researchers devoted themselves to study in this field using various modern techniques. Some valuable results have been found. For example, two infrared emission bands on the surface of acupoints were observed [3]. The gather of element Ca in acupoints was found by the X-ray fluorescence technique [4]. Therefore, the acupoints are also called Ca pools. However, despite considerable efforts for probing the anatomy of acupoints, the characterization of structures of acupoints remains elusive. In this work one kind of X-ray imaging technique, which has demonstrated enhanced image contrast for dense, highly absorbing materials was used to explore the structural character of acupoints.

2. Experimental and Samples

In order to find out whether there are some differences between acupoint regions and surrounding tissues out of the acupoint, then to identify the radiographic structure of the acupoints, the X-ray Dark Field Imaging method (DFI) [5] based on synchrotron radiation was used to provide detail images of acupoints. DFI is a kind of refraction based imaging [6-7]. When X-ray travels in a sample, the index of refraction n should be complex form. It can be written as

$$n = 1 - \delta - i\beta,$$

where, β describes the absorption term of X-ray, δ is the phase shift term caused by the refraction effect. The phase shift term can be up to 10^3 times greater than the absorption term for light elements. Therefore, one may observe the phase contrast when the absorption contrast is undetectable.

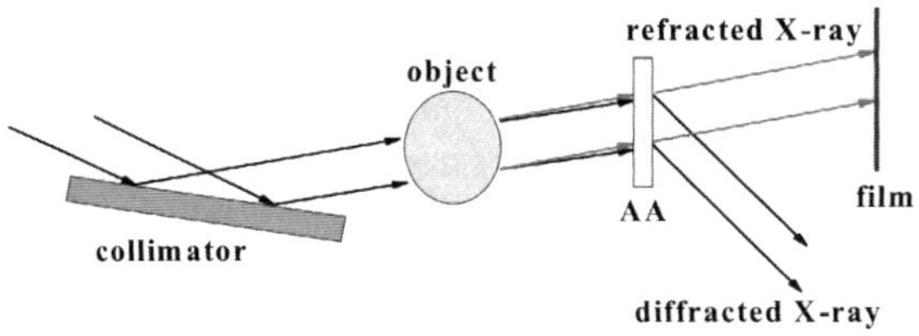

Figure 1. The sketch of the DFI setup at Photon Factory.

The DFI belongs to this kind of imaging technique, which can pick up the different phase information of the sample while suppressing the illumination light and the absorbed X-ray if any. The optics setup of DFI is shown in figure 1. A parallel incident beam emerges from a collimator. The sample is placed between a collimator and an analyzer (AA). When the beam transmitting the sample hits onto the analyzer, it is divided into two parts. One component transmits forward and does not change its direction. It will be diffracted by analyzer (shown with black lines in figure 1). Other component (indicated with red lines) will be refracted inside the sample due to the tiny refractive index variations and pass through the analyzer along original direction. So this analyzer just works as an angular filter. It removes the forward beam and at the same time keeps the refracted beam. Then, the phase information contained in the refracted beam is recorded in an imaging device, such as a film.

Rabbits are taken as the animal model and applied to do DFI experiments. The rabbit is a mammal and has acupoints too [8-9]. The names of rabbit acupoints are as same as in human being.

Table 1. The locations of the samples used for DFI experiment on rabbit's body.

Acupoint	Location on human (rabbit) body
Tianshu	Left and right side of the navel symmetrically on the belly, 2 Cun away from the navel
Neiguan	Lower side of the forearm, 2 Cun above the wrist
Zusanli	Lower part of the crus fibula, 3 Cun under the knee
Sanyinjiao	3 Cun above the ankle

According to the Chinese medicine, the proportional unit of the body, called "Cun" is used to locate the position of acupoint on human being. This method is also applied to rabbit body. The effective diameter of acupoints of rabbit is about 5mm. All samples were cut large enough to contain acupoints. We collected four acupoint samples from rabbit body and named these samples using corresponding names of acupoints. They are Tianshu, Neiguan, Zusanli and Sanyinjiao. Locations of these acupoints are shown in table 1. A pin was used to mark these acupoints positions approximately.

3. Results and discussion

The DFI experiment was performed at the endstation of BL14B in Photon Factory, KEK, where the synchrotron radiation was produced by a 5 Tesla wiggler. The photon energy is 36 keV. In our case, the beam size is approximately 19.7 mm (H)×33.5 mm (W). A Laue-case Si crystal was used as an analyzer, whose diffraction plane was (440) and thickness was about 1mm. The images were recorded on mammographic film of Kodak MinR 2000. The exposure time for each image was 2 minutes.

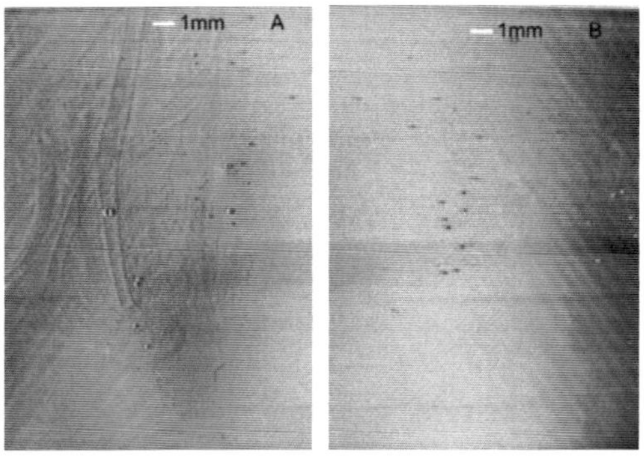

Figure 2. DFI images at Sanyinjiao acupoint region (A) and surrounding tissues (B).

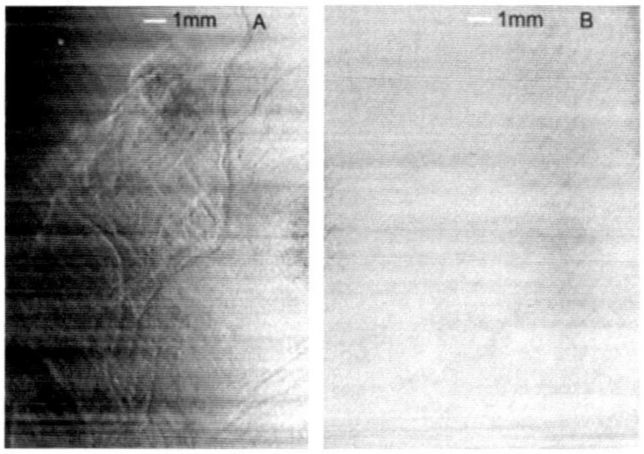

Figure 3. DFI images at Tianshu acupoint region (A) and surrounding tissues (B).

We took photos from the regions of acupoints and surrounding tissues far away from the acpoints. Figure 2 (A) is a photo taken at Sanyinjiao acupoint. A long blood vessel around with some fine branches is displayed clearly. Figure 3 (A) is a picture of Tianshu acupoint, where a short cluster of vessels can be seen. The micro-vessels gathering can also be observed at Neiguan and Zusanli acupoint regions. In the regions out of acupoints there are no such micro-vessels accumulations.

4. Conclusions

The synchrotron radiation based Dark Field Image method was used to study the topographic structures of acupoints. From pictures of four acupoints (Sanyinjiao, Tianshu, Neiguan and Zusanli) we have found that there exists the accumulation of micro-vessels, but in the surrounding tissues out of the acupoints one can not find such kind of structure. Therefore, the micro-vessels gathering are special structures of the acupoints. We suppose that these structures have certain relationship with the function of acupoints and play an important role in acupuncture.

Acknowledgments

This work was supported by National Basic Research Program of China (no.2005CB523306), Project of the State Key Program of National Natural Science Foundation of China (Grant No.10635060), and performed under the approval of the program Advisory Committee at the Photo Factory (2005G085).

References

[1] Cheng X 1987 *Chinese acupuncture and moxibustion* (Beijing: Foreign Language Press)
[2] O'Connor J and Bensky D 1981 *Acupuncture, a comprehensive text* (Seattle: Eastland Press)
[3] Fei L, Shen H S, Chen E Y, et al. 1998 *Sci. Forum* **43**(6) 658
[4] Chen E Y, Shen X Y, et al. 1998, *Shanghai Journal of Acupuncture and Moxibustion* **17** 47
[5] Ando M, Sugiyama H, et al. 2001 *Jpn. J. Appl. Phys.* **40** L844
[6] Davis T J, Gao D, Gureyev T E, Stevenson A W and Wilkins S W 1995 *Nature* **373** 595
[7] Gao D, Gureyev T E, Pogany A, Stevenson A W, Wilkins S W 1998 *Medical Applications of Synchrotron Radiation* eds. M. Ando and C. Uyama (Tokyo: Springer-Verlag) p 72
[8] Zheng L Y, et al. 2004 *Liaoning Journal of Traditional Chinese Medicine* **31** 63
[9] Zheng L Y, et al. 2004 *Liaoning Journal of Traditional Chinese Medicine* **31** 153

Observation of Bio-Fibres with a Hard X-ray Microscope

Hwa Shik Youn

PAL, POSTECH, 31 San, Hyoja-dong, Nam-ku, Pohang, KyungBuk 790-784, Korea

hsyoun@postech.ac.kr

Abstract. PLS (Pohang Light Source) hard x-ray microscope is briefly introduced. Zernike phase contrast is evaluated with images of polymer structures with various thicknesses. Images of bio-fibres taken with Zernike phase contrast are shown

1. The Microscope

The x-ray microscope has been at the 1B2 beam line of PLS since 2004. The optical system of this microscope is analogous to that of a visible light microscope in that there are a condenser lens and an objective lens, both of which are Fresnel zone plates.[1] The x-ray image of a sample which is magnified by the OZP (Objective Zone Plate) is converted into a visible one on a scintillation screen, which is further magnified by a microscope objective and captured by a digital CCD camera. The energy has been set at 6.95 keV by a crystal monochromator. We have added an annular aperture and phase plates for Zernike phase contrast.[2]

Since the last report [3] there has been upgrade in the visible light optics by replacing the microscope objective by a different make with a larger NA, which resulted in higher light intensity and more homogeneous background. Objective zone plates with the outer most zone width of either 100 nm or 70 nm have been employed with matching phase plates. There has been no image processing for all the images shown below except for normalization.

2. Results

We show x-ray images of a polymer structure and bio-samples. The former was taken to demonstrate the available contrast of our microscope. The latter consists of images of human hair shafts and silkworm silk, *Bombyx mori*.

2.1. Polymer samples

There have been attempts to take images of numerous biological samples with our microscope. However we didn't know the limit of the Zernike phase contrast. So we took images of known polymer structures with various thicknesses. Shown below in Figures 1-a and 1-b are two images of the same 200 nm thick polyimide structure taken with and without Zernike phase contrast, respectively. For these an OZP with the finest zone width of 70 nm was used together with a 2.1 μm thick Au phase plate, which gives negative contrast. Exposure time was 120 and 90 seconds for the image with and without Zernike phase contrast, respectively. It is clear that there is hardly any visibility without Zernike phase contrast while the image is clearly visible with the phase contrast when the polymer thickness gets 200 nm.

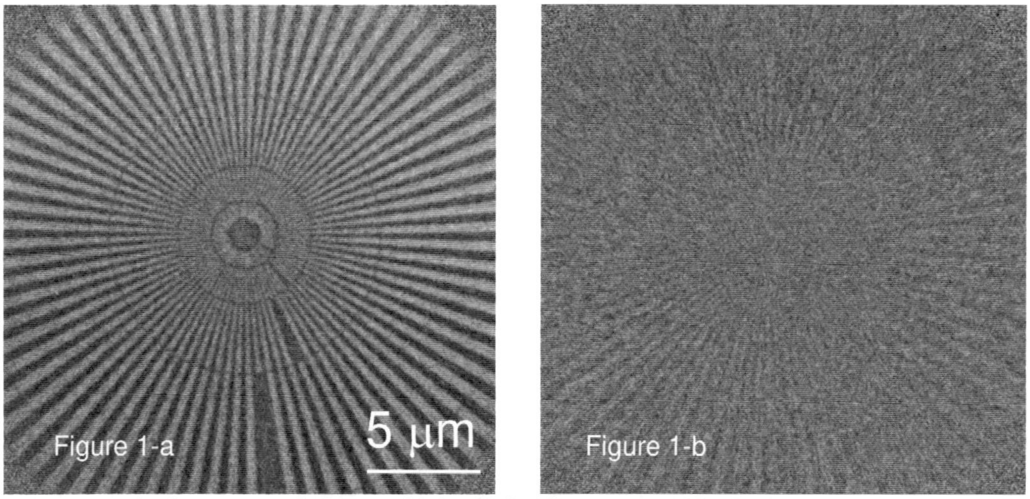

Figure 1-a and 1-b. Images of 200 nm thick polyimide structure. The four corners of the images are outside a field of view

2.2. Bio-fibres

2.2.1. *hair shaft.* Shown below in Fig. 2 and 3 are images of human hair shafts taken with a 100 nm OZP in the positive phase contrast. Fig. 2 is a white hair and Fig. 3 black one. The width of an Asian hair shaft is around 100 μm while the horizontal field of view is about 30 μm. So we took 4 images with scanning the sample and tiled them together to show the whole width of a shaft. Exposure time was 60 seconds a frame. We took this image with a new microscope objective which gives far superior contrast to the previous one. [3, 4]

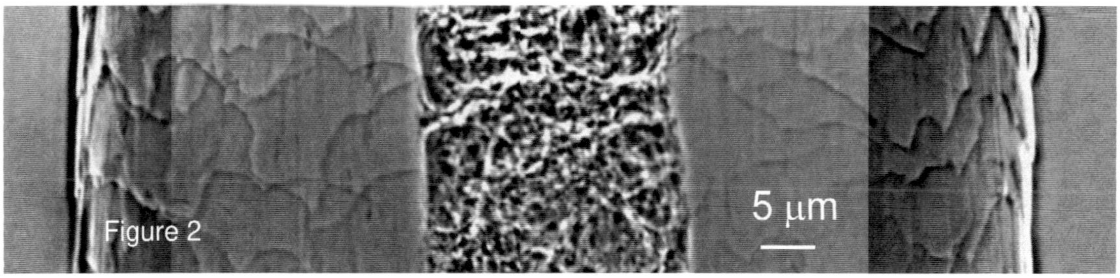

Figure 2. An image of a human hair shaft taken in the positive Zernike phase contrast. It is a white hair.

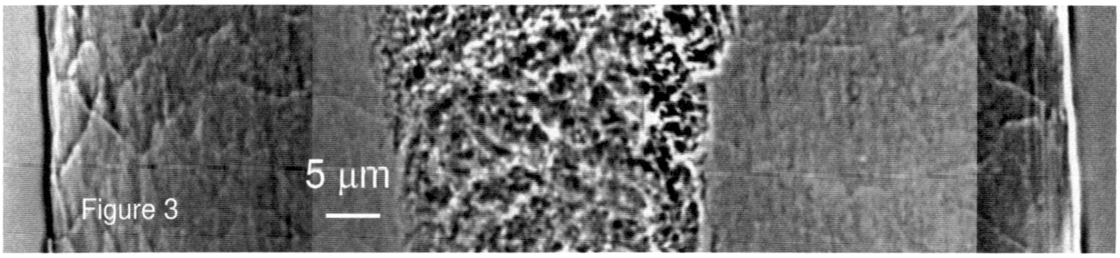

Figure 3. An image of human hair shaft taken. It is a black hair.

The black hair shaft looks somewhat more complicated than the white one. In the positive phase contrast any features with a higher density would look brighter than surrounding. Eumelanin granules of a black hair shaft are said to be 32 % denser than hair keratin [5], so there are numerous bright spots in the black hair shaft.

2.2.2. *Silkworm Silk.* There seems hardly any contrast in the soft x-ray microscope image of spider silk *N. Clavipes*. [6] Furthermore strong absorption of soft x-ray by the sample resulted in the severe radiation damage.

On the other hand silks are transparent to hard x-rays of 6.95 keV, so there is virtually no absorption by the samples. We looked at Silkworm silk, *Bombyx mori* to see the contrast of our microscope. Shown below in Fig. 4 is an x-ray image of Silkworm silk taken with a 70 nm OZP in the negative phase contrast and exposure time of 120 seconds. Nano-fibrils with a width of about 100 nm and bundles of them are said to be in the silkworm silk. [7] It was difficult to resolve nanofibrils in the projection image of the x-ray microscope. However there are clear indications of bundles in our images. Therefore we expect nano-tomogramphy can give a lot more interesting 3 D information.

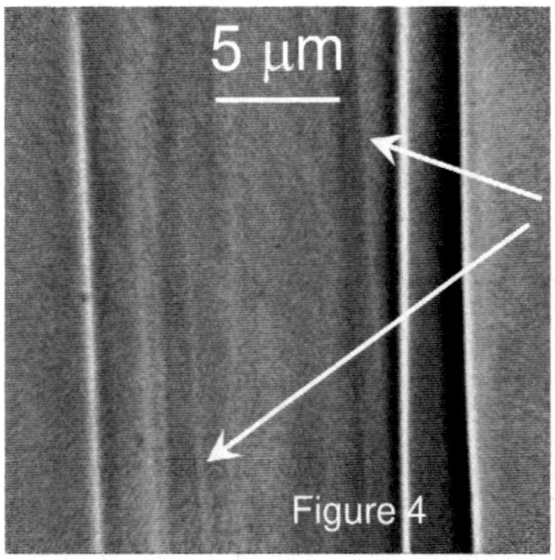

Figure 4 Image of Silkworm silk taken with a hard x-ray microscope where white arrows point to the suspected bundles of nano-fibrils.

I would like to thank Dr. Alan Swift for his illuminating comments and discussions on hair structures. Dr. Gene Ice has answered critical questions on x-ray optics and provided me with a nice pair of Ge crystals. Dr. Hyoung-Joon Jin of Inha University has sent us silkworm silk fibres. The excellent support of the staff at the Pohang Accelerator Laboratory especially, Dr. Tae Joo Shin is gratefully acknowledged. The research at the beam line 1B2 is supported by MOST and the POSCO (Pohang Iron and Steel Company).

References

[1] Hwa Shik Youn, Soo Yeun Baik, and Chang-Hwan Chang, Rev Sci Instrum **76**, 023702 (2005)
[2] Hwa Shik Youn and Suk-Won Jung, J Microsc **223**, 53 (2006)
[3] *ibid*, **228**, 107 (2007)
[4] Hwa Shik Youn and Suk-Won Jung, Phys Med Biol **50**, 5417 (2005)
[5] J. A. Swift, Ph.D. dissertation 'Some properties of pigment granules isolated from various animal fibres'. The University of Leeds, England (1963)
[6] Anja Glišović, Jürgen Thieme, Peter Guttmann, and Tim Salditt, Int J Biol Macromol **40**, 87 (2007)
[7] S. Putthanarat, N. Stribeck, S.A. Fossey, R.K. Eby, and W.W. Adams, Polymer **41**, 7735 (2000)

Three-dimensional cellular and subcellular structures of human brain tissue determined by microtomography

Ryuta Mizutani[1], Akihisa Takeuchi[2], Susumu Takekoshi[3], R Yoshiyuki Osamura[3], Kentaro Uesugi[2], Yoshio Suzuki[2]

[1]Department of Applied Biochemistry, School of Engineering, Tokai University, Kitakaname 1117, Hiratsuka, Kanagawa 259-1292, Japan

[2]Research and Utilization Division, JASRI/SPring-8, Kouto 1-1-1, Sayo, Hyogo 679-5198, Japan

[3]Department of Pathology, Tokai University School of Medicine, Shimokasuya 143, Isehara, Kanagawa 259-1193, Japan

E-mail: ryuta@tokai-u.jp

Abstract. We report here x-ray microtomographic studies of human cerebral cortex stained with high-Z elements. Brain tissues were stained with metal elements by the Golgi and Bodian impregnation methods and subjected to x-ray microtomographic analysis. Axons and dendrites arising from cell bodies were visualized as three-dimensional networks. Spherical structures of cellular nuclei were observed in the interiors of cell bodies, indicating that hard x-ray microtomography can reveal the intracellular structure. High-Z element microcontrasting in conjunction with microtomographic analysis can be applied to any soft tissues. Our results show that the metal contrasting facilitates the three-dimensional microtomographic visualization of cellular and subcellular structures of soft tissues.

1. Introduction

The transparency of biological tissue to hard x-rays enables radiographic analysis of the internal structure. However, soft tissues are composed of light elements, which produce little contrast in a hard x-ray transmission image. In clinical diagnosis, luminal structures of a living body are visualized by using x-ray contrast media. These contrast media contain high-Z elements, such as barium or iodine, that absorb x-rays efficiently. We have recently reported the three-dimensional visualization of the soft tissue microstructure of fruit fly [1,2], zebrafish [3], and human brain [3,4] by contrasting with high-Z elements. Element-specific structural analysis has also been achieved by using x-ray absorption edges of the contrasting elements [5].

Brain functions including creation, memory, and emotion are performed by three-dimensional neuronal networks composed of a huge number of neurons. The cellular functions of each neuron are achieved by the intracellular structure of neurons. Therefore, three-dimensional microstructural analysis of nerve tissues should reveal the functional mechanism of the brain. Here, we report on a computed microtomography (micro CT) analysis of the three-dimensional structure of human cerebral cortex at subcellular resolution.

2. Materials and Methods

2.1. Human brain tissue

Frontal cortex of normal brain tissue (44 years old, male) was dissected at autopsy and fixed with 10% formaldehyde for 7 days.

Golgi impregnation was performed using fixed tissues with dimensions of about 5 mm × 5 mm × 10 mm, as described previously [4]. The chromate and silver nitrate steps were repeated three times.

Bodian impregnation was performed using fixed tissues with typical dimensions of 0.5 mm × 0.5 mm × 2.5 mm. Defatting was performed by immersing the dehydrated tissues sequentially in acetone for 24 hr and xylene for 24 hr. After being rehydrated, tissues were subjected to reduced-silver Bodian impregnation, as described previously [2,3].

2.2. Micro CT

Stained samples were embedded in epoxy resin, as described previously [4]. Micro-CT analysis was performed at the BL20XU beamline [6] of SPring-8. Transmission radiographs were recorded with a CCD-based x-ray imaging detector (AA50 and C4880-41S, Hamamatsu Photonics, Japan) using 12.000-keV x-rays. The field of view and effective pixel size of the image detector were 1.00 mm × 0.65 mm and 0.50 μm × 0.50 μm, respectively. A total of 1800 images were acquired with a rotation step of 0.10° and an exposure time of 300 ms. The spatial resolution of the three-dimensional structure was estimated to be 1.0 μm. The convolution back projection method using a Chesler-type filter was used for tomographic reconstruction [7]. Volume-rendered figures of the obtained three-dimensional structures were produced using the program VG Studio MAX (Volume Graphics, Germany). CT densities were rendered by the maximum projection method.

3. Results and Discussion

Golgi impregnation is a conventional staining method for the optical observation of neural cells [8]. It has been reported that only a limited population of neurons is stained with this method. A three-dimensional structure of the internal pyramidal layer of the frontal cortex tissue is shown in Figure 1. Neurons were visualized as a three-dimensional distribution of linear absorption coefficients. Axons and dendrites arise from cell bodies and form network structures. Spherical structures were seen in the interiors of the cell bodies, indicating that the micro-CT analysis can reveal the intracellular microstructure. These microstructures were assigned as cellular nuclei. The surroundings of nuclei were stained deeper than the cytoplasm, while some of the nuclei displaced the metal dye and showed negatively stained images. Blood capillary vessels with an approximate diameter of 6–8 μm were observed as luminal structures, in which blood cells aligned along the lumen [3].

Reduced-silver Bodian impregnation is another conventional method used for the optical observation of neural tissues [9]. This method stains every neuropils with gold [2]. A layer of the three-dimensional structure of human frontal cortex stained with Bodian impregnation is shown in Figure 2. Spherical structures along with neuropil networks can be seen. These spherical structures have diameters of approximately 4–6 μm and should be cellular nuclei of smaller neural cells. Within these spherical structures, even denser granules having a typical dimension of 1.5 μm, were observed. This image suggests that the nucleolus can be visualized by this method.

4. Conclusion

High-Z element microcontrasting in conjunction with micro-CT analysis can be applied to any soft tissues. Our results indicate that the metal contrasting facilitates the three-dimensional visualization of cellular and subcellular structures.

The neuronal network visualized in the three-dimensional structure is essential for cerebral functions. The brain model should be built up as a circuit to reveal its functions, so it is essential to trace neuronal tracts in order to understand the functional mechanism of the brain.

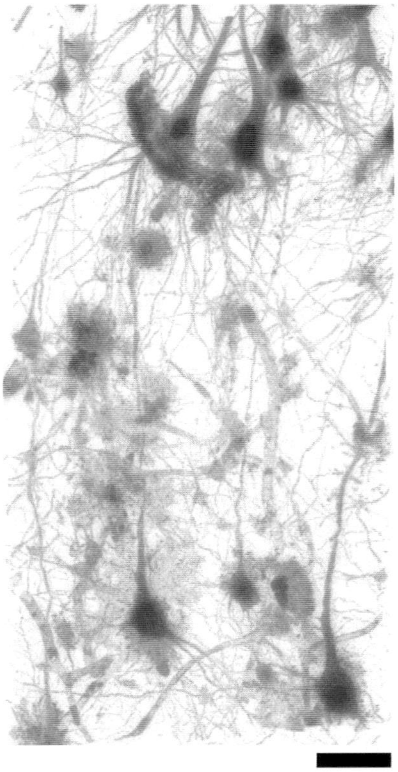

Figure 1. Three-dimensional structure of human frontal cortex stained using Golgi impregnation. CT densities are rendered from linear absorption coefficients of 0 to 180 cm^{-1}. Scale bar: 50 μm.

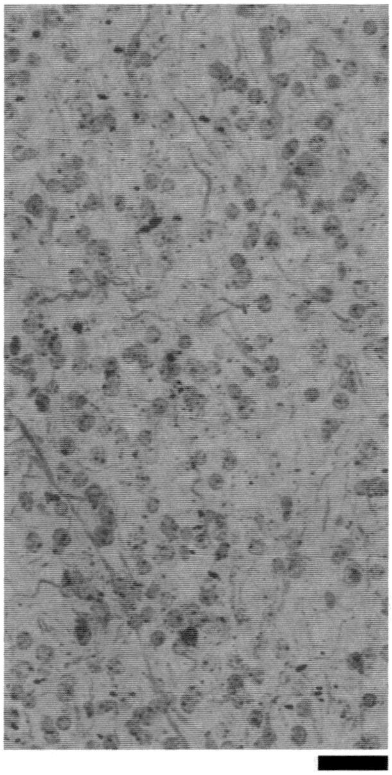

Figure 2. Three-dimensional structure of human frontal cortex stained using Bodian impregnation. CT densities are rendered from linear absorption coefficients of 0 to 40 cm^{-1}. Scale bar: 20 μm.

Acknowledgement
The synchrotron radiation experiments were performed at SPring-8 with the approval of the Japan Synchrotron Radiation Research Institute (JASRI) (proposal nos. 2006B1716, 2007A1844, 2007A2072, 2007B1102, 2007B1894, and 2008A1190).

References

[1] Mizutani R, Hara T, Takeuchi A, Uesugi K and Suzuki Y 2006 *The Fifth East Asian Biophysics Symposium, Abstracts* p S388
[2] Mizutani R, Takeuchi A, Hara T, Uesugi K and Suzuki Y 2007 *J. Synchrotron Radiat.* **14** 282
[3] Mizutani R, Takeuchi A, Uesugi K, Takekoshi S, Osamura R Y and Suzuki Y 2008 *Tissue Eng. Part C* **14** in press
[4] Mizutani R, Takeuchi A, Uesugi K, Ohyama M, Takekoshi S, Osamura R Y and Suzuki Y 2008 *Brain Res.* **1199** 53
[5] Mizutani R, Takeuchi A, Akamatsu G, Uesugi K and Suzuki Y 2008 *J Synchrotron Radiat.* **15** 374
[6] Suzuki Y *et al.* 2004 *AIP Conf. Proc.* **705** 344
[7] Mizutani R, Takeuchi A, Uesugi K and Suzuki Y 2008 *J. Synchrotron Radiat.* **15** in press
[8] Valverde F 1970 *Contemporary Research Methods in Neuroanatomy* (New York, Springer-Verlag) pp 12-31
[9] Tyrer N M, Shepherd D and Williams D W 2000 *J. Histochem. Cytochem.* **48** 1575

NEXAFS microscopy of polymeric materials: Successes and challenges encountered when characterizing organic devices

H. Ade[1,4], B. Watts[1], S. Swaraj[1], C. McNeill[2], L. Thomsen[3], W. Belcher[3], and P. C. Dastoor[3]

[1]Dept. of Physics, North Carolina State University, Raleigh, NC27685-8202, USA
[2]Cavendish Lab., Dept. of Physics, Univ. of Cambridge, Cambridge, CB3 0HE, UK
[3]Centre for Organic Electronics, Univ. of Newcastle, Callaghan, NSW 2308, AU

E-mail: harald_ade@ncsu.edu

Abstract. We summarize recent developments in x-ray microscopy of polymers by focusing on the characterization of organic electronic devices. The quantitative compositions of model polymer blends have been mapped at a resolution of ~35 nm. Since it could be inferred that these devices have structures smaller than 35 nm, quantitative compositional mapping at length scales below the present resolution limit of x-ray microscopy is required. Organic devices thus serve to both highlight the success of NEXAFS microscopy to date, but to also outline the very real need for higher spatial resolution. New approaches to create improved optics or different acquisition modalities are required if x-ray microscopy is to make sustained contributions to such an important area of research as organic devices.

1. Introduction

Over the past decade, x-ray microscopy characterization of organic and polymeric materials [1-3] has been very successful due to the high compositional sensitivity of Near Edge X-ray Absorption Fine Structure (NEXAFS) spectroscopy and the minor damage of soft X-rays [4]. We summarize recent efforts by focusing on NEXAFS microscopy of organic electronic devices. Organic photovoltaics based on solution processed conjugated polymers have the potential to meet the need for efficient and cheap solar cell technology [5]. In organic solar cells, the exciton created by the absorbed photon needs to be split at a donor-acceptor interface and the most common "bulk heterojunction" devices have these interfaces distributed throughout the active layer. Since solution processed films of donor and acceptor materials result in complex nonequilibrium morphologies with phase separation at multiple length scales, a quantitative compositional characterization tool is required to investigate the relationship between local composition and device performance. McNeill *et al.* have thus used NEXAFS microscopy to investigate a number of organic PV devices [6-9]. Here, we review the influence of annealing on the morphology of initially intimately mixed blends of the conjugated polymers poly(9,9′-dioctylfluorene-co-bis-N,N′-(4,butylphenyl)-bis-N,N′-phenyl-1,4-phenylene-diamine) (PFB) and poly(9,9′-dioctylfluorene-co-benzothiadiazole) (F8BT) and provide a context for higher resolution requirements in x-ray microscopy.

[4] To whom any correspondence should be addressed.

2. Experimental

~150 nm thick PFB/F8BT blends cast from chloroform onto poly(3,4-ethylenedioxythiophene):polystyrene sulfonic acid coated glass slides were annealed for 10 min on a hotplate, quenched to room temperature, and floated off onto TEM grids [9].

Measurements were performed with the 5.3.2 STXM at the Advanced Light Source [10]. Full details of the experiment and analysis can be found in a previous publication [6].

3. Results

Quantitative composition maps of an as-spun PFB/F8BT blend film and films annealed at 140, 160, 180 and 200 °C are displayed in Fig. 1, along with the photoluminescence quantum efficiency (PLQE) of these films and a F8BT reference film. Some compositional variations can be observed even for the as-cast films. For the 140 and 160 °C films, there is little difference compared to the as-spun film. In contrast, there is a pronounced coarsening of the morphology and corresponding increase in compositional variation for the 180 and 200 °C films. They also exhibit a bicontinuous morphology that evolves in a way reminiscent of spinodal decomposition. The quantitative analysis of feature size and compositional variation is summarized in Fig. 2, along with the domain size inferred from the PL data through the use of a model that takes the exciton diffusion length into account.

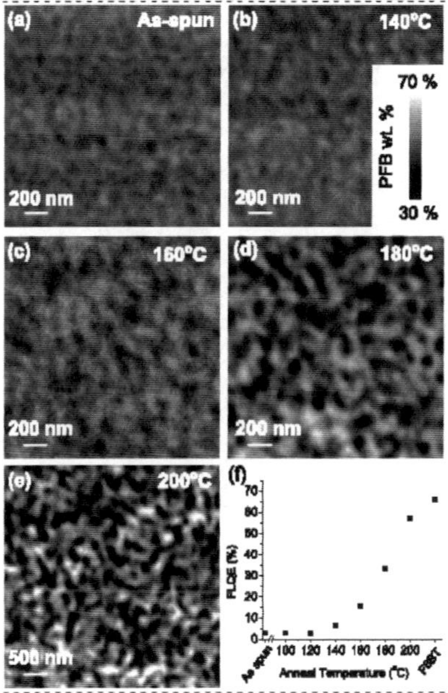

Figure 1. Composition maps of PFB/F8BT blend films: as-spun (a) and annealed at T=140 °C (b), 160 °C (c), 180 °C (d), and 200 °C (e). (f) Photoluminescence quantum efficiency (PLQE) of annealed blend films and a pure F8BT film.

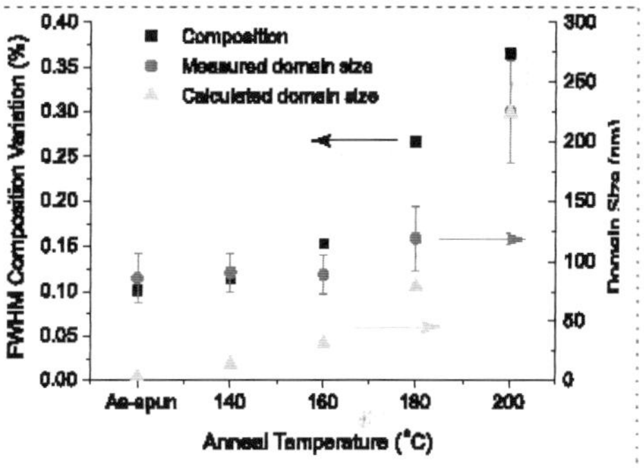

Figure 2. Compositional variation, measured feature size and calculated domain size from PLQE measurements versus annealing temperature.

4. Discussion

For the as-spun films, some structure is observed with dimension of ~ 85 nm. These structures are not pure phases and there must be significant intermixing with a hierarchy of phase separation existing on the length scale of less than 85 nm. With annealing, this finer scale of phase separation should evolve as pure (or very nearly pure) domains within these 85 nm structures. This hypothesis is supported by the fact that the length scales derived from PLQE data imply morphological evolution at annealing temperatures for which little variation can yet be observed with NEXAFS microscopy. At 160 °C, the size of the internal domains approaches that of the enclosing structures. Above 160 °C, the evolution of the morphology is directly recorded with x-ray microscopy and is characterized by a bicontinuous morphology favourable for charge transport. A Monte Carlo model of exciton diffusion and dissociation was able to achieve good agreement for T>140 °C between the domain size measured by STXM and those calculated by the model using an exciton diffusion length of 12 nm [9].

Clearly, real space characterization of these samples would greatly benefit from improved spatial resolution and tomography. Given that the exciton diffusion length in these organic materials is ~10 nm, the optimum devices will have to have phases ~10 nm in size. Consequently, characterization at that length scale is required for optimal understanding and significant improvements in spatial resolution are required to do. Unfortunately, advances in spatial resolution have become increasingly more difficult. Extrapolation of the best spatial resolution over the last two decades implies (see Fig. 3) that the 10 nm resolution limit might only be achieved in about a decade from now.

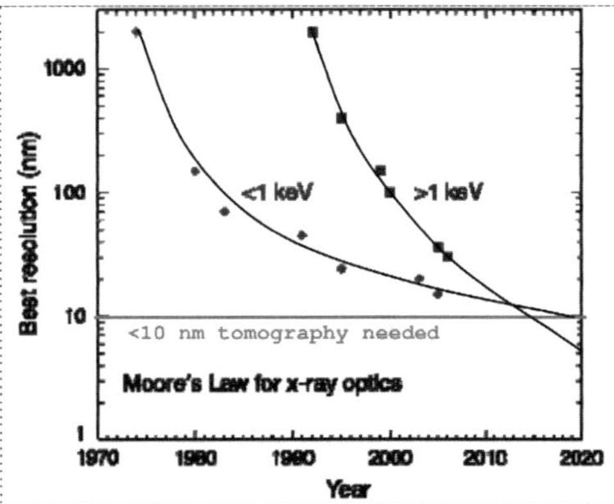

Figure 4. Best resolution of soft and hard x-ray microscopy during the last three decades. Lines are guides to extrapolate into the future. (Adopted from a figure courtesy of C. Jacobsen, Stony Brook University).

Organic devices thus serve to highlight the success of NEXAFS microscopy to date, but these devices also emphasize the very real need for higher spatial resolution. Clearly new approaches to zone plate fabrication or new methods are necessary to achieve faster progress. Recent zone plate developments [11,12] or x-ray ptychography [13] might be excellent ways forward.

Acknowledgments

The authors thank D. Kilcoyne and T. Tyliszczak for assistance and support. The work was supported by the EPSRC, U.K. (Advanced Research Fellowship EP/E051804/1), the Australian Research Council's Discovery funding scheme (DP0559417), and the U.S. Department of Energy under contracts DE-AC02-05CH11231 and DE-FG02-98ER45737.

References:

[1] Ade, H, Zhang, X, Cameron, S et al., 1992 *Science* **258** 972.
[2] Ade, H and Hsiao, B, 1993 *Science* **262** 1427.
[3] Ade, H and Hitchcock, A P, 2008 *Polymer* **49** 643.
[4] Rightor, E G, Hitchcock, A P, Ade, H et al., 1997 *J. Phys. Chem. B* **101** 1950.
[5] Mayer, A C, Scully, S R, Hardin, B E et al., 2007 *Mater Today* **10** 28.
[6] McNeill, C R, Watts, B, Thomsen, L et al., 2006 *Small* **2** 1432.
[7] McNeill, C R, Watts, B, Thomsen, L et al., 2007 *Macromol.* **40** 3263.
[8] McNeill, C R, Watts, B, Thomsen, L et al., 2008 *Nanotechnology* **19** 424015.
[9] McNeill, C R, Watts, B, Thomsen, L et al., 2008 *Macromol.* (submitted).
[10] Kilcoyne, A L D, Tyliszczak, T, Steele, W F et al., 2003 *J Synchrotron Radiat* **10** 125.
[11] Chao, W L, Harteneck, B D, Liddle, J A et al., 2005 *Nature* **435** 1210.
[12] Jefimovs, K, Vila-Comamala, J, Pilvi, T et al., 2007 *Phys. Rev. Lett.* **99** 264801.
[13] Thibault, P, Dierolf, M, Menzel, A et al., 2008 *Science* **321** 379.

Synchrotron-based in situ soft X-ray microscopy of Ag corrosion in aqueous chloride solution

B Bozzini[1], L D'Urzo[1], A Gianoncelli[2], B Kaulich[2], M Kiskinova[2], M Prasciolu[3], A Tadjeddine[4]
[1] Dipartimento di Ingegneria dell'Innovazione, Università del Salento, via Monteroni, 73100 Lecce - Italy
[2] Sincrotrone Trieste S.C.p.A., ELETTRA, s.s. 14 km 163.5 in Area Science Park, 34012 Basovizza.Trieste, Italy
[3] CNR-INFM TASC National Laboratory S.S.14 Km 163.5, Area Science Park, 34012 Basovizza, Trieste, Italy
[4] 3UDIL/CNRS, BP34 91898 Orsay Cedex - France
alessandra.gianoncelli@elettra.trieste.it

Abstract. In this paper we report an in situ X-ray microscopy study of a model metal electrochemistry system, incorporating faradaic reactivity: the anodic corrosion and cathodic electrodeposition of Ag in aqueous systems. The information at sub-μm scale about morpho-chemical evolution of the electrified interface, provided by this novel electroanalytical approach fosters fundamental understanding of important issues concerning material fabrication and stability, which are crucial in developing the next generation electrochemical technologies, such as fuel cells and biosensors. The key methodology challenge faced in this pilot electrochemical experiments is combining a three-electrode configuration and wet environment, which required metal electrodes suitable for transmitting soft X-rays and a sealed cell allowing working in high vacuum. This has been solved via lithographic fabrication route fabricating 75 nm thick Ag electrodes and using Si_3N_4 membranes as X-ray windows and electrode support. Imaging in the STXM mode with phase contrast allowed us to monitor the corrosion morphologies and metal outgrowth features. Localised thickness variation and the build-up of reaction products of electron density different from that of the starting material have been detected with high sensitivity.

1. Introduction

In situ electrochemical synchrotron-based soft X-ray transmission spectromicroscopy has been recently considered as an advanced electroanalytical tool combining nanoscopic capability with chemical and structural sensitivity, typical of X-ray spectroscopies. In addition, working with beam energies in the range encompassing the water window, opens up the investigation of dynamic processes in aqueous solution (see [1] and references therein contained). To the best of our knowledge - before our approach developed at the TwinMic beamline - the only electrochemical STXM work reported in the literature is [2], where the redox behaviour of polyaniline is studied in a two-electrode cell. This paper reports the first achievements of a long-term collaboration project between the X-ray Microscopy Section of ELETTRA and the Electrochemistry Group of Salento University, aimed at developing a unique in situ spectromicroscopy tool for the fundamental investigation of material degradation processes, that are critical for advanced electrochemical technologies: chiefly fuel cells. In order to prove the feasibility of the target electrochemical experiments, we focused on a widely-investigated - though still warranting research efforts - electrochemical system: the Ag/Ag^+ couple in aqueous Cl^- [1].

2. Experimental

The electrochemical cell was fabricated in collaboration between the TwinMic beamline at Elettra and TASC/INFM lithography group. The construction consists of two main parts, depicted in Figure 1.

Figure 1.1 - Electrode assembly, containing: four square Ag electrodes (5 nm Cr adhesion layer and 75 nm Ag top layer) *A*; cross-shaped electrolyte compartment *B*; X-ray transparent Si_3N_4 *C1* membrane onto which the electrodes *A* are deposited and confining the electrolyte present in the compartment *B*; Si frame *D1*, providing mechanical resistance of the cell body and support to the Si_3N_4 membrane. Figure 1.2 - Cell cover, consisting of the second optical window *C2*, supported by the Si frame *D2*. A cross-sectional view of the way the electrode-membrane (Figure 1.1) and the cover-membrane (Figure 1.2) are assembled is shown in Figures 1.3A and 1.3B. The process-flow for the fabrication of the electrodes comprised the following steps: (i) chemical vapor deposition of a 100 nm thick Si_3N_4 layer on both sides of a bare Si(100) wafer; (ii) opening of the optical window on the back side of the wafer by selective dry etching of the Si_3N_4 layer through a photoresist mask, removal of the mask by hot acetone, followed by wet etching in hot KOH of the Si down to the nitride layer on the opposite side, and thorough rinsing in deionised water; (iii) integration of four metal electrodes by properly masking the nitride side of the wafer and consecutive deposition of Cr and Ag. The X-ray beam crosses the electrode/electrolyte assembly normally to the optical windows *C1* and *C2* and the transmitted X-rays are monitored by the detector placed behind the cell. The electrolyte employed in this study was 0.02 M NaCl aqueous solutions in ultra-pure water (resistivity of 18.5 MΩ cm). Potentiostatic and galvanostatic experiments were performed in a conventional three-electrode configuration (see [1] for details).

The experiment was performed using the European twin X-ray microscopy station operating TwinMic at ELETTRA [3-4]. The performance of TwinMic matches well the requirements for this experiment, offering a broad-energy range from 0.25-2 keV with scanning and full-field imaging modes in a single instrument, and versatile contrast techniques, such as bright-field, differential phase [5] and differential interference contrast [6-7]. The lateral resolution better than 100 nm can be achieved over the entire photon energy range, by means of diffractive zone plate optics. In this experiment, TwinMic was operated in the scanning mode and the specimen was raster-scanned across the X-ray probe of 200 nm, provided by a Au zone plate of 250 μm in diameter. The STXM images were recorded by an Andor Ixon fast read-out camera, collecting absorption and phase information simultaneously. Dwell time per pixel of the raster scan have been adapted to follow dynamically the evolution of the corrosion processes of the electrodes. A photon energy of 477.5 eV was used.

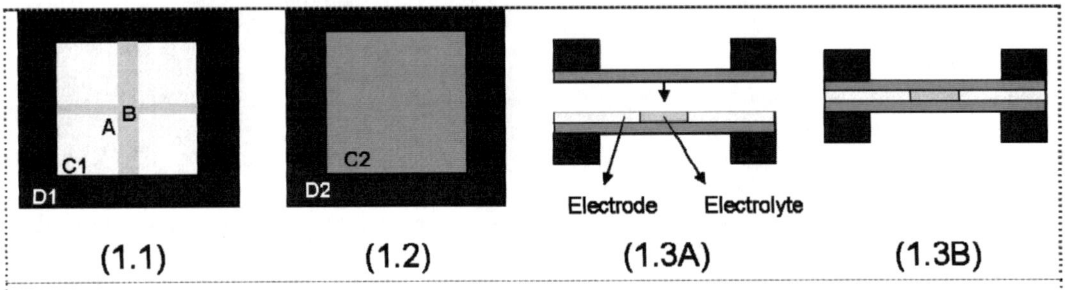

Figure 1. Schematic view of the electrochemical cell.

Results

The electrochemical experiments were performed by applying galvanostatic sequences across couples of electrodes. In this paper we report two typical examples of morphologies developing at the anodically and cathodically polarised electrodes. Complementary information regarding Ag behaviour in the chloride system - under different electrochemical conditions - as well as more detailed information about processes taking place in an ammonium electrolyte, can be found in [1].

Cathodic (Figures 2.1 and 2.2) and anodic and morphological changes were observed, brought about by the circulation of 0.2 mA cm^{-2}. The reduction process gives rise to the growth of Ag dendrites (location γ in Figure 2.1) in correspondence of high current density areas of the cathode: dynamic shape changes (overgrowth of peninsula δ in Figure 2.2) were found, typified by Figures 2.1 (after 20 min) and 2.2 (after 25 min).

Conclusions

For expanding the research fileds of the TwinMic beamline towards in-situ electrochemical studies an approach for the fabrication and utilization of sealed wet cells with electrochemical control has been developed and implemented. Its potential for electrochemical experiments has been demonstrated using a model system, highlighting non-trivial dynamic morphological aspects of corrosion and electrodeposition processes in a confined electrolyte. The unique feature of this approach is the possibility of observing subtle details of corrosion products and small electrodeposit features that are exclusively stabilised by the aqueous environment and would be either altogether destroyed or at least profoundly distorted by ex situ observations. In conclusion, very recently we have proved feasibility of applying the XAS spectroscopic facility to work with electrochemical wet cells, thus building the full capability of the TwinMic beamline into this class of experiments.

References

[1] Bozzini B, D'Urzo L, Gianoncelli A, Kaulich B, Kiskinova M, Prasciolu M and Tadjeddine A 2008 *Electr Comm*, **10**(11) 1680.

[2] Guay D, Stewert-Ornstein J, Zhang X, Hitchcock A P 2005 *Anal. Chem.* **77** 3479.

[3] Bianco A, Sostero G, Nelles B, Heidemann K F, Cocco D 2005 SPIE **5918** 591810.

[4] Kaulich B et al. 2006 *Proc. 8th Int. Conf. X-ray Microscopy* IPAP Conf. Series **7** 22.

[5] Gianoncelli A, Morrison G R, Kaulich B, Bacescu D, Kovac J 2006 *Appl. Phys. Lett.* **89** 251117.

[6] Wilhein T, Kaulich B, Fabrizio E, Romanato F, Cabrini S, and Susini J. 2001 *Appl. Phys. Lett.* **78**(14) 2082.

[7] Kaulich B, Wilhein T, Di Fabrizio E, Romanato F, Altissimo M, Cabrini S, Fayard B and Susini J 2002 *Journal of the Optical Society of America A* **19**(4) 797.

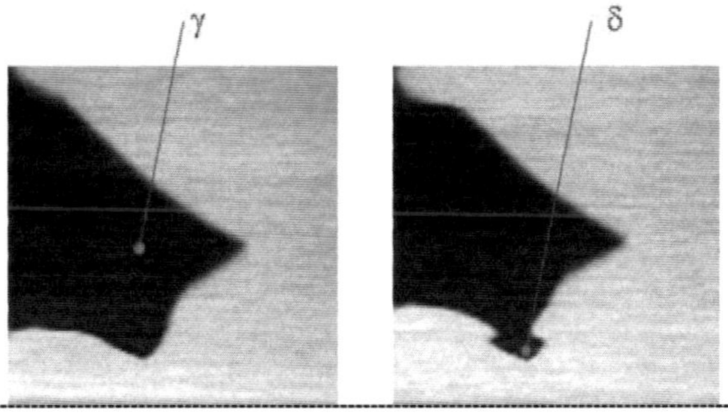

Figure 2. STXM Brightfield images (30μm by 30μm scan size, 100 by 100 pixels, 100ms CCD dwell time). (2.1) cathodic growth of Ag dendrite γ, (2.2) cathodic overgrowth of Ag peninsula δ on dendrite γ.

X-ray stereo microscopy for investigation of dynamics in soils

S-C Gleber[1], J Sedlmair[1], M Bertilson[2], O von Hofsten[2], S Heim[3], P Guttmann[1 c/o 3], H M Hertz[2], P Fischer[4], and J Thieme[1]

[1] Institut für Röntgenphysik, Georg-August-Universität Göttingen, Friedrich-Hund-Pl.1, 37077 Göttingen, Germany

[2] Biomedical and X-ray Physics, Department of Applied Physics, Royal Institute of Technology, AlbaNova, SE-10691 Stockholm, Sweden

[3] Berliner Elektronenspeicherring Gesellschaft für Synchrotronstrahlung m.b.H, Albert-Einstein-Str. 15, 12389 Berlin, Germany

[4] Center for X-Ray Optics, Lawrence Berkeley National Laboratory, 1 Cyclotron Road, Berkeley CA 94720, USA

E-mail: sgleber@gwdg.de

Abstract. The here presented combination of stereo imaging and elemental mapping with soft X-ray microscopy reveals the spatial arrangement of naturally aqueous colloidal systems, e. g. iron oxides in soil colloid clusters. Changes in the spatial arrangement can be induced by manipulating the sample in-situ and thus be investigated directly and as a function of time.

1. Introduction

Due to its high spatial and spectral resolution, X-ray microscopy is an important instrument for investigation of colloidal systems, e.g. from environmental sciences [1]. Taking advantage of the natural contrast mechanism, sample preparation is not necessary, it can be imaged in transmission within its original aqueous media up to $10\,\mu$m thickness with a resolution in the range of 20-50 nm. In soil science, the element distribution within soil colloid clusters is of great interest [2]. With images taken above and below an absorption edge, distribution maps of the corresponding element are achieved. Thus, the iron distribution in colloidal samples from the environment has been determined. By tilting the object, stereo pairs of images have been taken. The here presented combination of elemental mapping with X-ray microscopy and stereo imaging provides a tool for that and reveals the spatial arrangement of e. g. iron oxides in soil colloid clusters. Changing the chemical conditions of aqueous medium leads to changes in the spatial arrangement, which can be done directly in the X-ray microscope. Since many of the so induced morphological changes are much slower than the time required even for taking stereo pairs, the dynamical behaviour of these changes can be investigated.
Theory and results of such experiments at three different setups will be presented briefly.

2. Parallax equation based stereo reconstruction

Stereo calculations are based on the parallax equation [3]. It relates the parallax ΔY to the vertical distance h and the tilt angle 2θ (θ is the stereo angle). Thus, for a tilt by $\theta_2 - \theta_1 = \theta$

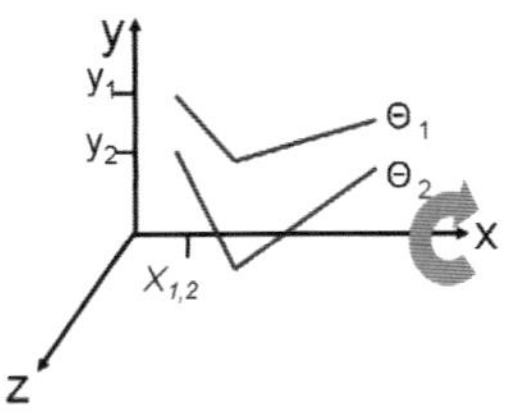

Figure 1. Rotation of a 3-d object around x-axis.

around the x-axis, the z-coordinate of a structure can be calculated by the difference between the y-coordinates (y_1 and y_2) in the respective images projecting the structure onto the x-y-plane (Figure 1). The spatial coordinates become:

$X = x_1 = x_2 \qquad Y = \frac{y_1+y_2}{2\cos\theta} \qquad Z = \frac{y_1-y_2}{2\sin\theta}$

The program xstereo has been written in IDL to mark or select significant structures in a set of tilted x-ray images and get information about the 3-d configuration, distances and lengths. Matching points recognizable in both projections are marked manually. In the 3-d plot, features like curvatures are displayed.

3. Experiments with XM-1, ALS

The soft X-ray microscope XM-1 is a full field microscope at a bending magnet [4]. For the experiments, a 25 nm micro zone plate was used to take high resolution images below and above the Fe L 3 edge at 707 eV. The magnification was about 2800, the spectral resolution E/ΔE≈500. Two kinds of tiltable sample holders for aqueous samples were used: glass capillaries with tips of about 3 μm diameter, and flat holders with two Si_3N_4 foils of 100 nm thickness each. Both holders were sealed, so the samples kept aqueous over several hours.

In Figure 2, images of an aqueous flock of montmorillonite with hematite between two Si_3N_4 foils taken at 704 eV (left) and 707 eV (right) are shown. The hematite particles are clearly appearing at the Fe L 3 absorption edge. Figure 3 presents a detail of the montmorillonite sample shown previously, viewed with an angular difference between both images of 14° around a horizontal tilt axis. Both images are taken at E = 707 eV. Lines, points, and edges are marked for analysis. The results from xstereo are given in Figure 4: view of a 3-d plot of points and edges marked in Figure 3. The unit of the axis is pixel, where 11 pixel relate to 100 nm. Distances had been determined e.g. between structure 6 (center of hematite) and 8 (clay edge) to 180 nm, or between structure 3 (center of hematite) and 7 (clay edge) to 630 nm, revealing results about the interrelation between pH value of the solution and the corresponding edge charge of the clay platelets [5].

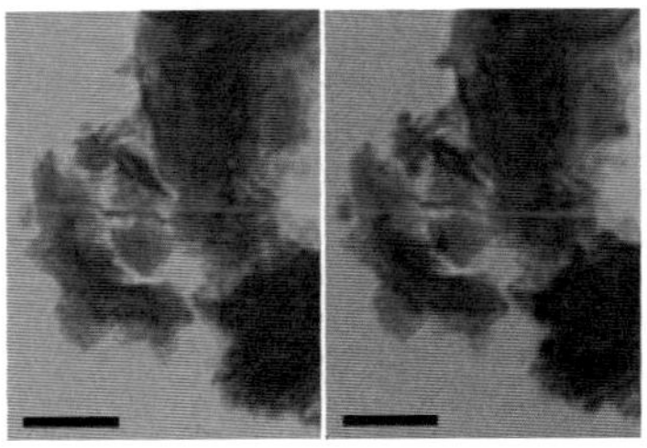

Figure 2. Aqueous sample of montmorillonite particles with hematite colloids. Left: E = 704 eV, right: E = 707 eV. Scale bar: 1 μm.

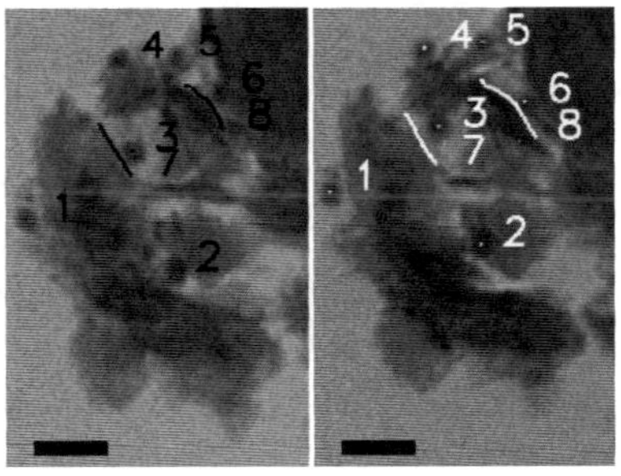

Figure 3. Detail of the sample shown in Figure 2, $\Delta\theta = 14°$, E = 707 eV, features marked. Scale bar: 500 nm.

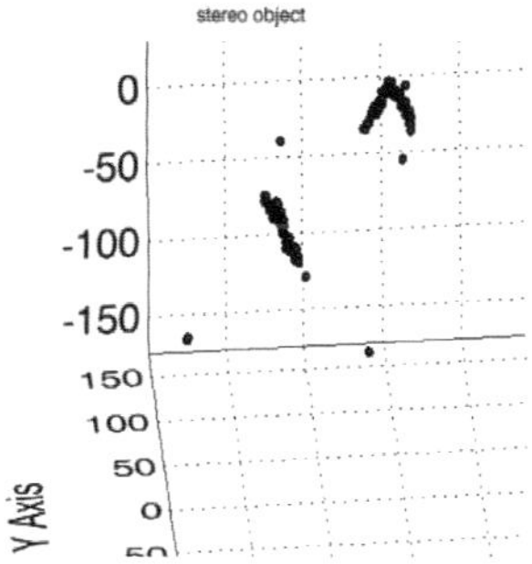

Figure 4. View of 3-d plot of marked features (Figure 3). Unit: pixels. 11 pix. = 100 nm.

4. Experiments with the compact X-ray microscope, KTH

The compact X-ray microscope at the KTH, Stockholm, Sweden, provides a table-top sub-30 nm resolution. It is based on a liquid-jet laser-plasma source, condenser optics and zone plates objectives [6]. The images were taken at $\lambda = 3.37$ nm wavelength with exposure times of 4 to 8 minutes. The image diameter is 20 μm. The vertical tilt given by the stereo mount was 14°, adapting the parallax to human stereo vision. Two pairs of images are shown (Figure 5 and 6).

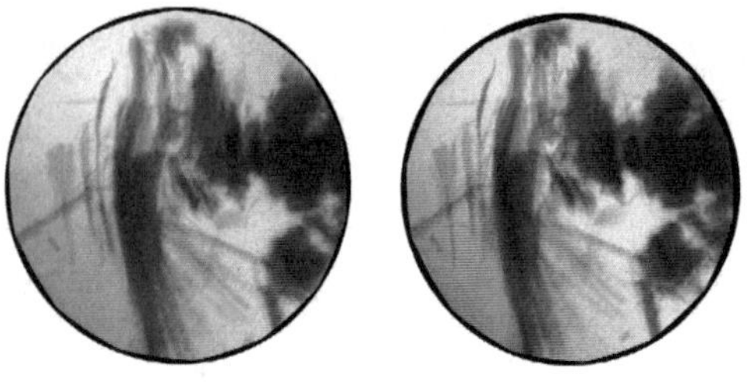

Figure 5. Aqueous nontronite particles, E = 368 eV, $\Delta\theta = 14°, \emptyset = 20\,\mu$m.

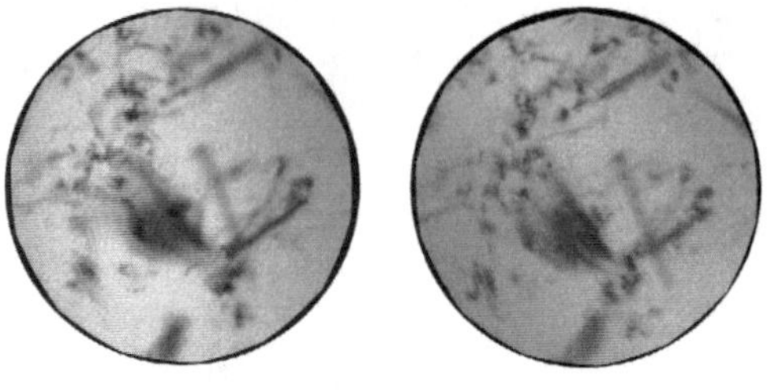

Figure 6. Aqueous nontronite with hematite particles, E = 368 eV, $\Delta\theta = 14°, \emptyset = 20\,\mu$m.

5. Experiments with the STXM, BESSY

The sample in the Scanning Transmission X-ray Microscope at the Undulator U41 at BESSY is scanned with a 50 nm spot, at best resolution a region of up to 40x40 μm^2 [7]. The sample stage has been adapted for tilting capillaries.

Figure 7 shows montmorillonite in a capillary. Two pairs of images were taken of the same sample region with equal image paramaters, one before (Figure 7 up) and one after (Figure 7 below) adding carbon nanotubes. Both image pairs were taken at 400 eV within 30 minutes, respectively, with a horizontal tilt of 15° of a sample region of 8x7 μm^2 onto 160x140 pixel^2 with 16 ms exposure time per pixel. The time between the pairs was 5 hours, resulting in a big change in structure.

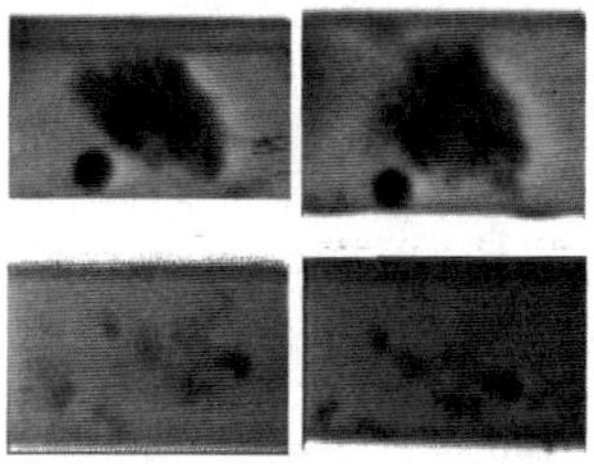

Figure 7. Montmorillonite. Below: after adding C nanotubes.

6. Conclusions

The suitability of a TXM, a STXM, and a TXM with a laboratory source for the investigation of samples from the environment in ambient conditions in combination with stereo imaging and elemental mapping could be shown. Manipulating the sample and tilting the sample was possible even under observation. Therefore, studies of the dynamical behavior of samples will be performed as a next step using both, stereo imaging and elemental mapping, to obtain a complete description of the system under investigation.

Acknowledgments

This work is supported by the DFG within the Collaborative Research Center SFB 755 Nanoscale Photonic Imaging and under contract numbers Th445/8-1 and Th445/8-2. The ALS microscope is supported by the Director, Office of Science, Office of Basic Energy Sciences, Materials Sciences and Engineering Division, of the U.S. Department of Energy. The KTH microscope is supported by the Swedish Science Research Council, the Göran Gustafsson Foundation, and the Wallenberg Foundation. We thank the staff of ALS and BESSY for providing excellent working conditions.

References

[1] Thieme J, McNulty I, Vogt S and Paterson D 2007 *Environ. Sci. Technol.* **41** 1027
[2] Mitrea G, Thieme J, Guttmann P, Heim S and Gleber S C 2008 *J. Synchrotron Rad.* **15** 26
[3] Gleber S C, Knöchel C, Thieme J, Rudolph D and Schmahl G 2003 *X-Ray Microscopy 2002* (*J. Phys. IV Proc.* vol 104) ed Susini J, Joyeux D and Polack F (EDP Sciences, Les Ulis) p 639
[4] Fischer P, Kim D H, Chao W, Liddle J A, Anderson E H and Attwood D 2006 *Materials Today* **9** 26
[5] Gleber S C, Thieme J, Chao W and Fischer P *J. Microsc.* submitted
[6] Bertilson M, von Hofsten O, Thieme J, Lindblom M, Holmberg A, Takman P, Vogt U and Hertz H *this volume*
[7] Wiesemann U 2003 *The Scanning Transmission X-Ray Microscope at BESSY II* Ph.D. thesis Göttingen

X-ray phase-contrast micro-tomography and image analysis of wood microstructure

Sheridan Mayo[1], Robert Evans[1], Fiona Chen[1] and Ryan Lagerstrom[2]
[1] CSIRO Materials Science and Engineering, Private Bag 33, Clayton, VIC 3169, Australia
[2] CSIRO Mathematical and Information Sciences, Building E6B, Macquarie University, North Ryde, NSW 2113, Australia

Abstract. A number of commercially important properties of wood depend on details of the wood micro- and nano- structure. CSIRO Forest Biosciences have developed SilviScan, an analytical instrument which uses a number of high-speed techniques for analyzing these properties. X-ray micro-tomographic analysis of wood samples provides detailed 3D reconstructions of the wood microstructure which can be used to validate results from SilviScan measurements. A series of wood samples was analysed using laboratory-based phase-contrast x-ray micro-tomography. Image analysis techniques were applied to the 3D data sets to extract significant features and statistical properties of the specimens. These data provide a means of verification of results from the more rapid SilviScan techniques, and will clarify the results of micro-diffraction studies of wood microfibrils.

1. Introduction

Important features of wood micro- and nano-structure include the presence and distribution of different wood cell types (tracheids, fibres, parenchyma, vessels) that constitute the structure and, on a smaller scale, the orientation of the cell wall cellulose microfibrils which give the wood its tensile strength and stiffness. A number of techniques can be used to analyse the micro-structural properties of wood, many of which are incorporated into the SilviScan[1] system developed by Robert Evans and his colleagues in the Forest Biosciences program at CMSE.

Two of the types of property that can be observed by SilviScan are the cross-sectional dimensions of wood cells, and microfibril angle variation obtained by x-ray scanning diffractometry.. Analysis of wood sections can be used to extract information about the distribution of vessels and different types of wood cells, but the method relies on 2D cut sections which can suffer from surface damage. The measurement of microfibril angle and angular variation is likewise complicated because information on the microfibril orientations is entangled with variation in orientation of the wood fibres and with the variation in wood density.

X-ray phase-contrast micro-tomography is very well suited to the analysis of wood microstructure and provides valuable data for validation of the above analyses from SilviScan. Tomography enables the analysis of 'sections' throughout the volume of the wood without physical sectioning. Coupled with image-processing techniques developed at CSIRO Mathematical and Information Sciences it also opens the door to image analyses in 3D that have been previously restricted to 2D optical sections. Analysis of wood fibre orientations can also be carried out for comparison with SilviScan diffractometry data on microfibrils.

2. Data collection

We have collected 3D data on different wood samples and used 3D image analysis methods to extract quantitative information about wood microstructure, such as cell wall orientations, and the distributions of different cell types.

Our micro-tomographic data were collected using the X-ray ultra-Microscope (XuM), an instrument developed by CSIRO and XRT Ltd. This instrument makes use of inline phase-contrast to improve contrast of low density materials[2]. It has imaging resolution <100nm at the highest magnification, and tomographic resolution down to 2μm or 0.5 μm with image alignment[3]. Polychromatic radiation was used, with an average energy of 8keV.

X-TRACT[4] software was used to process and reconstruct the data including phase retrieval[5] of the phase-contrast images prior to tomographic reconstruction. Figure 1 below shows a selection of reconstructed cross-sections from different samples with different wood structures.

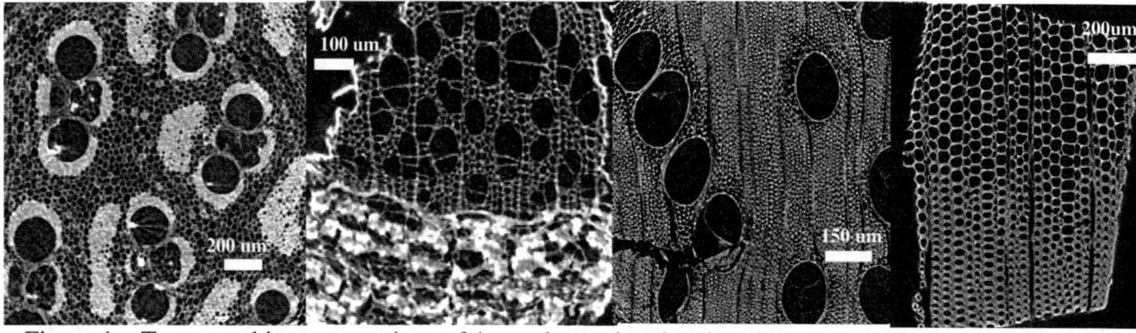

Figure 1 – Tomographic cross sections of 4 wood samples showing the variations in structure between different wood types. From left: bamboo; poplar; eucalypt and pine.

3. 3D Analysis

Two types of 3D analysis were performed on these datasets. The first was the extraction of gross features of the structure using mathematical morphological techniques and the second was extraction of variation in the cell orientations using a skeletonisation algorithm.

For the first of these analyses a low-magnification dataset of a eucalypt sample was used. Mathematical morphological methods[6] were used to extract specific features of the wood microstructure, namely vessels (large vertical channels), rays (narrow horizontal parenchyma cells) and longitudinal parenchyma cells (functional cells close to vessels in this species). The results are shown below in figure 2 including the rendered original 3D data and a rendered view of the segmented 3D data showing large vessels and horizontal rays (longitudinal parenchyma cells omitted for clarity). The rays are incompletely resolved in this example, and therefore appear to be discontinuous.

Figure 2. left: Rendered view of 3D tomographic reconstruction of a eucalypt sample; right: rendered view of segmentation of part of the dataset showing vessels (large vertical channels) and rays (narrow horizontal cells).

The next step was to extract information on the variation in orientation of the cell axes. The dominant cells (fibres) run vertically in the tree and it is the variations in the orientation of these cells that is potentially most significant in affecting microfibril diffractometry data. For this analysis we used two higher magnification datasets of eucalypt and pine respectively. The central axes of the fibres and ray cells were extracted from the reconstructed data using a skeletonisation algorithm[7] as shown for the pine specimen in figure 3.

The fibres (pine tracheids are also referred to as 'fibres' in this report) are the main body of cells in the wood and run approximately parallel to the trunk or branch axis and these can be seen running roughly vertically in the skeleton image. Analysis of the generated skeletons was carried out to extract the variation in the orientation of these cells in the tangential and radial directions in the tree. The angular variation in degrees for both wood types is shown in in figure 3 for both radial and tangential orientations. The standard deviations (sigma) values from these data (shown on plots) indicate that the variation in orientation is greater in the radial direction than in the tangential direction for both wood types, and that the variations are greater in the eucalypt sample relative to the pine sample.

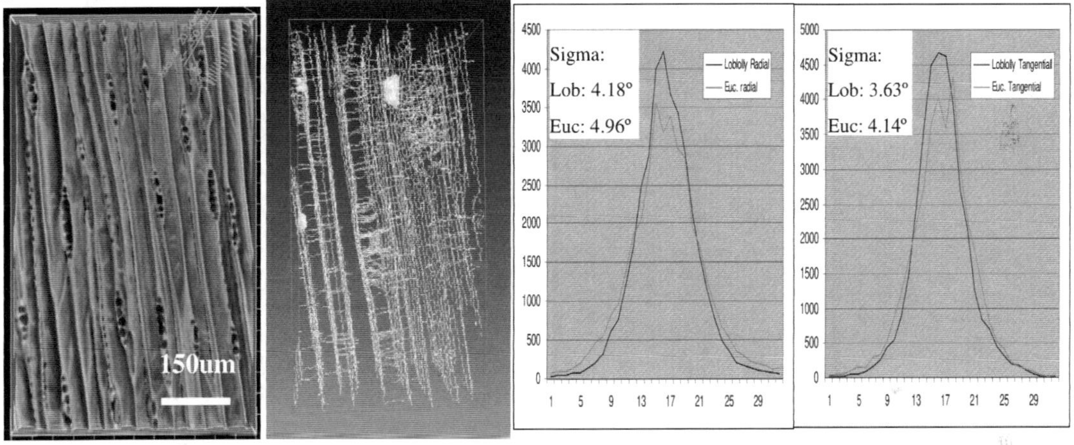

Figure 3 –from left: rendered view of pine sample; extracted pine cell-axis skeleton; normalised plots of fibre orientation variation for pine and eucalypt samples in the radial and tangential directions.

4. Conclusions

We have demonstrated the use of x-ray phase-contrast tomography to produce wood microstructure data of sufficient quality for extraction of details of gross and fine features.. Future work will focus on the use of angular variation data to refine the results of x-ray diffractometry studies of microfibril angle from SilviScan and the improvement of morphological and other image analysis methods for extracting and categorising a range of features in wood datasets.

References

[1] Evans, R 2006. Chapter 11. In: "Characterisation of the Cellulosic Cell Wall", Proceedings of the workshop 25-27 August 2003, Grand Lake, Colorado, USA. Southern Research Station, University of Iowa and the Society of Wood Science and Technology. D. Stokke and L. Groom, eds. Blackwell Publishing

[2] Mayo S C, Davis T J, Gureyev T E, Miller P R, Paganin D, Pogany A, Stevenson A W and Wilkins S W 2003 *Optics Express* **11** 2289

[3] Mayo S C, Miller P, Gao D and Sheffield-Parker J 2007 *J. Microscopy* **228** 257–263

[4] http://xrsi.cmit.csiro.au/Services/AppInfo/X-TRACT.aspx

[5] Paganin D, Mayo S C, Gureyev T E, Miller P R and Wilkins SW 2002 *J. Microscopy* **206** 33-40

[6] Soille P 1999 *Morphological Image Analysis* (Springer-Verlag Telos)

[7] Palágyi K and Kuba A 1998 *A Pattern Recognition Letters* **19** 612-627

Journal of Physics: Conference Series **186** (2009) 012106 doi:10.1088/1742-6596/186/1/012106

Scanning Transmission X-ray Microscopy of multi-walled carbon nanotubes

Ebrahim Najafi,[1] **Daniel Hernández Cruz,**[1] **Martin Obst,**[1] **Adam P. Hitchcock**[1], **Alexandre Felten,**[2] **Bastien Douhard,**[2] **Jean-Jacques Pireaux,**[2] **Konstantine Kaznatcheev**[3] and **Chithra Karunakaran**[3]

1. BIMR, McMaster University, Hamilton, Ontario L8S 4M1, Canada

2. LISE, University of Namur, 61 rue de Bruxelles, 5000, Namur, Belgium

3. CLS, University of Saskatchewan, Saskatoon, SK S7N 0X4 Canada

E-mail: aph@mcmaster.ca

Abstract. We have used the recently completed soft X-ray spectromicroscopy facility at the Canadian Light Source (CLS) to study individual multi-walled carbon nanotubes (MWCNTs) with scanning transmission X-ray microscopy (STXM). Specifically, spatially resolved C 1s X-ray absorption linear dichroism signals were used to map structural defects. The STXM on CLS 10ID1 beamline is particularly powerful for such studies since its elliptically polarized undulator provides fully linearly polarized X-rays with their E-vector orientation continuously tunable from -90 to +90°. We correlate the magnitude of linear dichroism of the C 1s →π* transition with the quality of the crystalline structure, in particular presence of sp^2 defects. We show that the dichroic magnitude along a MWCNT can be used to map local defect density.

1. Introduction

Carbon nanotubes (CNTs) have great potential for a wide range of applications such as channels in electronic devices, emitters in field emission displays, and reinforcements for polymer composites. Such applications often require the CNTs to be essentially perfect structures, often of well-defined chirality, or to have precisely controlled amount and types of defects or dopants to tailor their electronic and/or mechanical properties. A major challenge to technological applications of CNTs is that actual materials are far from perfect [1] and are often quite impure. Methods to characterize the quality of CNTs are thus desirable. A number of techniques have been developed that provide characterization of defect type and/or spatial distributions in carbon nanotubes. High resolution transmission electron microscopy can visualize defects directly [1]. Bockrath et al have developed a scanning probe electrical transport method (scanning gate microscopy) to visualize defects in metallic single walled nanotubes [2]. Selective electrochemical deposition followed by high resolution TEM [3] can detect and categorize defects over a wide range of concentrations. Methods to intentionally introduce structural defects at specific points in a carbon nanotubes have also been developed [4]. A method to rapidly survey a nanotube structure and evaluate its defect density at high resolution would be useful. Resonant Raman microscopy is very powerful since it can identify defect type as well as density [5], but in conventional implementations, its spatial resolution is limited to ~300 nm.

Recently we have shown [6,7] that scanning transmission X-ray microscopy (STXM) is a useful tool to characterize carbon nanotubes. The C 1s near edge X-ray absorption spectrum (NEXAFS) can be used to differentiate carbon nanotube materials from other types of graphitic-like carbon. In addition there is a very strong linear dichroism [7], first predicted by Banerjee et al. [8], which can be quantified and related to the local density of sp^2 defects. Here we report on the dichroic phenomena, show that the spectra without the dichroic response can be measured using circularly polarized light, and explore the use of maps of the dichroic amplitude as a means of mapping defect density in multi-walled carbon nanotubes (MWCNT).

2. Experimental

STXM spatial resolution is determined by the Fresnel zone plate (ZP) properties, specifically $1.22\delta_r$ where δ_r is the width of the outermost zone. The ZP used for this experiment had an outermost width of 25 nm, corresponding to a spatial resolution of 31 nm. This resolution has been verified using high contrast test structures. The source of CLS beamline 10ID1 is an elliptically polarizing undulator (EPU) in which all 4 quadrants can be scanned along the beam direction, thus allowing tuned X-rays with >99% linear polarization to be produced with arbitrary inclination of the E-vector orientation, from -90° to +90° relative the plane of the ring [9]. This is particularly convenient for STXM as a careful, complete and quantitative study of the dichroism can be performed without having to move the sample. Here we report the linear dichroism of individual MWCNTs of two different types– arc discharge (AD) and chemical vapour deposition (CVD).

3. Results

Figure 1 plots C 1*s* spectra extracted from image sequences recorded on an individual carbon nanotube using EPU phasing such that the E-vector is either circularly polarized (black) or linearly polarized with E parallel to the nanotube (red), or with E perpendicular to the nanotube (green). There is a very dramatic change in the intensity of the C 1*s* → π* transition at 285.2 eV, as well as smaller changes elsewhere in the spectrum. As expected, the spectrum measured with circularly polarized light gives a π* signal about half way between those for the parallel and perpendicular light, consistent with the linear dichroism effect being 'turned off' when the EPU is tuned to this mode. Adjacent carbon onion material or the holey carbon support did not show significant changes with changes in nature (circular/linear) or orientation (parallel, perpendicular) of the polarization of the light. **Figure 2** plots the intensity of the C 1s → π* transition of straight regions of an AD-MWCNT and a CVD MWCNT.

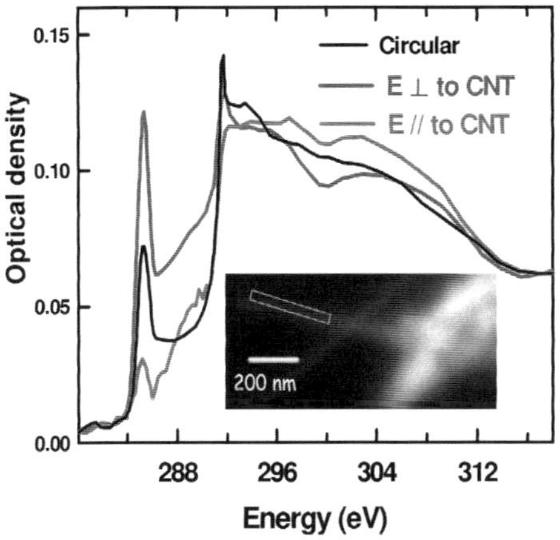

Figure 1. C 1*s* spectra, recorded with the indicated polarization, of a multi-walled carbon nanotube prepared by arc discharge.

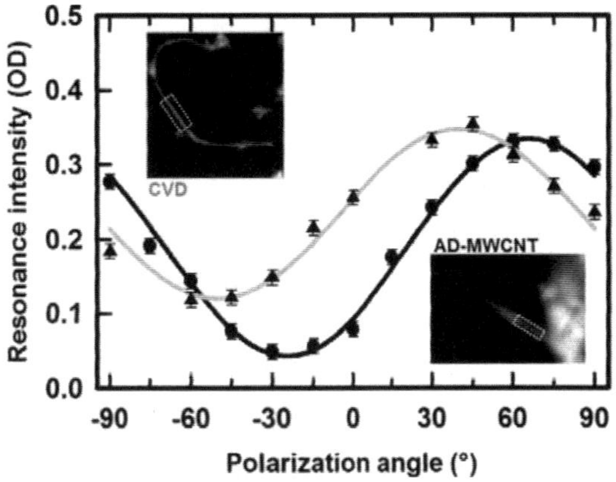

Figure 2 Fit of the intensity of the C 1s → π* transition of higher quality regions of an AD-MWCNT and a CVD-MWCNT.

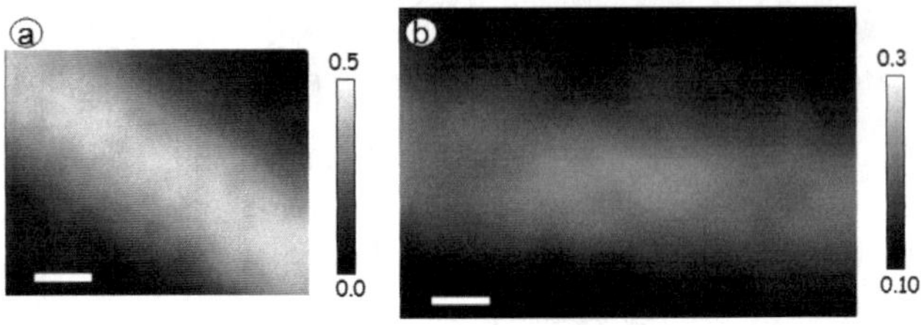

Figure 3. Map of the amplitude of the dichroic signal in a small region of (a) AD-MWCNT, and (b) CVD-MWCNT. Note different amplitude scales. Scale bar = 20 nm

Each curve is an excellent fit to the expected polarization dependence, $\{C + A\cos^2(\theta-\theta_{ref})\}$, where C represents the non-dichroic component of the signal, A is the amplitude of the dichroic component and θ_{ref} is the angle between the long axis of the MWCNT and horizontal. The position of the maximum agrees with the sample geometry and the fact that the intensity is maximum when the E-vector of the light is orthogonal to the tube axis. One sees that the amplitude of the dichroic signal is much larger for the AD-MWCNT than for the CVD-MWCNT, consistent with the expected relative quality and sp^2 defect density, since CVD-MWCNTs are known to be much more defective than AD-MWCNTs.

In order to look in more detail at the spatial distribution of sp^2 type defects along individual nanotubes, we have fit the dichroic signal at each pixel (5 nm x 5 nm) to the dichroic equation, $\{C + A\cos^2(\theta-\theta_{ref})$, and then assembled the dichroic amplitude coefficients from those fits into pseudo-images. Figure 3a plots this signal for the AD-MWCNT, while figure 3b plots that for the CVD-MWCNT on the same spatial scale, but different dichroic amplitude scales. It is clear that the dichroic amplitude from the AD-MWCNT is larger and more uniform than that for the CVD-MWCNT, indicating in a more detailed fashion the higher quality of the AD-MWCNT.

4. Summary

We have shown that the C 1s→π* transition measured by STXM for individual multi-walled carbon nanotubes exhibits a strong linear dichroism with maximum intensity when the E-vector is perpendicular to the carbon nanotube axis. Using a STXM with an EPU capable of inclining the E-vector over a full 180° greatly facilitates quantitative mapping of the dichroic signal. Comparison of AD- and CVD-MWCNTs showed that the π* dichroic magnitude correlates with the quality of the carbon nanotube. We have extended this approach to a full pixel-by-pixel fit of dichroic data in order to show how the amplitude of the dichroic effect could be used to map sp^3 defects along individual nanotubes. This method could be useful to help optimize preparation and purification procedures as well as the development of sensors or electronic devices based on single carbon nanotubes.

Acknowledgements

This research is supported by NSERC, CFI, CRC (Canada); PAI 6 (Belgium), Nanobeams EU Network of Excellence, Nano2hybrids (EU-STREP N°033311). STXM data was measured at CLS-SM (BL10ID1) (supported by NSERC, CIHR, NRC and U. Saskatchewan).

References

[1] Ebbesen T W and Takada T, 1995 *Carbon* **33** 973
[2] Bockrath M et al. 2000 *Science* **291** 283
[3] Fan Y, Goldsmith B R and Collins P G 2005 *Nature Materials* **4** 906
[3] Krasheninnikov A V and Banhart F 2007 *Nature Materials* **6** 723
[4] Jorio A et al. 2005 *Mater. Res. Soc. Symp. Proc.* **858E** HH11.2
[5] Felten A, et al. 2006 *Appl. Phys. Lett.* **89** 093123
[6] Felten A, et al. 2007 *Nanoletters* **7** 2435
[7] Najafi E, Felten A, Douhard B, Hernández Cruz D, Obst M, Hitchcock A P and Pireaux C J J 2008 *Small* **4** 2279
[8] Banerjee S, et al. 2005 *J. Phys. Chem. B* **109** 8489
[9] Kaznacheyev K V, et al. 2007 *Nucl. Inst. Meth. A* **582** 103

9th International Conference on X-Ray Microscopy IOP Publishing
Journal of Physics: Conference Series **186** (2009) 012107 doi:10.1088/1742-6596/186/1/012107

High-resolution imaging of soil colloids in aqueous media with a compact soft X-ray microscope

J Thieme*,[1], J Sedlmair[1], S-C Gleber[1], M Bertilson[2], O von Hofsten[2], P Takman[2], H Hertz[2]
[1]Institute for X-Ray Physics, University of Göttingen, Friedrich-Hund-Platz 1, D-37077 Göttingen, Germany
[2]Royal Institute of Technology/ Albanova, Biomedical and X-Ray Physics, KTH, SE-10691 Stockholm, Sweden

E-mail: jthieme@gwdg.de

Abstract. Colloids play an important role when describing parameters of and processes within soils, sediments or aquifers due to their abundance and their high specific surface area. It is of great importance to visualize the morphology of the structures formed by these particles as close as possible to environmental conditions. With X-ray microscopy colloids from the environment can be imaged directly in aqueous media with high spatial resolution. We demonstrate the first use of a compact laboratory x-ray microscope for studies of colloids from the environment, namely aqueous suspensions of clays and soils. The microscope is based on a high-brightness laser-produced-plasma X-ray source, a multilayer mirror and diffractive optics. The experiments show that such compact X-ray microscopes are reaching the image quality and operational maturity to make a significant impact in fields like environmental sciences.

1. Introduction to colloidal structures in soils

Soils are penetrated by atmosphere, hydrosphere and geosphere. Its fabric is expressed in the pore system of soils, where the distribution of the pore radii shows a wide range from mm to sub-μm. This pore system determines the inner surface of a soil. It affects many important soil parameters and processes as turnover of nutrients, adsorption of toxicants, bioavailability of these substances, water storage capacity and water flow. Pore radii of some μm or less are the most frequent, thus contributing to the largest part of the inner surface. These micro pores are mainly built up by inorganic colloids such as clays, quartz particles, iron and aluminium oxides, or organic colloids as bacteria, fungi, or humic substances. To better understand parameters and processes it is necessary to have a close look at the microstructure built up by these colloids. Due to their large specific surface area (e.g. clays with up to 800 m^2/g) and their small size many physical, chemical and biological processes in soils are influenced by these particles. Due to this particle size and as nearly all chemical reactions occur in aqueous phase, X-ray microscopy is a technique well suited for these purposes.

2. Compact X-ray microscopy

Successfully running X-ray microscopes have been installed at synchrotron radiation facilities, the most commonly used source for X-ray microscopes today. However, only few electron storage rings exist worldwide and the accessibility is limited, hampering the dissemination of X-ray microscopy as a useful technique. Using a high-brightness and minimum-debris regenerative liquid-jet laser-plasma source, a normal-incidence spherical multilayer condenser mirror and a micro zone plate, colloids from the environment in aqueous media have been imaged at λ=3.37nm. A micro zone plate with dr_n = 25 nm has been used as microscopic objective, and the resulting resolution in the images is of about the same value [1, 2]. Image and illustration in figure 1 show the arrangement of the compact laboratory soft X-ray microscope at the KTH. The images shown in this paper have been taken with exposure times of up to 600 seconds.

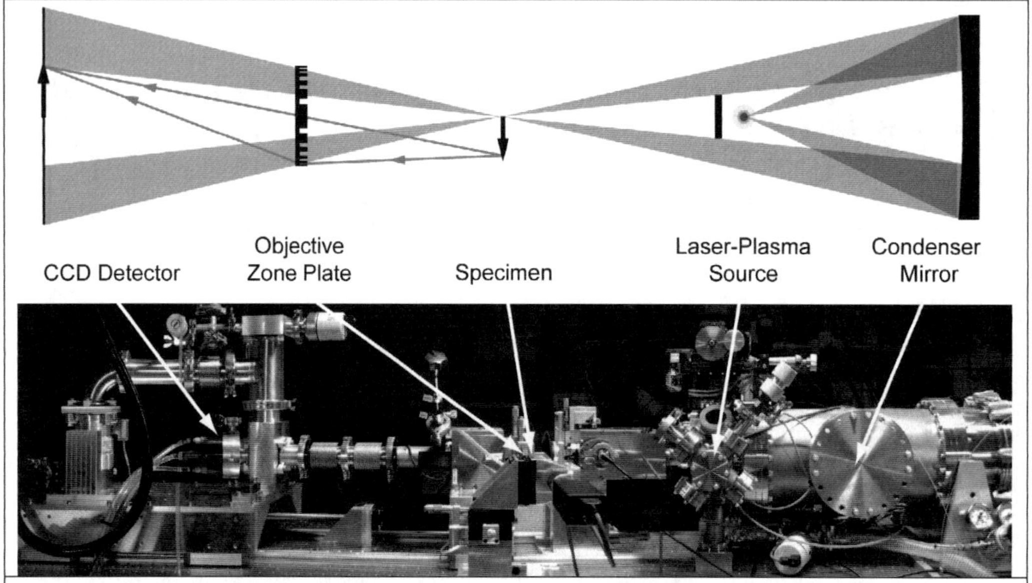

Figure 1. Arrangement of the compact X-ray microscope at the KTH, Stockholm.

3. Clay dispersion

A dispersion of nontronite clay has been chosen to demonstrate the necessity for imaging colloidal particles in aqueous suspension. Nontronite consists of small particles stacked together forming open structures within the soil solution. When falling dry, the delicate structures collapse. This change in morphology will make it impossible to assess the real internal surface under ambient conditions, i.e. when small pores within the soil are filled with water.

4. Soil colloids

Figure 2 shows images of samples from a chernozem soil show different appearances of colloidal particles, forming clusters. The trained eye quickly identifies in each sample particles of similar shapes. Small, optically dense clay particles are the majority of the particles visible. The cluster OF the center in the left image can be identified as a larger central clay particle with smaller particles attached. Due to the frayed edges of the smaller features they could possibly be of organic origin. The image diameter is ~2000 pixels, corresponding to 20 μm in the sample plane.

5. Discussion and Outlook

It has been clearly demonstrated that it is possible to visualize the morphology of colloidal structures from the environment with a compact laboratory X-ray microscope. The pictures shown here have been taken with exposure times of 600 seconds, which is of course a major drawback in comparison to exposure times in the range of seconds or less using X-ray microscopes at electron storage rings. The use of a state of the art laser with improved pulse energy, pulse duration and repetition rate will shorten the exposure time with about a factor 10. Equipping the microscope with a multilayer condenser mirror for the $\lambda = 2.48$ nm source, thereby increasing the numerical aperture and the collection efficiency of the condenser WILL shorten the exposure time even further. Finally, the transmission for thicker, ~10 μm, aqueous samples at this wavelength is ~5 times higher than for $\lambda = 3.37$ nm. This contributes also to shorter exposure times. All together these planned improvements lead to exposure times comparable to the time needed with synchrotron-based microscopes. Such short-exposure-time compact X-ray microscopes readily accessible in the laboratories of environmental sciences would make powerful tools for future studies.

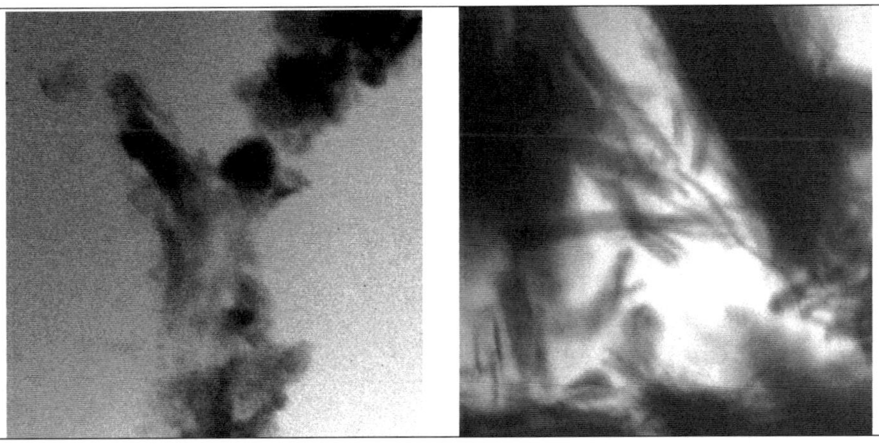

Figure 2. Cluster of nontronite particles in a dry (left) and an aqueous state (right).

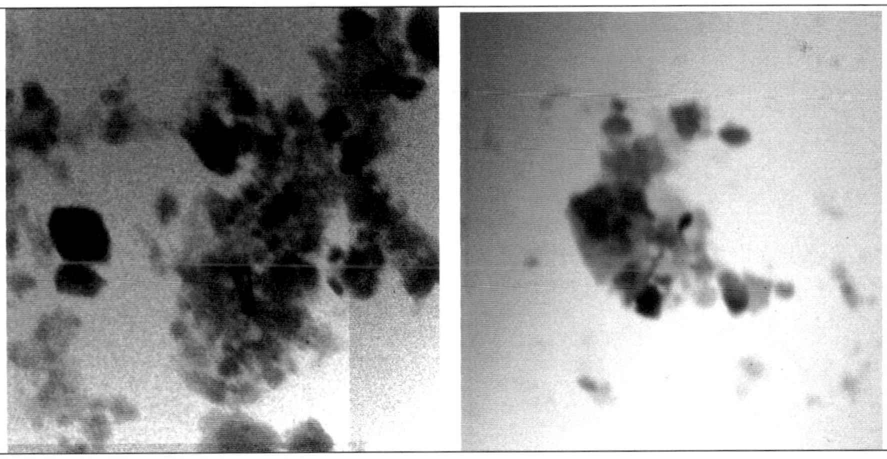

Figure 3. Colloid clusters from a Chernozem soil in aqueous suspension.

Acknowledgments
This work is supported by the DFG within the Collaborative Research Center SFB755 "Nanoscale Photonic Imaging" and under contract numbers Th445/8-1 and Th445/8-2; by the Swedish Science Research Council, the Göran Gustafsson Foundation, and the Wallenberg Foundation.

References

[1] Takman P, Stollberg H, Johansson G, Holmberg A, Lindblom M, Hertz H 2007 *J.Microscopy* **226** 175
[2] Bertilson M, von Hofsten O, Thieme J, Lindblom M, Holmberg A, Takman P, Vogt U, Hertz H, *this volume*

AUTHOR INDEX

AUTHOR INDEX

AUTHOR INDEX

AUTHOR INDEX

AUTHOR INDEX

Institute of Physics
Dirac House, Temple Back
Bristol BS1 6BE UK